Biomechanical Spectrum of Human Sport Performance

Biomechanical Spectrum of Human Sport Performance

Special Issue Editors

Redha Taiar

Mario Bernardo-Filho

MDPI • Basel • Beijing • Wuhan • Barcelona • Belgrade • Manchester • Tokyo • Cluj • Tianjin

Special Issue Editors
Redha Taiar
Université de Reims
Champagne Ardenne
France

Mario Bernardo-Filho
Institute of Biology Roberto
Alcantara Gomes, Rio de Janeiro
State University
Brazil

Editorial Office
MDPI
St. Alban-Anlage 66
4052 Basel, Switzerland

This is a reprint of articles from the Special Issue published online in the open access journal *Applied Sciences* (ISSN 2076-3417) (available at: https://www.mdpi.com/journal/applsci/special_issues/Sport_Biomechanical_Spectrum).

For citation purposes, cite each article independently as indicated on the article page online and as indicated below:

LastName, A.A.; LastName, B.B.; LastName, C.C. Article Title. *Journal Name* **Year**, *Article Number*, Page Range.

ISBN 978-3-03936-396-4 (Hbk)
ISBN 978-3-03936-397-1 (PDF)

Contents

About the Special Issue Editors

Redha Taiar, Ph.D. Biomechanics, is currently a Professor at the University of Reims Champaign, France. His research focuses on industry engineering for medicine and high-level sport. He is an engineer for different industries like Arena for high-level sport and Sidas, Medicapteur, for medical development. For industry workers, his last work was for the Notrax Society on the conception and validation of anti-fatigue mats. On the topic of sport, his last works focus on the development of swimsuits for triathlon and swimming in the Brazil Olympic Games (2016) and the suit fabrics for skiing in the Olympic Games at Sotchi in 2014. He is a specialist on the biomechanics of health disease and rehabilitation. The resume of his research can be found at www.redha-taiar.com; Orcid: https://orcid.org/0000-0002-0227-3884; Scopus: https://www.scopus.com/authid/detail.uri?authorId=15823162100; PubMed: https://www.ncbi.nlm.nih.gov/pubmed/?term=taiar+R.

Mario Bernardo-Filho is a Professor at the Rio de Janeiro State University, Brazil. He is a physiotherapist, and his research involves integrative and complementary medicine (auriculotherapy and acupuncture) and mechanical vibrations generated in an oscillating/vibratory platform that produces whole-body vibration exercises (WBVE) when a subject is in contact with the platform. Studies to evaluate the consequences of WBVE and extracts of medicinal plants in rats are ongoing. Investigations about the effects of WBVE in individuals with different diseases (metabolic syndrome, chronic obstructive pulmonary disease, arthrosis, and osteoporosis) and healthy people are also ongoing. He is a supervisor of various professionals who are preparing their Master of Sciences or Ph.D. thesis. He teaches graduate and undergraduate courses in the Rio de Janeiro State University. He was the Head of the First International Congress on Mechanical Vibration and Integrative and Complementary Practices, Cabo Frio, Brazil (2016). In this meeting, the World Association of Vibration experts (WAVex) was created, and 2018 marked the First Congress of the WAVex in Groningen, The Netherlands. He has more than 130 publications indexed in PubMed (https://www.ncbi.nlm.nih.gov/pubmed/?term=bernardo-filho).

applied sciences

MDPI

Editorial

Editorial "Biomechanical Spectrum of Human Sport Performance"

Redha Taiar [1,*] and Mario Bernardo-Filho [2]

1 Department of Physical Education and Sports (EPS), GRESPI, Université de Reims, 51100 Reims, France
2 Laboratório de Vibrações Mecânicas e Práticas Integrativas, Departamento de Biofísica e Biometria, Instituto de Biologia Roberto Alcantara Gomes e Policlínica Américo Piquet Carneiro, Universidade do Estado do Rio de Janeiro, RJ 20950-003, Brazil; bernardofilhom@gmail.com
* Correspondence: redha.taiar@univ-reims.fr

Received: 27 February 2020; Accepted: 7 March 2020; Published: 10 March 2020

Abstract: Several parameters can influence our health capital today and can have a negative impact on our performance, whether physiological or mechanical. Indeed, our health and wellbeing are influenced by a range of social, cultural, economic, psychological, and environmental factors across our lives. These change as we progress through the key transition points in life—from infancy and childhood through our teenage years to adulthood, working life, retirement and the end of life. Sport can be a vector that links many of these factors. Whether it is high-performance sport or sedentary practice, sport is very important for the improvement of psychological wellbeing and physical health. Our overarching aim was to increase quality of life. Sedentary practice can increase mobility and reduce the risk of disease, so changing adults' behavior through sedentary practice could reduce illness and decrease costs to society concerning health problems. Furthermore, a higher frequency of practice can lead to improvements in technique and optimized performance. Our objective is to summarize the latest research in sport science and to quantify the most important parameters influencing human performance related to the health sciences for all age groups, throughout their lives.

Keywords: modeling and simulation in sport science; strength and conditioning; mechanical analyses of sports; sport medicine; injury in sport; human behavior; quality of life; applied science in musculoskeletal disorders

---

The organization of a movement is regulated by the nervous system, which is subdivided into a central nervous system (CNS), composed of the brain (brain, cerebellum and brain stem) and the spinal cord, and a peripheral nervous system (PNS), composed of nerves that extend throughout the body. During voluntary movement, the cortical areas interact with the lower areas of the brain and spinal cord through the cortico-spinal motor pathway. The command is, therefore, generated in the CNS and then routed via the SNP to the muscle that generates the movement. The execution of a voluntary movement requires the coordination of several muscle contractions so that the movement performed corresponds to the desired movement and is adapted to the environmental situation in which it is performed. Motor control refers to the processes responsible for the preparation, organization and execution of this movement and refers to the coordinated organization of the individual's sensory-motor functions analysis. The analysis of movement by scientists back to antiquity (Hippocrates, 460-377; Aristotle, 384-322...) but the three-dimensional analysis of movement only began at the end of the last century with the work of the anatomist Wilhelm Braune and the mathematician Otto Fisher. These first works were devoted to the study of the march of the infantryman then required 8 to 10 hours of measurements and days of manual calculations, for the analysis of a movement. With the considerable developments in electronics and computer science, today's systems only take a few minutes to obtain the same type of results using biomechanical analysis. Biomechanics is, by definition, the study of the structure and functioning of living beings. It is based on the laws of mechanics and on the methods of the mechanic with the aim of knowing and understanding in order to exploit, improve or restore the functional capacities of humans. In fact, the results carried can be modelized with the aim to decrease experimentations and to understand better the complexity of this system. The complexity of the model required aims to replace the complicated visible with the simpler invisible depends on the aims of the analysis but also on the nature of the approach: kinematic or kinetic. Kinematics concerns the analysis of motion, whereas kinetics studies the forces that cause or result from it (for example, the reaction of the ground when walking). Different models can be considered, ranging from the human body represented by its center of gravity, to a model integrating both motor control and a musculoskeletal model of the human body. All the methods and the latest knowledge's have the same objective improve the human health, well-being and performance. Health is defined as 'a state of complete physical, mental and social wellbeing and not merely the absence of disease or infirmity' [1]. This definition links health clearly with wellbeing. Moreover, health is a human right requiring physical and social approaches to be reached and maintained. In addition, wellbeing is highly related to a positive rather than neutral state, leading the health in being a desired positive aspiration. Health and wellbeing are influenced by a range of social, economic, cultural, psychological and environmental conditions to be considered along of the life. Exercise is a responsible key factor in maintaining the functional autonomy of the body and can contribute to the protection against undesirable situations. There are strong scientific evidences that lifelong exercise is associated with a longer health span, favoring the delay to the onset of several chronic conditions/diseases [2,3]. The relevance of the exercise against age-related risks for commitment of the health and wellbeing that lead to disease and disability is unquestionable [4,5]. Furthermore, sedentarism is associated with an elevated incidence, in various stages of the life, in particular during aging, of chronic disease such as cardiovascular disease, chronic respiratory disease, cognitive decline, metabolic syndrome, type 2 diabetes, and cancer [6–14]. Regular exercise, additionally, improves health and decreases the incidence of oxidative-stress-related disease [6]. Sport, as an activity involving exercises, can be a vector that links many of these factors. Whether sport is high-performance sport or sedentary practice, it is very important for the improvement of psychological wellbeing and physical health. Indeed, our overarching aim was to increase quality of life. Sedentary practice can increase mobility and reduce the risk of disease, so changing adults' behavior through sedentary practice could reduce illness and decrease costs to society concerning health problems. Furthermore, a higher frequency of practice can lead to improvements in technique, and optimized performance. The objective of this Special Issue published by the Applied Sciences Journal is to summarize the most important biomechanical parameters influencing human performance related to the health sciences for all age groups, throughout their lives. The clinical and experimental studies presented here demonstrate the relevance of the exercises considering the Biomechanical Spectrum of Human Sport Performance.

In this Special Issue, 26 manuscripts [15–40] were published after the procedure of selection. Interesting manuscripts aimed on the quantification of human performance and his optimization. We will find studies permitting to determine the discriminate parameters of human performance as well as the latest technologies with the objective to analyze and understand the complexity of human mechanics and his performance in the different daily life tasks. This ranges from the improvement of sports performance to the rehabilitation of patients after injury.

Author Contributions: Investigation, R.T.; original draft preparation, R.T. and M.B.-F.; writing R.T. and M.B.-F.; review and editing, R.T. and M.B.-F.; visualization R.T. and M.B.-F.; supervision, R.T. and M.B.-F. All authors have read and agreed to the published version of the manuscript.

Funding: This research received no external funding.

Conflicts of Interest: The authors declare no conflict of interest.

References

1. World Health Organization. Available online: https://www.who.int/about/who-we-are/constitution (accessed on 27 February 2020).
2. Ruegsegger, G.N.; Booth, F.W. Health benefits of exercise. *Cold Spring Harb. Perspect. Med.* **2018**, *8*, 7. [CrossRef]
3. Ruegsegger, G.N.; Booth, F.W. Health Benefits of Exercise. *Cold Spring Harb. Perspect Med.* **2018**, *8*, a029694 [CrossRef] [PubMed]
4. Heine, M.; Lupton-Smith, A.; Pakosh, M.; Grace, S.L.; Derman, W.; Hanekom, S.D. Exercise-Based rehabilitation for major non-communicable diseases in low-resource settings: A scoping review. *BMJ Glob. Health* **2019**, *4*, e001833. [CrossRef] [PubMed]
5. Myers, J.; Nead, K.T.; Chang, P.; Abella, J.; Kokkinos, P.; Leeper, N.J. Improved reclassification of mortality risk by assessment of physical activity in patients referred for exercise testing. *Am. J. Med.* **2015**, *128*, 396–402
6. Norton, S.; Matthews, F.E.; Barnes, D.E.; Yaffe, K.; Brayne, C. Potential for primary prevention of Alzheimer's disease: An analysis of population-based data. *Lancet Neurol.* **2014**, *13*, 788–794.
7. Rea, I.M. Towards ageing well: Use it or lose it: Exercise, epigenetics and cognition. *Biogerontology* **2017**, *18*, 679–691. [CrossRef]
8. Gomes-Neto, M.; de Sá-Caputo, D.D.C.; Paineiras-Domingos, L.L.; Brandão, A.A.; Neves, M.F.; Marin, P.J.; Sañudo, B.; Bernardo-Filho, M. Effects of whole-body vibration in older adult patients with type 2 diabetes mellitus: A systematic review and meta-analysis. *Can. J. Diabetes* **2019**, *43*, 524–529. [CrossRef]
9. Paineiras-Domingos, L.L.; Sá-Caputo, D.D.C.; Francisca-Santos, A.; Reis-Silva, A.; Carvalho-Lima, R.P.; Neves, M.F.T.; Xavier, V.L.; Quinart, H.; Boyer, F.C.; Sartorio, A.; et al. Can whole-body vibration exercises promote improvement on quality of life and on chronic pain level of metabolic syndrome patients? A pseudo-randomized study. *J. Appl. Physiol.* **2020**. [CrossRef]
10. O'Donoghue, G.; Perchoux, C.; Mensah, K.; Lakerveld, J.; van der Ploeg, H.; Bernaards, C.; Chastin, S.F.M.; Simon, C.; O'Gorman, D.; Nazare, J.-A.; et al. A systematic review of correlates of sedentary behaviour in adults aged 18–65 years: A socio-ecological approach. *BMC Public Health* **2016**, *16*, 163. [CrossRef]
11. Honda, T.; Chen, S.; Yonemoto, K.; Kishimoto, H.; Chen, T.; Narazaki, K.; Kumagai, S. Sedentary bout durations and metabolic syndrome among working adults: A prospective cohort study. *BMC Public Health* **2016**, *16*, 888. [CrossRef]
12. Ekelund, U.; Ward, H.A.; Norat, T.; Luan, J.; May, A.M.; Weiderpass, E.; Sharp, S.S.; Overvad, K.; Nautrup, J.; Elio Riboli, E. Physical activity and all-cause mortality across levels of overall and abdominal adiposity in European men and women: The European prospective investigation into cancer and nutrition study (EPIC). *Am. J. Clin. Nutr.* **2015**, *101*, 613–621. [PubMed]
13. Kohl, H.W., III; Craig, C.L.; Lambert, E.V.; Inoue, X.; Alkandari, J.R.; Leetongin, G.; Kahlmeier, S. Lancet Physical Activity Series Working Group. The pandemic of physical inactivity: Global action for public health. *Lancet* **2012**, *380*, 294–305. [PubMed]
14. Lee, M.-L.; Shiroma, E.J.; Lobelo, F.; Puska, P.; Blair, S.N.; Katzmarzyk, P.T. Physical Activity Series Working Group. Impact of physical inactivity on the world's major non-communicable diseases. *Lancet* **2012**, *380*, 219–229. [PubMed]
15. Yu, P.; Xiang, L.; Liang, M.; Mei, Q.; Baker, J.; Gu, Y. Morphology-Related foot function analysis: Implications for jumping and running. *Appl. Sci.* **2019**, *9*, 3236. [CrossRef]
16. Akl, A.; Hassan, I.; Hassan, A.; Bishop, P. Relationship between Kinematic variables of jump throwing and ball velocity in elite handball players. *Appl. Sci.* **2019**, *9*, 3423. [CrossRef]
17. Collado-Mateo, D.; Dominguez-Muñoz, F.; Charrua, Z.; Adsuar, J.; Batalha, N.; Merellano-Navarro, E.; Raimundo, A. Isokinetic strength in peritoneal dialysis patients: A reliability study. *Appl. Sci.* **2019**, *9*, 3542. [CrossRef]
18. De Jesus, K.; Mourão, L.; Roesler, H.; Viriato, N.; de Jesus, K.; Vaz, M.; Fernandes, R.; Vilas-Boas, J. 3D device for forces in swimming starts and turns. *Appl. Sci.* **2019**, *9*, 3559. [CrossRef]
19. Sañudo, B.; Sánchez-Hernández, J.; Bernardo-Filho, M.; Abdi, E.; Taiar, R.; Núñez, J. Integrative neuromuscular training in young athletes, injury prevention, and performance optimization: A systematic review. *Appl. Sci.* **2019**, *9*, 3839. [CrossRef]
20. Zhang, X.; Luo, Z.; Wang, X.; Yang, Y.; Niu, J.; Fu, W. Shoe cushioning effects on foot loading and comfort perception during typical basketball maneuvers. *Appl. Sci.* **2019**, *9*, 3893. [CrossRef]

21. Schärer, C.; von Siebenthal, L.; Lomax, I.; Gross, M.; Taube, W.; Hübner, K. Simple assessment of height and length of flight in complex gymnastic skills: Validity and reliability of a two-dimensional video analysis method. *Appl. Sci.* **2019**, *9*, 3975.
22. Yang, C.; Xiao, S.; Yang, Y.; Zhang, X.; Wang, J.; Fu, W. Patellofemoral joint loads during running immediately changed by shoes with different minimalist indices: A cross-sectional study. *Appl. Sci.* **2019**, *9*, 4176. [CrossRef]
23. Caramenti, M.; Lafortuna, C.; Mugellini, E.; Abou Khaled, O.; Bresciani, J.; Dubois, A. No evidence that frontal optical flow affects perceived locomotor speed and locomotor biomechanics when running on a treadmill. *Appl. Sci.* **2019**, *9*, 4589. [CrossRef]
24. Dominguez-Muñoz, F.; Hernández-Mocholi, M.; Manso, L.; Collado-Mateo, D.; Villafaina, S.; Adsuar, J.; Gusi, N. Test-Retest reliability of kinematic parameters of timed up and go in people with type 2 diabetes. *Appl. Sci.* **2019**, *9*, 4709. [CrossRef]
25. Sousa-Gonçalves, C.; Paineiras-Domingos, L.; Teixeira-Silva, Y.; Amadeu, T.; Lírio, A.; Francisca-Santos, A.; De Souza, L.; Pereira, M.; Melo-Oliveira, M.; Meirelles, A.; et al. Evaluation of whole-body vibration exercise on neuromuscular activation through electromyographic pattern of vastus lateralis muscle and on range of motion of knees in metabolic syndrome: A quasi-randomized cross-over controlled trial. *Appl. Sci.* **2019**, *9*, 4997. [CrossRef]
26. Figueiredo Azeredo, C.; de Castro de Paiva, P.; Azeredo, L.; Reis da Silva, A.; Francisca-Santos, A.; Paineiras-Domingos, L.L.; Pereira da Silva, A.L.; Bernardes-Oliveira, C.L.; Pessanha-Freitas, J.; Moura-Fernandes, M.C.; et al. Effects of whole-body vibration exercises on parameters related to the sleep quality in metabolic syndrome individuals: A clinical trial study. *Appl. Sci.* **2019**, *9*, 5183. [CrossRef]
27. Ribeiro Kütter, C.; Moreira-Marconi, E.; Teixeira-Silva, Y.; Moura-Fernandes, M.C.; Gonçalves de Meirelles, A.; dos Santos Pereira, M.J.; Chang, S.; Bachur, J.A.; Paineiras-Domingos, L.L.; Taiar, R.; et al. Effects of the whole-body vibration and auriculotherapy on the functionality of knee osteoarthritis individuals. *Appl. Sci.* **2019**, *9*, 5194. [CrossRef]
28. Bodini, B.; Lucenteforte, G.; Serafin, P.; Barone, L.; Vitale, J.; Serafin, A.; Sansone, V.; Negrini, F. Do grade II ankle sprains have chronic effects on the functional ability of ballet dancers performing single-leg flat-foot stance? An observational cross-sectional study. *Appl. Sci.* **2020**, *10*, 155. [CrossRef]
29. Krajewski, K.; McCabe, C.; Sinnott, A.; Moir, G.; Lamont, H.; Brown, S.; Connaboy, C. Inter-Segmental coordination during a unilateral 180° jump in elite rugby players: Implications for prospective identification of injuries. *Appl. Sci.* **2020**, *10*, 427. [CrossRef]
30. Alves, K.; Ferreira, A.; Parente, L.; Rodrigues, F.; Marques, T.; Antonino, G.; Melo, L.; Villela, D.; Guerino, M.; Leite, W.; et al. Immediate effect of whole-body vibration on skin temperature and lower-limb blood flow in older adults with type 2 diabetes: Pilot study. *Appl. Sci.* **2020**, *10*, 690. [CrossRef]
31. Basalp, E.; Bachmann, P.; Gerig, N.; Rauter, G.; Wolf, P. Configurable 3D rowing model renders realistic forces on a simulator for indoor training. *Appl. Sci.* **2020**, *10*, 734. [CrossRef]
32. Gao, Y.; Kristensen, L.; Grøndberg, T.; Murray, M.; Sjøgaard, G.; Søgaard, K. Electromyographic evaluation of specific elastic band exercises targeting neck and shoulder muscle activation. *Appl. Sci.* **2020**, *10*, 756. [CrossRef]
33. Yamashita, D.; Murata, M.; Inaba, Y. Effect of landing posture on jump height calculated from flight time. *Appl. Sci.* **2020**, *10*, 776. [CrossRef]
34. Pietraszewska, J.; Struzik, A.; Burdukiewicz, A.; Stachoń, A.; Pietraszewski, B. Relationships between body build and knee joint flexor and extensor torque of polish first-division soccer players. *Appl. Sci.* **2020**, *10*, 783. [CrossRef]
35. Hébert-Losier, K.; Hanzlíková, I.; Zheng, C.; Streeter, L.; Mayo, M. The 'DEEP' landing error scoring system. *Appl. Sci.* **2020**, *10*, 892. [CrossRef]
36. Silva, H.; Ferreira, H.; Rocha, C.; Monteiro Rodrigues, L. Texture analysis is a useful tool to assess the complexity profile of microcirculatory blood flow. *Appl. Sci.* **2020**, *10*, 911. [CrossRef]
37. Rúbio, G.; Martins Ferreira, F.; Brandão, F.; Machado, V.; Tonelli, L.; Martins, J.; Kozan, R.; Vimieiro, C. Evaluation of commercial ropes applied as artificial tendons in robotic rehabilitation orthoses. *Appl. Sci.* **2020**, *10*, 920. [CrossRef]
38. Olchowik, G.; Czwalik, A. Effects of soccer training on body balance in young female athletes assessed using computerized dynamic posturography. *Appl. Sci.* **2020**, *10*, 1003. [CrossRef]

39. Stein, K.; Mombaur, K. Whole-Body dynamic analysis of challenging slackline jumping. *Appl. Sci.* **2020**, *10*, 1094. [CrossRef]
40. King, M.; Towler, H.; Dillon, R.; McErlain-Naylor, S. A correlational analysis of shuttlecock speed kinematic determinants in the badminton jump smash. *Appl. Sci.* **2020**, *10*, 1248. [CrossRef]

applied sciences

MDPI

Article

A Correlational Analysis of Shuttlecock Speed Kinematic Determinants in the Badminton Jump Smash

Mark King [1], Harley Towler [1,*], Romanda Dillon [1] and Stuart McErlain-Naylor [1,2]

1 School of Sport, Exercise, and Health Sciences, Loughborough University, Loughborough LE11 3TU, UK; M.A.King@lboro.ac.uk (M.K.); R.Dillon@lboro.ac.uk (R.D.); s.mcerlain-naylor@uos.ac.uk (S.M.-N.)

2 School of Health and Sports Sciences, University of Suffolk, Ipswich IP3 8AH, UK

* Correspondence: H.Towler@lboro.ac.uk; Tel.: +44-0795-792-7465

Received: 31 December 2019; Accepted: 10 February 2020; Published: 13 February 2020

check for updates

Featured Application: The findings suggest that players and/or coaches should focus on proximal segment movements: specifically, producing greater pelvis-thorax separation during the retraction phase and greater shoulder internal rotation at shuttle contact to increase shuttlecock speed.

Abstract: The forehand jump smash is an essential attacking stroke within a badminton player's repertoire. A key determinate of the stroke's effectiveness is post-impact shuttlecock speed, and therefore awareness of critical technique factors that impact upon speed is important to players/coaches. Three-dimensional kinematic data of player, racket and shuttlecock were recorded for 18 experienced players performing maximal effort forehand jump smashes. Joint angles and X-factor (transverse plane pelvis-thorax separation) were calculated at key instants: preparation, end of retraction, racket lowest point, turning point and shuttlecock contact. Peak shoulder, elbow, and wrist joint centre linear velocities, phase durations and jump height were also calculated. Correlational analyses were performed with post-impact shuttlecock speed, revealing significant correlations to peak wrist joint centre linear velocity ($r = 0.767$), acceleration phase duration ($r = -0.543$), shoulder internal/external rotation angle at shuttlecock contact ($r = 0.508$) and X-factor at the end of retraction ($r = -0.484$). Multiple linear regression analysis revealed 43.7% of the variance in shuttlecock speed could be explained by acceleration phase duration and X-factor at the end of retraction, where shorter acceleration phase durations and more negative X-factor at end of retraction caused greater shuttlecock speeds. These results suggest that motions of the proximal segments (shoulder and pelvis–thorax separation) are critical to developing greater distal linear velocities, which subsequently lead to greater post-impact shuttlecock speed.

Keywords: velocity; technique; overhead; racket; swing; stroke

---

1. Introduction

The forehand smash is an effective attacking shot in badminton, accounting for 54% of "unconditional winner" and "forced failure" shots in international matches [1]. Success of the stroke is dependent upon two components: speed and direction, where speeds as high as $89.3 \pm 7.2\ m{\cdot}s^{-1}$ have been reported in the literature for elite Malaysian players [2], whereas the competition world record is $118\ m{\cdot}s^{-1}$ [3]. A shuttlecock with a greater post-impact speed will give an opponent less reaction time, while directing the shuttlecock away from the opponent requires them to make fast reactive movements in order to return the shuttlecock.

Several studies have determined that both linear and angular velocities of distal segments (hand and racket) are strong positive correlates with shuttlecock speed [2,4], which is unsurprising due to higher velocities causing greater transfer of linear momentum to the shuttlecock. Explaining what kinematic parameters determine this greater distal velocities may be more useful for players, coaches and practitioners, as the greater joint powers associated with distal segments are not fully generated by the muscles associated with these segments, but transferred between segments through reaction forces of a proximal-to-distal nature, e.g., shoulder–elbow–wrist [5]. Varied experimental set-ups and methodologies of quantifying joint contributions have generated conflicting results, where Rambely et al. [4] suggested that a distal-to-proximal order of wrist, followed by the elbow and shoulder are the major contributors to racket head speed (26.5%, 9.4% and 7.4%, respectively). In contrast, Liu et al. [6] reported that a proximal-to-distal order of shoulder internal rotation, forearm pronation and wrist palmar flexion contribute 66%, 17% and 11%, respectively, towards racket head speed. Rambely et al. [5] captured video data at 50 Hz and calculated the resultant linear velocities of the shoulder, elbow and wrist joint centres at impact, and expressed the segment contributions as these velocities represented as percentages of the racket head centre resultant velocity at impact. Conversely, Liu et al. [6] used a three-dimensional kinematic method developed by Sprigings et al. [7] and calculated the relative angular velocities of distal segments by removing contributions from the adjacent proximal segment. Data were captured using high-speed video (200 Hz) and processed using direct linear transformation [8]. Teu et al. [9] found no specific proximity order using a dual Euler angles method with high-speed video (200 Hz), and reported torso rotation, shoulder internal rotation, forearm pronation and wrist abduction contributed 57%, 3%, 27% and 10% to resultant racket head velocity, respectively.

In summary, previous research has reported that linear velocities of the distal segments best explain variation in shuttlecock speed/racket head speed, however it is unclear how the distal segment velocities and subsequent racket head and shuttlecock speeds are generated. Additionally, use of low frame rates and unclear methodology for defining both shuttlecock speed and racket head speed mean that it is difficult to compare results. The present study therefore aims to identify full-body kinematic parameters that best explain the generation of post-impact shuttlecock velocities in the badminton jump smash, such that coaches/practitioners can advise players how best to increase smash speeds through technique and/or strength training. It is hypothesised that positions (joint angles) of proximal segments and linear velocities of more distal segments will best explain variation in post-impact shuttlecock speed.

2. Materials and Methods

2.1. Participants

Eighteen male badminton players (mean ± SD: age 24.3 ± 7.1 years, height 1.84 ± 0.08 m, mass 79.6 ± 8.8 kg) of regional (n = 9), national (n = 4) and international (n = 5) standards participated in this study, each performing a series of twelve forehand jump smashes from a racket-fed lift via an international coach/player, representative of match conditions. A range of abilities were used to provide a variety of maximal smash speeds, facilitating an investigation into causal factors associated with this variation. Testing procedures were explained to each participant, and informed written consent was obtained in accordance with the guidelines of the Loughborough University Ethical Advisory Committee (SSEHS-1959).

2.2. Data Collection

An eighteen-camera Vicon Motion Analysis System (400 Hz; OMG Plc, Oxford, UK) was used to collect three-dimensional kinematic data of the participant, racket and shuttlecock on a mock badminton court within a hall of sufficient height. Forty-seven 14 mm retroreflective markers were attached to the participant (Figure 1), where joint centres were calculated from a pair of markers placed across the joint so that their midpoint coincided with the joint centre [10]. Hip, thorax, neck and head joint centres were calculated according to Worthington et al. [11]. A further marker was placed on the bottom of the racket handle, seven pieces of 3M Scotchlite reflective tape were attached to the racket frame and a single piece of reflective tape was attached around the base of the cork of the shuttlecock (Figure 2). Participants used their own racket and new Yonex AS40 shuttlecocks throughout, where misshapen or broken shuttlecocks were removed.

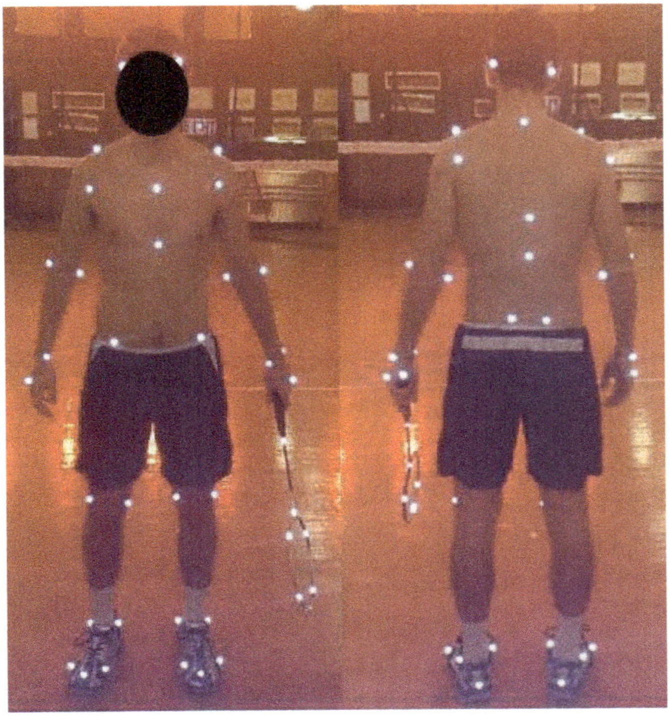

Figure 1. Participant marker locations.

Figure 2. Racket and shuttlecock marker locations.

2.3. Data Reduction

Position data were labelled within Vicon Nexus 1.7.1 where gaps within marker trajectories were filled using the "pattern-filled" function where possible, and "spline-filled" function thereafter. All position data were then imported into Matlab v.2018b (The MathWorks Inc., Natick, MA, USA) for all further processing. Position data of all body markers were filtered using a fourth-order, zero-phase, low-pass Butterworth filter with a cut-off frequency of 30 Hz, determined through residual analysis [12]. Racket and shuttlecock markers remained unfiltered to avoid double-filtering during subsequent curve-fitting methodologies.

Joint angles were calculated using three-dimensional rotation matrices, defining the rotation applied to the proximal segment coordinate system to bring it into coincidence with the coordinate system of the distal segment [12]. An XYZ rotation sequence was used, representing flexion–extension, adduction–abduction and longitudinal axis rotation, respectively. When describing humerothoracic motion, a YZY rotation sequence was used as recommended by ISB [13], using a different coordinate system for the humerus segment where xyz represented adduction–abduction, longitudinal axis rotation and flexion–extension, respectively. Wrist angles were normalised based on the player adopting their normal grip within a static trial, which was considered the neutral position. The mean offset was 28.8 ± 8.4° and 7.1 ± 7.4° of palmar extension and ulnar deviation, respectively. X-factor referred to the separation angle between vectors connecting the right and left shoulder joint centres and, the right and left hip joint centres, respectively, in the transverse plane [14]. Centre of mass was calculated using segment inertial values from de Leva [15], where the body was modelled as fourteen segments. Jump height was defined as the difference between the maximum centre of mass height and the height of that during a static standing trial. Table 1 details the joint angles calculated.

Table 1. Calculated joint angles, with their relative zero positions and positive directions.

Joint	Motion	Anatomical Position (°)	Positive Direction
Shoulder	Internal/External Rotation	†	Internal Rotation
Elbow	Flexion/Extension	0	Extension
	Pronation/Supination	0	Supination
Wrist	Palmar Flexion/Extension	180	Extension
	Ulnar/Radial Deviation	0	Ulnar Deviation
Trunk	X-Factor (Transverse Plane)	0	‡

† For further details on motion of the shoulder joint as a result of the ISB rotation sequence recommendations, see Wu et al. [13]. ‡ Positive direction relates to the direction of rotation during the forward swing of the movement, i.e., anticlockwise for a right-handed player. Note transverse plane is viewed from above.

Instantaneous post-impact shuttlecock speed and racket–shuttlecock contact timing were determined using a logarithmic curve-fitting methodology [16] with minor adjustments for the application to the badminton smash. The time of impact was derived from the intersection of pre- and post-impact shuttlecock displacement curves in the global anterior–posterior direction (dominant direction of the smash), and an intermediate 1 ms contact period [17] was added between the pre- and post-impact shuttlecock curves, where the racket face and shuttlecock velocity were assumed equal. Post-impact shuttlecock speed was determined via differentiation of the post-impact logarithmic shuttlecock displacement curve. Racket head speed was the component of the linear racket head centre linear velocity acting perpendicular to the racket stringbed. The pre-impact racket head speed data was then interpolated to the calculated time of initial shuttlecock contact.

Based upon previous literature within badminton, the movement was defined around two phases, defined as backswing and acceleration phases [18]. Five discrete instants were identified such that trials could be compared appropriately: preparation (P) was defined as the point at which centre of mass height was minimal [2]; end of retraction (ER) was defined as the point at which the racket was most medio-laterally positioned towards the non-dominant side of the participant within the global coordinate system; racket lowest point (RLP) when the racket tip was at its lowest vertical point [19]; and turning point (TP) was defined as the point, after minimum (most negative), at which the racket head speed became positive [4] and shuttlecock contact (SC) was defined as the closest motion capture frame to the previously defined instant of racket-shuttlecock contact. The backswing phase (BSP) was defined as the time between PP and TP, whereas the acceleration phase (AP) was defined as the time between TP and SC.

Five joint angles (Table 1) were calculated for each trial, describing the elements of badminton smash technique which have previously been linked to shuttlecock velocity in literature or thought to be linked to shuttlecock velocity. Joint angles were defined at each key instant and their maximum range of motion through to SC calculated, e.g., maximum external rotation angle through to angle at contact. Furthermore, post-impact shuttlecock speed; racket head speed at impact; jump height; and peak shoulder, elbow and wrist joint centre linear velocities were calculated for each trial. Length of phases (BSP and AP), as well as total swing time, were also calculated for each trial. All kinematic variables for each player's trial with the greatest shuttlecock speed were entered into subsequent correlation analyses.

2.4. Statistical Analysis

All statistical correlational analyses were performed in Matlab v.2018b (The MathWorks Inc., Natick, MA, USA). Pearson product moment correlation analyses were performed between each kinematic (independent) variable and shuttlecock speed. Pearson product moment correlations (r) and their 95% confidence intervals (CI) were interpreted as negligible < 0.3; $0.3 \leq$ low < 0.5; $0.5 \leq$ moderate < 0.7; $0.7 \leq$ high < 0.9; very high ≥ 0.9 [20]. An alpha value of 0.05 was used to determine significance. Correlates were then entered as "candidate" variables into a forwards stepwise multiple linear regression model to identify key kinematic parameters that best explain variation in shuttlecock speed. Entry requirements for inclusion of a parameter were $p < 0.05$, with a removal coefficient of $p > 0.10$. The regression model was rejected if the 95% CI coefficients included zero, the residuals of the predictor were heteroscedastic or if the bivariate correlations, tolerance statistics or variance inflation factors showed any evidence of multicollinearity [21–25]. The normality of the standardised residuals in the regression model was also confirmed using the Shapiro–Wilk test. The percentage of variance in the dependent variable explained by the independent variables (predictors) within the regression equation was determined by Wherry's adjusted R^2 value [26]. Multiple linear regression analysis was performed in IBM SPSS Statistics 23 (IBM, Armonk, NY, USA).

3. Results

Maximal shuttlecock speeds for the cohort were 89.6 ± 5.3 m·s^{-1} (range: 80.1–99.8 m·s^{-1}). Racket head speeds at SC achieved during these trials were 56.3 ± 4.0 m·s^{-1} (range: 46.7–64.6 m·s^{-1}). Five kinematic variables were significantly correlated with shuttlecock speed, where a greater racket head speed, greater peak wrist joint centre linear velocity and shorter acceleration phase duration were associated with greater shuttlecock speeds. Likewise, greater shoulder internal rotation at SC and more negative X-factor at ER produced greater shuttlecock speeds. No other variables were significantly correlated to shuttlecock speed (Tables 2–4). Means and standard deviations, as well as a time-normalized comparison between the "fastest" and "slowest" participant for shoulder internal rotation angle, X-factor and racket head speed are shown in Figures 3–5, respectively.

Table 2. Pearson product moment correlation (r) between each of racket head speed, jump height, phase durations, joint centre linear velocities and ranges of motion, and shuttlecock speed.

Kinematic Variable	Key Instant/Phase	Mean (SD)	r	95% CI	p
Racket Head Speed (m·s^{-1})	SC	56.3 (4.0)	0.903	0.753, 0.964	<0.001 *
Jump Height (cm)		31.6 (9.2)	0.454	−0.017, 0.760	0.059
Phase duration (ms)	BSP	509.0 (94.8)	0.412	−0.067, 0.737	0.089
	AP	38.1 (5.2)	−0.543	−0.805, −0.101	0.020 *
	TS	547.1 (92.5)	0.392	−0.092, 0.726	0.108
Peak Shoulder JC LV (m·s^{-1})		3.5 (0.4)	0.177	−0.316, 0.595	0.482
Peak Elbow JC LV (m·s^{-1})		8.3 (0.8)	0.353	−0.136, 0.704	0.151
Peak Wrist JC LV (m·s^{-1})		14.2 (1.7)	0.767	0.467, 0.908	<0.001 *
Shoulder IR ROM (°)	Peak−SC	98.1 (20.7)	0.403	−0.079, 0.732	0.097
Elbow Flexion ROM (°)	Peak−SC	106.0 (7.9)	0.016	−0.454, 0.479	0.950
Elbow PRO ROM (°)	Peak−SC	21.2 (14.9)	0.298	−0.197, 0.671	0.230
Wrist Flexion ROM (°)	Peak−SC	40.4 (12.5)	−0.242	−0.254, 0.637	0.333
X−Factor ROM (°)	Peak−SC	46.0 (9.4)	0.208	−0.287, 0.615	0.408

Abbreviations: CI: confidence interval, BSP: backswing phase, AP: acceleration phase, TS: total swing, JC LV: joint centre linear velocity, IR: internal rotation, ROM: range of motion, PRO: pronation. * Significant ($p < 0.05$) .

Table 3. Pearson product moment correlation (r) between joint angles of the shoulder and elbow at key instants and shuttlecock speed.

Kinematic Variable	Key Instant/Phase	Mean (SD)	r	95% CI (Lower, Upper)	p
Shoulder INT/EXT Rotation Angle (°)	P	102.4 (29.8)	0.324	−0.168, 0.678	0.189
	ER	1.9 (8.8)	0.013	−0.457, 0.477	0.959
	RLP	−33.6 (8.2)	0.161	−0.331, 0.584	0.523
	TP	−29.2 (13.7)	0.299	−0.195, 0.672	0.228
	SC	58.7 (20.0)	0.508	0.054, 0.788	0.031 *
Elbow Flexion Angle (°)	P	88.4 (19.8)	0.329	−0.163, 0.690	0.183
	ER	64.9 (8.0)	−0.206	−0.614, 0.289	0.413
	RLP	74.2 (9.8)	0.161	−0.648, 0.236	0.298
	TP	120.6 (10.5)	0.225	−0.271, 0.626	0.370
	SC	166.1 (4.3)	0.243	−0.252, 0.638	0.331
Elbow Pronation Angle (°)	P	−61.6 (17.5)	0.398	−0.084, 0.730	0.102
	ER	−99.4 (20.0)	−0.180	−0.597, 0.313	0.475
	RLP	−84.0 (13.2)	0.120	−0.556, 0.368	0.636
	TP	−95.4 (15.4)	−0.376	−0.717, 0.111	0.125
	SC	−101.5 (19.0)	−0.420	−0.741, 0.059	0.083

Abbreviations: KI: key instant, CI: confidence interval, INT: internal, EXT: external, P: preparation, ER: end of retraction, RLP: racket lowest point, TP: turning point, SC: shuttlecock contact. * Significant ($p < 0.05$).

Table 4. Pearson product moment correlation (r) between joint angles of the wrist and X-factor at key instants and shuttlecock speed.

Kinematic Variable	Key Instant/Phase	Mean (SD)	r	95% CI (Lower, Upper)	p
Wrist Flexion Angle (°)	P	191.0 (13.0)	0.368	−0.120, 0.712	0.133
	ER	208.2 (10.6)	−0.134	−0.565, 0.356	0.598
	RLP	225.4 (9.9)	0.146	−0.344, 0.574	0.563
	TP	225.8 (9.3)	0.021	−0.451, 0.483	0.935
	SC	187.8 (8.9)	0.190	−0.304, 0.603	0.451
X−factor (°)	P	0.8 (8.6)	0.024	−0.448, 0.485	0.926
	ER	−33.3 (5.8)	−0.484	−0.775, −0.022	0.042 *
	RLP	−23.5 (7.1)	−0.362	−0.709, 0.126	0.140
	TP	−5.2 (7.6)	−0.175	−0.593, 0.318	0.489
	SC	7.9 (7.9)	−0.077	−0.525, 0.404	0.760

Abbreviations: KI: key instant, CI: confidence interval, P: preparation, ER: end of retraction, RLP: racket lowest point, TP: turning point, SC: shuttlecock contact. * Significant ($p < 0.05$).

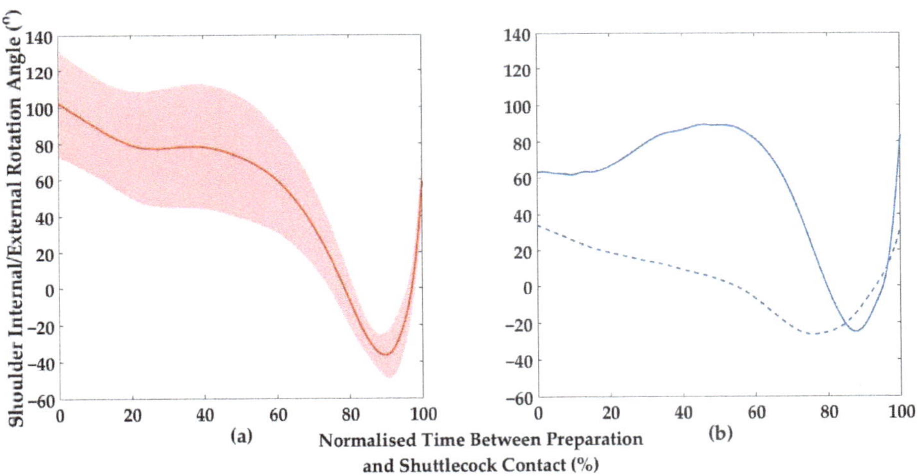

Figure 3. (**a**) Mean (solid line) ± SD (shaded area) for the normalised racket arm shoulder internal/external rotation angle between preparation and shuttlecock contact key instants. (**b**) Normalised racket arm shoulder internal/external rotation angle between preparation and shuttlecock contact key instants, for participants with the fastest (solid line) and slowest (dashed line) participant's smash. Internal rotation (positive); external rotation (negative). A significant correlation between shuttlecock speed and shoulder internal rotation angle was present at shuttlecock contact.

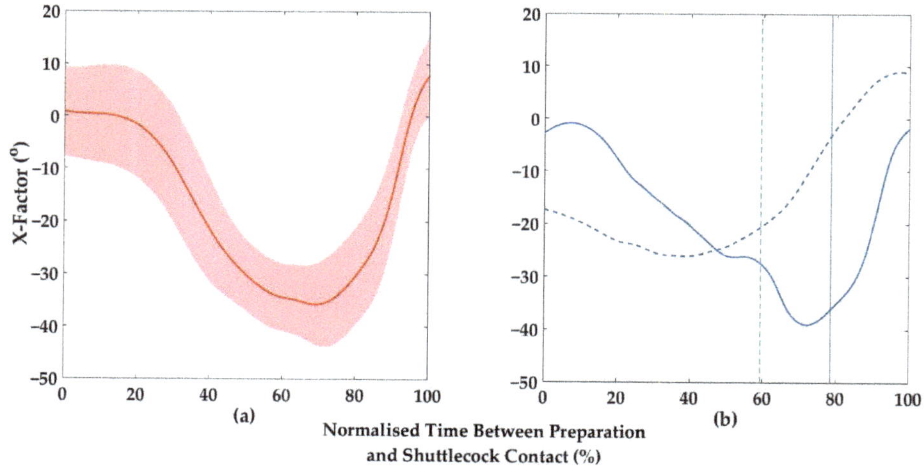

Figure 4. (**a**) Mean (solid line) ± SD (shaded area) for normalised X-factor between preparation and shuttlecock contact key instants for all participants; (**b**) Normalised X-factor between preparation and shuttlecock contact key instants for participants with the fastest (solid line) and slowest (dashed line) smash. The end of retraction (ER) key instant for each participant is represented by a vertical line. X-factor at ER was significantly correlated with shuttlecock speed.

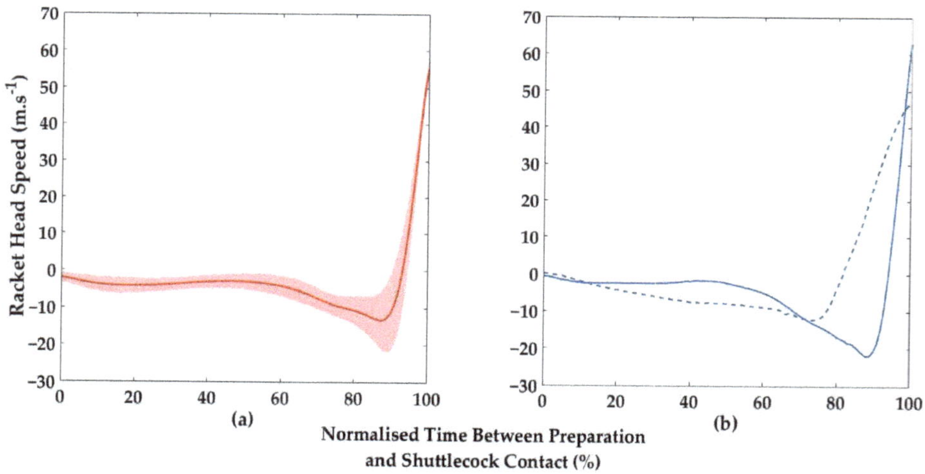

Figure 5. (**a**) Mean (solid line) ± SD (shaded area) for the normalised racket head speed between preparation and shuttlecock contact key instants for all participants; (**b**) Normalised racket head speed (normal to the stringbed plane) between preparation and shuttlecock contact key instants for participants with the fastest (solid line) and slowest (dashed line) smash. The acceleration phase begins at the point at which the racket head speed becomes positive, following the initial negative peak, and the duration between this turning point and shuttlecock contact was significantly negatively correlated with shuttlecock speed.

Multiple regression analysis revealed that the duration of AP alone explained 25.0% of the variation in shuttlecock speed, with a standard error of the estimate (SEE) of 4.4 m·s^{-1}. The addition of X-factor at ER explained 43.7% of the variance in post-impact shuttlecock speed (SEE = 3.8 m·s^{-1}), where a shorter acceleration phase and a more negative X-Factor at ER caused greater shuttlecock speeds (Table 5; Figure 6).

Table 5. Multiple regression equations explaining variance in shuttlecock speed.

Model	Kinematic Parameters	Coefficient	95% CI		Variable, *p*	Model		
			Lower Bound	Upper Bound		Percent Explained	*p*	SEE
a	(Constant)	110.1	93.2	127.0	<0.001	25.0	0.02	4.4
	AP Duration	−536.5	−976.8	−96.2	0.020			
b	(Constant)	95.8	76.7	114.9	<0.001	43.7	0.05	3.8
	AP Length	−514.0	−898.1	−129.9	0.012			
	X−Factor at ER	−0.4	−0.7	−0.1	0.024			

Abbreviations: CI: confidence intervals, SEE: standard error of the estimate, AP: acceleration phase, ER: end of retraction.

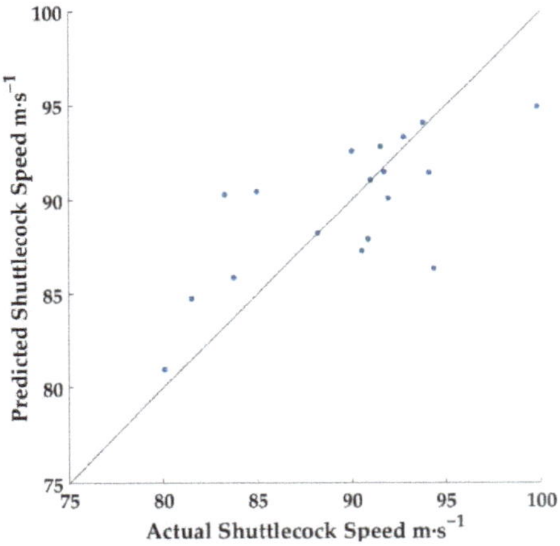

Figure 6. Predicted shuttlecock speed against actual shuttlecock speed for the two-parameter stepwise regression equation (Table 5; Model b). With a higher percentage of the variation in shuttlecock speed explained the closer the data points lie to the line y = x (predicted shuttlecock speed = actual shuttlecock speed).

4. Discussion

Racket head speeds and shuttlecock speeds achieved by the participants showed good agreement with previously reported values by elite players [2,27,28]. Shuttlecock speed was greatest in the international players (94.4 ± 3.2 $m{\cdot}s^{-1}$), followed by national players (91.2 ± 2.6 $m{\cdot}s^{-1}$) and then regional players (86.4 ± 4.6 $m{\cdot}s^{-1}$), suggesting that the ability to achieve greater post-impact shuttlecock speeds is a good indication of playing level [29]. A one-way ANOVA revealed a significant difference in shuttlecock speed between groups ($F_{(2,15)} = 7.29$, $p = 0.006$). Bonferroni post-hoc tests revealed the international group smashed significantly faster than the regional group (mean difference = 8.06 $m{\cdot}s^{-1}$, CI: 2.22, 13.90; $p = 0.006$). The national group were not significantly different to either the international or regional groups. Five kinematic variables correlated significantly with shuttlecock speed: racket head speed at impact, peak wrist joint centre linear velocity, length of the acceleration phase, shoulder internal rotation angle at SC and X-factor (transverse) at ER. These results are indicative of proximal joint motions (shoulder and trunk) causing greater linear velocities in the distal segments (wrist and racket).

Racket head speed at impact, normal to the stringbed plane, correlated very highly (r = 0.903; CI: 0.753, 0.964; $p < 0.001$) with instantaneous post-impact shuttlecock speed. This is a larger effect size than those previously reported [2,27], which may be due to the more accurate calculation of post-impact shuttlecock speed i.e., at a precise time of impact, and an appropriate racket head speed (normal to the stringbed plane) interpolated to the time of initial contact with the shuttlecock in the present study. The non-perfect relationship between racket head and shuttlecock speed may be due to differences in racket specifications used by each player, where impact efficiency (location and coefficient of restitution), as well as mass properties of the racket, will affect the amount of linear momentum transferred to the shuttlecock, for a given impact velocity. Longitudinal impact locations, calculated using the previously described methodology of Peploe et al. [16], ranged between −29.4 and 24.9 mm from the racket head centre, whereas medio-lateral impact locations ranged between −33.2 and 17.1 mm from the racket head centre (medial–negative, lateral–positive; relative to the player). Modelling the racket as two segments (handle and frame) and assuming a frame transverse angular velocity of 80 rad·s^{-1} at impact, consistent with experienced players [28], the difference in racket head speed at the most proximal and distal longitudinal impact locations would be 4.3 m·s^{-1}, suggesting that impact location can have a large effect on subsequent shuttlecock speed, even given identical input in terms of racket angular velocities.

As well as racket head speed, peak linear velocity of the wrist joint centre, was found to be a strong predictor of shuttlecock speed (r = 0.767; CI: 0.467, 0.908; $p < 0.001$), indicating that players should ultimately aim to achieve high linear velocities within distal components of the kinetic chain. This correlation was stronger than previous reported by Rambely et al. [4] in elite international players, who reported a low positive relationship (r = 0.454) despite mean values in the present study showing good agreement (14.2 vs. 11.7 m·s^{-1}) with differences attributable to the variety in player ability in the present study and differences in capture frequency (400 vs. 50 Hz), where 50 Hz may be inadequate for the badminton smash [28]. Greater shuttlecock speeds were also produced when AP was shorter in duration (r = -0.543; CI: -0.805, −0.101; $p = 0.020$), demonstrated in Figure 5b, where the racket head speed becomes positive much closer to impact, when comparing the "fastest" to the "slowest" participant.

Furthermore, two kinematic technique factors significantly correlated with shuttlecock speed. First, players who were in a position of greater shoulder internal rotation at impact produced greater shuttlecock speeds (r = 0.508, moderate; CI: 0.054, 0.788; $p = 0.031$), where a clear difference in position between the "fastest" and "slowest" player is shown in Figure 3b, despite very similar minimum positions. This suggests that greater shoulder internal rotation allows more work done by this joint rotation, which Liu et al. [6] found contributed 66% towards racket head velocity. However, correlational analysis revealed a non-significant relationship between shoulder internal rotation range of motion and shuttlecock speed (r = 0.403, low; CI: −0.079, 0.732; $p = 0.097$). When accounting for the time in which this range was completed, i.e., average angular velocity, the correlation came closer to significance (r = 0.444, low; CI: −0.029, 0.755; $p = 0.065$). Second, a low negative correlation was found between X-factor at ER and shuttlecock speed (r = 0.484; CI: −0.755, −0.022; $p = 0.042$), where players who produced a greater pelvis–thorax separation angle in the transverse plane at ER produced greater shuttlecock speeds. Again, a clear difference is seen between the "fastest" and "slowest" participants in Figure 4b. X-factor (maximum pelvis-thorax separation) has previously been found to be a strong correlate (r = 0.60, $p < 0.01$) with shuttlecock speed [12]. A greater rotational countermovement of the torso presumably allows greater trunk rotation contribution to racket head speed, which Teu et al. [9] found to contribute 57% to racket head speed.

It has been previously reported that both peak vertical ground reaction force and jump height significantly correlate with shuttlecock speed (r = 0.548 and 0.508, respectively) [2]. The present study found a non-significant correlation, with CI marginally crossing zero, between jump height and shuttlecock speed (r = 0.454; CI: −0.017, 0.760; p = 0.059); however, a similar effect size. Perhaps greater jump heights are a characteristic of more able players, who typically produce greater shuttlecock speeds [12,29], and serves as a tactical factor allowing players to produce steeper smash strokes as opposed to producing greater shuttlecock speeds. Note that jump height was significantly correlated with racket head speed in the present study (r = 0.494; CI: 0.035, 0.781; $p = 0.037$).

The fact that distal linear velocities (wrist and racket) best explain variation in shuttlecock speed, yet proximal angles of the trunk and shoulder best explain variation in shuttlecock speed, is suggestive of the kinetic link principle whereby movements of a proximal-to-distal nature generate and conserve angular momentum to produce high distal end-point velocities [5,30–32]. Important longitudinal axis rotations, typically difficult to measure and observe, may not always follow this strict sequence with regards to timing, however the proximal-to-distal nature of overhead strokes provides a good general understanding of how high distal end-point velocities can be generated [33,34]. A more negative X-factor at ER causing greater shuttlecock speed endorses the idea of the stretch–shortening cycle, whereby more elastic energy is stored and recovered to enhance the concentric phase when X-factor is more negative at ER. The stretch–shortening cycle has been linked to greater velocities in throwing actions due to enhancement of the concentric phase [35,36]. Finally, no ranges of motion were found to significantly correlate with shuttlecock speed. Lees et al. [32] previously suggested that increasing the range of motion can improve performance (racket head speed) by increasing the acceleration path of the racket, allowing more muscular force to be generated and applied to accelerate the racket.

A further relevant biomechanical principle, not explored in this study, is the velocity lever principle [32]. The angle between the forearm and racket longitudinal axes is of importance for generating racket head speed. For example, if the racket is held at 90° to the forearm, then the racket head will move through the greatest distance when the forearm pronates, increasing the racket head linear speed for any given angular velocity of pronation [31]. Likewise, if the elbow is flexed at 90°, the contribution from shoulder internal rotation is maximised [30]. This principle is difficult to analyse within a complex motion such as the badminton smash, as multiple segmental rotations are responsible for producing the racket motion, as well as ensuring racket head orientation is optimal at impact, which may make certain joint angles, such as a racket–forearm angle of 90°, undesirable. Tang et al. [37] reported that within their cohort of four elite players, the average racket–forearm angle was 147°, which may represent a suitable compromise between the height and the speed at contact in practical play.

The multiple linear regression analysis found that two predictor variables were able to explain 43.7% of the variance in shuttlecock speed (Table 5; Figure 6). Participants with the fastest smashes were found to have a shorter acceleration phase duration and more negative (greater separation) X-factor at ER. From a practical standpoint, this would suggest that players attempt to delay the onset of their forward swing (when the racket head velocity normal to the stringbed becomes positive) such that the forward swing can be completed in the shortest possible time. This would ultimately achieve a greater velocity of the racket, i.e., the same acceleration path of the racket for a given player but completed in a shorter time period [31]. Additionally, players should seek to utilise as much counter-rotation of the trunk as possible before reversing this rotation within the acceleration phase.

A potential limitation to the present study was use of each participant's own racket causing a lack of experimental control over the impact mechanics between racket and shuttlecock, including effective mass of the "racket particle" within the collision between racket and shuttlecock, and thus the momentum transferred [27]. The effect of racket properties on player kinematics is also a potential source of limitation, where Whiteside et al. [38] reported that increasing the "swingweight" of the racket caused a reduction in peak angular velocities of shoulder internal rotation and wrist flexion during the tennis serve, where both of these joint rotations have been reported to be the two greatest contributors to racket head speed at impact [39]. Conversely, players are accustomed to their own racket, and introducing a "control" racket may require a great amount of familiarization or cause suboptimal performances due to unfamiliarity with the racket. It must also be acknowledged that the present study relies on investigation of joint kinematics at discrete time points (i.e., key instants). Future studies may therefore extend the current work using methodologies to investigate the continuous time series of kinematic data, such as statistical parametric mapping [40], vector coding [41] or principal component analysis [42], as well as considering other biomechanical principles not explored within this study. Additionally, differences in anthropometric data were not accounted for, where the same angular velocities and joint angles may lead to different linear velocities. Retrospective power analysis revealed that for the lowest significant correlation coefficient ($r = 0.484$), with 80% power and a significance threshold ($p = 0.05$), a sample size of 31 would be required. The study was therefore underpowered; however, recruiting more participants meeting the minimum standard criteria was not achievable.

5. Conclusions

In conclusion, proximal kinematics explained the greatest proportion of variation in shuttlecock speed during the forehand jump smash stroke in a cohort of experienced male badminton players. From a practical standpoint, it is suggested that players and/or coaches attempt to produce high shuttlecock speeds by increasing racket and distal joint centre linear velocities. This should be achieved primarily by having a greater internal rotation angle at shuttlecock contact. Furthermore, greater pelvis–thorax separation during the backswing phase is likely to aid the concentric phase of the swing via the stretch–shortening cycle.

Author Contributions: Conceptualisation, methodology and data collection: all authors; data processing: H.T. and R.D.; formal analysis and writing—original draft preparation: H.T. and S.M.-N.; writing—review and editing: all authors; supervision, M.K. and S.M.-N. All authors have read and agreed to the published version of the manuscript.

Funding: This research received no external funding.

Conflicts of Interest: The authors declare no conflict of interest.

References

1. Tong, Y.-M.; Hong, Y. The playing pattern of world's top single badminton players. In Proceedings of the 18th International Symposium on Biomechanics in Sports, Hong Kong, China, 25–30 June 2000; pp. 825–830.
2. Ramasamy, Y.; Osman, J.; Sundar, V.; Joseph, S. Ground reaction force and kinematics of forehand jumping smash among elite Malaysian badminton players. In Proceedings of the 37th International Society of Biomechanics in Sport Conference, Oxford, OH, USA, 21–25 July 2019.
3. Guinness World Records, Fastest Badminton Hit in Competition (Male). Available online: https://www.guinnessworldrecords.com/world-records/fastest-badminton-hit-in-competition-(male)/ (accessed on 19 June 2019).
4. Rambely, A.S.; Osman, N.A.A.; Usman, J.; Abas, W.A.B.W. The contribution of upper limb joints in the development of racket velocity in the badminton smash. In Proceedings of the 23rd International Symposium on Biomechanics in Sports, Beijing, China, 22–27 August 2005.
5. Rasmussen, J.; Kwan, M.; Andersen, M.S.; De Zee, M. Analysis of segment energy transfer using musculoskeletal models in a high speed badminton stroke. In Proceedings of the 9th International Symposium on Computer Methods in Biomechanics and Biomedical Engineering, Valencia, Spain, 24–27 February 2010.
6. Liu, X.; Kim, W.; Tan, J. An Analysis of the Biomechanics of Arm Movement during a Badminton Smash. Nanyang Technology University: Singapore, 2002.
7. Sprigings, E.; Marshall, R.; Elliott, B.; Jennings, L. A three-dimensional kinematic method for determining the effectiveness of arm segment rotations in producing racquet-head speed. *J. Biomech.* **1994**, *27*, 245–254. [CrossRef]
8. Abdel-Aziz, Y.I.; Karara, H.M. Direct linear transformation from comparator coordinates into object coordinates in close-range photogrammetry. In Proceedings of the ASP Symposium on Close-Range Photogrammetry, Urbana, IL, USA, 26–29 January 1971.
9. Teu, K.K.; Kim, W.; Tan, J.; Fuss, F.K. Using dual Euler angles for the analysis of arm movement during the badminton smash. *Sports Eng.* **2005**, *8*, 171–178. [CrossRef]
10. McErlain-Naylor, S.; King, M.; Pain, M.T.G. Determinants of countermovement jump performance: A kinetic and kinematic analysis. *J. Sports Sci.* **2014**, *32*, 1805–1812. [CrossRef] [PubMed]

11. Worthington, P.; King, M.; Ranson, C. The influence of cricket fast bowlers' front leg technique on peak ground reaction forces. *J. Sports Sci.* **2013**, *31*, 434–441. [CrossRef] [PubMed]
12. Winter, D.A. *Biomechanics and Motor Control of Human Movement*, 4th ed.; Wiley: New York, NY, USA, 2009; pp. 70–73.
13. Wu, G.; Van Der Helm, F.C.T.; Veeger, H.E.J.; Makhsous, M.; Van Roy, P.; Anglin, C.; Nagels, J.; Karduna, A.R.; McQuade, K.; Wang, X.; et al. ISM recommendation on definitions of joint coordinate systems of various joints for the reporting of human joint motion-Part II: Shoulder, elbow, wrist and hand. *J. Biomech.* **2005**, *38*, 981–992. [CrossRef]
14. Zhang, Z.; Li, S.; Wan, B.; Visentin, P.; Jiang, Q.; Dyck, M.; Shan, G. The influence of X-factor (trunk rotation) and experience on the quality of the badminton forehand smash. *J. Hum. Kinet.* **2016**, *53*, 9–22. [CrossRef]
15. De Leva, P. Adjustments to Zatsiorsky-Seluyanov's segment inertia parameters. *J. Biomech.* **1996**, *29*, 1223–1230. [CrossRef]
16. Peploe, C.; McErlain-Naylor, S.A.; Harland, A.R.; Yeadon, M.R.; King, M.A. A curve fitting methodology to determine impact location, timing, and instantaneous post-impact ball velocity in cricket batting. *J. Sports Eng. Technol.* **2018**, *232*, 185–196. [CrossRef]
17. Cohen, C.; Texier, B.D.; Quéré, D.; Clanet, C. The physics of badminton. *New J. Phys.* **2015**, *17*, 063001. [CrossRef]
18. Brahms, B.V. *Badminton Handbook: Training, Tactics, Competition*, 2nd ed.; Meyer & Meyer Sport: Maidenhead, UK, 2014.
19. Martin, C.; Kulpa, R.; Delamarche, P.; Bideau, B. Professional tennis players; serve: Correlation between segmental angular momentums and ball velocity. *Sports Biomech.* **2012**, *12*, 2–14. [CrossRef]
20. Hinkle, D.; Wiersma, W.; Jurs, S.G. *Applied Statistics for the Behavioural Sciences*, 5th ed.; Houghton Miffin: Boston, MA, USA, 2003.
21. Bowerman, B.L.; O'Connell, R.T. *Linear Statistical Models: An Applied Approach*; Duxbery Press: Belmont, CA, USA, 1990.
22. Draper, N.R.; Smith, H. *Fitting a Straight Line by Least Aquares, in Applied Regression Analysis*; John Wiley & Sons, Inc.: Hoboken, NJ, USA, 1998.
23. Field, A.P. *Discovering Statistics Using IBM SPSS Statistics*, 4th ed.; Sage: London, UK, 2013.
24. Menard, S. *Applied Logistic Regression Analysis: Sage University Series on Quantitative Applications in the Social Sciences*; Sage: Thousand Oaks, CA, USA, 1995.
25. Myers, R. *Classical and Modern Regression with Applications*, 2nd ed.; Duxbury Press: Boston, MA, USA, 1990.
26. Wherry, R. A new formula for predicting the shtrinkage of the coefficient of multiple correlation. *Ann. Math. Stat.* **1931**, *2*, 440–457. [CrossRef]
27. Kwan, M. Designing the World's Best Badminton Racket. Ph.D. Thesis, Aalborg University, Aalborg, Denmark, 2010.
28. Kwan, M.; Andersen, M.S.; Cheng, C.-L.; Tang, W.-T.; Rasmussen, J. Investigation of high-speed badminton racket kinematics by motion capture. *Sports Eng.* **2011**, *13*, 57–63. [CrossRef]
29. Phomsoupha, M.; Laffaye, G. Shuttlecock velocity during a smash stroke in badminton evolves linearly with skil level. *Comput. Methods Biomech. Biomed. Eng.* **2014**, *17* (Suppl. 1), 140–141. [CrossRef] [PubMed]
30. Waddell, D.B.; Gowitzke, B.A. Biomechanical principles applied to badminton power strokes. In Proceedings of the 18th International Symposium on Biomechanics in Sports, Hong Kong, China, 25–30 June 2000.
31. Lees, A. Technique analysis in sports: A critical review. *J. Sports Sci.* **2002**, *20*, 813–828. [CrossRef] [PubMed]
32. Lees, A.; Cabello, D.; Torres, G. *Science and Racket Sports IV*; Routledge: London, UK, 2008.
33. Marshall, R.N.; Elliott, B.C. Long axis rotation: The missing link in proximal-to-distal sequencing. *J. Sports Sci.* **2000**, *18*, 247–254. [CrossRef]
34. Fleisig, G.; Nicholls, R.; Elliott, B.; Escamilla, R. Kinematics used by world class tennis players to produce high-velocity serves. *Sports Biomech.* **2003**, *2*, 51–64. [CrossRef]
35. Elliott, B. Biomechanics and tennis. *Br. J. Sports Med.* **2006**, *40*, 392–396. [CrossRef]
36. Elliott, B.C.; Baxter, K.G.; Besier, T.F. Internal rotation of the upper-arm segment during a stretch-shorten cycle movement. *J. Appl. Biomech.* **1999**, *15*, 381–395. [CrossRef]
37. Tang, H.P.; Abe, K.; Ae, M.; Katoh, L. 3-D cinematographic analysis of the badminton forehand smash: Movement of the forearm and hand. In *Science & Racket Sports*; Reilly, T., Hughes, M., Lees, A., Eds.; E & FN SPON: Cambridge, UK, 1995; pp. 113–118.

38. Whiteside, D.; Elliott, B.; Lay, B.; Reid, M. The effect of racquet swing weight on serve kinematics in elite adolescent female tennis players. *J. Sci. Med. Sport* **2014**, *17*, 124–128. [CrossRef]
39. Elliott, B.C.; Marshall, R.N.; Noffal, G.J. Contributions of upper limb segment rotations during the power serve in tennis. *J. Appl. Biomech.* **1995**, *11*, 433–442. [CrossRef]
40. Serrien, B.; Coossens, M.; Baeyens, J.P. Statistical parametric mapping of biomechanical one-dimensional data with Bayesian inference. *Int. Biomech.* **2019**, *6*, 9–18. [CrossRef]
41. Floria, O.; Sánchez-Sixto, A.; Harrison, A.J.; Ferber, R. The effect of running speed on joint coupling coordination and its variability in recreational runners. *Hum. Mov. Sci.* **2019**, *66*, 449–458. [CrossRef] [PubMed]
42. Cushion, E.J.; Warmenhoven, J.; North, J.; Cleather, D.J. Principal component analysis reveals the proximal to distal pattern in vertical jumping is governed by two functional degrees of freedom. *Front. Bioeng. Biotechnol.* **2019**, *7*, 193. [CrossRef] [PubMed]

applied sciences

MDPI

Article

Whole-Body Dynamic Analysis of Challenging Slackline Jumping

Kevin Stein * and Katja Mombaur

Optimization, Robotics and Biomechanics (ORB), Institute of Computer Engineering (ZITI), Heidelberg University, 69117 Heidelberg, Germany; katja.mombaur@ziti.uni-heidelberg.de
* Correspondence: Kevin.Stein@ziti.uni-heidelberg.de; Tel.: +49-6221-54-14869

Received: 31 December 2019; Accepted: 1 February 2020; Published: 6 February 2020

check for updates

Abstract: Maintaining balance on a slackline is a challenging task in itself. Walking on a high line, jumping and performing twists or somersaults seems nearly impossible. Contact forces are essential to understanding how humans maintain balance in such challenging situations, but they cannot always be measured directly. Therefore, we propose a contact model for slackline balancing that includes the interaction forces and torques as well as the position of the Center of Pressure. We apply this model within an optimization framework to perform a fully dynamic motion reconstruction of a jump with a rotation of approximately 180°. Newton's equations of motions are implemented as constraints to the optimization, hence the optimized motion is physically feasible. We show that a conventional kinematic analysis results in dynamic inconsistencies. The advantage of our method becomes apparent during the flight phase of the motion and when comparing the center of mass and angular momentum dynamics. With our motion reconstruction method all momentum is conserved, whereas the conventional analysis shows momentum changes of up to 30%. Furthermore, we get additional and reliable information on the interaction forces and the joint torque that allow us to further analyze slackline balancing strategies.

Keywords: slackline balancing; dynamics reconstruction; contact force modeling; optimal control; subject-specific modeling

1. Introduction

Slackline balancing is a recreational sport where the athlete tries to stand, walk or jump on a spring-like elastic ribbon band that is mounted between two anchor points as shown in Figure 1. Unlike balancing on a stiff beam or when performing a tandem walk on a regular surface, the slackline can swing both sideways and vertically, which increases the difficulty of maintaining an upright position [1]. In ongoing work, we investigate how humans maintain balance on a slackline and, to date, motion captured over 20 subjects of different skill levels. This led to a big data set containing standing, walking and jumping motions from beginners, intermediate and expert subjects. We previously analyzed different performance indicators for standing and walking based on kinematic measurements and subject-specific rigid body modeling [2] and found that experts show reduced angular momentum around the vertical axis and reduced Center of Mass (CoM) acceleration due to adjusted stance leg compliance. The analysis, however, lacks information on the interaction forces with the slackline. They are fundamental for understanding human locomotion [3] and would allow us to further investigate the data. The measurement of Ground Reaction Forces (GRF) enables us to compute joint torques, assess stability parameters in balancing tasks and is therefore part of many gait analysis protocols [4,5]. Pressure sensors and insoles are undergoing rapid development and are applied in various setups to measure interaction forces [6], however, it is not always possible to measure all forces directly for every task. Slackline balancing is usually done barefoot as experts report higher contact friction with the slackline which is crucial when landing from jumping. Beginners also prefer barefoot balancing over wearing shoes because they get a better feedback and control of the contact force. In the literature there are no reports on instrumented slacklines that can measure all interaction forces between the slackline and the subject. Previous work by Karatsidis et al. [7] estimated GRF for walking based only on inertial measurement data, a dynamic subject model and a heuristic on how the GRF are distributed during the double support phase. In previous work, we applied optimization methods and rigid body modeling to reconstruct GRF from kinematic data for sprinting motions with prostheses [8] and for a gymnastic cartwheel motion [9]. A dynamic model of the subject and a specific model of the interaction was crucial to obtain meaningful results. We now propose a new contact model specifically for balancing on a slackline and demonstrate its application by formulating and solving an optimal control problem (OCP) that reconstructs joint torques and forces of the slackline from marker-based motion capture and purely kinematic data. Newton's equations of motions are implemented as constraints to the optimization and therefore the optimized motion is physically feasible. When analyzing jumping motions with the conventional kinematic method we found inconsistencies in the CoM and angular momentum dynamics during the flight phase. In this work, we show the advantage of the proposed analysis for a jump motion with rotation as an example.

Figure 1. Left: A subject balancing on a slackline. An elastic ribbon band is mounted between two anchor points. The foot contact is able to move sideways and in up and down direction. Right: Kinematic motion capture data of slackline balancing. The coordinate system definition is used throughout this work.

2. Materials and Methods

2.1. Experimental Approach to the Problem

We present an overview of the slackline measurements and the data acquisition in Section 2.2. The conventional way of analyzing the data is described in Section 2.3. In a pre-study we investigated the differences between a regular planar contact and the contact with the slackline using sensor insoles (Section 2.4). The new analysis method is described in Section 2.5. It is based on the new contact model (Section 2.5.1) that takes the contact force, the contact torque and the position of the Center of Pressure (CoP) position into account. The subject-specific modeling and the formulation of the rigid body system dynamics with external contact forces are described in Section 2.5.2. In Section 2.5.3 we formulate and solve an OCP that tracks the recorded kinematic data by employing the proposed model. As a result we obtain a fully dynamic description of the motion, including joint torques and contact forces. We apply our approach to reconstruct the dynamics of one example motion where the subject jumps on the slackline and performs a rotation of approximately 180°. A visualization of the resulting motion and the contact forces and torques is shown in Section 3. We validate the OCP result by computing the CoM and angular momentum dynamics to show that the motion is physically feasible and follows Newton's mechanics (Section 2.5.4). We then compare the proposed method to the results of a conventional analysis that is often used in the literature.

2.2. Slackline Study and Data Acquisition

The slackline was installed using the Gibbon Slackrack 300 (ID Sports GmbH, Gibbon Slacklines, Stuttgart, Germany) as shown in Figure 1. The slackline was 3 m in length, 5 cm in width and mounted 31 cm above the ground. The motions were recorded using the marker-based motion capture system Qualisys (Qualisys, Goeteborg, Sweden) consisting of 8 Oqus 500 cameras at a frame rate of 150 Hz. The subject was prepared with 49 spherical infrared-reflective markers of 14 mm diameter, following the bone landmarks of the Gait-IOR marker set [10]. This marker set has originally been designed for gait analysis and was extended by two additional markers on the Medial Epicondyle of the Humerus to allow for better upper arm tracking and shoulder angle reconstruction. A static pose was recorded to create the subject-specific rigid body model. Six static markers were removed after this pose following the Gait-IOR guidelines, since the static markers were located at the medial side of the leg segments and are often occluded, thus tending to collide with other body parts and eventually hinder free movement. All dynamic motions were recorded using the remaining 45 markers. The slackline motion capture experiments were approved by the ethics committee of the Faculty of Behavioral and Cultural Studies of Heidelberg University according to the Helsinki Declaration (AZ Mom 2016 1/2-A1, 2016 with amendment 2019). Written informed consent was obtained from the subject. before the measurements. The subject was 1.9 m tall, weight 86 kg and was 27 years old and had considerable experience in slackline balancing.

2.3. Conventional Analysis

In conventional motion analysis the joint angles of the rigid body model are computed using whole-body inverse kinematics for every frame individually. We used the open source sofware tool Puppeteer by Martin Felis [11] for this task. It employs a damped Levenberg-Marquardt algorithm, described by Sugihara [12], to solve for the whole-body posture on a frame-by-frame basis. Afterwards, a low-pass filter was applied to generate smooth and differentiable joint trajectories. Joint velocities and joint accelerations are computed using finite central differences. The resulting trajectories represent the measured kinematics and we can compute different mechanical properties such as CoM acceleration or angular momentum when employing a dynamic subject model. The model is usually based on anthropomorphic measurements, e.g., by de Leva [13] or Jensen [14]. This kind of analysis is widely spread in the sports and biomechanics community and is also part of commercial motion analysis software like Visual3d (C-Motion, Germantown, MD). However, a frame-by-frame analysis in combination with an approximated model does not necessarily follow the laws of physics. Noise in the measurement, filtering and numerical derivation of joint velocities and accelerations. can result in violation of Newton's Equations of Motion (EoM). Additionally, it does not provide information on the acting joint torques or any interaction forces. We used a conventional analysis of the example motion as a comparison to show the advantages of the proposed method.

2.4. Experimental Foot Contact Analysis and Modeling—A Pre-Study

The Center of Pressure (CoP) is often used to analyze balance capabilities [5]. In this pre-study we performed CoP measurements using Moticon Pressure Insoles (Moticon GmbH, Munich, Germany) as we expected additional insight into the interaction from these measurements. We recorded standing on one leg in three different situations and for each foot individually. The following results are visualized in Figure 2:

- **Left:** Single leg balancing on regular surface
- **Middle:** Balancing on the slackline with aligned stance foot

- **Right:** Balancing on the slackline with turned stance foot

We see a clear difference between the cases: On flat ground the CoP moves in medial-lateral and in anterior-posterior direction, as is established in the literature. It is different on the slackline: we observe that the CoP does not move within the whole contact plane, but only on a single line. From the aligned and turned foot positioning we can conclude that the direction of this line is determined by the direction of the slackline. Even though the slackline measures 5 cm in width, we see from the measurement data that the CoP does not deviate from the very center of the slackline. We explain this with the fact that the ribbon band of the slackline is able to freely rotate around the center axis between the anchor points. This rotation is barely damped and therefore highly sensitive to shifting the CoP away from the slackline center. Rotating a contact surface, however, greatly reduces the normal force and therefore reduces the non-slipping threshold. We conclude that a small shift of the CoP away from the center of the slackline can already lead to the contact foot slipping off. The very narrow constraints on the CoP together with the continuous vertical and sideways movement of the stance foot explains the difficulty of slackline balancing especially in contrast to walking on a beam.

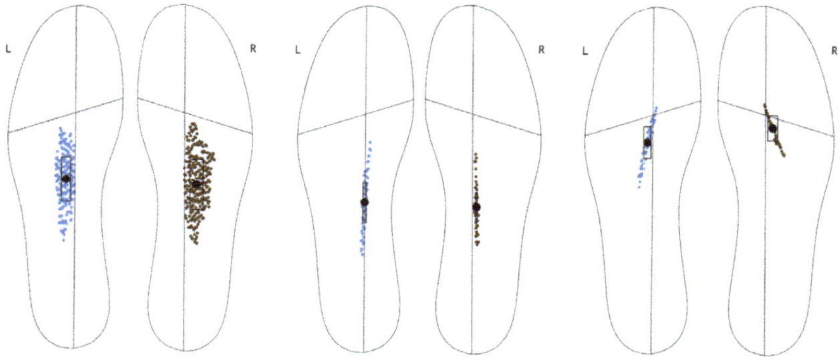

Figure 2. Center of Pressure recording of different single leg balancing tasks. We recorded both legs individually using Moticon Pressure Insoles. Left: Single leg balancing on regular surface. The CoP is moving in the anterior-posterior and in medial-lateral direction. Middle: Balancing on the slackline with aligned stance foot. Right: Balancing on the slackline with turned stance foot. The CoP is constrained by the direction of the slackline and to a single line.

2.5. Proposed Analysis Method

2.5.1. Slackline Contact Modeling

A foot contact model is required to perform a dynamic motion analysis. A regular planar, non-slipping contact can be described using a total of 8 variables [15]: The contact force $\boldsymbol{F}$ acting in three directions, two variables for the CoP position $\boldsymbol{p}$ in the contact plane and three contact torques $\boldsymbol{M}$. The forces are constrained by the friction cone to prevent slipping. Taking the measurements presented in Section 2.4 into account, we propose a new contact model for slackline balancing that includes 6 variables. Based on the slackline coordinate system shown in Figure 1 they are:

$$\boldsymbol{F} = [F_x, F_y, F_z]\text{: The contact forces} \tag{1}$$

$$\boldsymbol{p} = [p_x, 0]\text{: The CoP position in along the slackline} \tag{2}$$

$$\boldsymbol{M} = [0, M_y, M_z]\text{: The contact torques in the slackline coordinate system} \tag{3}$$

Zeros are placed where the regular contact model would have had an additional variable. The contact force can still act in all three directions and is still subject to the friction cone. As a consequence of the measurements presented in Figure 2, the CoP position is described by only one free variable instead of two. The Y-coordinate is always zero. The position can no longer be defined in the local coordinate system of the foot, but has to be transformed in the coordinate system of the slackline. Furthermore, we allow only two instead of three contact torques. Again, these are defined by the global position of the slackline and cannot be modeled locally. They act around the vertical axis and are perpendicular to the slackline. As described before, the slackline can freely rotate around the X-Axis and therefore no torque can be applied around this axis.

2.5.2. Subject Modeling and Dynamics Computations

The subject model consists of 17 segments. The length of each segment and the joint locations were estimated from the marker positions of the static trial recording following the Gait-IOR marker set guidelines. Figure 3 shows the static pose recording of the subject at the left and the specific rigid body model in the middle. Joint center positions are visualized in white. Virtual markers are placed on each segment of the model according to the static pose. These virtual markers are used to track the recorded marker data as shown at the right. For the dynamic properties of the model, such as segment mass, inertia or the relative segment CoM position, we refer to the measurements by de Leva [13] and linearly adjust them to the individual segment length. The degrees of freedom (DoF) and the kinematic structure is described in Table 1.

Table 1. Description of the kinematic structure of the model with all degrees of freedom.

Segment	Parent	DoF
Pelvis	Root	6d Floating Base
Lower Trunk	Pelvis	RY, RZ
Upper Trunk	Lower Trunk	RX, RY
Head	Upper Trunk	RZ
Upper Arm	Upper Trunk	RX, RY, RZ
Lower Arm	Upper Arm	RY, RZ
Hand	Lower Arm	Fixed
Thigh	Pelvis	RX, RY, RZ
Shank	Thigh	RY
Foot	Shank	RY

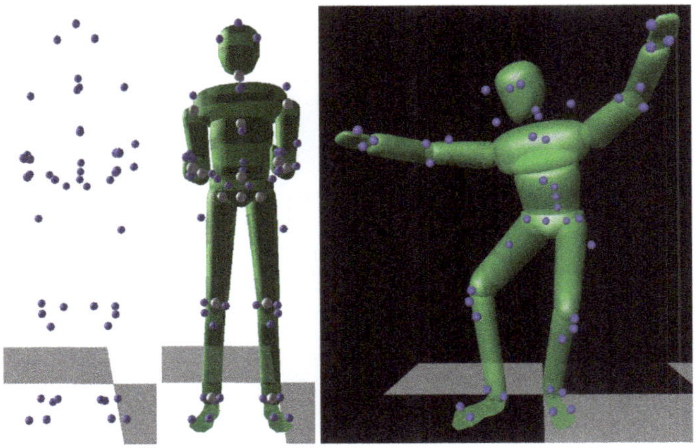

Figure 3. Left: The static pose of the subject. Middle: The rigid body model in Puppeteer. Joint centers are visualized in white. Virtual markers were placed on the model following the static pose. Right: Inverse kinematic fit of the model to a slackline pose.

The dynamics of a rigid body system with n_{dof} DoF using generalized coordinates q are described by the following equation [16]:

$$H(q)\ddot{q} + C(q,\dot{q}) = \tau \tag{4}$$

where the matrix $H(q) \in \mathbb{R}^{n_{dof} \times n_{dof}}$ is the generalized inertia matrix that is constructed from the current configuration of the model and the inertia and CoM positions of each segment, $C(q,\dot{q}) \in \mathbb{R}^{n_{dof}}$ is the generalized bias force (e.g., gravity or the Coriolis force) and τ are the generalized forces applied at the joints. This equation holds when the subject is in the air and no external forces are applied. When the model is in contact with the slackline, a contact force F and a contact torque M are acting at the contact point p as derived in Section 2.5.1. We can compute the generalized forces τ_c resulting from an external force using:

$$\tau_c = G(q,p)^T * \begin{bmatrix} M \\ F \end{bmatrix} \tag{5}$$

where $G(q,p)$ is the 6D Jacobian for a point on a body that when multiplied with $\dot{q}$ gives a 6-D vector that has the global angular velocity as the first three entries and the global linear velocity as the last three entries. With external contact forces Equation (4) becomes:

$$H(q)\ddot{q} + C(q,\dot{q}) = \tau + \sum_{contacts} \tau_c \tag{6}$$

For known joint angles q, joint velocities $\dot{q}$ and joint torques τ we are able to compute the joint acceleration $\ddot{q}$. This is known as Forward dynamics and implemented in RBDL under the `ForwardDynamics` function. The contact Jacobean was computed using the `CalcPointJacobian6D` function. The exact recursive implementation is described in [16].

2.5.3. Optimal Control Problem Formulation

We formulate the dynamic reconstruction of the recorded motion as an OCP. The general multi-phase OCP formulation is derived for example in [17]. We only describe the formulation that was used for the specific problem of motion reconstruction. Unlike our previous work [8,9] where we tracked joint trajectories that were computed beforehand, we now formulate an OCP that tracks the marker positions directly. In the following cost function we minimize the distance between the positions $\vec{m}_i$ of the virtual markers on the model and the 45 recorded marker trajectories $\vec{m}_i^*$ over the time $t \in [0,T]$ of the motion:

$$\min_{\mathbf{x},\mathbf{u}} \int_0^T \sum_{i=0}^{45} \|\vec{m}_i(q(t)) - \vec{m}_i^*(t)\|^2 dt \tag{7}$$

with respect to the state vector x and control vector u. They are the same for all phases. The state vector consists of joint angles and joint velocities: $\mathbf{x}(t) = [\mathbf{q}, \dot{\mathbf{q}}]$. The control vector represents the torques of the actuated joints and all variables of the slackline contact model: $\mathbf{u}(t) = [\tau(t), \lambda_L, \lambda_R]$ These are for each foot: three contact forces, two contact torques and the CoP position. All are given in the coordinate system of the slackline as it is shown in Figure 1.

$$\lambda_{R/L} = [F_x, F_y, F_z, T_y, T_z, p] \tag{8}$$

Controls are approximated to be piecewise linear continuous. To retrieve a feasible motion, the EoM must be satisfied and are therefore handled as constraints in the OCP. We reformulate Equation (4) and (6) as first order differential equations for the state vector:

$$\dot{x}(t) = f_j(t, q(t), \tau(t)) \tag{9}$$

where $j \in \mathbb{N}$ is the phase index. At this point we implemented two formulations: In the first implementation, we used the joint trajectories from the conventional inverse kinematics fit to determine the three different phases of the motion. Figure 4 shows the height of the feet above the ground plotted over time. The slackline is mounted at a rest height of 31 cm. Therefore, we can define a flight phase whenever both feet are above this height. This is indicated in red for the motion at hand. Hence, phases 1 and 3 are contact phases and subject to Equation (6), Phase 2 is a flight phase and subject to Equation (4).

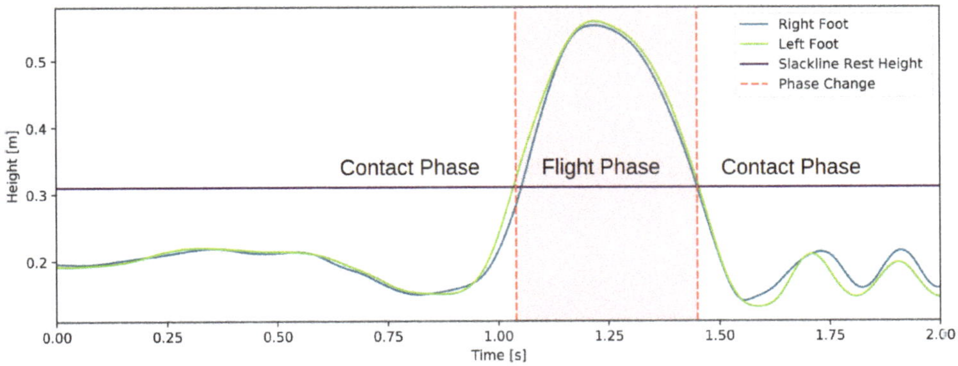

Figure 4. Feet positions are plotted against time. The rest height of the slackline is 31 cm. We determine the times of contact and flight phases when both feet are higher than the slackline. The flight phase is indicated in red.

In the second implementation, we defined the whole motion as one phase subject to the contact dynamics (Equation (6)). Additionally, we formulated the following discontinuous path constraint throughout the motion:

$$r^{eq}(\mathbf{u}(0),..,\mathbf{u}(T)) = \begin{cases} \lambda_R, & \text{if Right Foot Height } > 0.31 \text{ m} \\ \lambda_L, & \text{if Left Foot Height } > 0.31 \text{ m} \quad != 0 \\ 0, & \text{otherwise} \end{cases} \tag{10}$$

This allows the optimizer to determine the exact timing of contact which is fixed within the other implementation. Additionally, this formulation allows for different contact timings, such as jumping or landing with one foot after the other. This should enable for better marker positions tracking. On the other hand, such a formulation results in non-differentiabilities in the model which might cause numerical problems.

In both implementations, further boundary constraints $g(\cdot)$ are implemented for joint angles, velocities and torques and must be respected throughout the motion. The friction cone is implemented as an inequality constraint that requires the normal force to be larger than the horizontal forces. To determine reasonable upper limits for contact forces and torques, we performed a similar jumping motion on two force plates (Bertec, Columbus, OH, USA) and recorded the acting GRF. Limits were set to 1.5 times the measured maximum values. We solve both implementations of the OCP numerically using MUSCOD-II [17,18]. It was developed at the Interdisciplinary Center for Scientific Computing, IWR, Heidelberg University. The state variables are parameterized by the direct multiple-shooting method as it is derived in [19]. The control variables were discretized by piecewise linear continuous functions. On all multiple-shooting intervals, the dynamics of the system are computed in parallel. The same intervals are apply for states and controls. We ensure a continuous solution by imposing continuity constraints at the shooting interval transitions for all state variables. This way, a large but structured nonlinear programming problem (NLP) is obtained. It is solved by an adapted sequential quadratic programming (SQP) method. Further detail can be found in [17]. We expect the first implementation to show better convergence, since discontinuities in the dynamics are supposed to be formulated as phase changes and constraints should be differentiable throughout one phase. However, in practical tests we observed that formulating phase changes as constraints also works in the present case. We did not use a regularization term, which is often used to account for possible redundancies in the contact forces during the double support phases.

2.5.4. Validation

We validate our method by showing that important mechanical properties of the overall system are satisfied due to the fact that Newton's EoM have been formulated as constraints to the OCP. This includes:

- Horizontal momentum is conserved during the flight phase
- Gravity is the only acceleration acting on the CoM during the flight phase
- Angular momentum is conserved during the flight phase
- The change of momentum is proportional to the sum of external forces
- The change of angular momentum is equal to the sum of all acting torques.

With the change of momentum being equal to the CoM acceleration $\ddot{c}$ times the subject mass m and $\boldsymbol{L}$ being the angular momentum, we reformulate Newton's EoM:

$$\ddot{c} * m = \sum_{\lambda_{L/R}} \boldsymbol{F} \tag{11}$$

$$\dot{\boldsymbol{L}} = \sum_{\lambda_{L/R}} (\boldsymbol{M} + (\boldsymbol{p}_i - \boldsymbol{c}) \times \boldsymbol{F}) \tag{12}$$

where $\boldsymbol{p}$ is the point where the force is acting. During all phases Equations (11) and (12) must hold. For the contact phases we can compute the left hand side of Equations (11) and (12) using the `CalcCenterOfMass` function of RBDL and the right hand side from result values of $\lambda_{L/R}$ of the OCP result.

3. Results and Discussion

Both implementations converged. Due to the problem's complexity, we were numerically limited to ≈80 shooting nodes. We chose 40 multiple-shooting intervals per 1 s of motion and reconstructed around 2 s of motion. Other choices are possible, but we found that the computation time drastically increases with more shooting nodes and that the solver is not always able to find solutions for less shooting nodes per second of motion. As expected, the single-phase implementation resulted in a slightly lower tracking error. Therefore we present the results for the single-phase implementation. The resulting motion and the interaction forces (visualized as yellow arrows) are shown in Figure 5. Contact and flight phases are clearly distinguished. We see that the subject equally used both feet to initiate and land the jump and that contact torques were acting to build the necessary angular momentum for the rotation. The feet were aligned with the Slackline during the first contact phase and were turned perpendicular for landing.

We present the fitting error of our model to the measured marker data: Figure 6 shows the average marker residuum on the top and the frame by frame error on the bottom. Overall we achieved an average tracking error of 3 cm per marker. This accuracy is similar to the least squares kinematic fit of the conventional analysis that was used to initialize the OCP and to what is reported in the literature. The largest deviations occur for the two shoulder markers (L_SAE and R_SAE). One reason for this is that the arms were aligned to the upper body during the static pose on which the model is based. During slacklining, however, the arms are turned 90° compared to this pose and mainly parallel to the ground which results in skin and marker movement relative to the bone and shoulder joint. This offset is visible throughout the motion. Additionally, the shoulder is modeled as a spherical joint with only three DoF. In reality this joint is much more complex and also has translational DoF. This currently limits the tracking accuracy; however, a more precise kinematic shoulder model could be used.

Figure 5. The result of this work: A fully dynamic reconstruction of a jumping motion with rotation. The contact forces and torques are visualized.

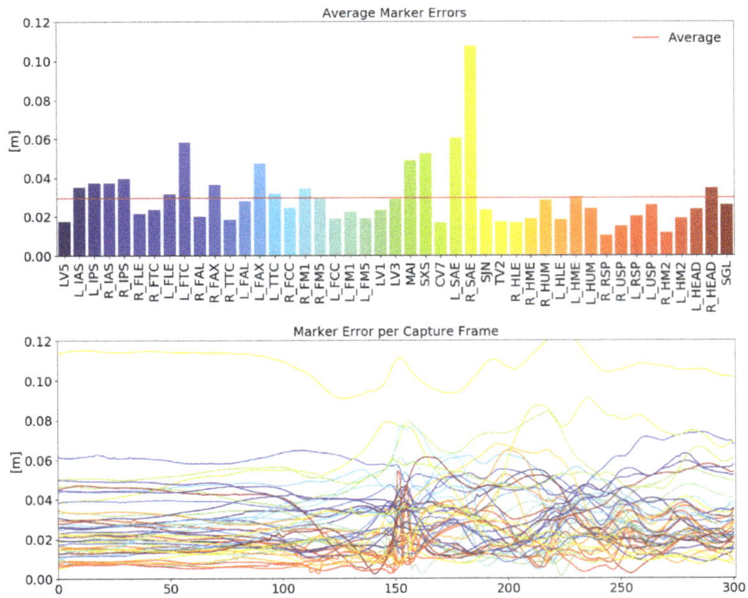

Figure 6. Top: Average marker error for all 45 markers. Largest deviations occur for the shoulder markers. This is due to the fact that the model is based on a N-Pose capture and the slackline motion mainly had a T-Pose like arm positioning. Bottom: Frame by frame marker error.

Figure 7 shows the CoM velocity and acceleration throughout the motion. The conventional motion analysis is plotted in green, the OCP result of the proposed method in purple. The most apparent differences occur during the flight phase when no forces are acting. As derived in Section 2.5.4, we expect constant accelerations and conservation of momentum in horizontal direction. Gravity should be the only force acting in vertical direction. This is indeed the case for the OCP result but is not given for the conventional analysis. Looking at the CoM acceleration we see values of up to 0.5 $\frac{m}{s^2}$ in X direction and 0.7 $\frac{m}{s^2}$ in Y direction during the flight phase. With the maximum values during the whole motion being 2.5 $\frac{m}{s^2}$ and 3.5 $\frac{m}{s^2}$, respectively, this leads to an estimated relative error of up to about 20% for the conventional analysis.

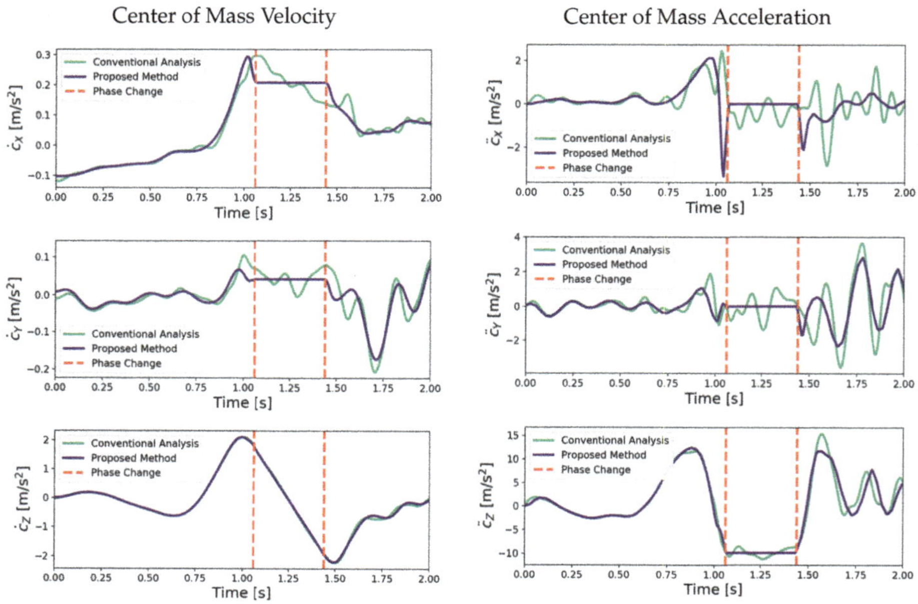

Figure 7. Center of Mass velocity and acceleration in global coordinates during the motion. Results from the conventional method are plotted in green, the proposed method is plotted in purple. During the flight phase all horizontal accelerations are supposed to be zero and the according velocities should be constant. We see inconsistencies and changes in the range of 20% for the conventional method whereas the proposed method is physically feasible.

Angular momentum is plotted in Figure 8. It should also be conserved during the flight phase and the change should be equal to zero when no external torques are acting. Again, we observe the desired properties for the optimized motion. The conventional analysis shows high variability during the flight phase and errors in the range of 30% in the horizontal plane.

On the left of Figure 9 we plotted the contact forces for each foot stacked on top of each other and the CoM acceleration times the subject mass as a dashed line. We can see that they exactly match and that Equation (11) holds. On the right we have the same result for the contact torques. Again, Equation (12) is satisfied throughout the motion. This shows that the optimization result follows Newton's EoM also during the contact phases. The resulting forces are consistent with the CoM acceleration and the resulting contact torques and CoP position are consistent with the change of angular momentum. We have established that the dynamics of our optimization framework are physically feasible and can further analyze the motion itself.

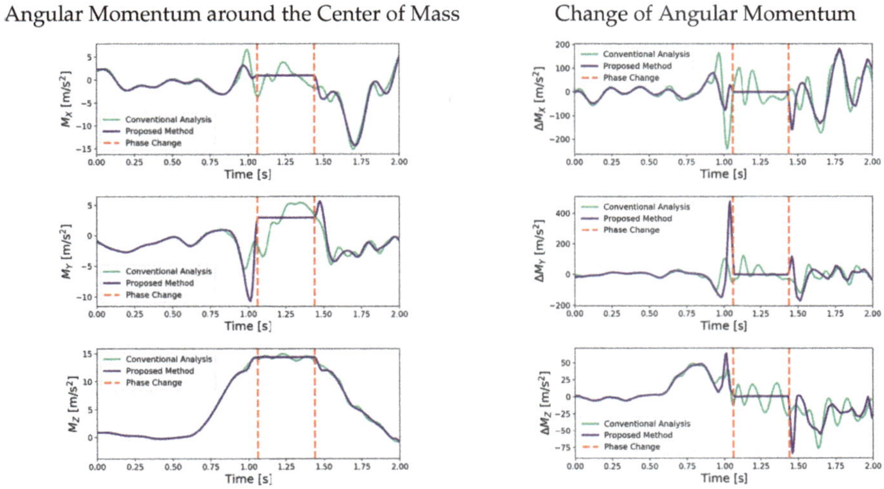

Figure 8. Angular Momentum and Change of Angular Momentum around the Center of Mass. Similar to Figure 7 we see constant values during the flight phase and zero values of the derivative for the proposed method. Values computed with the conventional method are not respecting Newtons EoM.

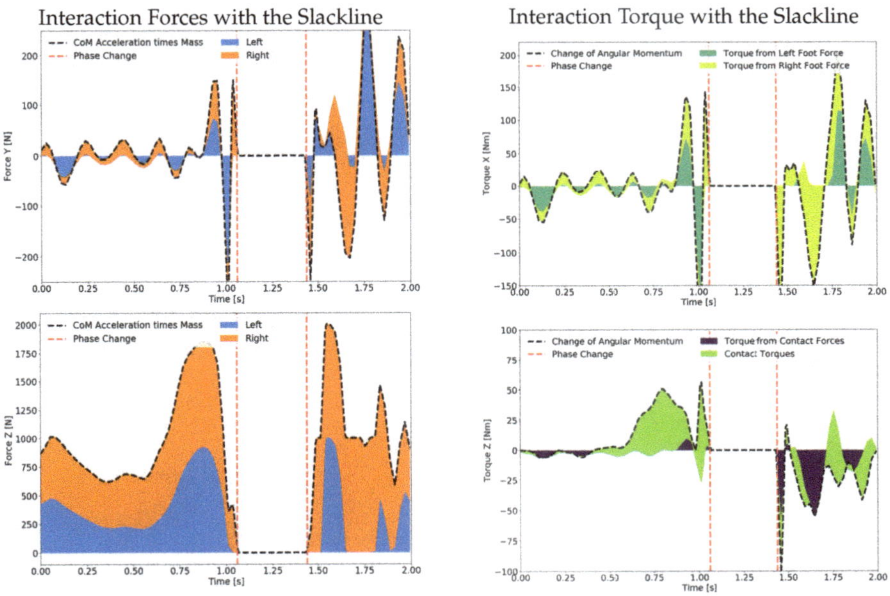

Figure 9. Left: Sum of contact forces and Center of Mass acceleration times subject mass. Right: Sum of contact torques and torques produced by the contact forces plotted against the change of angular momentum. Equations (11) and (12) are satisfied throughout the motion.

We do so by looking at the sideways and vertical direction and forces and torques individually. As shown in Figure 1 the X-Axis is aligned with the slackline, the Y-Axis perpendicular and the Z-Axis in vertical direction. The Y direction is particularly interesting since the subject is very unstable due to the slackline and torques can only be applied controlling the contact force. Looking at the CoM acceleration we see spring-like behavior before and after the jump. The subject is constantly swinging sideways. Amplitudes are small when the subject is in balance, larger after the landing and decreasing during the stabilization process. The oscillation frequency appears to be the same throughout the motion. The interaction forces of both feet point into the same direction, back towards the center of the anchor points of the slackline. We also observe spring spring-like properties before and after the jump in the vertical direction. The slackline model proposed by Paoletti and Mahadevan [1] models the force of the slackline also as a spring with the direction towards the resting position. However, they only considered sideways movement and did not take vertical movement of the subject into account. The lower left plot of Figure 9 shows the sum of contact torques and the sum of the torque generated from the contact forces. We see that the subject initiates the rotation applying contact torques when the feet are parallel during the first contact phase. After landing, the feet are perpendicular to the slackline and the contact forces are used to decrease the rotation.

In the future, we plan to apply the method to several other slackline tricks such as front or backflips, chest or pelvis bounces. For this task we plan to introduce other contact points for example at the hands, the chest or the pelvis. Another possibility would be a thorough gait analysis of slackline walking and the comparison to flat ground walking regarding stability parameters. We intend to reconstruct a full gait cycle and gain more insight in human balancing strategies. Furthermore, we want to analyze the resulting contact forces relative to the foot positions and develop a thorough spring model of the slackline. We showed that the current model by Paoletti and Mahadevan [1] is oversimplified as it does not take the whole range of motion of the slackline into account.

4. Conclusions

In this work we derived a general contact model for slackline balancing and demonstrated how it can be applied to analyze jumping on slackline. We employed the model inside an optimization framework to reconstruct the dynamics of a slackline jump with 180° rotation. We successfully implemented a multi-phase and a single-phase formulation leading to almost equivalent results. We found that the resulting motion has similar fitting error to the markers when compared to the inverse kinematics approach, but is physically feasible and time consistent. The advantage of our approach becomes apparent during the flight phase of the motion and when comparing the CoM and angular momentum dynamics. We found variations of up to 30% for the conventional method. Due to numeric complexity and the high amount of variables necessary to formulate the OCP, this method is limited to a few seconds of motion. This might limit the application to compare longer slackline motions of beginners and experts as initially intended. However, we can apply it to interesting parts of motion and analyze specific movements connected to balance recovery.

Author Contributions: Conceptualization, K.S. and K.M.; Formal analysis, K.S.; Funding acquisition, K.M.; Investigation, K.S.; Project administration,K.M.; Supervision, K.M.; Writing—original draft, K.S.; Writing—review & editing, K.M. All authors have read and agreed to the published version of the manuscript.

Funding: Funding by the Carl Zeiss Foundation within the "Heidelberg Center for Motion Research" is gratefully acknowledged.

Acknowledgments: We want to thank the Simulation and Optimization research group of the IWR at Heidelberg University for giving us the possibility to work with MUSCOD-II. We also want to acknowledge financial support by the Baden-Württemberg Ministry of Science, Research and the Arts and by Ruprecht-Karls-Universität Heidelberg.

Conflicts of Interest: The authors declare no conflict of interest.

Abbreviations

The following abbreviations are used in this manuscript:

CoM	Center of Mass
CoP	Center of Pressure
DoF	Degrees of Freedom
EoM	Equation of Motion
GRF	Ground Reaction Forces
L_SAE	Left Scapula Acromial Edge
n_{dof}	Number of Degrees of Freedom
NLP	Nonlinear Programming Problem
RBDL	The Rigid Body Dynamics Library
R_SAE	Right Scapula Acromial Edge
OCP	Optimal Control Problem
SQP	Sequential Quadratic Programming
ZMP	Zero Moment Point

References

1. Paoletti, P.; Mahadevan, L. Balancing on tightropes and slacklines. *J. R. Soc. Interface R. Soc.* **2012**, *9*, 2097–108. [CrossRef] [PubMed]
2. Stein, K.; Mombaur, K. Performance indicators for stability of slackline balancing. In Proceedings of the IEEE/RAS International Conference on Humanoid Robots (Humanoids 2019), Toronto, ON, Canada, 15–17 October 2019.
3. Winter, D. Human balance and posture control during standing and walking. *Gait Posture* **1995**, *3*, 193–214. [CrossRef]
4. Sutherland, D. The evolution of clinical gait analysis: Part II Kinematics. *Gait Posture* **2002**, *16*, 159–179. [CrossRef]
5. Thompson, L.; Badache, M.; Cale, S.; Behera, L.; Zhang, N. Balance performance as observed by center-of-pressure parameter characteristics in male soccer athletes and non-athletes. *Sports* **2017**, *5*, 86. [CrossRef] [PubMed]
6. Xu, F.; Li, X.; Shi, Y.; Li, L.; Wang, W.; He, L.; Liu, R. Recent Developments for Flexible Pressure Sensors: A Review. *Micromachines* **2018**, *9*, 580. [CrossRef] [PubMed]
7. Karatsidis, A.; Bellusci, G.; Schepers, H.M.; de Zee, M.; Andersen, M.S.; Veltink, P.H. Estimation of ground reaction forces and moments during gait using only inertial motion capture. *Sensors* **2016**, *17*, 75. [CrossRef] [PubMed]
8. Emonds, A.L.; Funken, J.; Potthast, W.; Mombaur, K. Comparison of Sprinting with and without Running-Specific Prostheses Using Optimal Control Techniques. *Robotica* **2019**, *37*, 2176–2194. [CrossRef]
9. Stein, K.; Mombaur, K. Optimization-Based Analysis of a Cartwheel. In Proceedings of the 7th IEEE International Conference on Biomedical Robotics and Biomechatronics (Biorob), Enschede, The Netherlands, 26–29 August 2018; pp. 909–915
10. Leardini, A.; Biagi, F.; Merlo, A.; Belvedere, C.; Benedetti, M.G. Multi-segment trunk kinematics during locomotion and elementary exercises. *Clin. Biomech.* **2011**, *26*, 562–71. [CrossRef] [PubMed]
11. Felis, M. Modeling Emotional Aspects in Human Locomotion. PhD Thesis, Heidelberg University, Heidelberg, Germany, 2015.
12. Sugihara, T. Solvability-Unconcerned Inverse Kinematics by the Levenberg- Marquardt Method. *IEEE Trans. Robot.* **2011**, *27*, 984–991. [CrossRef]
13. De Leva, P. Adjustments to Zatsiorsky-Seluyanov's segment inertia parameters. *J. Biomech.* **1996**, *29*, 1223–1230. [CrossRef]
14. Jensen, R.K. Body segment mass, radius and radius of gyration proportions of children. *J. Biomech.* **1986**, *19*, 359–368. [CrossRef]
15. Elftman, H. Forces and Energy Changes in the Leg During Walking. *Am. J. Physiol.* **1939**, *125*, 339–356. [CrossRef]
16. Felis, M.L. RBDL: An efficient rigid-body dynamics library using recursive algorithms. *Autonomous Robots* **2016**, 1–17. [CrossRef]

17. Kuhl, P.; Ferreau, J.; Albersmeyer, J.; Kirches, C.; Wirsching, L.; Sager, S.; Potschka, A.; Schulz, G.; Diehl, M.; Leineweber, D.; et al. *MUSCOD-II Users' Manual*; Interdisciplinary Center for Scientific Computing (IWR): Heidelberg, Germany, 2001.
18. Leinweber, D.; Bauer, I.; Bock, H.; Schloeder, J. An efficient multiple shooting based reduced SQP strategy for large-scale dynamic process optimization. Part I: Theoretical aspects. *Comput. Chem. Eng.* **2003**, *27*, 157–166. [CrossRef]
19. Bock, H.; Plitt, K. *A Multiple Shooting Algorithm for Direct Solution of Ooptimal Control Problems*; Pergamon Press: Oxford, UK, 1984; pp. 243–247.

applied sciences

MDPI

Article

Effects of Soccer Training on Body Balance in Young Female Athletes Assessed Using Computerized Dynamic Posturography

Grażyna Olchowik and Agata Czwalik *

Department of Biophysics, Medical University of Lublin, K. Jaczewskiego 4, 20-090 Lublin, Poland; grazyna.olchowik@umlub.pl

* Correspondence: agata.czwalik@umlub.pl; Tel.: +48-81-448-6330

Received: 18 December 2019; Accepted: 23 January 2020; Published: 3 February 2020

Abstract: The aim of this study was to determine the effect of regular soccer training on the balance system for young women. Computerized dynamic posturography of female footballers (n = 25) and control group (n = 50) was assessed during three tests: Sensory Organization Test, Motor Control Test, and Adaptation Test. Statistically significant differences between the groups was found in Composite Equilibrium Score with higher values, indicating better postural stability, for footballers. Regular trainees also showed better usefulness of vestibular system while maintaining balance. Weight symmetry of the lower limbs during Motor Control Test also showed statistically significant differences between the groups. This study shows that female footballers have better postural stability than their inactive peers and that regular workouts may improve the balance system.

Keywords: posture stability; balance; football; exercise; training

1. Introduction

Football is the most popular sport discipline with around 200,000 professional players and 240 million amateur players. It is responsible for almost 10% of sports injuries requiring medical attention in adolescents [1,2]. This discipline requires players to have unprecedented coordination to cope with rapidly changing external conditions. With regular practice a player acquires precision in movement and more muscle mass. Frequent training helps the player acquire the ability to execute appropriate strategic movements to effectively target his/her opponent's goal and to prevent serious injuries. To evaluate motor coordination a test was adopted many years ago to measure maximum rotation to the left and right when jumping in the air with both feet [3]. In modern training it is necessary to constantly monitor the sportsperson's coordination and balance [4,5]. Presently there are many different methods to assess training progress, physical activity, and body postural stability, but the most important conditions in which football players should be assessed are dynamic conditions. Postural stability in athletes has been reported widely in several sports disciplines [6,7]. That suggest that the type of sport and repetitive training may affect the balance system control. Some researchers found that football players were better that other athletes [6,8] and that the level of playing experience influences postural control performance and adopted motor strategies [9].

The human body's static and dynamic balance is maintained by the posture control system, which coordinates the stimuli received by the vestibular system, the visual system, and proprioceptors, and also provides a selection of optimal postural responses aimed at preventing falls [10,11]. The posture control system tracks the position of the center of gravity (COG) over an area defined by the outline of the human feet, thus guaranteeing a stable posture [12]. The body's multi-segment construction, the height of the COG above the base of support (BOS), and a relatively small BOS, results in the human body being unstable when in an upright position. The balance control system therefore needs to constantly analyze and counterbalance all the destabilizing factors through proper stimulation of the relevant muscle groups [12,13]. The central nervous system watches over the appropriate choice of reaction and the stimulation of the appropriate muscle reflexes. To correct posture, it adapts a movement strategy based on its analysis of linear and angular accelerations of individual body segments. Stability control of body posture depends on many factors, which includes a range of movements in the joints, muscle strength, speed and precision of movement, and the ability to perceive body positioning in space ("body sense") [14].

Computerized Dynamic Posturography (CDP) allows the individual components of the human balance system to be evaluated. During CDP, signals from appropriate senses involved in maintaining balance are evaluated. The postural response time and the reaction time to unexpected support platform changes, with an appropriate motor response, is also determined as well as the efficiency of adaptive mechanisms [15]. CDP is the gold standard to differentiate between sensory, motor, and central adaptive impairments to postural control [16]. In football it can be an opportunity to track the rehabilitation of postural control impairment after anterior cruciate ligament injury, which is one of the most common in this sport [17].

The aim of this study was to compare the behavior of the balance system between two groups of young women and to determine the effect of regular practice. One group consisted of young female footballers who train regularly and the other, a group of peers who do not participate in any regular sporting activity. We hypothesize that football players would show better postural balance performance.

2. Materials and Methods

2.1. Participants

The study group consisted of 25 young (age = 18.9 ± 4.5 year) female footballers from AZS-PSW Biała Podlaska, a club in the Polish Women's Football League. Each footballer in this study group has been attending training sessions for at least 6 years with five training sessions per week of 90 min duration. The control group consisted of 50 students (age = 20.7 ± 1.2 year) from the Medical University of Lublin, who had declared a lack of sporting activity. Both groups were chosen such that there were no statistically significant differences between them in height, weight, or body mass index (BMI) (Table 1). All subjects participating in the study had the same functional preferences (right-sided handedness, footedness, and eyedness) and same postural lateral preferences (hand-clasping, arm-folding, leg-crossing, and stair climbing) [18].

Table 1. Characteristics of female footballers and the control group.

	Footballers (n = 25)		Control Group (n = 50)		*p*
	M	SD	M	SD	
Height [cm]	167.47	4.28	168.04	5.33	0.726
Body weight [kg]	62.00	5.17	61.72	5.55	0.873
BMI [kg/m^2]	22.06	1.65	21.86	1.51	0.672

M = mean, SD—standard deviation, *p*—probability value.

2.2. Procedure

The study was conducted with the approval of the Bioethics Committee at the Medical University of Lublin (KE 0254/195/2011). At the beginning, participants were informed about the purpose of the study and asked to complete a questionnaire and to sign the consent for the study. The questionnaire included questions about general health: all diseases or health problems, surgeries or hospitalizations, brain injuries, loss of consciousness episodes, bone/muscle/joint injuries as well as any medication or drugs (including alcohol) taken in the past month. Another part of the questionnaire was about physical activity: what sport or physical activity the participant undertakes, for how long, and how many times per week. Afterwards each participant was accurately weighed and measured, and a short lateralization preference test was performed (e.g., which hand the participant uses for drawing/throwing a small object, which leg for kicking a football or stepping onto a chair, and which eye for looking into a bottle or a door viewer). The final stage of the study was a posturographic examination during which the participant stood barefoot on the posturographic platform and was secured with a special harness.

2.3. *Instruments*

Posturographic tests were performed using an Equitest posturograph manufactured by NeuroCom International®. The device consists of a dynamic force plate, visual surround, and a computer with software. Both the force plate and visual surround are moveable (±10° rotation for both and a maximum angular velocity of 50°/s and 15°/s, respectively). The study protocol included the following tests: Sensory Organization Test, Motor Control Test, and the Adaptation Test.

The Sensory Organization Test (SOT) evaluates the usefulness of signals coming from the different senses involved in maintaining body balance. This test is performed using six sensory stimulation conditions, during which visual stimuli are changed and a rotation of the foot support platform, or movements of the visual surround, are introduced. During the first three tests (SOT1–SOT3) the foot support platform is stationary while the visual information is varied: SOT1—eyes open, SOT2—eyes closed, SOT3—moving visual surround. During these tests, analyzing postural stability determines the usefulness of the visual signal and the patient's ability to suppress visual stimuli which are contrary to reality. The next three tests (SOT4–SOT6) are performed with a moving foot support platform, which interferes with the information received by the proprioceptive system. As with the previous three tests, visual information is varied: SOT4—eyes open, SOT5—eyes closed, SOT6—moving visual surround. Two final trials evaluate the usefulness of the vestibular system, whose role increases significantly in the case of incorrect or missing stimuli from the remaining systems involved in posture control. During SOT, the Equilibrium Score (ES) that quantifies the COG sway or postural stability under each of the 3 trials of the 6 sensory conditions is evaluated. A score of 100 represents perfect balance (no sway) and a score of 0 indicates a fall. During SOT, Composite Equilibrium Score (CES)—a weighted average of all 6 individual scores ES—the body's COG displacement in the anterior–posterior direction, as well as the motor strategy (MS)—correctness of the selected postural strategy—is also assessed. A score near 100 indicates a full ankle strategy while a score near 0 indicates a full hip strategy with maximum shear force. In addition the Sensory Analysis (SRS) that determines a patient's ability to use input from the somatosensory (SOM), visual (VIS), vestibular (VES) system to maintain balance as well as the degree to which the patient relies on visual information whether it is correct or not (PREF) is assessed [19,20].

The Motor Control Test (MCT) is performed for 6 conditions using the foot support platform capable of forward and backward movements through small, medium, and large displacements. A series of three trials is performed for each condition during which the patient's ability to perform corrective movements is evaluated in response to unexpected changes in the foot support platform. The MCT analysis the Latency Response (L)—the time between start of platform movement and start of postural response—the Amplitude (A) of the postural response and the Weight Symmetry (WS)—a nondimensional quantity with a score of 100 indicating that weight is borne equally by the two legs. The WS score decreases to zero or increases to 200 when all the weight is borne by the left or right leg, respectively.

In the Adaptation Test (ADT), the patient is subjected to two series of sudden platform perturbations, one of which causes dorsiflexion (ATU) while the other causes plantarflexion (ATD) in the ankles. Each series consists of five trials. During subsequent trials, the patient should maintain a vertical posture, minimizing with every subsequent trial the amount of energy required to rebalance the body. This sway energy (SE) is determined after each platform perturbation and indicates the amount of COG displacement during each trial.

2.4. Data Analysis

Statistical analysis was performed using the STATISTICA 10 (StatSoft) computer program. To verify the normality of the data the Shapiro–Wilk test was used. Because normality tests failed to confirm that all the parameters variables studied had normal distribution (most probable reason for not normal distribution in footballers group was the small number of participants), all variables were analyzed with non-parametric statistics—the Mann–Whitney U-test—for differences between groups. Statistically significant changes were those with a statistical significance level of $p < 0.05$.

3. Results

The SOT results are shown in Figures 1–3. ES analysis (Figure 1) shows statistical significance ($p < 0.05$) for the CES, with higher values for footballers. Of the six SOT conditions, COG displacement differed significantly between the groups only for conditions ES5 and ES6 which provided conflicting information to the sense organs, carried out on a moving platform with eyes closed (ES5) or a moving visual surround (ES6). These results reflect significant differences in the use of vestibular stimuli (VES) between people training regularly and those not performing any regular sport.

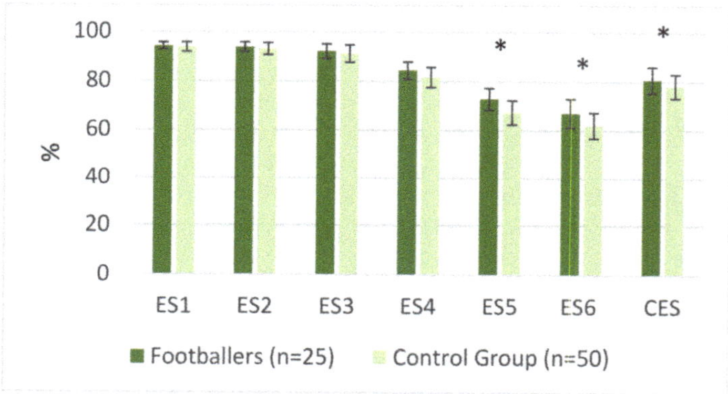

Figure 1. Sensory Organization Test—Composite Equilibrium Score (CES) and Equilibrium Score (ES) results (where the ES suffix refers to a particular SOT condition) for footballers and control group. The symbol * refers to a significant difference at $p < 0.05$.

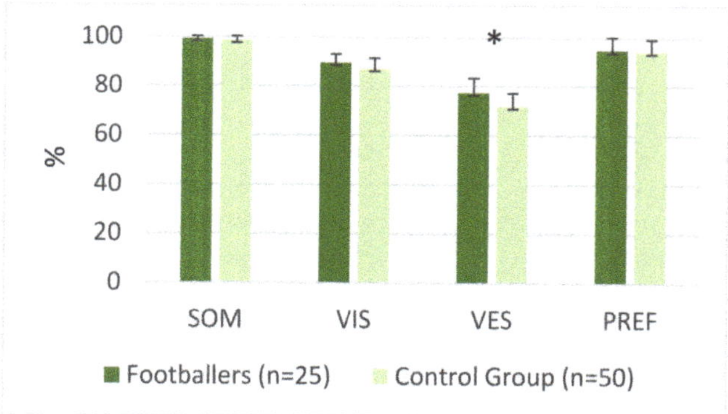

Figure 2. Sensory Organization Test—sensory analysis (SRS) results for the somatosensory system (SOM), the visual system (VIS), the vestibular system (VES), and visual preference (PREF) for footballers and control group. The symbol * refers to a significant difference at $p < 0.05$.

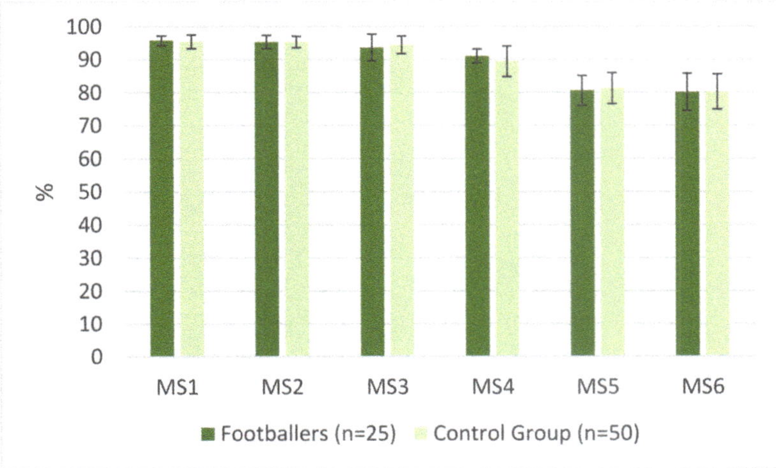

Figure 3. Sensory Organization Test—motor strategy (MS) for particular SOT conditions for footballers and control group.

The usefulness of the signals from the other sensory organs involved in the control of body balance (VIS, SOM, and PREF) did not highlight any significant differences between the two groups (Figure 2).

There was no impact on the selection of an appropriate motor strategy (MS1–MS6) by people playing football (Figure 3).

During MCT, statistically significant differences between the two groups were found in the symmetry of loading of the lower limbs during all the trials (Figure 4). Football players were characterized by disproportionate distribution of body weight with a predominance of left leg.

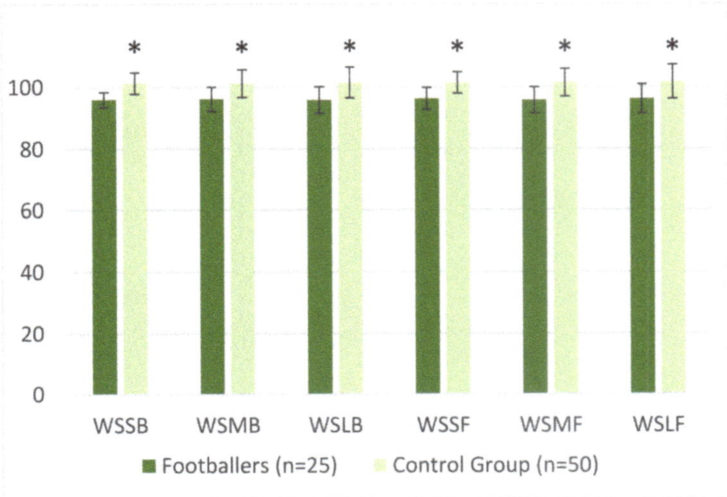

Figure 4. Motor Control Test—weight symmetry (WS) results for the lower limbs for small, medium and large platform transitions in backward (SB, MB, LB) and forward (SF, MF, LF) direction for footballers and control group. The symbol * refers to a significant difference at $p < 0.05$.

Postural response latencies (Figure 5) and their amplitudes (Figure 6) did not reveal any statistically significant differences between the two groups, which means that neither the reaction time nor the angular velocity of COG during the trials depends on the physical activity.

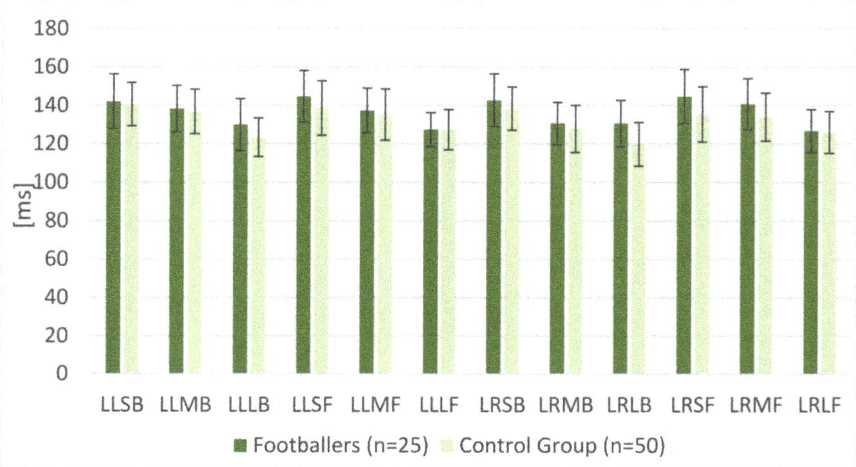

Figure 5. Motor Control Test—postural latencies for left (LL) and right (LR) lower limb for small, medium and large platform transitions in backward (SB, MB, LB) and forward (SF, MF, LF) direction for footballers and control group.

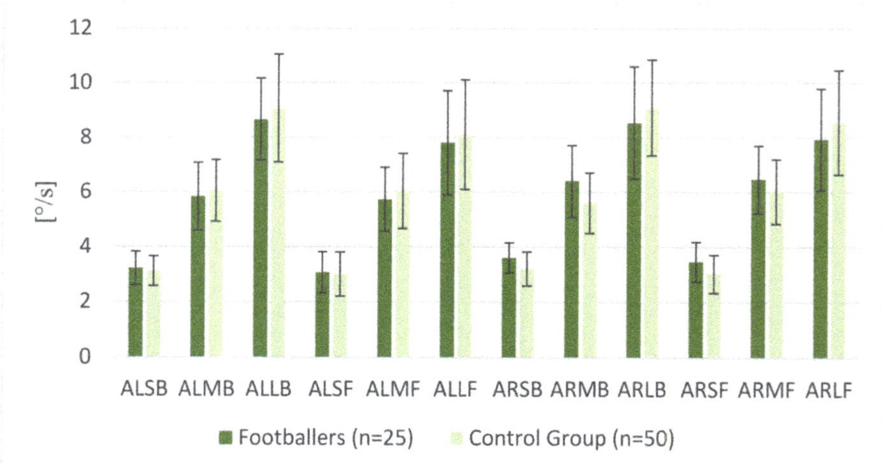

Figure 6. Motor Control Test—amplitudes of the postural responses for left (AL) and right (AR) lower limb for small, medium and large platform transitions in backward (SB, MB, LB) and forward (SF, MF, LF) direction for footballers and control group.

In the ADT there were no statistically significant differences between the measured parameters for both groups, which means that the adaptive postural response system is independent of physical activity. The results are shown in Figure 7.

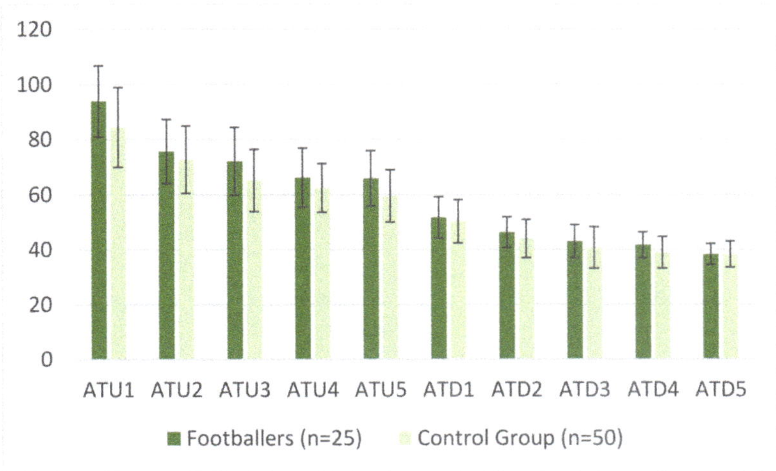

Figure 7. Adaptation Test—sway energy for 5 dorsiflexion (ATU) and 5 plantarflexion (ATD) condition for footballers and control group.

4. Discussion

Football is a sport discipline which, in addition to unprecedented coordination, requires exceptionally divided attention. The players must focus not only on the ball, but must also control their positions, the actions of the players in their team as well as those in the opposing team. They must analyze and foresee subsequent situations.

This study shows that professional footballers achieve significantly better results in the CES. The smaller COG displacements observed during all SOTs provides the evidence for the higher CES value for footballers who train regularly. The footballers also showed smaller COG displacements in the anterior–posterior direction for conditions which were inconsistent with the real signals from the somatosensory and visual systems. The footballers made better use of the signals from the vestibular system, which is the most reliable source of information regarding the positioning of the body's COG [15]. An analysis of the symmetry of the distribution of body weight gives a higher load asymmetry in the lower limbs in the footballers, which arises from using the dominant leg during the game. This is probably the result of more symmetrical actions performed during training and matches, which are due to unexpected situations arising during the game. For both male [21] and female footballers [22], in particular when shooting at goal, which is an important technical skill, there is a clear asymmetry with the right leg being dominant. Barone et al. [23] suspects that sportspeople prefer to use one leg (dominant), because they have better standing balance on non-dominant one.

The regularity of exercising in the context of improving balance has been written about repeatedly. One of the few studies relating to female footballers showed that doing dynamic exercises during warm-up improves both static and dynamic balance [24]. Jakobsen et al. [25] observed postural control improvement after 12 weeks of football training and high-intensity interval running. Paillard et al. [9] paid attention to the fact that the players with more frequent and intensive training sessions showed better postural control, which may be due to greater sensitivity of sensory receptors or better integration of information. Both the general muscular exercises (that mobilizes the whole body) and local muscular exercises (that concentrates on a particular muscle group) can disturb postural control. General muscular exercises (especially running and walking, which are a big part of a football training) contribute to changing the effectiveness of sensory inputs: vestibular—decreases the sensitivity of orthotic organs—and proprioceptive—disturbs the senses of the force and limb position [26]. Söderman et al. [27] Baghbaninaghadehi et al. [28] have drawn attention to the importance of maintaining proper body stability in team sports. This is an important issue, because footballers must have adequate stability, which is especially important when jumping up to head a ball. Brito et al. [29] found that the postural sway (represented by sway velocity) increased and the ankle proprioception decreased after a 45 min of a match. The authors suggest that prophylactic balance training should be performed following rather than before the practice session. Also, Gioftsidou et al. [30] observed improvement in the balance ability, especially in non-dominant lower limbs when the balance training was performed after the football training.

Mohammadi et al. [31] observed significant changes in both the static and dynamic balance parameters among young sportspeople who undertook a 6-week appropriately selected training schedule. According to the authors, an increase in the sportsperson's muscle mass was responsible for the changes. Results of this work also show better postural stability of young football players in dynamic conditions, which are more valuable than static ones, because they are the closest conditions to these on the football field. Interestingly, Pau et al. [32] observed dynamic balance impairment in young football players due to fatigue after the first half on football match. The impairment was observed in the increase in the time necessary to stabilize after landing from a single-leg jump.

On the basis of tracking professional players' COG trajectory while they were standing on one leg and then monitoring the time taken to stabilize their posture, Pau et al. [32] showed that they achieved significantly lower stabilization times while their center of pressure (COP) displacement surface area was significantly less than for their novice colleagues.

The assessment of the balance and dizziness following sports-related concussions is very important for making decisions about further training or removal from the team [33]. In addition, the study authors [34,35] have shown that the use of exercises to improve balance also has a positive effect on coping with stress and reduces the risk of injury during physical activity, which is extremely important in the effective use of a sportsperson during the game.

The findings of this study must be seen in light of some limitations. The first limitation is the sample size: because of the number of football players in the world it is difficult to deduce a general conclusion based on a small sample. The second limitation concerns the sample profile—using purely student sampling is also extremely limiting on the population. Another limit regarding the student population is the questionnaire and their truthfulness, especially in questions about alcohol, drugs, etc. Gribble et al. [36] observed that males were more adversely affected by fatigue than females. This may suggest that for future research the study group should consist of both sexes. In our research, only dynamic conditions were tested; however, Hrysomallis et al. [37] and Pau et al. [38] paid attention to the fact that assessment of balance in football players should be performed with both dynamic and static tests, because postural control performance is not related in these two cases.

5. Conclusions

In conclusion, postural stability in young female footballers was found to be better than their inactive peers. The study showed better usefulness of the vestibular system as well as asymmetrical weight-bearing in regularly trainees.

Further research to explore the possibilities of use CDP in assessing the progress of training is needed. The Sensory Organization Test assesses the quality of the human balance system, which could help with evaluation of training effects. Also, the results of the WS may be useful for determination of the dominant leg, which is important when selecting the role the player can play on the field. More studies are needed to specify which exercises contribute to improving the balance and perhaps apply them in rehabilitation techniques.

Author Contributions: Conceptualization, G.O.; methodology, G.O.; validation, G.O. and A.C.; formal analysis, A.C.; investigation, A.C..; resources, G.O.; writing—original draft preparation, G.O. and A.C.; writing—review and editing, G.O. and A.C. All authors have read and agreed to the published version of the manuscript.

Funding: This research received no external funding.

Acknowledgments: We would like to pay our gratitude and our respects to our co-organizer Jozef Bergier, deceased March 2019.

Conflicts of Interest: The authors declare no conflict of interest.

References

1. Arliani, G.G.; Almeida, G.P.L.; dos Santos, C.V.; Venturini, A.M.; da Costa Astur, D.; Cohen, M. The effects of exertion on the postural stability in young soccer players. *Acta Ortop. Bras.* **2013**, *21*, 155–158. [CrossRef] [PubMed]
2. Gstöttner, M.; Neher, A.; Scholtz, A.; Millonig, M.; Lembert, S.; Raschner, C. Balance Ability and Muscle Response of the Preferred and Nonpreferred Leg in Soccer Players. *Mot. Control.* **2009**, *13*, 218–231. [CrossRef] [PubMed]
3. Starosta, W.; Karpińska, A.; Podciechowska, K. Lateral differentiation of global movement coordination results in girls and women depending in their age, kind of physical activity and dominant hand. *Pol. J. Sports Med.* **2011**, *27*, 289–298. [CrossRef]
4. Kraemer, W.J.; Adams, K.; Cafarelli, E.; Dudley, G.A.; Dooly, C.; Feigenbaum, M.S.; Fleck, S.J.; Franklin, B.; Fry, A.C.; Hoffman, J.R.; et al. American College of Sports Medicine position stand. Progression models in resistance training for healthy adults. *Med. Sci. Sports Exerc.* **2002**, *34*, 364–380. [PubMed]
5. Randers, M.B.; Nielsen, J.J.; Krustrup, B.R.; Sundstrup, E.; Jakobsen, M.D.; Nybo, L.; Dvorak, J.; Bangsbo, J.; Krustrup, P. Positive performance and health effects of a football training program over 12 weeks can be maintained over a 1-year period with reduced training frequency. *Scand. J. Med. Sci. Sports* **2010**, *20*, 80–89. [CrossRef] [PubMed]
6. Bressel, E.; Yonker, J.C.; Kras, J.; Heath, E.M. Comparison of Static and Dynamic Balance in Female Collegiate Soccer, Basketball, and Gymnastics Athletes. *J. Athl. Train.* **2007**, *42*, 42–46. [PubMed]
7. Liang, Y.; Hiley, M.; Kanosue, K. The effect of contact sport expertise on postural control. *PLoS ONE* **2019**, *14*, e0212334. [CrossRef]
8. Matsuda, S.; Demura, S.; Uchiyama, M. Centre of pressure sway characteristics during static one-legged stance of athletes from different sports. *J. Sports Sci.* **2008**, *26*, 775–779. [CrossRef]
9. Paillard, T.; Noé, F.; Rivière, T.; Marion, V.; Montoya, R.; Dupui, P. Postural Performance and Strategy in the Unipedal Stance of Soccer Players at Different Levels of Competition. *J. Athl. Train.* **2006**, *41*, 172–176.
10. Biegański, P.; Pyskir, M.; Pyskir, J.; Trela, E.; Hagner, W. Postural stability of young football players against their physically inactive peers. *J. Health Sci.* **2013**, *3*, 477–488.
11. Maranesi, E.; Merlo, A.; Fioretti, S.; Zemp, D.D.; Campanini, I.; Quadri, P. A statistical approach to discriminate between non-fallers, rare fallers and frequent fallers in older adults based on posturographic data. *Clin. Biomech.* **2016**, *32*, 8–13. [CrossRef] [PubMed]
12. Horak, F.B. Postural orientation and equilibrium: What do we need to know about neural control of balance to prevent falls? *Age Ageing* **2006**, *35*, ii7–ii11. [CrossRef] [PubMed]
13. Maranesi, E.; Fioretti, S.; Ghetti, G.G.; Rabini, R.A.; Burattini, L.; Mercante, O.; Di Nardo, F. The surface electromyographic evaluation of the Functional Reach in elderly subjects. *J. Electromyogr. Kinesiol.* **2016**, *26*, 102–110. [CrossRef]
14. Błaszczyk, J.; Cieślinska-Swider, J.; Plewa, M.; Zahorska-Markiewicz, B.; Markiewicz, A. Effects of excessive body weight on postural control. *J. Biomech.* **2009**, *42*, 1295–1300. [CrossRef]
15. Nashner, L.M. Computerized dynamic posturography. In *Handbook of Balance Function Testing*; Delmar: Clifton Park, NY, USA, 1993; pp. 280–334.
16. Trueblood, P.R.; Rivera, M.; Lopez, C.; Bentley, C.; Wubenhorst, N. Age-based normative data for a computerized dynamic posturography system that uses a virtual visual surround environment. *Acta Oto Laryngol.* **2018**, *138*, 597–602. [CrossRef]
17. Mohieldin, A.; Douaa, M.; Sherif, K.; Thabat, I.; Fawzy, H.; Walid, A.-B. Evaluation of dynamic posturography in anterior cruciate ligament injury patients. *Maced. J. Med. Sci.* **2011**, *4*, 167–173.
18. Dittmar, M. Functional and Postural Lateral Preferences in Humans: Interrelations and Life-Span Age Differences. *Hum. Biol.* **2002**, *74*, 569–585. [CrossRef]

19. Vanicek, N.; King, S.A.; Gohil, R.; Chetter, I.C.; Coughlin, P.A. Computerized dynamic posturography for postural control assessment in patients with intermittent claudication. *J. Vis. Exp.* **2013**, *82*, e51077. [CrossRef]
20. NeuroCom International Clinical Interpretation Guide. In *Balance Manager Systems*; NeuroCom International, Inc.: Clakamas, OR, USA, 2008.
21. Carey, D.P.; Smith, G.; Smith, D.T.; Shepherd, J.W.; Skriver, J.; Ord, L.; Rutland, A. Footedness in world soccer: An analysis of France '98. *J. Sports Sci.* **2001**, *19*, 855–864. [CrossRef]
22. Bergier, J. In the Search of the Tendency to Symmetrize the Shots in Female Soccer at Top Level Competitions. *Afr. J. Phys. Act. Health Sci.* **2015**, *21*, 1024–1029.
23. Barone, R.; Macaluso, F.; Traina, M.; Leonardi, V.; Farina, F.; Di Felice, V. Soccer players have a better standing balance in nondominant one-legged stance. *Open Access J. Sports Med.* **2010**, *2*, 1. [CrossRef]
24. Amiri-Khorasani, M.; Gulick, D.T. Acute effects of different stretching methods on static and dynamic balance in female football players. *Int. J. Ther. Rehabil.* **2015**, *22*, 68–73. [CrossRef]
25. Jakobsen, M.D.; Sundstrup, E.; Krustrup, P.; Aagaard, P. The effect of recreational soccer training and running on postural balance in untrained men. *Eur. J. Appl. Physiol.* **2011**, *111*, 521–530. [CrossRef]
26. Paillard, T. Effects of general and local fatigue on postural control: A review. *Neurosci. Biobehav. Rev.* **2012**, *36*, 162–176. [CrossRef]
27. Söderman, K.; Werner, S.; Pietilä, T.; Engström, B.; Alfredson, H. Balance board training: Prevention of traumatic injuries of the lower extremities in female soccer players? A prospective randomized intervention study. *Knee Surg. Sports Traumatol. Arthrosc.* **2000**, *8*, 356–363. [CrossRef]
28. Baghbaninaghadehi, F.; Ramezani, A.R.; Hatami, F. The effect of functional fatigue on static and dynamic balance in female athletes. *Int. Sportmed J.* **2013**, *14*, 77–85.
29. Brito, J.; Fontes, I.; Ribeiro, F.; Raposo, A.; Krustrup, P.; Rebelo, A. Postural stability decreases in elite young soccer players after a competitive soccer match. *Phys. Ther. Sport* **2012**, *13*, 175–179. [CrossRef]
30. Gioftsidou, A.; Malliou, P.; Pafis, G.; Beneka, A.; Godolias, G.; Maganaris, C.N. The effects of soccer training and timing of balance training on balance ability. *Eur. J. Appl. Physiol.* **2006**, *96*, 659–664. [CrossRef]
31. Mohammadi, V.; Alizadeh, M.; Gaieni, A. The Effects of six weeks strength exercises on static and dynamic balance of young male athletes. *Procedia Soc. Behav. Sci.* **2012**, *31*, 247–250. [CrossRef]
32. Pau, M.; Mereu, F.; Melis, M.; Leban, B.; Corona, F.; Ibba, G. Dynamic balance is impaired after a match in young elite soccer players. *Phys. Ther. Sport* **2016**, *22*, 11–15. [CrossRef]
33. Doettl Steven, M. Sports Concussions (TBI), Imbalance, and Dizziness. *Perspect. Neurophysiol. Neurogenic Speech Lang. Disord.* **2015**, *25*, 36–41. [CrossRef]
34. Iacono, A.D.; Martone, D.; Alfieri, A.; Ayalon, M.; Buono, P. Core Stability Training Program (CSTP) effects on static and dynamic balance abilities. *Gazz. Med. Ital. Arch. Per Le Sci. Med.* **2014**, *173*, 197–206.
35. Cankaya, S.; Gokmen, B.; Tasmektepligil, M.Y.; Con, M. Special Balance Developer Training Applications on Young Males' Static and Dynamic Balance Performance. *Anthropologist* **2015**, *19*, 31–39. [CrossRef]
36. Gribble, P.A.; Robinson, R.H.; Hertel, J.; Denegar, C.R. The Effects of Gender and Fatigue on Dynamic Postural Control. *J. Sport Rehabil.* **2009**, *18*, 240–257. [CrossRef] [PubMed]
37. Hrysomallis, C.; McLaughlin, P.; Goodman, C. Relationship between static and dynamic balance tests among elite Australian Footballers. *J. Sci. Med. Sport* **2006**, *9*, 288–291. [CrossRef]
38. Pau, M.; Arippa, F.; Leban, B.; Corona, F.; Ibba, G.; Todde, F.; Scorcu, M. Relationship between static and dynamic balance abilities in Italian professional and youth league soccer players. *Phys. Sport* **2015**, *16*, 236–241. [CrossRef]

applied sciences

MDPI

Article

Evaluation of Commercial Ropes Applied as Artificial Tendons in Robotic Rehabilitation Orthoses

Guilherme de Paula Rúbio [1], Fernanda Márcia Rodrigues Martins Ferreira [1], Fabrício Henrique de Lisboa Brandão [2], Victor Flausino Machado [2], Leandro Gonzaga Tonelli [2], Jordana Simões Ribeiro Martins [3], Renan Fernandes Kozan [4] and Claysson Bruno Santos Vimieiro [1,2,3,*]

1 Graduate Program in Mechanical Engineering, Universidade Federal de Minas Gerias, Belo Horizonte, MG, Brazil; guilhermeprubio@gmail.com (G.d.P.R.); fernandaferreira.to@gmail.com (F.M.R.M.F.)

2 Department of Mechanical Engineering, Universidade Federal de Minas Gerias, Belo Horizonte, MG, Brazil; fabriciohlisboa@gmail.com (F.H.d.L.B.); victor.fmachado72@gmail.com (V.F.M.); ltonellig@gmail.com (L.G.T.)

3 Graduate Program in Mechanical Engineering, Pontifícia Universidade Católica de Minas Gerais, Belo Horizonte, MG, Brazil; martinsjsr@gmail.com

4 Department of Electrical Engineering, Universidade Federal de Minas Gerias, Belo Horizonte, MG, Brazil; renankozan@gmail.com

* Correspondence: claysson@pucminas.br; Tel.: +55-31-99161-1320

Received: 28 December 2019; Accepted: 23 January 2020; Published: 31 January 2020

Featured Application: This study can be used as a reference for works that develop active orthoses or prostheses that use artificial tendons to move the fingers, helping in the process of defining these tendons.

Abstract: This study aims to present the design, selection and testing of commercial ropes (artificial tendons) used on robotic orthosis to perform the hand movements for stroke individuals over upper limb rehabilitation. It was determined the load applied in the rope would through direct measurements performed on four individuals after stroke using a bulb dynamometer. A tensile strength test was performed using eight commercial ropes in order to evaluate the maximum breaking force and select the most suitable to be used in this application. Finally, a pilot test was performed with a user of the device to ratify the effectiveness of the rope. The load on the cable was 12.38 kgf (121.4 N) in the stroke-affected hand, which is the maximum tensile force that the rope must to supports. Paragliding rope (DuPont™ Kevlar®) supporting a load of 250 N at a strain of 37 mm was selected. The clinical test proved the effectiveness of the rope, supporting the requested efforts, without presenting permanent deformation, effectively performing the participant's finger opening.

Keywords: orthosis; biomechanics; tensile strength test; robotic therapy

1. Introduction

Stroke is the world's leading cause of death and disability [1]. with 6.7 million deaths per year, estimated to be the second leading cause of death by the year 2030 [2,3].

Stroke is a clinical syndrome caused by a reduction in blood perfusion in brain structures and is characterized by disturbances in brain function [4]. About 57% of individuals, after having a stroke, have some of limitations in daily activities and more than 50% of these individuals have functional impairment, thus requiring some external help to perform functional tasks [5]. The prevalent motor deficits presented are paralysis (hemiplegia) or weakness (hemiparesis) of body contralateral half to stroke injury [6]. Clinically, muscle weakness, muscle tone abnormality, movement control deficiency, body composition changes of the extremities with loss of muscle and bone mass [7], postural alteration, lack of mobility, abnormal synergistic patterns and loss or reduction of motor coordination can also be observed [8].

An innovative and promising rehabilitation alternative capable of enhancing the motor and functional capacity of individuals after stroke is the robotic therapy [9]. It uses robotic orthoses, which is mechatronic equipment ranging from the creation of artificial limbs, to robots that assist in rehabilitation or hospital or residential care [10].

The great advantage of using these devices is the high degree of repeatability and the performance of intensive activities with less professional supervision, using a simple routine of pre-programmed robot rehabilitation activities [8,11–13]. Numerous systematic reviews have been performed showing the efficiency in using these devices to rehabilitate individuals [12,14–18]. An Improvement in short and long-term proximal end (shoulder and elbow) motor control has been observed in acute and chronic post-stroke patients, using these devices [12,17]. A combination of traditional methods and robotic therapy in some stages of stroke recovery can produce a significant improvement in elbow and shoulder motor recovery [16,18]. An improvement in daily life activity, arm function and muscle strength has also been observed [15], as well as minor effects on motor control and mild effects on muscle strength compared with other short-term interventions using robotic therapy [14]. A combination of these devices with brain–computer interfaces has been also used to improve the rehabilitation ability of individuals after stroke, thus proving the effectiveness of these robots during therapy sessions [19].

Several devices have also been developed to rehabilitate or assist upper limb function, such as MIT-Manus [20], ARMin III [21], MIME [22], BI-MANU-TRACK [23], ARM Guide [24] and NeReBot [25]. In Brazil, the Laboratório de Bioengenharia da Universidade Federal de Minas Gerais (LabBio-UFMG), located in Belo Horizonte/MG, developed a robotic orthosis for upper limb rehabilitation of post stroke individuals [26,27].

Currently is desire the development of soft robots, made by tissue, or elastic polymers [28]. In this way, mechanical tendons or soft actuators act like a human tendon performing the exoskeleton actuation function, while the hand and fingers perform the strutural function [29]. These devices often use springs, cables, elastics, ropes or elastic wires, making the same fuction of artificial tendons, to transmit movement to the paralyzed muscles, through traction and relaxation of them. In the Exo-Glove [29] (Figure 1b) a Bowden cable, frequently used like a bicycle brake cable, was used to perform the artificial tendon function and control the fingers movement, your choice was due to a teflon coating which protects the steel cable. In the Xiloyannis et al. [30,31] work a Bowden cable was used to transmit de motor force to the artificial tendon, but in the first a Kevlar rope was used in the artifical tendon function and in the second another cable steel was used for this. In the Hero system [32] (Figure 1a) a tie up, made by plastic, was used and a simple rope was used in the previously version of the LabBio hand orthesis [33] (Figure 1c).

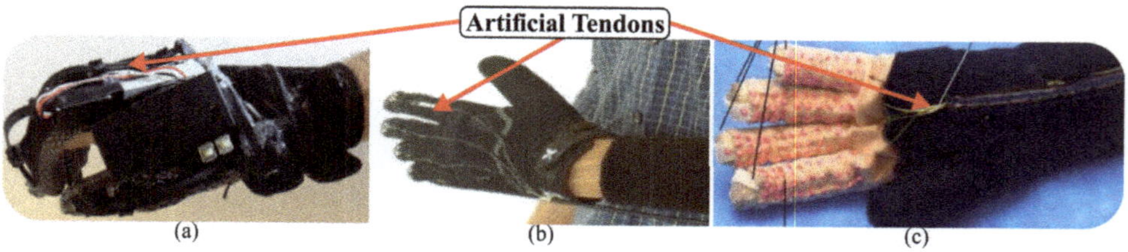

Figure 1. Use of artificial tendon in: (**a**) Hero orthosis, (**b**) hand exoskeleton and (**c**) LabBio orthosis. Adapted from Cherian et al. [29], Yurkewich et al. [32], Rocha [33].

These artificial tendons must be able to withstand heavy loads, have low strain, ease of handling and good fit for the user limb, ensuring safe use of the equipment. To achieve this a correct analysis of each cable must be made, through tests that prove the correct functioning of the tendons. Thus, this paper aims to present the design, selection and testing of commercial ropes that were used to generate the hand movements of a robotic orthosis for upper limb rehabilitation of individuals after stroke.

2. Methodology

2.1. Device

The orthosis developed in (LabBio-UFMG) is portable, low cost and low weight composed by a module with glove and artificial fingers and phalanxes, witch are connected to ropes (artificial tendons) that are able to open the user's fingers. The rope is pulled through a power screw system coupled to a motor, which executes and controls the motions. The rope was attached to the transmission system through a support fixed by a bolt (Figure 2).

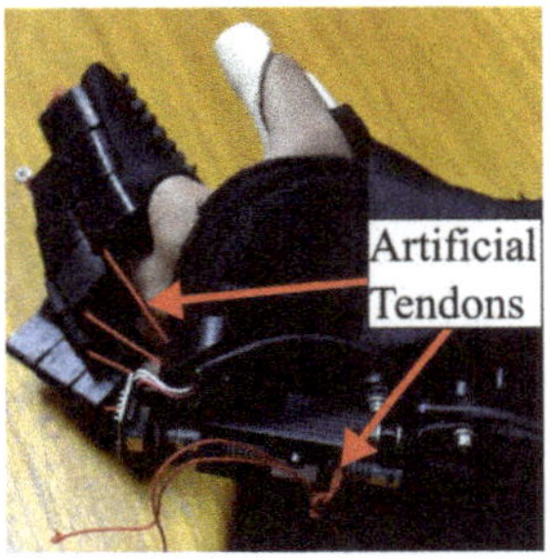

Figure 2. Orthosis with artificial tendons use.

The artificial phalanges limit the users' phalanges movement avoiding anything greater than physiological amplitudes. For fingers' opening movement, the artificial tendon was tensioned and the distal artificial phalanx was pulled, in that way, performing a rotation about the distal articulation axis. This movement was performed until the distal artificial phalanx collided with the middle artificial phalanx. Keeping the traction in the artificial tendon, a rotation about the middle articulation axis was performed until the collision between middle and proximal artificial phalanges. The rotation now occurred about the proximal articulation axis until the transmission system nut actuated the travel limiter sensor present in the system, so the fingers' opening completed movement was performed (Figure 3). The closing movement of the fingers was performed passively, taking advantage of the user's ability to perform this movement. Due to this, the artificial tendon was designed to support the traction during the fingers opening movement.

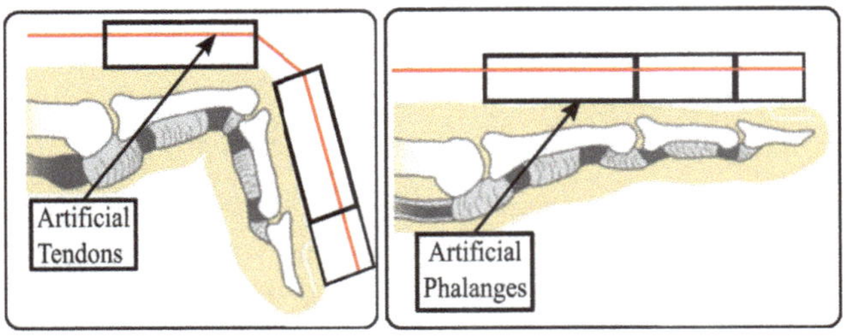

Figure 3. Schematic representation of the (**a**) opening and (**b**) closing fingers movement by the device. Adapted from Rúbio et al. [27].

Before to the operation started, with the orthosis dressed up, the operator commanded the actuator to the fully fingers opening setup with the bracket screw loose.Upon reaching the position, the operator pulled the ropes, with the user's hand fully opening, until they were on full traction, and tightened the fastener screw, thus ensuring the correct performance of the system. All position and speed control was made through an application developed for the Android® system and a microcontrolled electronic circuit by a STM32F103.

2.2. Volunteers

The volunteers were recruited, for a pilot test with the device, through a public call issued by stroke associations, rehabilitation centers, hospitals and social media in the city of Belo Horizonte, Minas Gerais in December 2016. They were selected according to the inclusion and exclusion criteria and informed about the study objectives. Those who agreed to participate signed the informed consent form. The study project was approved by the Universidade Federal de Minas Gerais Research Ethics Committee (CAAE Registry: 22207213.5.0000.5149).

Inclusion criteria were: age greater or equal than 18 years with left unilateral stroke and chronic impairment diagnosis (minimum six months after stroke) [34]; present hemiparesis with reduced upper limb motor function (elbow and fingers incomplete flexion and extension movements, presenting arm and hand in the 4th Brunnstrom phase) [35]; present at least 45° of the shoulder flexion and abduction, complete passive movement of the elbow, hand and finger joints without compromised sensitivity measure by the Fugl–Meyer scale [36]; present muscle tone alteration from mild to moderate, measured through the Modified Ashworth Scale [37]; present absence of severe cognitive deficits, evaluated by the Brazilian version of the Mini-Mental State Examination [38]. The exclusion criteria in the study were: flaccidity in the affected upper limb; severe neurological, orthopedic or rheumatologic impairment before the stroke that may interfere with the performance task; severe cognitive impairment (global aphasia, attention deficit, neglect) that limits understanding of commands or conclusion of experimental tasks; severe pain in the affected upper limb, measured using the Visual Analogue Scale (VAS) (> 8 on a scale from 0 to 10); opened skin injuries where the device would be attached; having used Botox in the last three months for spasticity or other medicines known to increase motor recovery; having participated in the last three months of another research study to improve upper limb function.

Four individuals were selected to participate in the study. Their characteristics are presented in Table 1.

Table 1. Volunteers characteristics.

Participant	Sex	Age	Dominance	Post Stroke Time	Finger Spasticity Level *
1	Male	73	Right	53 months	1
2	Male	38	Right	113 months	2
3	Female	25	Left	48 months	2
4	Female	48	Right	24 months	1

(*) Spasticity assessed using the Ashworth Scale.

2.3. Artificial Tendons Design and Selection

For the artificial tendons design and selection used in the hand module, some design steps were required. The first was the determination of the loads to which the ropes were subjected; thereafter a tensile strength test using various commercial ropes to evaluate and determine what would best fit the use; and finally a pilot test with the volunteers to confirm the effectiveness of the rope as an artificial tendon.

2.3.1. Applied Loads Determination

To determine the traction force that the rope would be subjected, direct measurements were performed on the four volunteers with a New Saehan Squeeze Dynamometer-SH5008 bulb-type dynamometer. Normative anthropometric data were not used for this determination because individuals after stroke present abnormality in tone, which implies hypotonia in the development and subsequent hypertonia in these individuals. This hypertonia can lead to spasticity, which implies speed-dependant increase in muscle tone, thereby increasing myotactic reflex, postural changes and stereotyped movements, which leads to a reduction in the range of joint motion, pain, muscles limb activities limitation and consequently hinders the performance of daily functional activities [39,40]. Because of that, these individuals present a lower grip strength than normative anthropometric data, which present an average of 44.2 kgf for men and 31.6 kgf for women when using the dominant hand [35].

As users of this equipment present difficulties only in the fingers opening movement, the maximum force required to the rope was greater or equal than the maximum grip strength of the hand affected, that is, pulling the rope. The movement of finger closure was given passively, taking advantage of the spasticity presented by the individuals and the ability to close the fingers preserved.

To measure the grip strength, each participant squeezed the dynamometer body as tightly as possible for three consecutive times and an average of the strength measured was calculated. The traction force applied to the selected cable was determined to be the highest of the average forces measured, thus ensuring that the rope safely supports the required loads.

2.3.2. Tensile Strength Test

After defining the applied traction force, the rope used was selected. For this, a set of eight different commercial ropes were subjected to a tensile strength test, until their rupture, to determine the maximum supported force and the strain of each one. The first guarantees the system's operation, and must have a value greater or equal than the volunteers highest average grip force, which was the maximum traction load defined for the project. Already the strain ensures the device a degree of repeatability, since one of the great advantages of robotic therapy is the ability to perform the same movements over and over again on the patient [8,11,12], the selected rope must not have permanent strain after performing one or more actions, to ensure the fingers correct position.

The selected ropes to the tests should have a diameter equal or less than 1.25 mm to easily couple the rope in the nut and artificial phalanges without raising the device components size and weight. Furthermore, they usually used in hight loads applications and could be easily found in the regional market due to the fabrication costs. Their characteristics are presented in the Table 2.

Table 2. Ropes characteristic.

Manufacturer	Product	Material	Diameter [mm]
D'Addarío	Guitar String D (85/15)	Brass	0.64
D'Addarío	Guitar String E (85/15)	Brass	0.23
D'Addarío	Guitar String B (85/15)	Brass	0.30
D'Addarío	Guitar String A (85/15)	Brass	0.89
D'Addarío	Guitar String G (85/15)	Brass	0.38
D'Addarío	Guitar String E (85/15)	Brass	1.14
TufLine XP	Fishing line	Spectra®	0.48
SOL Paragliders	Paragliding rope	DuPont™ Kevlar®	1.00

The tests were performed at the Structural Analysis Laboratory of the Pontifícia Universidade Católica de Minas Gerais, using the EMIC 23-5D tensile testing machine, which has a maximum load capacity of up 5 KN and clamps for specimen fixation. For the attachment of the ropes, a tie was used, where the cable was tied to the machine's claws ends, using a simple knot. With the cable properly fixed, the upper clamp was adjusted to an initial height of 300 mm, thus allowing a usable wire clearance of 255 mm and a maximum test height of 600 mm.

For control and data acquisition, Tesc Version 3.04 software was used. The test parameters were then defined in the software, which were the "Rectangular Tie Pull" method, the type of material tested and the rope diameter analyzed. The test was initiated by the operator, and the machine gradually increased the applied force automatically until the sample rupture was reached. With the test finished, the software generated the strain and force plots with the value measured. Several tests were performed in order to stabilize the values of force and strain obtained, since the desired curve for each rope should have similar behaviors, without great variability in their force values by strain. However a great variability in the values found were observed, mainly due to the way the ropes were attached to the claw, which often fails to fix them sometimes loosening the knot made which altered the obtained data and thus several tests were needed. The results presented took into account the tests that behaved with greater similarity for each rope.

2.3.3. Pilot Test

With the rope selected, a device functionality test was performed with the four volunteers to approve this rope like artificial tendons. Only the results of one of the volunteers (participant 3) was shown, because this participant presented one of the highest hand spasticity values, so if the tendons could open their fingers efficiently, all of the individuals, with a spasticity lower of equal than his, would be also able to open their fingers.

The opening and closing movement of the fingers (complete extensions and flexions) were performed several times to verify if the artificial tendon supported the loads during the whole movement. Partial openings were also made in order to verify the degree of precision of the system by measuring the opening angle of the hand using a plastic PVC Carci finger goniometer. In addition, it was verified that the cable did not suffer any permanent strain during its operation.

3. Results

3.1. Applied Loads Determination

With the data of volunteers grip strength (Table 3), the highest average grip strength observed was equal to 17.83 kgf (174,91 N) in the hand afected by the stroke. Therefore this is the maximum traction force that the rope used should support.

Table 3. Unaffected and affected limb grip strength of volunteers.

Participant	Measured Unaffected [kgf]	Measured Affected [kgf]	Average Unaffected [kgf]	Average Affected [kgf]	Standard Deviation Unaffected [kgf]	Standard Deviation Affected [kgf]
1	36 39 41	9 12 12	38.67	11.00	2.52	1.73
2	54 49 44	18 18 17.5	49	17.83	5	0.29
3	38 38 40	15 10 15	38.67	13.33	1.15	2.89
4	34.5 37 30	6 9 7	33.83	7.33	3.55	1.53
Average			40.04	12.38	6.45	4.28

3.2. Tensile Strength Test

The TufLine XP fishing line behaved heterogeneously, as shown in the three force x strain curves below (Figure 4). It presented the maximum strength range between 45 and 140 N and a strain between 14 and 39 mm approximately.

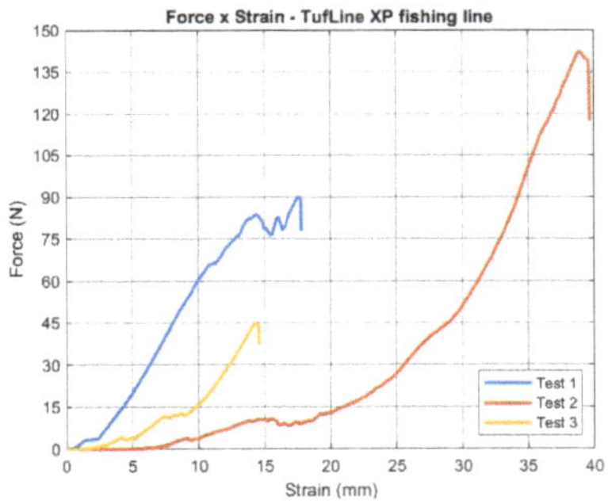

Figure 4. Force x strain curves TufLine XP fishing line.

The guitar strings tested, showed a more consistent behavior, and were analyzed as shown in Figure 5. Only one curve force x strain for each string was shown, due to close behavior of the curves obtained by the tests. Among the guitar strings submitted to the tensile strength test, the one that presented the best result was the D string (0.64 mm in diameter), it presents approximately a 37 mm of strain, when it is pulled at 200 N force. This value is higher than the global average grip force in the volunteers (12.38 kgf or 121.45 N) and the maximum traction force determined in the project (174.91 N). The ropes did not perform as expected, there was no proportional increase in the maximum tensile strength relative to the increase in the diameter of the rope. This is because, in order to change the musical notes (frequencies), in addition to increasing the diameter, changes in the string structures are necessary.

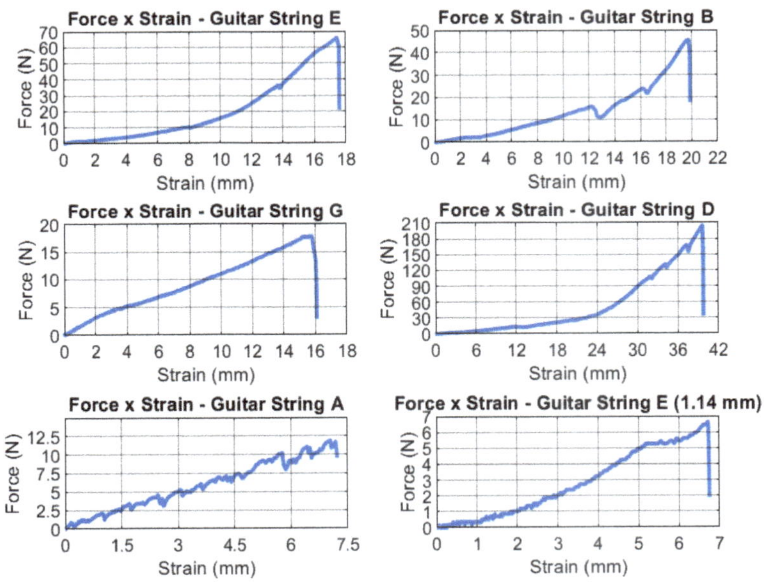

Figure 5. Force x strain curves of: **Guitar String E, Guitar String B, Guitar String G, Guitar String D, Guitar String A, Guitar String E (1.14 mm)**.

The last rope analyzed was the SOL Paragliders paragliding rope. The force x strain curves shown (Figure 6) to demonstrate the more stable behavior of the paragliding rope than the fishing line. It withstood high tensile loads between 225 N and 280 N and maximum strain between 33 mm and 35 mm. In addition, the generated curves present greater congruence of values for the elastic, plastic and rupture limit regions compared to the other lines submitted to the test.

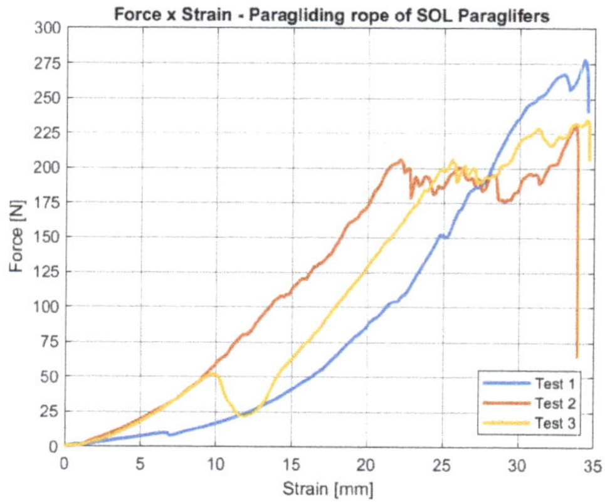

Figure 6. Force x strain curves of paragliding rope of SOL Paragliders.

3.3. Pilot Test

With the artificial tendon selected, it was then possible to use it with a post stroke individual to validate its application. As shown in Figure 7, the participant's finger opening, using the orthosis, was satisfactory, the full opening was performed, which was not possible naturally, through voluntary control. As shown in the Figure 7a, the spasticity prevented the volunteer's hand opening movement, which at the same time performed wrist pronation due to stroke sequelae. With the orthosis (Figure 7b), pronation was no longer performed and the full opening movement of the hand was performed, proving the functionality of the device for this individual.

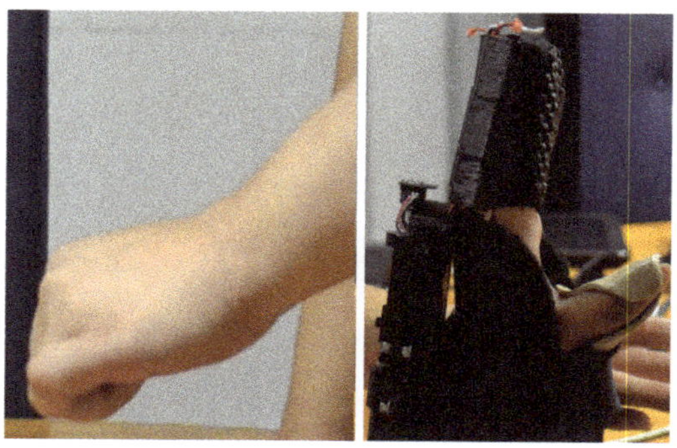

(a) (b)

Figure 7. Participant's finger openings of (**a**) naturally form and (**b**) using the orthosis.

Making repeated partial openings of 80°, 60° and 40° (Figure 8), it was observed that the fingers presented a consistency and precision in the opening angle, showing that the use of this type of rope can satisfactorily perform as an artificial tendon in robotic orthoses for individuals after stroke. It is important to emphasize that the repeated use of the equipment did not generate permanent strains in the rope, being able to execute without variations the same amount of opening requested.

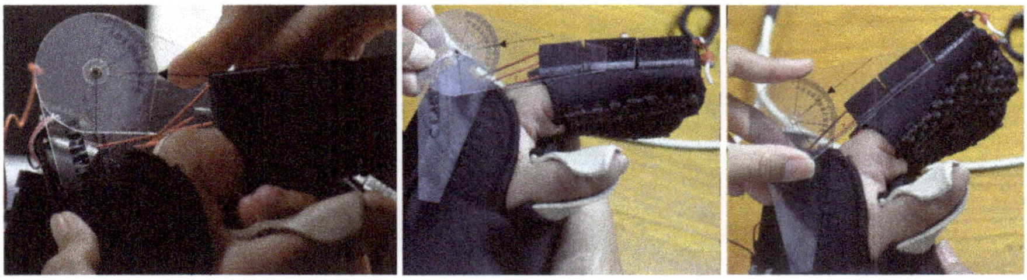

(a) (b) (c)

Figure 8. Fingers aperture at (a) 80°, (b) 60° and (c) 40°.

4. Discussion

Results of the volunteers grip force measures confirm that they have a strength lower than that of typical individuals. Normative data indicate that healthy individuals had an average dominant hand grip force of 28.4 ± 9.7 kgf for women and 40.3 ± 14.3 kgf for men [41]. The change in post-stroke individuals grip force is related to the motor deficits found after the brain injury, which result from the injury to the upper motor neurons that control the distal and proximal muscles, generating a decrease in the activation of some muscle groups [42]. In addition, there is the presence of spasticity, could cause postural changes in the upper limb, which, if not corrected it generate deformities, affecting even the joints and muscles viscoelastic properties and the tendons integrity. Therefore, based on the studies of Wu et al. [43,44], we can use a high resolution ultrasound as an alternative to check these phenomena and select the most suitable cable to be used in robotic orthoses.

It can be noted in the fishing line tensile strength test (Figure 4), that there is a variability in its behavior, presented maximum traction strength below than the desired. In addition to inconsistent behavior, it is difficult to handle and could difficulty in properly adjusting tendons before orthosis operation.

The D'Addarío (85/15 bronze) guitar strings (Figure 5), despite their smallest strain, suffered some damage with the application of low force, once their constitution was related to the sound and not to the material resistance. Its structure broke internally or suffered cracks with low loads, which can be seen with a simple visual inspection.

The paragliding rope showed the highest tensile strength among the tested ropes (Figure 6), breaking through the external structure destruction, with a smoothly break.

Thus, given the analysis of data obtained in the tests performed with different ropes, the conclusion was that the most suitable for use in robotic orthosis is the paragliding rope which meets the pre-established criteria of: strength, diameter and cost. (DuPont™ Kevlar®). The rope presented an easy handling and greater security due to values of resistance to traction superior to the other ropes, and due to its form of rupture, since the others present a whip when broken.

The pilot test confirmed the consistency and precision in the opening angle using this rope, showing the satisfactory use as an artificial tendon in robotic orthoses. This system makes it possible to efficiently open and close the post-stroke individuals fingers, since a significant proportion of individuals with physical sequel resulting from stroke, remains with problematic or unsatisfactory manual function return [45]. With the use of artificial tendons in the orthosis, there is an increase in functional skills, facilitating the daily activities performance, such as reaching, picking up and holding objects, using tools such as a cell phone, eating, dressing and performing personal care in a general way. This significantly impacts the patient's life, improving his independence, self-esteem and life quality.

5. Conclusions

The use of artificial tendons applied to upper limb orthoses is one of the ways to effectively perform fingers opening of the individual. As shown in this study, commercial ropes can be used for this purpose as long as they support the requested loads without presenting a permanent strain.

We analyzed eight ropes to act as artificial tendon and withstand a tensile load of 121.4 N. Of these, the paragliding rope manufactured by SOL Paragliders showed the best performance for this purpose, withstanding a load of 250 N at approximately 30 to 35 mm of strain. When used in the device, in an individual after stroke, it was effective for application, being able to perform the task without breaking, and maintaining the degree of repeatability, i.e., did not suffer permanent strain, which would affect the equipment degree of repetition, validating its use as an artificial tendon.

Author Contributions: Conceptualization: G.P.R., F.M.R.M.F. and C.B.S.V.; Data curation: F.H.d.L.B., V.F.M., L.G.T.; Writing and editing: G.d.P.R., F.M.R.M.F., J.S.R.M. and C.B.S.V.; Supervision: R.F.K. and C.B.S.V.; Project administration: C.B.S.V. All authors have read and agreed to the published version of the manuscript.

Funding: This research was funded by Financiadora de Estudos e Projetos (FINEP: 01.12.0476.00), Fundação de Amparo à Pesquisa do Estado de Minas Gerais (FAPEMIG) and Coordenação de Aperfeiçoamento de Pessoal de Nível Superior (Capes): finance code 001.

Acknowledgments: The authors would like to thank the Universidade Federal de Minas Gerais, Pontifícia Universidade Católica de Minas Gerais and the Graduate Program in Mechanical Engineering for the support available to carry out this project.

Conflicts of Interest: The authors declare no conflict of interest.

References

1. GBD 2015 Neurological Disorders Collaborator Group. Global, regional, and national burden of neurological disorders during 1990–2015: A systematic analysis for the Global Burden of Disease Study 2015. *Lancet Neurol.* **2017**, *16*, 877–897. [CrossRef]
2. WHO. Health Statistics and Information Systems—Projections of Mortality and Causes of Death, 2015 and 2030. 2013. Available online: http://www.who.int/healthinfo/global_burden_disease/projections/em (accessed on 22 November 2018).
3. WHO. The top 10 causes of death. 2014. Available online: http://www.who.int/mediacentre/factsheets/fs310/em (accessed on 22 November 2018).
4. Royal College of Physicians. *National Clinical Guidelines for Stroke*, 5 ed.; Intercollegiate Stroke Working: London, UK, 2016.
5. Mayo, N.E.; Wood-Dauphinee, S.; Côté, R.; Durcan, L.; Carlton, J. Activity, participation, and quality of life 6 months poststroke. *Arch. Phys. Med. Rehabil.* **2002**, *83*, 1035–1042. [CrossRef]
6. O'Sullivan, S.B.; Schmitz, T.J. *Physical Rehabilitation*, 5 ed.; F.A. Davis Company: Philadelphia, PA, USA, 2006; p. 1383.
7. Chang, K.V.; Wu, W.T.; Huang, K.C.; Han, D.S. Segmental body composition transitions in stroke patients: Trunks are different from extremities and strokes are as important as hemiparesis. *Clin. Nutr.* **2019**. [CrossRef]
8. Poli, P.; Morone, G.; Rosati, G.; Masiero, S. Robotic Technologies and Rehabilitation: New Tools for Stroke Patients' Therapy. *BioMed Res. Int.* **2013**, *2013*, 1–8. [CrossRef]
9. Liepert, J.; Uhde, I.; Graf, S.; Leidner, O.; Weiller, C. Motor cortex plasticity during forced-use therapy in stroke patients: A preliminary study. *J. Neurol.* **2001**, *248*, 315–321. [CrossRef]
10. Maciejasz Pawełand Eschweiler, J.; Gerlach-Hahn, K.; Jansen-Troy, A.; Leonhardt, S.; Maciejasz, P.; Eschweiler, J.; Gerlach-Hahn, K.; Jansen-Troy, A.; Leonhardt, S. A survey on robotic devices for upper limb rehabilitation. *J. NeuroEng. Rehabil.* **2014**, *11*, 3. [CrossRef]
11. Duret, C.; Grosmaire, A.G.; Krebs, H.I. Robot-Assisted Therapy in Upper Extremity Hemiparesis: Overview of an Evidence-Based Approach. *Front. Neurol.* **2019**, *10*. [CrossRef]
12. Kwakkel, G.; Kollen, B.J.; Krebs, H.I. Effects of Robot-Assisted Therapy on Upper Limb Recovery After Stroke: A Systematic Review. *Neurorehabil. Neural Repair* **2008**, *22*, 111–121. [CrossRef] [PubMed]
13. Yozbatiran, N.; Francisco, G.E. Robot-assisted Therapy for the Upper Limb after Cervical Spinal Cord Injury. *Phys. Med. Rehabil. Clin. North Am.* **2019**, *30*, 367–384. [CrossRef] [PubMed]
14. Ferreira, F.M.R.M.; Chaves, M.E.A.; Oliveira, V.C.; Van Petten, A.M.V.N.; Vimieiro, C.B.S. Effectiveness of robot therapy on body function and structure in people with limited upper limb function: A systematic review and meta-analysis. *PLoS ONE* **2018**, *13*, e0200330. [CrossRef] [PubMed]

15. Mehrholz, J.; Pohl, M.; Platz, T.; Kugler, J.; Elsner, B. Electromechanical and robot-assisted arm training for improving activities of daily living, arm function, and arm muscle strength after stroke. *Cochrane Database Syst. Rev.* **2018**. [CrossRef] [PubMed]
16. Norouzi-Gheidari, N.; Archambault, P.S.; Fung, J. Effects of robot-assisted therapy on stroke rehabilitation in upper limbs: Systematic review and meta-analysis of the literature. *J. Rehabil. Res. Develop.* **2012**, *49*, 479. [CrossRef] [PubMed]
17. Prange, G.B.; Jannink, M.J.A.; Groothuis-Oudshoorn, C.G.M.; Hermens, H.J.; IJzerman, M.J. Systematic review of the effect of robot-aided therapy on recovery of the hemiparetic arm after stroke. *J. Rehabil. Res. Develop.* **2006**, *43*, 171. [CrossRef] [PubMed]
18. Veerbeek, J.M.; Langbroek-Amersfoort, A.C.; van Wegen, E.E.H.; Meskers, C.G.M.; Kwakkel, G. Effects of Robot-Assisted Therapy for the Upper Limb After Stroke. *Neurorehabil. Neural Repair* **2017**, *31*, 107–121. [CrossRef] [PubMed]
19. Lee, J.; Mukae, N.; Arata, J.; Iihara, K.; Hashizume, M. Comparison of Feature Vector Compositions to Enhance the Performance of NIRS-BCI-Triggered Robotic Hand Orthosis for Post-Stroke Motor Recovery. *Appl. Sci.* **2019**, *9*, 3845. [CrossRef]
20. Krebs, H.; Hogan, N.; Aisen, M.; Volpe, B. Robot-aided neurorehabilitation. *IEEE Trans. Rehabil. Eng.* **1998**, *6*, 75–87. [CrossRef]
21. Klamroth-Marganska, V.; Blanco, J.; Campen, K.; Curt, A.; Dietz, V.; Ettlin, T.; Felder, M.; Fellinghauer, B.; Guidali, M.; Kollmar, A.; et al. Three-dimensional, task-specific robot therapy of the arm after stroke: A multicentre, parallel-group randomised trial. *Lancet Neurol.* **2014**, *13*, 159–166. [CrossRef]
22. Lum, P.; Burgar, C.; Van Der Loos, M.; Shor, P.; Majumbar, M.; Yap, R. MIME robotic device for upper-limb neurorehabilitation in subacute stroke subjects: A follow-up study. *J. Rehabil. Res. Dev.* **2006**, *45*, 631,642. [CrossRef]
23. Hesse, S.; Schulte-Tigges, G.; Konrad, M.; Bardeleben, A.; Werner, C. Robot-assisted arm trainer for the passive and active practice of bilateral forearm and wrist movements in hemiparetic subjects. *Arch. Phys. Med. Rehabil.* **2003**, *84*, 915 920. [CrossRef]
24. Kahn, L.; Lum, P.; Rymer, W.; Reinkensmeyer, D. Robot-assisted movement training for the stroke-impaired arm: Does it matter what the robot does? *J. Rehabil. Res. Dev.* **2006**, *43*, 619–630. [CrossRef]
25. Masiero, S.; Celia, A.; Rosati, G.; Armani, M.; Masieiro, S.; Celia, A.; Rosati, G.; Armani, M. Robotic-Assisted Rehabilitation of the Upper Limb After Acute Stroke. *Arch. Phys. Med. Rehabil.* **2007**, *88*, 142–149. [CrossRef]
26. Araújo, R.C.D. Desenvolvimento e Avaliação de Sistema de Auxílio à Reabilitação Motora do Membro Superior Após Acidente VasculaR EncefáLico. Doctoral Dissertation, Universidade Federal de Minas Gerais, Belo Horizonte - MG, Brazil, 2011.
27. Rúbio, G.P.; Ferreira, F.; de Lisboa Brandão, F.H.; Machado, V.; Tonelli, L.; Kozan, R.F.; Vimieiro, C. Design of Actuators Applied to a Upper Limb Orthosis. In Proceedings of the 25th International Congress of Mechanical Engineering, ABCM, Uberlandia, MG, Brazil, 20–25 October 2019. [CrossRef]
28. Radder, B.; Prange-Lasonder, G.B.; Kottink, A.I.R.; Holmberg, J.; Sletta, K.; van Dijk, M.; Meyer, T.; Melendez-Calderon, A.; Buurke, J.H.; Rietman, J.S. Home rehabilitation supported by a wearable soft-robotic device for improving hand function in older adults: A pilot randomized controlled trial. *PLoS ONE* **2019**, *14*, e0220544. [CrossRef]
29. Cherian, B.; Dominic, C.; G, V.; Vishakh, K.R. Exo-Glove: A Soft Wearable Robotic Hand for Stroke Survivors. *Inte. Res. J. Eng. Technol.* **2018**, *05*, 3358–3364.
30. Xiloyannis, M.; Galli, L.; Chiaradia, D.; Frisoli, A.; Braghin, F.; Masia, L. A Soft Tendon-Driven Robotic Glove: Preliminary Evaluation. In *Converging Clinical and Engineering Research on Neurorehabilitation III. ICNR 2018. Biosystems & Biorobotics*; Masia, L.; Micera, S.; Akay, M.; Pons, J.L., Eds.; Springer: Cham, Switzerland, 2019; pp. 329–333. [CrossRef]
31. Xiloyannis, M.; Cappello, L.; Dinh Binh Khanh.; Shih-Cheng Yen.; Masia, L. Modelling and design of a synergy-based actuator for a tendon-driven soft robotic glove. In Proceedings of the 2016 6th IEEE International Conference on Biomedical Robotics and Biomechatronics (BioRob), Singapore, 26–29 June 2016; pp. 1213–1219. [CrossRef]
32. Yurkewich, A.; Hebert, D.; Wang, R.H.; Mihailidis, A. Hand Extension Robot Orthosis (HERO) Glove: Development and Testing With Stroke Survivors With Severe Hand Impairment. *IEEE Trans. Neural Syst. Rehabil. Eng.* **2019**, *27*, 916–926. [CrossRef] [PubMed]

33. Rocha, D.N. Desenvolvimento de um sistema de controle para a órtese funcional de mão da UFMG. Master's Thesis, Universidade Federal de Minas Gerais, Belo Horizonte, Brazil, 2007.
34. Wechsler, L.R.; Bates, D.; Stroemer, P.; Andrews-Zwilling, Y.S.; Aizman, I. Cell Therapy for Chronic Stroke. *Stroke* **2018**, *49*, 1066–1074. [CrossRef] [PubMed]
35. Freitas, P.P. *Reabilitação da Mão*, 1st ed.; Atheneu: São Paulo, Brazil, 2006; p. 578.
36. Fugl-Meyer, A.; Jaasko, L.; Leyman, I.; Olsson, S.; Steglind, S. The post-stroke hemiplegic patient: 1. A method for evaluation of physical performance. *Scand. J. Rehab. Med.* **1975**, *7*, 13–31.
37. Bohannon, R.W.; Smith, M.B. Interrater Reliability of a Modified Ashworth Scale of Muscle Spasticity. *Phys. Ther.* **1987**, *67*, 206–207. [CrossRef] [PubMed]
38. Bertolucci, P.H.; Brucki, S.M.; Campacci, S.R.; Juliano, Y. O Mini-Exame do Estado Mental em uma população geral: Impacto da escolaridade. *Arquivos de Neuro-Psiquiatria* **1994**, *52*, 01–07. [CrossRef]
39. Raghavan, P. Upper Limb Motor Impairment After Stroke. *Phys. Med. Rehabil. Clin. North Am.* **2015**, *26*, 599–610. [CrossRef]
40. Thompson, A.J.; Jarrett, L.; Lockley, L.; Marsden, J.; Stevenson, V. Clinical management of spasticity. *J. Neurol. Neurosurg. Psychiatry* **2005**, *76*, 459–463. [CrossRef]
41. Chagas, H.M.A. Determinação de valores de referência para a força de preensão palmar e força muscular respiratória em adultos saudáveis. Master's Thesis, Universidade de São Paulo, Brazil, 2018.
42. Bohannon, R. Muscle strength and muscle training after stroke. *J. Rehabil. Med.* **2007**, *39*, 14–20. [CrossRef]
43. Wu, W.T.; Chang, K.V.; Mezian, K.; Naňka, O.; Lin, C.P.; Özçakar, L. Basis of Shoulder Nerve Entrapment Syndrome: An Ultrasonographic Study Exploring Factors Influencing Cross-Sectional Area of the Suprascapular Nerve. *Front. Neurol.* **2018**, *9*. [CrossRef] [PubMed]
44. Wu, W.T.; Chang, K.V.; Mezian, K.; Naňka, O.; Yang, Y.C.; Hsu, Y.C.; Hsu, P.C.; Özçakar, L. Ulnar Wrist Pain Revisited: Ultrasound Diagnosis and Guided Injection for Triangular Fibrocartilage Complex Injuries. *J. Clin. Med.* **2019**, *8*, 1540. [CrossRef] [PubMed]
45. Faria-Fortini, I.; Michaelsen, S.M.; Cassiano, J.G.; Teixeira-Salmela, L.F. Upper Extremity Function in Stroke Subjects: Relationships between the International Classification of Functioning, Disability, and Health Domains. *J. Hand Ther.* **2011**, *24*, 257–265. [CrossRef] [PubMed]

applied sciences

MDPI

Article

Texture Analysis is a Useful Tool to Assess the Complexity Profile of Microcirculatory Blood Flow

Henrique Silva [1,2,*], Hugo A. Ferreira [3], Clemente Rocha [1] and Luís Monteiro Rodrigues [1,2]

1. CBIOS (Research Center for Biosciences and Health Technologies), School of Health Sciences and Technologies, Lusófona University, 1749-024 Lisboa, Portugal; c.rocha.dr@gmail.com (C.R.); monteiro.rodrigues@ulusofona.pt (L.M.R.)
2. Department of Pharmacological Sciences, Faculty of Pharmacy, University of Lisboa, 1649-003 Lisboa, Portugal
3. IBEB Institute of Biophysics and Biomedical Engineering, Faculty of Sciences, University of Lisboa, 1749-016 Lisboa, Portugal; hatdferreira@gmail.com

* Correspondence: henrique.silva@ulusofona.pt

Received: 31 December 2019; Accepted: 28 January 2020; Published: 30 January 2020

check for updates

Abstract: The quantitative assessment of cardiovascular functions is particularly complicated, especially during any physiological challenge (e.g., exercise), with physiological signals showing intricate oscillatory properties. Signal complexity is one of such properties, and reflects the adaptability of the physiological systems that generated them. However, it is still underexplored in vascular physiology. In the present study, we calculate the complexity of photoplethysmography (PPG) signals and their frequency components obtained with the wavelet transform (WT), with two analytical tools—(i) texture analysis (TA) of WT scalograms, and (ii) multiscale entropy (MSE) analysis. PPG signals were collected from twelve healthy young subjects (26.0 ± 5.0 y.o.) during a unilateral leg lowering maneuver to evoke the venoarteriolar reflex (VAR) while lying supine, with the contralateral leg remaining stationary. Results showed that TA was able to detect a decrease in complexity, viewed as an increase in texture entropy (TE), of the PPG scalograms during VAR, similarly to MSE, suggesting that a decrease in the competence of vascular regulation mechanisms might be present during VAR. Nonetheless, TA showed lower sensitivity than MSE for low frequency spectral regions. TA seems to be a promising and straightforward analytical tool for the assessment of the complexity of PPG perfusion signals.

Keywords: signal complexity; texture analysis; multiscale entropy analysis; wavelet transform; photoplethysmography

1. Introduction

The regulation of the cardiovascular system results from the adjustment of several biophysical phenomena, both electrical and mechanical. Their coordination is highly complex, depending on the continuous feedback and cross-talk between effector organs and controlling systems [1]. The heart pump and respiration are the most notorious "central" processes governing the performance of the cardiovascular system, while on a more "peripheral" level, the phenomena that control vascular tone, such as the myogenic activity of the vessel wall, the sympathetic activity and the endothelial release of vasoactive substances, play crucial roles [2–4]. Microcirculation signals are composed of both these central and peripheral components and therefore provide an "integrated" view of cardiovascular function [4,5]. The contributions from these multiple regulation systems and their intricate interplay explain the oscillatory properties of microcirculation signals. From an analytical standpoint, these oscillatory properties are increasingly considered as a means to extract more sensible information not apparent from the general analysis of the raw signals, in particular their spectral origin and the significance of their fractal and chaotic profiles [5–7]. In recent years, much attention has been given to the study of the complexity of physiological signals, a property that reflects the adaptability of the systems that generated them. Intuitively, the term complexity is often associated with "structural richness" [8], which can be applied to the study of biological signals and images. Complexity is typically assessed as "entropy," a general measure of "disorganization" in physical, chemical and biological phenomena [9]. In biomedical research, entropy is used as a quantitative parameter, and is typically assessed in continuous, but temporally unpredictable, physiological signals, such as electrocardiography [10], electroencephalography [11], blood pressure [12], laser Doppler flowmetry (LDF) [13], and photoplethysmography (PPG) [14]. Physiological signals may be regular (e.g., more periodic) or irregular. The higher entropy measured in irregular signals reflects the adaptation capability to changing internal and external conditions of the biological system that generated them [15]. Overall, the entropy of physiological signals tends to decrease whenever the systems that generated them lose adaptability, which can result in a compromise of their function or even in disease. Therefore, a suitable metric of complexity should assign higher values to the output signal of a "healthier" system with a rich and meaningful structure and lower values to either random dynamics or predictable systems, which are often associated with disease [8,16,17]. This has been observed in the aging process [18,19] and in cardiovascular, metabolic and neurological diseases [18], among several others. In addition to pathological states, some experimental procedures are sufficiently "challenging" to temporarily provoke tissue dysfunction, especially those that reduce perfusion [6]. For continuous signals, the multiscale entropy analysis (MSE) is considered a robust analytical tool, having shown superiority over other measures for its ability to assess entropy over different time scales [19]. Entropy can also be assessed in texturally rich biological images, reflecting their textural "disorganization." Texturally rich images can be assessed with texture analysis (TA), an analytical tool allowing the detection of various features on a gray-level image to discriminate textural differences, such as texture entropy (TE), contrast, correlation, homogeneity, and energy. Entropy calculated from TA of images may reflect different information than that conveyed by the entropy assessment of continuous numerical series. When applied to a grayscale image, high entropy means that the pixels can adopt a high number of gray levels, expressing richer information [20]. TA has been used in medical imaging with the purpose of improving diagnostic capacity, given that human visual inspection constitutes a process that is expensive, time-consuming and prone to interpretation errors [21]. Thus far, TA in medical imaging has been more frequently applied to mammography [22], ultrasonography, computerized tomography (CT) [23], and magnetic resonance imaging (MRI) [24] techniques [25], as well as to dermatoscopy [26] and microscopy images [27]. In particular, TE has been used to explore immune competence of lymphoid tissue in microscopic images [28], to assess the onset organ (liver) disease and tumor (colorectal cancer) evolution in CT scans [29–31], and to assess bone regeneration in fractures [32] and osteoarthritis [33], among others. In medical imaging, TE does not necessarily decrease when images display abnormal features. In fact, since abnormal or dysmorphic features often increase the coarseness of smooth or uniform images from healthy subjects, increases in TE are often seen in pathological images, such as in the case of aged tissues and organs [34] as well as the case of neoplastic formations [30,31]. In these situations, a richer texture and therefore, higher entropy, is associated with dysfunction. Although currently underexplored in vascular physiology, the complexity assessment of microcirculation signals, both raw and decomposed, may deepen our knowledge of the mechanisms underlying perfusion regulation. Microcirculation signals are easily decomposed into their frequency components by the wavelet transform (WT) [4]. One of the WT main outputs is a scalogram, i.e., a multi-patterned image representation of the time evolution of all components of the decomposed signal. Given their textural richness, WT scalograms seem suitable candidates for TE analysis, however, to our knowledge this analysis has not been previously attempted. Since WT scalograms are not medical images, the interpretation of TE in this context must be necessarily different from the one performed in medical diagnosis. Given the fact that a WT scalogram is a 2D projection of a 3D frequency spectrum, (i.e., the 3D spectrum seen from above), it is only logical to assume that a change in the entropy of the numerical series that constitute the 3D spectrum would also be observed in corresponding WT scalograms. Therefore, the assumption that similar trends in complexity would be observed with MSE and TA seems legitimate. Our objective was to compare the complexity of microcirculation PPG signals calculated with TE against MSE during a typical venoarteriolar reflex (VAR).

2. Materials and Methods

2.1. Experimental

Twelve healthy young adult subjects (26.0 ± 5.0 y.o., seven females, five males) participated in this study after giving informed written consent. All subjects were healthy, with no cardiovascular diseases, were non-smokers, and abstained from consuming alcohol and caffeine-containing beverages 24 hours prior to the experimental procedure. The protocol was approved by the School of Health Sciences and Technologies' Ethical Commission. All subjects gave written informed consent, and the study was conducted in accordance with the Declaration of Helsinki and subsequent amendments [35]. After 20 minute acclimatization to room conditions (temperature: 22 ± 1 °C, humidity: 40–60%), a postural challenge to elicit the VAR was applied, as previously described [4]—10 minutes lying supine with both legs extended (baseline, Phase I); 10 minutes with one foot lowered 50 cm from heart level (challenge, Phase II); and 10 minutes resuming the initial posture (recovery, Phase III). The contralateral foot remained stationary and served as control. Blood flow signals, expressed in arbitrary units (AU), were acquired from the first toe of both feet at a 100 Hz sampling rate with a reflection photoplethysmography Blood Pulse Volume sensor (Biosignals Plux, Lisboa, Portugal) connected to a BITalino Plugged microprocessor board (Biosignals Plux).

2.2. Numerical

The raw PPG signals were imported to Matlab (Mathworks R2012, Natick, MA, USA) and smoothed with a moving average filter.

The WT (http://noc.ac.uk/using-science/crosswavelet-wavelet-coherence) was then applied to the smoothed signals, which allowed the decomposition into their main frequency components. Wavelets are a family of functions constructed from translations and dilations of a single function called the "mother wavelet" $\psi(t)$, collectively defined by:

$$\psi_{a,b}(t) = \frac{1}{\sqrt{|a|}}\psi\left(\frac{t-b}{a}\right),\ a, b \in R,\ a \neq 0 \tag{1}$$

The parameter *a* is the scaling parameter or scale, measuring the degree of compression, while parameter *b* is the translation parameter, determining the time location of the wavelet. If $|a| < 1$, then the wavelet in the above equation is the compressed version of the mother wavelet and corresponds mainly to higher frequencies. If $|a| > 1$, then $\psi_{a,b}(t)$ has a larger time-width than $\psi(t)$ and corresponds to lower frequencies. Thus, wavelets have time-widths adapted to their frequencies.

In this case, the central frequency of the wavelet was 6 Hz, with a discretization factor of 10 scales per octave. The detected components of the PPG signals occurred at the following frequency ranges: cardiac [1.6–0.7 Hz], respiratory [0.7–0.26 Hz], myogenic [0.26–0.1 Hz], neurogenic/sympathetic [0.1–0.045 Hz], endothelial NO-dependent (NOd) [0.045–0.015 Hz] and NO-independent (NOi) [0.015–0.007 Hz], which were in line with the ranges previously described for LDF [36]. From the WT, two main outputs were generated for each subject: (1) a scalogram and (2) a 3D periodogram. The WT scalogram (Figure 1) shows the time (here converted to sample number) evolution of the amplitude with a factor of 2^n (here in color scale) for each signal scale. The approximate frequency of a component is obtained by dividing the central wavelet frequency by the scale of the component. From a visual inspection, the WT scalogram also presents an oscillatory pattern in pixel distribution on different scales, making it a suitable candidate for TA. The scalogram was then used as a grayscale image, where several regions of interest (ROI) were marked, each per component per phase of the protocol. Each ROI was then converted to a gray-level co-occurrence matrix (GLCM), a matrix where the number of rows and columns is equal to the number of gray levels (G) in the image, from which the TE was calculated as follows [37]:

$$\text{Entropy} = -\sum_{i=0}^{G-1}\sum_{j=0}^{G-1} P(i,j).\log[P(i,j)] \tag{2}$$

where $P(i, j)$ denotes the probability of occurrence of a given element in the matrix.

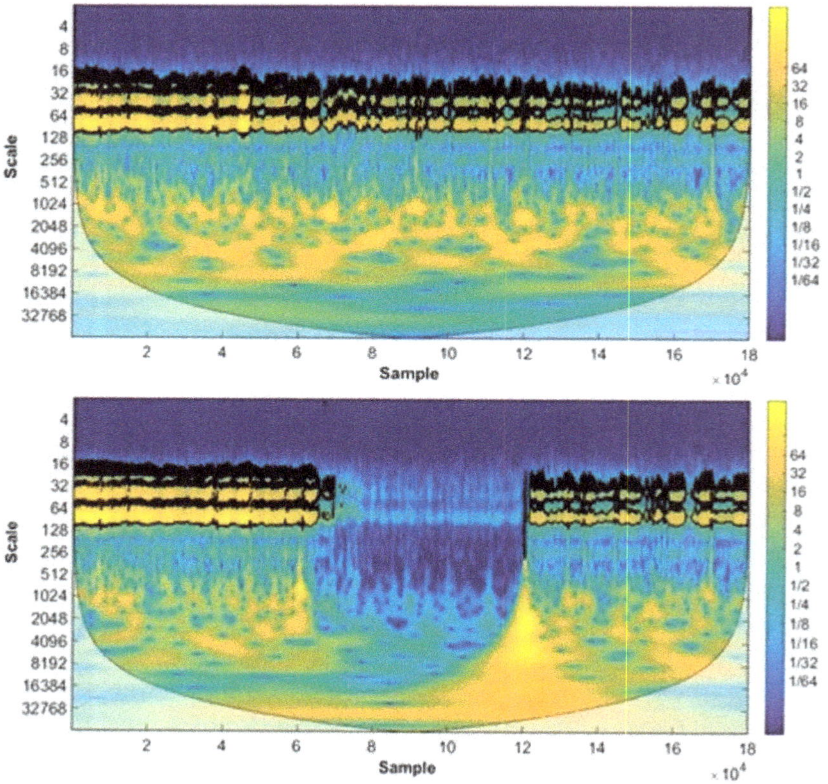

Figure 1. WT scalogram of the PPG signal from the control (top) and test (bottom) feet from a representative subject (20 y.o.).

To correctly select each ROI, a control sine wave signal was generated with multiple components occurring at known frequencies (f_1 = 3.2 Hz; f_2 = 1.6 Hz; f_3 = 0.7 Hz, f_4 = 0.26 Hz; f_5 = 0.1 Hz; f_6 = 0.045 Hz; f_7 = 0.015 Hz; f_8 = 0.007 Hz), where each frequency is close to the ones that define the borders of the PPG components' detected frequency ranges. The sine wave signal was then deconstructed with the WT, and a reference scalogram was generated (Figure 2).

From the 3D periodogram (Figure 3), a 2D time evolution of the amplitude of each component was constructed by averaging the amplitude at each time point for each spectral interval (Figure 4). Signals were acquired at a 100 Hz sampling rate, which translates to a total number of 180,000 samples for the entire 30 minute protocol. The wavelet period is a natural logarithmic representation of the wavelet scale, with the frequency in Hz exponentially related to this period. The time evolution was analyzed for both the raw PPG signal and its components with MSE, an algorithm that quantifies the randomness/unpredictability of a signal over different time scales as follows [19]:

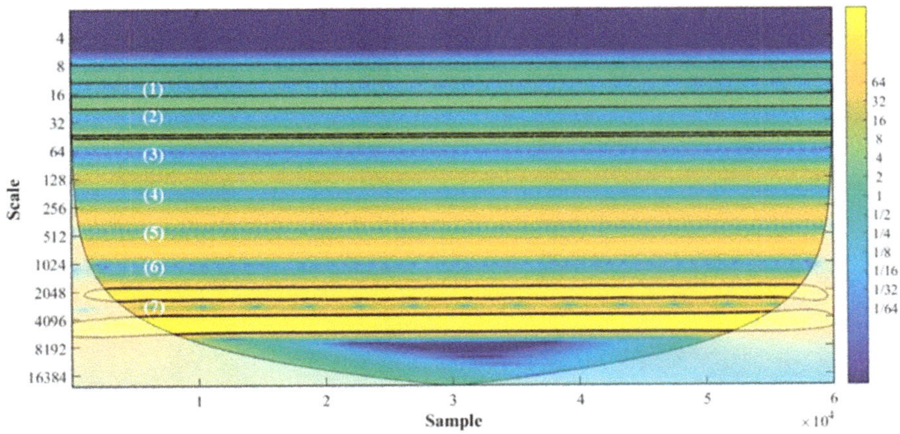

Figure 2. WT scalogram of the reference sine wave signal. The region corresponding to each component frequency interval is shown (1—first harmonic of the cardiac component; 2—cardiac; 3—respiratory; 4—myogenic; 5—sympathetic; 6—endothelial NO-dependent; 7—endothelial NO-independent).

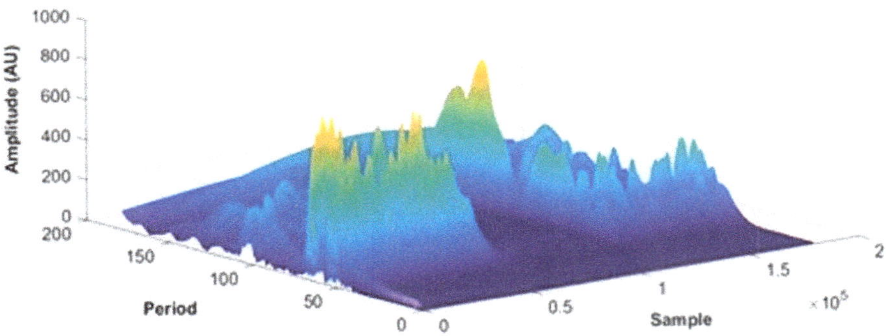

Figure 3. PPG 3D frequency spectrum (period vs. sample vs. amplitude) for the test foot of a representative subject (20 y.o.). Signals were acquired at a 100 Hz sampling rate, which translates to a total of 180,000 signal samples for a 30 minute acquisition period. Period is a natural logarithmic representation of the wavelet scale.

Given a time series $x_i = 1,..., N$, a consecutive coarse-grained time series $y(\tau)$ is constructed:

$$y_j^{(\tau)} = \frac{1}{\tau} \sum_{i=(j-1)\tau+1}^{j\tau} x_i \quad (3)$$

where τ represents the scale factor and $1 \leq j \leq N/\tau$. The sample entropy (*SampEn*) of each coarse-grained time series is then computed. *SampEn(m, r, N)* is the negative natural logarithm of the conditional probability that a dataset of length N, having repeated itself with a tolerance of r points, will also repeat itself for $m + 1$ points, without allowing self matches:

$$\text{SampEn}(\text{m}, \text{r}, \text{N}) = -\ln\frac{\text{A}^{\text{m}}(\text{r})}{\text{B}^{\text{m}}(\text{r})} \tag{4}$$

where $A^m(r)$ is the probability that two sequences will match for $m + 1$ points and $B^m(r)$ is the probability that two sequences will match for m points. The more regular and predictable a time series is, the lower the value of *SampEn*. The more random a time series is, the higher the value of *SampEn*. Plotting the *SampEn* over the scale factor yields the MSE curve, which gives insight into the integrated complexity of the system over the time scales of interest, which can be of interest when comparing groups where differences in specific time scales are probable. The randomness/unpredictability of the signal can finally be straightforwardly summarized as the complexity index (CI), which corresponds to the area under the MSE curve [38]. The CI and TE were statistically compared between each phase of the protocol with the Wilcoxon signed-rank test, and were compared between feet for each phase with the Mann-Whitney independent sample test, adopting a 95% confidence interval.

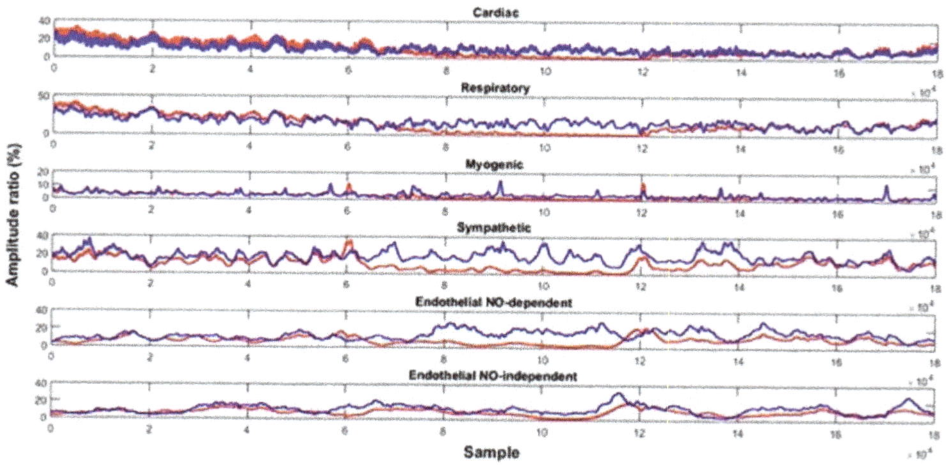

Figure 4. Time evolution of the PPG signal components' amplitude ratio for the control (blue) and test (red) feet throughout the protocol (baseline: signal samples 0–60,000; challenge: signal samples 60,000–120,000; recovery: signal samples 120,000–180,000) for a representative subject (20 y.o.).

3. Results

As previously published, during the postural challenge, perfusion decreased significantly on both the test and control feet [4]. The mean and standard deviation (SD) of the entropy parameter (CI) calculated with MSE is presented in Table 1. For the test foot, the CI of the raw PPG signal and all its components decreased significantly during the challenge (Phase II, raw signal: $p = 0.005$; cardiac: $p = 0.012$; respiratory: $p = 0.002$; myogenic: $p = 0.003$; sympathetic: $p = 0.002$; NOd: $p = 0.002$; NOi: $p = 0.002$). For the control foot, the CI of the raw PPG signal showed an increase without statistical significance. All the components' CI decreased, with the exception of the cardiac, although not so pronounced as in the test foot (respiratory: $p = 0.002$; myogenic: $p = 0.003$; sympathetic: $p = 0.002$; NOd: $p = 0.002$; NOi: $p = 0.002$). During recovery (Phase III), significant differences were still detected on both feet regarding baseline for the respiratory (test: $p = 0.002$; control: $p = 0.002$), myogenic (test: $p = 0.008$; control: $p = 0.008$), sympathetic (test: $p = 0.002$; control: $p = 0.002$), NOd (test: $p = 0.002$; control: $p = 0.002$) and NOi (test: $p = 0.002$; control: $p = 0.003$). The mean and SD of the entropy parameter (TE) calculated for the sine wave and PPG signals with TA is shown in Table 2. For the sine wave signal scalogram, TE decreased from the cardiac component towards the low frequency components, where it increased again for the NOi. For the test foot PPG scalogram, TE decreased significantly in the cardiac (p = 0.004), respiratory (p = 0.013), and myogenic (p = 0.008) components, increased significantly in the NOd (p = 0.012) and NOi (p = 0.010), and showed no change in the sympathetic component. During recovery, all components' TE returned except for the NOd, which showed a significant difference regarding baseline (p = 0.007). For the control foot PPG scalogram, changes in TE were not so pronounced as in the test foot, with only the cardiac component showing a significant difference regarding baseline (p = 0.001). During recovery, only the cardiac component showed a significant difference (p = 0.022) regarding baseline. No significant differences in entropy were found between feet at baseline (Tables 1 and 2) either with CI or TE. During the challenge phase, significant differences were found for the raw signal ($p < 0.001$) and all components (p=0.001 for the respiratory; $p < 0.001$ for the remaining) with CI. Similarly, TE showed significant differences in all components (cardiac: p = 0.001; respiratory: p = 0.045; myogenic: p = 0.028; NOd: p = 0.001; NOi: p = 0.001), except the sympathetic. During recovery, the CI of the sympathetic (p = 0.012), NOd (p = 0.002) and NOi ($p < 0.001$) were significantly different, while with TE no statistical differences were found.

Table 1. Mean and standard deviation (SD) of the MSE complexity index for each foot on each phase of the protocol (card—cardiac, resp—respiratory, myo—myogenic, sym—sympathetic, NOd—endothelial NO-dependent, NOi—endothelial NO-independent). Statistical comparison to Phase I is shown (*—$p < 0.05$).

Complexity Index		Test Foot		Control Foot		p-Value (T vs. C)
		Mean ± SD	p-Value	Mean ± SD	p-Value	
raw signal	Phase I	34.8 ± 6.1	-	34.2 ± 8.2	-	0.843
	Phase II	15.0 ± 11.5	0.005 *	35.9 ± 8.7	0.182	<0.001 *
	Phase III	31.4 ± 7.3	0.182	37.0 ± 6.7	0.239	0.089
card	Phase I	25.9 ± 6.1	-	29.3 ± 7.2	-	0.347
	Phase II	15.6 ± 9.0	0.012 *	28.7 ± 6.5	0.754	<0.001 *
	Phase III	23.8 ± 7.3	0.224	29.7 ± 7.8	0.875	0.089
resp	Phase I	25.9 ± 6.1	-	29.3 ± 7.2	-	0.347
	Phase II	6.7 ± 4.2	0.002 *	14.3 ± 4.8	0.002 *	0.001 *
	Phase III	10.1 ± 5.5	0.002 *	13.7 ± 4.9	0.002 *	0.160
myo	Phase I	23.2 ± 8.7	-	25.9 ± 7.5	-	0.378
	Phase II	3.9 ± 3.1	0.003 *	12.0 ± 2.8	0.003 *	<0.001 *
	Phase III	7.9 ± 5.4	0.008 *	10.9 ± 5.8	0.008 *	0.160
sym	Phase I	25.9 ± 6.1	-	29.3 ± 7.2	-	0.347
	Phase II	2.6 ± 2.4	0.002 *	10.5 ± 1.9	0.002 *	<0.001 *
	Phase III	6.8 ± 2.2	0.002 *	10.7 ± 4.0	0.002 *	0.012 *
NOd	Phase I	25.9 ± 6.1	-	29.3 ± 7.2	-	0.347
	Phase II	2.5 ± 3.5	0.002 *	11.5 ± 2.6	0.002 *	<0.001 *
	Phase III	5.5 ± 2.1	0.002 *	11.0 ± 4.8	0.002 *	0.002 *
NOi	Phase I	25.9 ± 6.1	-	29.3 ± 7.2	-	0.347
	Phase II	1.9 ± 0.9	0.002 *	10.2 ± 3.6	0.002 *	<0.001 *
	Phase III	6.2 ± 3.4	0.002 *	13.6 ± 5.0	0.003 *	<0.001 *

Table 2. Mean and standard deviation (SD) of the texture entropy for each foot on each phase of the protocol (card—cardiac, resp—respiratory, myo—myogenic, sym—sympathetic, NOd—endothelial NO-dependent, NOi—endothelial NO-independent). Statistical comparison to Phase I and between the test and control limbs are shown (*—$p < 0.05$).

Texture Entropy		Sine Wave Signal	Test Foot		Control Foot		p-Value (T vs. C)
			Mean ± SD	p-Value	Mean ± SD	p-Value	
card	Phase I	6.1	6.6 ± 0.5	-	6.8 ± 0.3	-	0.198
	Phase II		6.2 ± 0.7	0.004 *	6.9 ± 0.3	0.011 *	0.001 *
	Phase III		7.0 ± 0.2	0.563	7.0 ± 0.2	0.022 *	0.713
resp	Phase I	5.4	6.5 ± 0.2	-	6.5 ± 0.2	-	0.551
	Phase II		5.5 ± 1.3	0.013 *	6.3 ± 0.2	0.286	0.045 *
	Phase III		6.5 ± 0.2	0.638	6.4 ± 0.2	0.530	0.478
myo	Phase I	5.2	6.6 ± 0.3	-	6.7 ± 0.1	-	0.089
	Phase II		6.3 ± 0.4	0.008 *	6.6 ± 0.1	0.433	0.028 *
	Phase III		6.5 ± 0.2	0.289	6.7 ± 0.2	0.754	0.266
sym	Phase I	5.3	6.2 ± 0.3	-	6.2 ± 0.4	-	0.671
	Phase II		6.2 ± 0.2	0.147	6.2 ± 0.2	0.814	0.887
	Phase III		6.0 ± 0.2	0.594	6.1 ± 0.3	0.239	0.630
NOd	Phase I	5.3	5.8 ± 0.5	-	5.9 ± 0.4	-	0.630
	Phase II		6.4 ± 0.2	0.012 *	5.7 ± 0.5	0.272	0.001 *
	Phase III		6.1 ± 0.3	0.007 *	5.9 ± 0.3	0.307	0.060
NOi	Phase I	6.2	6.0 ± 0.4	-	5.9 ± 0.4	-	0.551
	Phase II		6.2 ± 0.3	0.010 *	5.5 ± 0.7	0.753	0.001 *
	Phase III		6.2 ± 0.3	0.594	5.8 ± 0.8	0.126	0.060

4. Discussion

In this study, the acquired PPG signals and an artificial sine wave signal were decomposed with the WT into their respective components so that TE and CI could be calculated. The relative contributions of each of the components to the overall signal, expressed as amplitude ratios, are typical WT outputs but were not considered for this study as these results were previously published [4]. During leg lowering, perfusion decrease in the test foot is explained by a constriction of arterioles secondary to the increased venous distension when the foot is in a pendent posture, which constitutes the VAR [39]. The fast perfusion decreases in the control foot (not so pronounced as in the test foot) is thought to be due to a centrally mediated neurogenic reflex initiated to maintain the vascular homeostasis in the lower limb [4]. The sine wave signal scalogram showed TE values generally invariant regarding the component, which reflects the regularity of these signals. The highest TE values were recorded for both the highest (cardiac) and lowest (NOi) frequency components, meaning a richer texture was found in the corresponding ROIs, likely explained by the particular overlap of the spectral waves that define them. Contrarily, the PPG signal scalogram showed that the higher frequency (HF) components (cardiac, respiratory, myogenic) displayed higher TE values in either foot, meaning the corresponding ROIs are more texturally rich, i.e., more complex in comparison to the lower frequency components. This suggests that TE assessment of PPG scalograms is sensitive to the underlying phenomena that explain the physiological complexity of these signals. In our data, both TE and MSE analyses showed an overall decrease in the PPG signal and in the respective components' entropies during leg lowering in both the test and control feet. This result seems to indicate a bilateral decrease in the vascular adaptation capacity to the lower perfusion levels observed during leg lowering. This apparent decrease in adaptation capacity is further reinforced by the observation that several PPG components' entropies show significant differences between the recovery and baseline phases, suggesting that full recovery was not attained at the end of the protocol. In our view, these results indicate that that VAR creates a state of reduced perfusion that, while not compromising tissue viability, destabilizes the evoked regulatory mechanisms. When considering the test foot, MSE detected a significant decrease in the entropy of all components during VAR. TE, however, detected an entropy decrease in the cardiac, respiratory, and myogenic components, an increase in both endothelial components and no change in the sympathetic. Regarding the control foot, an overall entropy decrease was observed with both tools. MSE recorded a decrease in the complexity of all components, although not so pronounced as in the test foot, which again is likely attributable to the magnitude of the perfusion decrease. Again, TE showed more subtle changes, only significant for the respiratory and NOi components. Overall, MSE and TE showed similar responses in the high frequency components (cardiac, respiratory, myogenic), and dissimilar responses in the low frequency components (sympathetic, NOd, NOi), suggesting that TE loses analytical sensitivity in the latter regions. This loss of sensitivity may be attributed to an incomplete visual resolution of the scalogram bands of the low frequency components, leading to a less precise identification of the respective ROIs. Nevertheless, TE was able to differentiate the entropies of most components during VAR between the test and control limbs. Overall, although both TE and MSE were able to detect differences in entropy, MSE was consistently more sensitive. One should keep in mind that MSE and TE ensure different scale evaluations—frequency for TE and time for MSE, which are two aspects of the same reality (i.e., perfusion signal). Thus, both analytical strategies provide different views into the same physiological event, which can also help to explain the differences in sensitivity found. MSE detected changes in entropy between limbs during both VAR and recovery phases, while TE was only able to detect the more pronounced changes that occurred during VAR. Furthermore, MSE was sensitive to changes in entropy for a greater number of components in comparison to TE. Although the performance of the reference MSE method was superior under the conditions studied, these results suggest that TE is an interesting and suitable tool for assessing the complexity of PPG microcirculation signals considered as WT scalograms, in particular for the analysis of high frequency components.

5. Conclusions

In this paper we present for the first time the use of texture entropy (TE) for the quantification of complexity in microcirculatory perfusion signals, using MSE as a reference analytical tool. Our results show that TE was able to detect a decrease in complexity of the PPG scalograms during VAR, similarly to MSE. This complexity decrease suggests that a decrease in the competence of vascular regulation mechanisms might be present during VAR and should be further investigated. Results from both tools were aligned for the high frequency components, but not for the low frequency components, suggesting a decrease in sensitivity of TE in the latter spectral regions. Recognizing the value of increasing the sample size (and heterogeneity) in future studies, our results have shown Texture Analysis (TA) to be a promising method to assess the complexity of PPG perfusion signals.

Author Contributions: Conceptualization, H.S.; Methodology, H.S., C.R. and H.A.F.; Software, H.S. and H.A.F.; Validation, H.A.F., and L.M.R.; Formal Analysis, H.S.; Investigation, H.S. and C.R.; Resources, H.A.F. and L.M.R.; Data Curation, H.S. and H.A.F.; Writing—Original Draft Preparation, H.S.; Writing—Review & Editing, H.A.F.; Visualization, H.A.F. and L.M.R.; Supervision, L.M.R.; Project Administration, L.M.R.; Funding Acquisition, L.M.R. All authors have read and agreed to the published version of the manuscript.

Funding: This work was supported by national funds from FCT—Fundação para a Ciência e a Tecnologia, I.P, within the project UID/DTP/04567/2019.

Acknowledgments: The authors would like to express their thanks to all the volunteers for their participation in this study.

Conflicts of Interest: The authors declare that the research was conducted in the absence of any commercial or financial relationships that could be construed as a potential conflict of interest.

Ethics Statement: This study was carried out in accordance with the recommendations of Ethical Principles for Medical Research Involving Human Subjects, Declaration of Helsinki. The protocol was approved by the School of Health Sciences and Technologies' Ethical Commission. All subjects gave written informed consent in accordance with the Declaration of Helsinki.

References

1. Stefanovska, A. Coupled Oscillators: Complex but Not Complicated Cardiovascular and Brain Interactions. *IEEE Eng. Med. Biol.* **2007**, *26*, 25–29. [CrossRef] [PubMed]
2. Stefanovska, A.; Lotric, M.B.; Strle, S.; Haken, H. The cardiovascular system as coupled oscillators? *Physiol. Meas.* **2001**, *22*, 535–550. [CrossRef]
3. Bernjak, A.; Stefanovska, A. Importance of wavelet analysis in laser Doppler flowmetry time series. In Proceedings of the 29th Annual International Conference of the IEEE Engeniring in Medicine and Biology Society, Lyon, France, 22–26 August 2007; pp. 4064–4067.
4. Silva, H.; Ferreira, H.A.; Da Silva, H.P.; Rodrigues, L.M. The Venoarteriolar Reflex Significantly Reduces Contralateral Perfusion as Part of the Lower Limb Circulatory Homeostasis in vivo. *Front. Physiol.* **2018**, *9*, 1–9. [CrossRef] [PubMed]
5. Silva, H.; Bento, M.; Vieira, H.; Rodrigues, L.M. Comparing the spectral components of laser Doppler flowmetry and photoplethysmography signals for the assessment of the vascular response to hyperoxia. *J. Biomed. Biopharm. Res.* **2017**, *14*, 187–194. [CrossRef]
6. Silva, H.; Ferreira, H.; Rodrigues, L.M. Studying the Oscillatory Components of Human Skin Microcirculation. In *Measuring the Skin*; Springer Science and Business Media LLC: Berlin, Germany, 2015; pp. 1–15.
7. Mizeva, I.; Di Maria, C.; Frick, P.; Podtaev, S.; Allen, J. Quantifying the correlation between photoplethysmography and laser Doppler flowmetry microvascular low-frequency oscillations. *J. Biomed. Opt.* **2015**, *20*, 37007. [CrossRef] [PubMed]
8. Costa, M.; Goldberger, A.L.; Peng, C.-K. Multiscale entropy analysis of biological signals. *Phys. Rev. E* **2005**, *71*, 021906. [CrossRef]
9. Rangel, J.A.O. The Systemic Theory of Living Systems and Relevance to CAM Part I: The Theory. *Evid-Based Compl. Alt.* **2005**, *2*, 13–18. [CrossRef]
10. Zhang, Y.; Wei, S.; Long, Y.; Liu, C. Performance Analysis of Multiscale Entropy for the Assessment of ECG Signal Quality. *J. Electr. Comput. Eng.* **2015**, *2015*, 1–9. [CrossRef]
11. Chung, C.-C.; Kang, J.-H.; Yuan, R.-Y.; Wu, D.; Chen, C.-C.; Chi, N.-F.; Chen, P.-C.; Hu, C.-J. Multiscale Entropy Analysis of Electroencephalography During Sleep in Patients With Parkinson Disease. *Clin. EEG Neurosci.* **2013**, *44*, 221–226. [CrossRef]
12. Trunkvalterova, Z.; Javorka, M.; Tonhajzerova, I.; Javorkova, J.; Lazárová, Z.; Javorka, K.; Baumert, M. Reduced short-term complexity of heart rate and blood pressure dynamics in patients with diabetes mellitus type 1: multiscale entropy analysis. *Physiol. Meas.* **2008**, *29*, 817–828. [CrossRef]
13. Humeau, A.; Buard, B.; Mahé, G.; Rousseau, D.; Chapeau-Blondeau, F.; Abraham, P. Multiscale entropy of laser Doppler flowmetry signals in healthy human subjects. *Med. Phys.* **2010**, *37*, 6142. [CrossRef] [PubMed]
14. Schack, T.; Harb, Y.S.; Muma, M.; Zoubir, A.M. Computationally efficient algorithm for photoplethysmography-based atrial fibrillation detection using smartphones. In *2017 39th Annual International Conference of the IEEE Engineering in Medicine and Biology Society (EMBC)*; Institute of Electrical and Electronics Engineers (IEEE): Piscataway, NJ, USA, 2017; pp. 104–108.
15. Zhang, X.D.; Shen, B. Entropy for the Complexity of Physiological Signal Dynamics. *Adv. Exp. Med. Biol.* **2017**, *1028*, 39–53. [PubMed]
16. Yang, D.; Luo, Z.; Ma, S.; Wong, W.T.; Ma, L.; Zhong, J.; He, H.; Zhao, Z.; Cao, T.; Yan, Z.; et al. Activation of TRPV1 by dietary capsaicin improves endothelium-dependent vasorelaxation and prevents hypertension. *Cell Metab.* **2010**, *12*, 130–141. [CrossRef] [PubMed]
17. Liu, Y.; Lin, Y.; Wang, J.; Shang, P. Refined generalized multiscale entropy analysis for physiological signals. *Phys. A: Stat. Mech. its Appl.* **2018**, *490*, 975–985. [CrossRef]
18. Vargas, B.; Cuesta-Frau, D.; Ruiz-Esteban, R.; Cirugeda, E.; Varela, M. What Can Biosignal Entropy Tell Us About Health and Disease? Applications in Some Clinical Fields. *Nonlinear Dyn. Psychol. Life Sci.* **2015**, *19*, 419–436.
19. Costa, M.; Goldberger, A.L.; Peng, C.-K. Multiscale Entropy Analysis of Complex Physiologic Time Series. *Phys. Rev. Lett.* **2002**, *89*, 068102. [CrossRef]
20. Tsai, D.-Y.; Lee, Y.; Matsuyama, E. Information Entropy Measure for Evaluation of Image Quality. *J. Digit. Imaging.* **2008**, *21*, 338–347. [CrossRef]

21. Larroza, A.; Bodí, V.; Moratal, D. Texture Analysis in Magnetic Resonance Imaging: Review and Considerations for Future Applications. In *Assessment of Cellular and Organ Function and Dysfunction using Direct and Derived MRI Methodologies*; IntechOpen: London, UK, 2016; pp. 75–106.
22. Pharwaha, A.P.S.; Singh, B. Shannon and Non-Shannon Measures of Entropy for Statistical Texture Feature Extraction in Digitized Mammograms. In Proceedings of the World Congres on Engineering nd Computer Science, San Francisco, CA, USA, 20–22 October 2009.
23. Al-Kadi, O.; Watson, D. Texture Analysis of Aggressive and Nonaggressive Lung Tumor CE CT Images. *IEEE Trans. Biomed. Eng.* **2008**, *55*, 1822–1830. [CrossRef]
24. Gibbs, P.; Turnbull, L.W. Textural analysis of contrast-enhanced MR images of the breast. *Magn. Reson. Med.* **2003**, *50*, 92–98. [CrossRef]
25. Castellano, G.; Bonilha, L.; Li, L.; Cendes, F. Texture analysis of medical images. *Clin. Radiol.* **2004**, *59*, 1061–1069. [CrossRef]
26. Shrestha, B.; Bishop, J.; Kam, K.; Chen, X.; Moss, R.H.; Stoecker, W.V.; Umbaugh, S.; Stanley, R.J.; Celebi, M.E.; Marghoob, A.A.; et al. Detection of atypical texture features in early malignant melanoma. *Skin. Res. Technol.* **2010**, *16*, 60–65. [CrossRef] [PubMed]
27. Gao, Y.; Helgeson, M.E. Texture analysis microscopy: quantifying structure in low-fidelity images of dense fluids. *Opt. Express* **2014**, *22*, 10046–10063. [CrossRef] [PubMed]
28. Pantic, I.; Pantic, S. Germinal Center Texture Entropy as Possible Indicator of Humoral Immune Response: Immunophysiology Viewpoint. *Mol. Imaging. Biol.* **2012**, *14*, 534–540. [CrossRef] [PubMed]
29. Mir, A.; Hanmandlu, M.; Tandon, S. Texture Analysis of CT. *IEEE Eng. Med. Biol.* **1995**, *14*, 781–786. [CrossRef]
30. Ng, F.; Ganeshan, B.; Kozarski, R.; Miles, K.A.; Goh, V. Assessment of Primary Colorectal Cancer Heterogeneity by Using Whole-Tumor Texture Analysis: Contrast-enhanced CT Texture as a Biomarker of 5-year Survival. *Radiology* **2013**, *266*, 177–184. [CrossRef] [PubMed]
31. Ng, F.; Kozarski, R.; Ganeshan, B.; Goh, V. Assessment of tumor heterogeneity by CT texture analysis: Can the largest cross-sectional area be used as an alternative to whole tumor analysis? *Eur. J. Radiol.* **2013**, *82*, 342–348. [CrossRef]
32. El Hassani, A.; El Hassouni, M.; Jennane, R.; Rziza, M.; Lespessailles, E. Texture Analysis for Trabecular Bone X-Ray Images Using Anisotropic Morlet Wavelet and Rényi Entropy. In *Image and Signal Processing, Proceedings of the 5th International Conference on Image and Signal Processing, Agadir, Morocco, 28–30 June 2012*; Elmoataz, A., Mammass, D., Lezoray, O., Nouboud, F., Aboutajdine, D., Eds.; Springer: Berlin, Germany, 2012; pp. 290–297.
33. Kołaciński, M.; Kozakiewicz, M.; Materka, A. Textural entropy as a potential feature for quantitative assessment of jaw bone healing process. *Arch. Med. Sci.* **2015**, *11*, 78–84. [CrossRef]
34. Shamir, L.; Wolkow, C.A.; Goldberg, I.G. Quantitative measurement of aging using image texture entropy. *Bioinformatics* **2009**, *25*, 3060–3063. [CrossRef]
35. WMA. World Medical Association Declaration of Helsinki: Ethical principles for medical research involving human subjects. *J. Am. Med. Assoc.* **2013**, *310*, 2191–2194. [CrossRef]
36. Kvandal, P.; Landsverk, S.A.; Bernjak, A.; Stefanovska, A.; Kvernmo, H.D.; Kirkebøen, K.A. Low-frequency oscillations of the laser Doppler perfusion signal in human skin. *Microvasc. Res.* **2006**, *72*, 120–127. [CrossRef]
37. Gonzalez, R.; Woods, R.E.; Eddins, S.L. *Digital Image Processing Using MATLAB*, 2nd ed.; Gatesmark Publishing: USA, 2009.
38. Ferreira, H.; Rodrigues, F.; Meyer, M.; Santos-Ribeiro, A.; Gonçalves-Pereira, P.; Manaças, R.; Andrade, A. Complexity analysis of resting-state networks. *MAGMA* **2013**, *26*, 2013.
39. Gabrielsen, A.; Norsk, P. Effect of spaceflight on the subcutaneous venoarteriolar reflex in the human lower leg. *J. Appl. Physiol.* **2007**, *103*, 959–962. [CrossRef] [PubMed]

applied sciences

MDPI

Article

The 'DEEP' Landing Error Scoring System

Kim Hébert-Losier [1,*], Ivana Hanzlíková [1], Chen Zheng [2], Lee Streeter [3] and Michael Mayo [2]

[1] Te Huataki Waiora School of Health, Division of Health, Engineering, Computing and Science, University of Waikato, Tauranga 3116, New Zealand; xhanzlikova@gmail.com

[2] School of Computing and Mathematical Sciences, Division of Health, Engineering, Computing and Science, University of Waikato, Hamilton 3216, New Zealand; zhengchenloot@gmail.com (C.Z.); michael.mayo@waikato.ac.nz (M.M.)

[3] School of Engineering, Division of Health, Engineering, Computing and Science, University of Waikato, Hamilton 3216, New Zealand; lee.streeter@waikato.ac.nz

* Correspondence: kim.hebert-losier@waikato.ac.nz; Tel.: +64-7-837-9476

Received: 20 December 2019; Accepted: 24 January 2020; Published: 29 January 2020

check for updates

Featured Application: The Landing Error Scoring System, an injury-risk screening tool used in sports to detect high risk of anterior cruciate ligament injury, can be automated using deep-learning-based computer vision on 2D videos combined with machine learning methods. The successful application of this method paves the way for the automatic detection of individuals at high risk of injury using smartphone-based applications and opens doors to addressing other related injury prevention problems.

Abstract: The Landing Error Scoring System (LESS) is an injury-risk screening tool used in sports; but scoring is time consuming, clinician-dependent, and generally inaccessible outside of elite sports. Our aim is to evidence that LESS scores can be automated using deep-learning-based computer vision combined with machine learning and compare the accuracy of LESS predictions using different video cropping and machine learning methods. Two-dimensional videos from 320 double-leg drop-jump landings with known LESS scores were analysed in OpenPose. Videos were cropped to key frames manually (clinician) and automatically (computer vision), and 42 kinematic features were extracted. A series of 10 × 10-fold cross-validation experiments were applied on full and balanced datasets to predict LESS scores. Random forest for regression outperformed linear and dummy regression models, yielding the lowest mean absolute error (1.23) and highest correlation ($r = 0.63$) between manual and automated scores. Sensitivity (0.82) and specificity (0.77) were reasonable for risk categorization (high-risk LESS ≥ 5 errors). Experiments using either a balanced (versus unbalanced) dataset or manual (versus automated) cropping method did not improve predictions. Further research on the automation would enhance the strength of the agreement between clinical and automated scores beyond its current levels, enabling quasi real-time scoring.

Keywords: anterior cruciate ligament; automation; drop jump; injury risk; deep learning; machine learning; movement screen; OpenPose

---

1. Introduction

Lower-extremity injuries due to physical activities have devastating short-term and long-term consequences to the health and wellbeing of individuals [1,2] and burden societies worldwide [3,4]. Non-contact injuries account for approximately 20% of injuries in game situations and 37% of injuries in training situations [5]. Non-contact injuries in sport and recreation are the ones of most practical interest to coaches and clinicians as preventable through neuromuscular training programs [6].

The mechanism of non-contact lower-extremity injuries and their underlying risk factors have been linked with 'risky' movement patterns [7,8], such as knee valgus and stiff landings. 3D motion analysis systems, which provide gold-standard measures for the objective quantification of human motion noninvasively, can readily identify altered movement patterns and biomechanical control. However, conventional 3D motion analysis using infrared systems requires a considerable financial outlay and an expert-user, in addition to time and space to perform the analysis. These constraints limit its practical application and use for large-scale screening of injury risk factors in physically active individuals.

As a countermeasure and to reduce technological requirements, various clinician-led movement screens have been developed [9]. Even though these clinician-led screens reduce the financial costs and space requirements compared to 3D motion analysis, they nonetheless require expert clinicians and dedicated time for testing and scoring, limiting their widespread use. For instance, the Functional Movement Screen^{TM} takes 12 to 15 min and the Tuck jump assessment takes 12 min to administer and score for one individual [9].

The Landing Error Scoring System (LESS) is one movement screen with demonstrated reliability [10,11] and validity [11,12]. Clinicians evaluate 2D video recordings from three double-leg drop-jump landing tasks per individual to detect 'movement errors' linked to non-contact anterior cruciate ligament (ACL) and other lower-extremity injury mechanisms [10]. The LESS consists of 17 items (Table 1), with the total number or possible errors ranging from 0 (best) to 17 (worst). Greater scores hence indicate more movement errors, poorer landing biomechanics, and greater relative risk of sustaining non-contact lower-extremity injuries. In a prospective study, Padua et al. [12] determined that scoring 5 or more errors on the LESS was associated with a 10.7 times greater relative risk of sustaining a non-contact ACL injury in youth soccer players (sensitivity 0.86, specificity 0.64). The total testing time (including set up) takes ~5 min with 3 to 4 min for a trained rater to score the three drop-jump landing trials of one individual once downloaded to a computer [10].

A few of the drawbacks of the LESS is the subjective nature of the assessment, requirement for an expert-rater, and need to view videos at a later stage [13,14]. In recent years, researchers have striven to automate the LESS to streamline the process using depth sensor cameras [13,15]. Dar, Yehiel, and Cale' Benzoor [13] introduced the PhysiMax system (PhysiMax Technologies Ltd., Tel Aviv, Israel) to automate LESS scoring using a personal computer, 3D Microsoft Kinect, and motion analysis software that requires limited clinical input. Their results indicated high consensus between clinician and PhysiMax LESS scores (intra-class correlation, ICC = 0.80, mean absolute difference 1.13 errors), although the clinician manually inputted the overall impression item (no. 17, Table 1). Despite the automated quantification of the LESS using markerless motion capture using depth cameras provides time-cost saving benefits, there are still additional hardware-software expenditures to consider.

Deep-learning-based computer vision technologies enable the automatic identification and quantification of human motion without the need for depth sensor cameras. Numerous such systems are currently being developed. For example, OpenPose [16] is a system enabling real-time multi-person pose estimation in video streams captured by a camera. The system tracks both body pose as well as keypoints associated with joints and anatomical features. The same technology is also being deployed for solving other related problems, such as tracking lab animal motion in laboratory settings [17,18]. In this work, we aim to apply deep-learning techniques to LESS score estimation. Applying these approaches to 2D video recordings would improve the accessibility to end-users and pave the way to smartphone-based applications for injury risk screening. Our aim is to evidence that LESS scores can be automated from 2D videos using deep-learning-based computer vision with machine learning and compare the accuracy of LESS predictions using different video cropping and machine learning methods. Our work substantiates that: LESS automation is possible without the need for 3D motion analysis or depth sensor cameras, random forest leads to more accurate predictions than linear or dummy (ZeroR) regression models, and that cropping method (manual versus automated) does not affect predictions.

Table 1. Landing Error Scoring System operational definitions of errors. (Adapted from Padua et al. [10].)

No	Item	Definition of Error
1.	Knee flexion IC	Knee flexion < 30°
2.	Hip flexion IC	Thigh is in line with the trunk (hips not flexed)
3.	Trunk flexion IC	Trunk is vertical or extended at the hips (trunk not flexed)
4.	Ankle plantar flexion IC	Heel-to-toe or flat foot landing
5.	Knee valgus IC	The centre of the patella is medial to the midfoot
6.	Lateral trunk flexion IC	The midline of the trunk is flexed to the left or right
7.	Stance width (wide)	Feet are greater than shoulder width apart
8.	Stance width (narrow)	Feet are less than shoulder width apart
9.	Foot (toe-in)	Foot is externally rotated > 30° between IC and KF_{max}
10.	Foot (toe-out)	Foot is internally rotated > 30° between IC and KF_{max}
11.	Symmetric foot contact IC	One foot lands before the other One foot lands heel-toe and the other foot lands toe-heel
12.	Knee flexion displacement	Knee flexes < 45° between IC and KF_{max}
13.	Hip flexion at KF_{max}	Thigh does not flex more on trunk from IC to KF_{max}
14.	Trunk flexion at KF_{max}	Trunk does not flex more from IC to KF_{max}
15.	Knee valgus displacement	At max medial knee position, centre of the patella is medial to the midfoot
16.	Joint displacement	Soft, average, stiff
17.	Overall impression	Excellent, average, poor

Abbreviations: IC, initial contact; KF_{max}, maximal knee flexion.

2. Materials and Methods

2.1. Participants

A sample of 144 individuals (45 males and 99 females) volunteered to participate in this study. Age, height, mass, and body mass index (mean ± standard deviation) for males were 21.0 ± 5.9 years (range 17 to 42 years), 179.1 ± 7.2 cm, and 82.2 ± 13.6 kg; and for females were 17.1 ± 3.7 years (range 12 to 31 years), 169.2 ± 6.1 cm, and 64.8 ± 9.6 kg. All participants were involved in physical activity (34% participated in netball, 19% in rugby, 9% in field hockey, 9% in soccer, and 29% in other sports). On average, participants were involved in physical activity four times per week, 6 h a week. Participants had to be free from injury, pain, or any other issue that would limit physical activity participation. Previous injuries were not an exclusion criterion. Participants were recruited via word-of-mouth, research contacts, social media, and emails sent to local sports clubs. The study protocol was approved by our institution's health research ethics committee [HREC(Health)#41] and adhered to the Declaration of Helsinki. All participants and their legal guardian when younger than 16 years of age signed a written informed consent document that explained the potential risks associated with testing prior to participation.

2.2. Data Collection

We used the original LESS protocol for testing [10]. Participants jumped horizontally from a 30 cm high box to a line placed at 50% of their body height, and immediately jumped upward for maximal vertical height. We placed an emphasis on jumping off the box with both feet, landing in front of the designated line, jumping as high as possible straight up in the air once they landed from the box, and completing the task in a fluid motion. We did not provide any feedback on participants landing technique unless they were performing the task incorrectly. Participants used their own footwear for testing.

After task instructions and practice jumps for familiarization (typically 1), each participant performed three successful trials of the double-leg drop-jump landing task in front of two standard video cameras capturing at 120 Hz (Sony RX10 II, Sony Corporation, Tokyo, Japan) with an actual focal length of 8.8 to 73.3 mm (35 mm equivalent focal length of 24–200 mm). We mounted the cameras on tripods placed 3.5 m in front of and to the right side of the landing area with a lens-to-floor distance of 1.3 m. We allowed participants to rest until they felt ready to perform the task again to limit fatigue between the three trials. Total testing time was typically 2 min per participant.

2.3. Clinical LESS

A qualified physiotherapist who completed over 400 LESS evaluations (IH) replayed the videos using the Kinovea software (version 0.8.15, www.kinovea.org), identified the two key frames of initial ground contact (IC) and maximal knee flexion (KF_{max}), and scored all trials using the 17-item LESS scoring sheet (Table 1). The clinician was blinded to the results from the automated computer-vision scoring. A total of 320 double-leg drop-jump landings from the potential 432 trials (3 jumps × 144 participants) were retained for analysis because of certain participants not completing three trials, one or both video files being not usable, or a clear misidentification of time events from the automatic cropper described in the following subsection (i.e., more than 100 ms difference with the clinician).

2.4. Automated LESS

The LESS score prediction algorithm we developed was a multistage process. Generally, the first stage consisted of processing the videos to detect the IC and KF_{max} key frames, which involved running the frontal and lateral videos for each jump through OpenPose v.1.21 [16], and then using a heuristic method to identify the key frames. Once that stage was complete, we extracted measurements from the key frames to use as features for machine learning. The final stage was the score prediction for the drop-jump landing trial from the features using a machine learning algorithm. The entire process is depicted in Figure 1. We further evaluated the predictive accuracy of the final machine learning stage using cross validation.

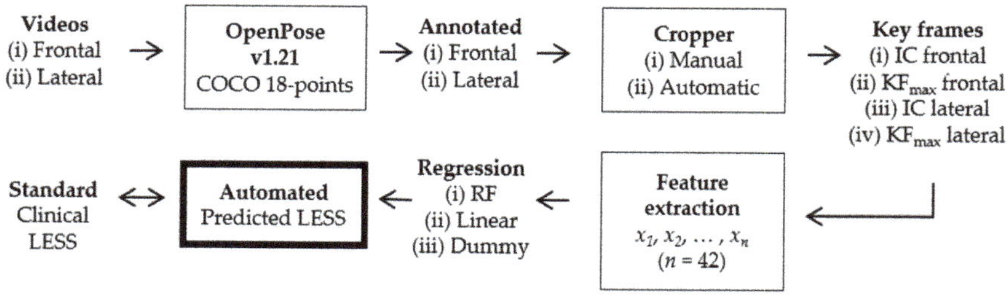

Figure 1. Flow diagram of data processing leading to comparing 'gold standard' clinical LESS scores from an expert rater to 'automated' predicted LESS scores from the automation process. *Abbreviations*: IC, initial contact; KF_{max}, maximal knee flexion; LESS, Landing Error Scoring System; RF, random forest.

In more detail, the algorithm used to detect key frames in the first stage is described in Table 2. The input to the algorithm are the frontal and lateral videos for a single drop-jump landing trial, and the output are cropped versions of the same videos where the first and last frames correspond to the IC and KF_{max} key frames, respectively.

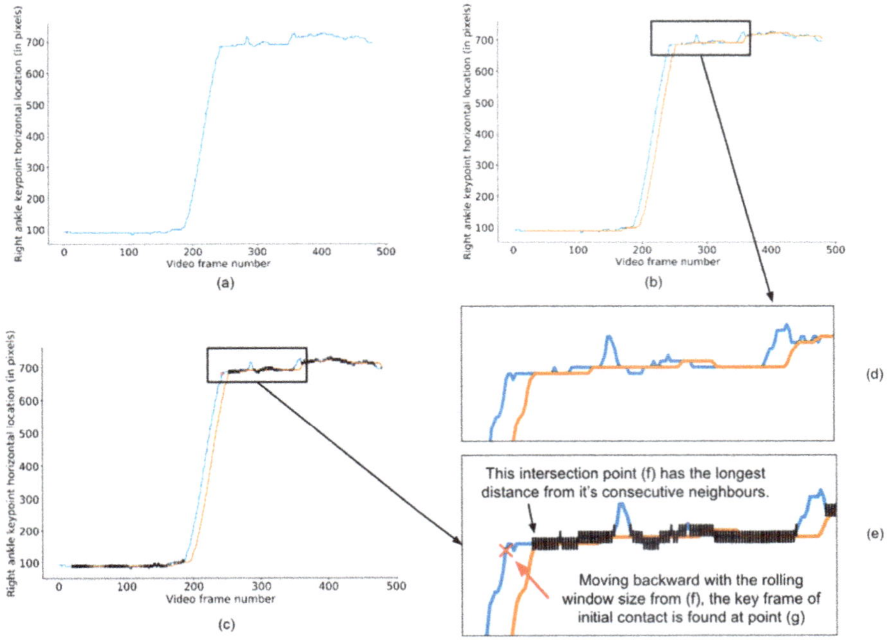

Figure 2. This figure is an example of the original (blue line) plot and rolling window (orange line) plot for the right ankle keypoint of one individual during a drop-jump landing trial taken from the lateral view video. More specifically, (**a**) the blue line depicts the distance of the right ankle to the left boarder (y-axis) in each video frame (x-axis); (**b**) the orange line is the 20-frame rolling median of the original blue line; (**c**) the black bars indicate the intersections of the two lines, whereas the red dotted line represents the distance between two consecutive intersection points. Figure (**d**) is a zoomed-in view of the intersections around the initial contact key frame. Figure (**e**) highlights the points (f) and (g) as the initial contact key frame on the rolling window plot and original plot, respectively.

The basic method is to track the location of the ankles (using OpenPose and COCO 18-points model [16]) across the frames to detect the frame in which landing occurs based on the original and rolling window plots (Figure 2), and additionally to track the body and knee keypoints so that the ankle/knee/body angle can be calculated and used to identify the point of maximum knee flexion. Once these two points are identified in both videos, then the frames before and after the key frames are cropped away. This stage generally reduces the length of the original videos from several seconds down to less than 250 ms.

Once cropping is complete, two videos in which the first frame corresponds to IC and the last frame corresponds to KF_{max} pass to the second stage. In the second stage of processing, features are extracted from both videos and merged into a single 'example' that will be used for machine learning. A total of 42 kinematic features from the two key frames in each video were generated. The features are a mixture of angles between specific OpenPose keypoints (shown in Figure 3) and ratio between distances. The specific features are listed in Table 3. A total of six angles were extracted from all four key frames with an additional eight features (mixture of angles, distances, and distance ratios) being extracted from the two frontal key frames only, for a total of 40 measurements. Two further features, being the length in frames of the cropped frontal and lateral videos, were also included.

Table 2. Algorithm used to detect key frames from the two input videos.

Step	Description
	Input: F, Frontal view video; L, Lateral view video
1.	Obtain the body part keypoints in each frame in both F and L videos using OpenPose
2.	Impute keypoint positions using linear interpolation when not recognized by OpenPose
3.	Find F key frames IC and KF_{max}
3.1.	Based on the coordinates of the left and right ankle (both visible in F), find the intersections of the original and rolling window plots [a] for each ankle
3.2.	Calculate the distances between each consecutive intersection point pairs
3.3.	Find the first point of the pair of intersection points with the longest distances for each ankle
3.4.	Identify the first point of the pair of intersection points that has the lowest x value (i.e., number of frames) as IC
3.5.	Based on the coordinates of the body keypoint, find the intersections of the original and rolling window plots [a]
3.6.	Calculate the distances between each consecutive intersection point pairs
3.7.	Identify the first point of the pair of intersection points with the longest distances as KF_{max}
4.	Find L key frames IC and KF_{max}
4.1.	Based on the coordinates of the individual's right ankle (which is closest to the camera L), find the intersections of the original and rolling window plots
4.2.	Calculate the distances between each consecutive intersection point pairs
4.3.	Identify the first point of the pair of intersection points with the longest distances as IC
4.4	Based on the coordinates of the body keypoint, find the intersections of the original and rolling window [a] plots
4.5.	Calculate the distances between each consecutive intersection point pairs
4.6.	Identify the first point of the pair of intersection points with the longest distances with upper/positive trend as KF_{max}
5.	Crop the videos (F and L) according to IC and KF_{max} key frames.
	Output: F', cropped version of frontal view video; L, cropped version of lateral view video

Notes. [a] Rolling window plot, plot of median values from a rolling 20-frame window. See Figure 2. *Abbreviations.* F, Frontal view video; IC, initial contact; L, Lateral view video KF_{max}, maximal knee flexion.

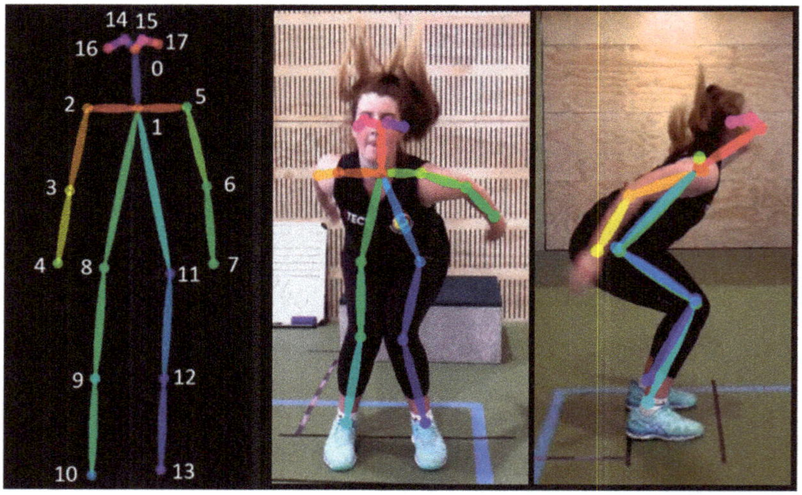

Figure 3. OpenPose's COCO 18-points model keypoint positions (left image) [16] and example of a frontal (middle image) and lateral (right image) view processed video at the maximal knee flexion key frame.

Table 3. Measurements extracted from key frames and used as kinematic features.

Key Frames and Views	Measurement (OpenPose Numbers [a])	Kinematic Features
All four key frames (two frontal key frames and two lateral key frames)	Angle (8,9,10) Angle (9,8,1) Angle (2,1,8) Angle (3,2,1) Angle (4,3,2) Angle (2,1,0)	Right knee angle Right hip angle Right trunk angle Right shoulder angle Right elbow angle Right neck angle
Two key frames (two frontal key frames)	Angle (11,12,13) Angle (1,11,12) Angle (11,1,5) Angle (1,5,6) Angle (5,6,7) Distance (9,12) Distance (2,5) Distance (9,12)/Distance (2,5)	Left knee angle Left hip angle Left trunk angle Left shoulder angle Left elbow angle Knee distance Shoulder distance Knee distance/Shoulder distance

Notes. Key frames are: (i) initial contact, (ii) maximal knee flexion. [a] Refer to Figure 3 for keypoints.

Following feature extraction, we then used a machine learning algorithm to predict the LESS score associated with the drop-jump landing videos. To evaluate the predictive effectiveness of the various machine learning algorithms, we generated features for all 320 drop-jump landings in the dataset using the approach described above. It was also noticed that the distribution of the LESS scores in the dataset was imbalanced, with the majority of LESS scores falling in the range 4–6. Given that unbalanced datasets can potentially affect the accuracy of machine learning techniques, we additionally generated a balanced version of the dataset consisting of 153 drop-jump landing trials with at most 20 trials per LESS score. All evaluations of machine learning techniques were applied to both datasets.

The machine learning techniques chosen to be evaluated were random forest regression, because it is a state-of-the-art machine learning approach and generally performs well 'out of the box' on most problems in practice; and linear regression, which is a widely understood linear modelling technique. Unlike random forest regression, linear regression produces an interpretable model, but it has the disadvantage of being unable to model interactions between features. Given that the full dataset was imbalanced, we also evaluated a dummy regressor (ZeroR) that simply predicts the mean LESS score from the training data. For the original dataset, this method was expected to have reasonably high accuracy, but lower accuracy for the balanced dataset. All machine learning methods implemented were available in WEKA 3.8.0 [19], and returned floating point numbers (i.e., decimals) that added granularity to the data.

2.5. Statistical Method

As noted in Section 2.3, 320 double-leg drop-jump landings were analysed. A series of 10 × 10-fold cross validation experiments were applied on full (320 videos) and balanced (153 videos, ≤ 20 videos per LESS score) to predict the scores using random forest for regression, linear regression, and dummy regression (ZeroR) models in WEKA [20]. To assess the effectiveness of the automated cropping algorithm in the context of the overall system, we additionally ran the entire pipeline with crops generated by the clinician. Mean absolute error and Pearson correlation coefficient (r) were calculated to assess the accuracy of the predictions. Predictions were then converted to a binary category and sensitivity-specificity for categorising individuals at high risk of non-contact ACL injury (LESS ≥ 5 errors [12]) were assessed for each method. The outcomes of the models were compared using paired corrected t-tests in WEKA [20], and the timestamps of the key frames IC and KF_{max} respectively compared between manual (clinician) and automated (OpenPose) cropping methods using unpaired t-tests assuming homoscedasticity. Since the LESS score was treated as a regression problem, actual (clinical LESS) versus predicted (automated LESS) and Bland-Altman [21] plots were used to allow for a visual inspection of the models. Statistical significance was set at $p \leq 0.05$.

3. Results

The mean LESS score from the 320 drop-jump landings was 5.5 ± 1.8 errors (range 0 to 12 errors) as rated by the clinician. The absolute time difference between manually identified IC and KF_{max} was 26.5 ± 17.0 ($p = 0.484$) and 32.8 ± 18.0 ms ($p = 0.445$) for the frontal videos, and 53.5 ± 16.2 ($p = 0.125$) and 20.8 ± 16.3 ms ($p = 0.827$) for the sagittal videos.

Random forest yielded the lowest mean absolute error (1.23) and greatest correlation ($r = 0.63$) between actual and predicted scores based on results from the cross validation experiments (Table 4). Sensitivity (0.82) and specificity (0.77) were reasonable for high (LESS ≥ 5 errors) and low (LESS < 5 errors) injury risk categorisation. Experiments using a balanced (versus unbalanced) dataset or manually (versus automated) cropping methods did not improve predictions. An actual versus predicted plot from the random forest regression is depicted in Figure 4, and two Bland-Altman plots on the same dataset in Figure 5. Note that both conventional (mean difference ± 1.96 standard deviation) and regression-based (regressed difference between methods on the mean of the two methods ± 2.46 standard deviation of the residual) Bland-Altman plots were generated given the non-uniform differences in mean [21].

Table 4. Results from machine learning experiments.

Cropper	Dataset	Mean Absolute Error (n Errors)			Correlation (*r*)		
		RF	**Linear**	**Dummy**	**RF**	**Linear**	**Dummy**
(i) Manual	(i) Full	1.23 ± 0.18	1.39 ± 0.20 *	1.44 ± 0.20 *	0.52 ± 0.15	0.39 ± 0.14 *	0.0 ± 0.0 *
	(ii) Balanced	1.57 ± 0.27	1.90 ± 0.61	2.08 ± 0.34 *	0.60 ± 0.15	0.48 ± 0.21 *	0.0 ± 0.0 *
(ii) Automatic	(i) Full	1.23 ± 0.18	1.32 ± 0.20 *	1.44 ± 0.20 *	0.53 ± 0.15	0.44 ± 0.16 *	0.0 ± 0.0 *
	(ii) Balanced	1.56 ± 0.29	1.63 ± 0.32	2.08 ± 0.32 *	0.63 ± 0.17	0.51 ± 0.20 *	0.0 ± 0.0 *
Cropper	**Dataset**	**Sensitivity** [a]			**Specificity** [a]		
		RF	**Linear**	**Dummy**	**RF**	**Linear**	**Dummy**
(i) Manual	(i) Full	0.80 ± 0.09	0.75 ± 0.09	1.0 ± 0.0 *	0.50 ± 0.18	0.51 ± 0.18	0.0 ± 0.0 *
	(ii) Balanced	0.77 ± 0.13	0.73 ± 0.13	1.0 ± 0.0 *	0.73 ± 0.18	0.63 ± 0.21	0.0 ± 0.0 *
(ii) Automatic	(i) Full	0.82 ± 0.07	0.77 ± 0.09 *	1.0 ± 0.0 *	0.52 ± 0.19	0.52 ± 0.18	0.0 ± 0.0 *
	(ii) Balanced	0.76 ± 0.15	0.77 ± 0.13	1.0 ± 0.0 *	0.77 ± 0.19	0.70 ± 0.21	0.0 ± 0.0 *

Notes. Values are means ± standard deviations. *Abbreviations.* RF, random forest. * Significant difference versus random forest ($p \leq 0.05$) using paired-corrected t-tests. [a] Categorising high (LESS ≥ 5 errors) and low (LESS < 5 errors) injury risk individuals [12].

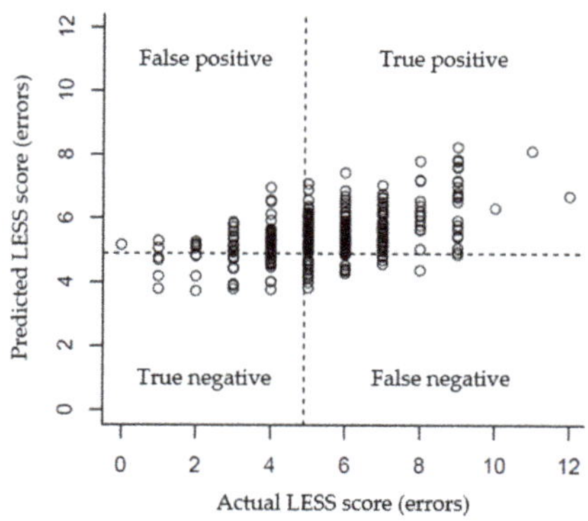

Figure 4. Actual (clinical) versus predicted (automated) LESS score plots from the random forest regression using full dataset (n = 320) and automatic cropping method. Dashed lines represent the 5-error threshold that defines high risk of injury (i.e., scoring 5 or more errors during LESS has been associated with a 10.7 times greater relative risk of sustaining a non-contact anterior cruciate ligament injury [12]). Note that the clinical scores are integers and predicted scores are decimals, which adds granularity. *Abbreviations*: LESS, Landing Error Scoring System.

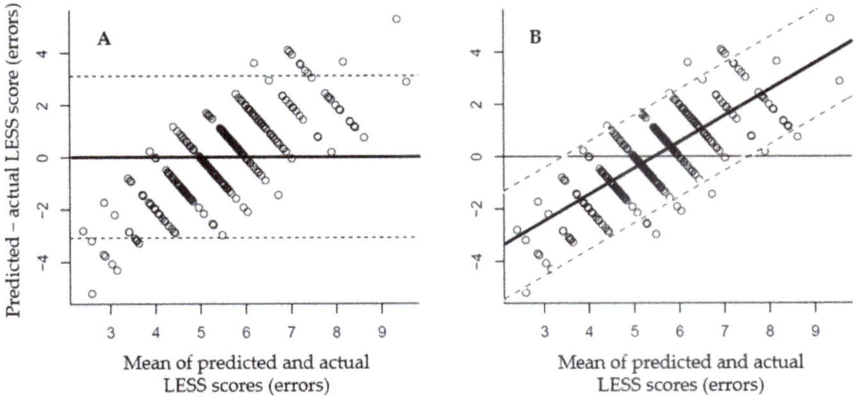

Figure 5. Bland-Altman [21] plots depicting the difference in predicted (automated) and actual (clinical) LESS scores versus the mean scores with (**A**) conventional 95% limits of agreement (mean difference ± 1.96 standard deviation), and (**B**) regression-based limits of agreement (regressed difference between methods on the mean of the two methods ± 2.46 standard deviation of the residual).

4. Discussion

The use of the LESS to assess injury risk is common in sport science and clinical practice [9,22], but scoring is time consuming, clinician-dependent, and generally inaccessible for large-scale screening outside of elite sports. This study provides evidence that the LESS can be automated using deep-learning-based computer vision combined with machine learning methods without the need for 3D motion analysis or depth sensor cameras. A clear benefit of automating LESS scoring is immediate feedback to end-users. The successful application of this method paves the way for the automatic detection of individuals at high risk of injury using smartphone-based applications of LESS videos (Video S1: https://youtu.be/q1wiGt4K8MU).

The characteristics of an ideal injury risk screening tool are good reliability, validity, and predictive value for injury incidence. In practical or field settings, an ideal screening method is easy to administer without an expert, and has minimal financial, spatial, and temporal requirements. Ideally, the screening tool provides immediate results and is accessible to everyone, from the recreational to elite athlete, as well as novice to expert rater. Overall, the LESS responds to most of these stated requirements. The test demonstrates acceptable reliability and validity [10,11,23], as well as predictive value for non-contact ACL injury using a threshold of 5 errors [12]. The inter-rater reliability of the total LESS score is good to excellent, with ICC ranging from 0.83 to 0.92 [10,11,23] and typical errors at 0.71 LESS errors [10]. The results from the current study indicate that the typical errors from the automated processing and scoring of the LESS through computer vision when applying the random forest model (Table 2) are less than half an error greater than scores taken from two expert clinicians. In fact, certain individual LESS items yield suboptimal psychometric properties between raters and 3D motion analysis [23]. More specifically, no significant agreement between raters was found for knee and trunk flexion at IC, and poor agreement between rater and 3D motion capture analysis was found for knee flexion at IC, lateral trunk flexion at IC, and symmetric foot contact at IC [23]. As such, a certain level of disagreement between clinical ratings and computerised ratings is expected.

As seen in Figures 4 and 5, the estimated error is not uniform across the range of LESS scores, but depends on the target. For example, trials with a low actual LESS score tend to have a positive error (the prediction is an overestimation) and trials with a higher actual LESS score tend to have a negative error (the prediction is an underestimation). If these biases stemmed from the over representation of the mid-range LESS values (i.e., majority of LESS scores falling in the range 4–6), the balanced dataset should have provided more accurate predictions, which was not the case. It might be possible to attempt correcting predictions to improve accuracy in future work using probability calibration methods, such as Platt Scaling and Isotonic Regression. The large errors in LESS score predictions were attributed to inaccurate foot and IC key frame detection. The newest body model in OpenPose (Body 25) contains 25 points, including coordinates that define the feet and enable computations of angles at the ankles [24]. Improving the LESS score automation relies on either refining body part detection or training a new system specifically to solve this problem.

In previous research, depth sensor technology has been used to automate LESS scoring [13,15]. Comparisons between automated and expert clinicians indicate a mean difference of 1.20 errors [15], mean absolute difference of 1.13 errors [13], intra-class correlation of 0.80 [16], and percentage agreement of the individual items ranging from 55–100% [13,15]. These research findings are comparable to our lowest mean absolute error (1.23), greatest correlation ($r = 0.63$), and agreement in risk classification (sensitivity 0.82, specificity 0.77) between actual and predicted scores from the cross validation experiments using random forest regression. In contrast to the PhysiMax system [13,15], our approach did not require the clinician to add the overall impression manually (no. 17 in Table 1) given that the LESS items were not scored one-by-one. Although the lack of individual-item scores might be perceived as a limitation of the deep LESS approach; no subjective rating from the clinician or hardware other than a handheld camera or smart portable device are required. Furthermore, only the final LESS score has shown predictive value in terms of injury risk [12]; hence, the individual items are of lesser clinical value.

The better accuracy achieved by random forest can be explained by the fact that the features (angles, distances, and ratios) are likely correlated and related in a non-linear manner. Decision tree ensembles in general are better able to cope with correlated variables and model non-linear patterns [25]. Linear regression, on the other hand, achieves optimal results when the predictor variables are independent and do not interact. We also foresee a possibility of processing the raw video images themselves and attempting direct deep learning-based classification with minimal pre-processing. Such an approach would obviate the need to use OpenPose or a similar pose-tracking tool. However, taking such an approach would be challenging because of the lack of training data relative to size of datasets usually used to train deep image recognisers. Another significant disadvantage of the proposed approach is that deep learning needs GPU-based acceleration hardware, and is therefore currently unable to process videos independently on consumer smartphones. That said; the rapidly increasing computational power of consumer smartphones and the current trend in research of compressing deep models [26] so that they run efficiently on mobile devices should solve this problem in the next few years.

One of the main concerns in clinical screening tools are their subjective nature and reliance on visual observations to estimate angles, which are challenging to quantify accurately [27,28]. During the LESS, a small kinematic difference (e.g., knee angle 29°, 1—error present; knee angle 30°, 0—error absent) can result in poor agreement between raters and between clinical LESS scores and motion capture scores. Recent technological advances have allowed the more objective quantification of human motion using wearable technology [29,30]. Inertial measurement units are able to measure linear and angular motion of individual body segments and centre of mass, and are proposed as more accurate means of identifying risky movement patterns than through visual observations [31]. Although inertial measurement units are relatively inexpensive; they are not commonly used in clinical environments and an expert is still needed to process and interpret data signals. The automated scoring process here developed using standard video recordings offers an alternative solution that can possibly improve consistency of LESS ratings, removing the subjective interpretation of the task. Moving forward, the reliability of deep LESS scores, validity of OpenPose derived data during the dynamic double-leg drop-landing task, and predictive ability of the method need empirical support.

An indisputable advantage of automated scoring using deep-learning-based computer vision combined with machine learning methods or markerless methods from depth sensor cameras is immediate results and feedback to patients, athletes, coaches, or healthcare professionals.

Our developed method that automates LESS scores provides a viable solution to decreasing scoring time, increasing accessibility to non-expert raters, and delivering immediate results without any additional expenditure other than conventional video recordings. Conventional 2D video recordings are adequate for quantifying kinematics [32–34] and are readily accessible through tablets or smartphones. The successful application of this method would pave the way for the automatic detection of individuals at high risk of injury using smartphone-based applications of LESS and 2D video footage (Video S1: https://youtu.be/q1wiGt4K8MU). Other than expediting mass injury risk screening initiatives in youth or team sports, LESS automation could be a valuable and convenient tool to track injury risk factors over time and to assess the effectiveness of intervention programs at improving landing mechanics (Video S2: https://youtu.be/Ve_QJu0fuLs). The proposed method could be extended to other injury risk screening methods based on 2D camera recordings to decrease manual labour and time required for screening initiatives; e.g., the Cutting Movement Assessment Scale [35] and Tuck jump assessment [36].

This preliminary investigation provides evidence that it is feasible to automate the LESS from 2D video recordings alone. Further research could lead to improved automation outcomes and enhance the strength of the agreement between clinical and automated LESS scores beyond its current levels. The newest body model in OpenPose (Body 25) contains 25 points, including coordinates that define the feet and enable computations of angles at the ankles [24]. Although the timestamped IC key frame in frontal and sagittal videos were comparable between the clinician and scripted process (mean difference: 32.8 ms, $p = 0.445$ and 20.8 ± 16.3 ms, $p = 0.827$), using the foot coordinates rather than ankle and body coordinates would certainly enhance precision. A number of videos from the available dataset were not used because of a clear misidentification of time events from the automatic cropper (i.e., more than 100 ms). We were unable to determine the reason underlying the mislabelling of these videos upon visual inspection. We speculate that rerunning the current experiment using the COCO + Foot model might lead to the correct identification of key events in a greater number of our database videos, increasing the number of eligible videos for analysis. The increased number of coordinates from the 25-point Body rather than 18-point COCO model would also allow us to extract a greater number of features from the processed videos and use these as input in the subsequent regression experiments.

5. Conclusions

We provide evidence that the Landing Error Scoring System (LESS)—an injury-risk screening tool—can be automated using deep-learning-based computer vision combined with machine learning methods. Further research on the automation would enhance the strength of the agreement between clinical (gold standard) and automated (predicted) LESS scores, and risk classification beyond its current levels. Automation of the LESS using standard 2D recordings would facilitate mass injury-risk screening initiatives with quasi real-time feedback, without the need of depth cameras or expert clinicians. The successful application of this method would pave the way for the automatic detection of individuals at high risk of injury using smartphone-based applications of LESS and 2D video footage (Video S1: https://youtu.be/q1wiGt4K8MU), increasing accessibility of injury-risk assessment methods beyond elite athletes and removing depth-sensor camera requirements. It may also open doors to other related injury prevention problems. Future work includes updating the framework using the newest body model in OpenPose (Body 25) to extract a greater number of features and more accurately detect key frames.

Supplementary Materials: The following are available online: https://youtu.be/q1wiGt4K8MU. Video S1: LESS demonstration. https://youtu.be/Ve_QJu0fuLs. Video S2: The 'DEEP' Landing Error Scoring System.

Author Contributions: Conceptualization, K.H.-L.; methodology, K.H.-L.; formal analysis, K.H.-L., L.S., M.M.; investigation, K.H.-L., I.H.; data curation, I.H., C.Z.; writing—original draft preparation, K.H.-L., I.H.; writing—review and editing, K.H.-L., I.H., C.Z., L.S., M.M.; supervision, K.H.-L., M.M.; project administration, K.H.-L.; funding acquisition, K.H.-L., L.S., M.M. Authorship must be limited to those who have contributed substantially to the work reported. All authors have read and agreed to the published version of the manuscript.

Funding: This research was funded by a University of Waikato Strategic Investment Fund 2018 Medium research Grant.

Acknowledgments: We would like to acknowledge Ruili Wang for expert advice and Christopher Martyn Beaven for research support.

Conflicts of Interest: The authors declare no conflict of interest.

References

1. Hébert-Losier, K.; Pini, A.; Vantini, S.; Strandberg, J.; Abramowicz, K.; Schelin, L.; Häger, C.K. One-leg hop kinematics 20years following anterior cruciate ligament rupture: Data revisited using functional data analysis. *Clin. Biomech.* **2015**, *30*, 1153–1161. [CrossRef] [PubMed]
2. Hébert-Losier, K.; Schelin, L.; Tengman, E.; Strong, A.; Hager, C.K. Curve analyses reveal altered knee, hip, and trunk kinematics during drop-jumps long after anterior cruciate ligament rupture. *Knee* **2018**, *25*, 226–239. [CrossRef] [PubMed]
3. Mather, R.C., 3rd; Hettrich, C.M.; Dunn, W.R.; Cole, B.J.; Bach, B.R., Jr.; Huston, L.J.; Reinke, E.K.; Spindler, K.P. Cost-effectiveness analysis of early reconstruction versus rehabilitation and delayed reconstruction for Anterior Cruciate Ligament tears. *Am. J. Sports Med.* **2014**, *42*, 1583–1591. [CrossRef] [PubMed]
4. Sutherland, K.; Clatworthy, M.; Fulcher, M.; Chang, K.; Young, S.W. Marked increase in the incidence of anterior cruciate ligament reconstructions in young females in New Zealand. *ANZ J. Surg.* **2019**, *89*, 1151–1155. [CrossRef]
5. Hootman, J.M.; Dick, R.; Agel, J. Epidemiology of collegiate injuries for 15 sports: Summary and recommendations for injury prevention initiatives. *J. Athl. Train.* **2007**, *42*, 311–319.
6. Al Attar, W.S.A.; Alshehri, M.A. A meta-analysis of meta-analyses of the effectiveness of FIFA injury prevention programs in soccer. *Scand. J. Med. Sci. Sports* **2019**, *29*, 1846–1855. [CrossRef]
7. Hewett, T.E.; Myer, G.D.; Ford, K.R. Anterior cruciate ligament injuries in female athletes: Part 1, mechanisms and risk factors. *Am. J. Sports Med.* **2006**, *34*, 299–311. [CrossRef]
8. Leppänen, M.; Pasanen, K.; Kujala, U.M.; Vasankari, T.; Kannus, P.; Ayramo, S.; Krosshaug, T.; Bahr, R.; Avela, J.; Perttunen, J.; et al. Stiff landings are associated with increased ACL injury risk in young female basketball and floorball players. *Am. J. Sports Med.* **2017**, *45*, 386–393. [CrossRef]
9. Chimera, N.J.; Warren, M. Use of clinical movement screening tests to predict injury in sport. *World J. Orthop.* **2016**, *7*, 202–217. [CrossRef]
10. Padua, D.A.; Marshall, S.W.; Boling, M.C.; Thigpen, C.A.; Garrett, W.E., Jr.; Beutler, A.I. The Landing Error Scoring System (LESS) is a valid and reliable clinical assessment tool of jump-landing biomechanics: The JUMP-ACL study. *Am. J. Sports Med.* **2009**, *37*, 1996–2002. [CrossRef]
11. Hanzlíková, I.; Hébert-Losier, K. Is the Landing Error Scoring System reliable and valid? A systematic review. *Sports Health* **2020**. [CrossRef]
12. Padua, D.A.; DiStefano, L.J.; Beutler, A.I.; de la Motte, S.J.; DiStefano, M.J.; Marshall, S.W. The Landing Error Scoring System as a screening tool for an anterior cruciate ligament injury-prevention program in elite-youth soccer athletes. *J. Athl. Train.* **2015**, *50*, 589–595. [CrossRef] [PubMed]
13. Dar, G.; Yehiel, A.; Cale'Benzoor, M. Concurrent criterion validity of a novel portable motion analysis system for assessing the landing error scoring system (LESS) test. *Sports Biomech.* **2019**, *18*, 426–436. [CrossRef] [PubMed]
14. Markbreiter, J.G.; Sagon, B.K.; Valovich McLeod, T.C.; Welch, C.E. Reliability of clinician scoring of the landing error scoring system to assess jump-landing movement patterns. *J. Sport Rehabil.* **2015**, *24*, 214–218. [CrossRef] [PubMed]
15. Mauntel, T.C.; Padua, D.A.; Stanley, L.E.; Frank, B.S.; DiStefano, L.J.; Peck, K.Y.; Cameron, K.L.; Marshall, S.W. Automated quantification of the Landing Error Scoring System with a markerless motion-capture system. *J. Athl. Train.* **2017**, *52*, 1002–1009. [CrossRef] [PubMed]
16. Cao, Z.; Simon, T.; Wei, S.-E.; Sheikh, Y. Realtime multi-person 2d pose estimation using part affinity fields. *arXiv* **2017**, arXiv:1611.08050v2.
17. Mathis, A.; Mamidanna, P.; Cury, K.M.; Abe, T.; Murthy, V.N.; Mathis, M.W.; Bethge, M. DeepLabCut: Markerless pose estimation of user-defined body parts with deep learning. *Nat. Neurosci.* **2018**, *21*, 1281–1289. [CrossRef]

18. Mathis, M.W.; Mathis, A. Deep learning tools for the measurement of animal behavior in neuroscience. *Curr. Opin. Neurobiol.* **2020**, *60*, 1–11. [CrossRef]
19. Frank, E.; Hall, M.A.; Witten, I.H. The WEKA Workbench. In *Data Mining: Practical Machine Learning Tools and Techniques*, 4th ed.; Witten, I.H., Frank, E., Hall, M.A., Pal, C.J., Eds.; Morgan Kaufmann: Cambridge, MA, USA, 2016.
20. Nadeau, C.; Bengio, Y. Inference for the Generalization Error. *Mach. Learn.* **2003**, *52*, 239–281. [CrossRef]
21. Bland, J.M.; Altman, D.G. Measuring agreement in method comparison studies. *Stat. Methods Med. Res.* **1999**, *8*, 135–160. [CrossRef]
22. Dallinga, J.M.; Benjaminse, A.; Lemmink, K.A.P.M. Which screening tools can predict injury to the lower extremities in team sports? A systematic review. *Sports Med.* **2012**, *42*, 791–815. [CrossRef] [PubMed]
23. Onate, J.; Cortes, N.; Welch, C.; Van Lunen, B.L. Expert versus novice interrater reliability and criterion validity of the landing error scoring system. *J. Sport Rehabil.* **2010**, *19*, 41–56. [CrossRef] [PubMed]
24. Cao, Z.; Hidalgo, G.; Simon, T.; Wei, S.-E.; Sheikh, Y. OpenPose: Realtime multi-person 2D pose estimation using Part Affinity Fields. *arXiv* **2018**, arXiv:1812.08008v1. [CrossRef] [PubMed]
25. Breiman, L. Random forests. *Mach. Learn.* **2001**, *45*, 5–32. [CrossRef]
26. Nan, K.; Liu, S.; Du, J.; Liu, H. Deep model compression for mobile platforms: A survey. *Tsinghua Sci. Technol.* **2019**, *24*, 677–693. [CrossRef]
27. Ekegren, C.L.; Miller, W.C.; Celebrini, R.G.; Eng, J.J.; Macintyre, D.L. Reliability and validity of observational risk screening in evaluating dynamic knee valgus. *J. Orthop. Sports Phys. Ther.* **2009**, *39*, 665–674. [CrossRef]
28. Whatman, C.; Hing, W.; Hume, P. Physiotherapist agreement when visually rating movement quality during lower extremity functional screening tests. *Phys. Ther. Sport* **2012**, *13*, 87–96. [CrossRef]
29. Willy, R.W. Innovations and pitfalls in the use of wearable devices in the prevention and rehabilitation of running related injuries. *Phys. Ther. Sport* **2018**, *29*, 26–33. [CrossRef]
30. Iqbal, M.H.; Aydin, A.; Brunckhorst, O.; Dasgupta, P.; Ahmed, K. A review of wearable technology in medicine. *J. R. Soc. Med.* **2016**, *109*, 372–380. [CrossRef]
31. Whelan, D.F.; O'Reilly, M.A.; Ward, T.E.; Delahunt, E.; Caulfield, B. Technology in rehabilitation: Evaluating the single leg squat exercise with wearable inertial measurement units. *Methods Inf. Med.* **2017**, *56*, 88–94. [CrossRef]
32. McLean, S.G.; Walker, K.; Ford, K.R.; Myer, G.D.; Hewett, T.E.; van den Bogert, A.J. Evaluation of a two dimensional analysis method as a screening and evaluation tool for anterior cruciate ligament injury. *Br. J. Sports Med.* **2005**, *39*, 355–362. [CrossRef] [PubMed]
33. Willson, J.D.; Davis, I.S. Utility of the frontal plane projection angle in females with patellofemoral pain. *J. Orthop. Sports Phys. Ther.* **2008**, *38*, 606–615. [CrossRef] [PubMed]
34. De Oliveira, F.C.L.; Fredette, A.; Echeverria, S.O.; Batcho, C.S.; Roy, J.S. Validity and reliability of 2-Dimensional video-based assessment to analyze foot strike pattern and step rate during running: A systematic review. *Sports Health* **2019**, *11*, 409–415. [CrossRef] [PubMed]
35. Dos'Santos, T.; McBurnie, A.; Donelon, T.; Thomas, C.; Comfort, P.; Jones, P.A. A qualitative screening tool to identify athletes with 'high-risk' movement mechanics during cutting: The cutting movement assessment score (CMAS). *Phys. Ther. Sport* **2019**, *38*, 152–161. [CrossRef]
36. Myer, G.D.; Ford, K.R.; Hewett, T.E. Tuck jump assessment for reducing Anterior Cruciate Ligament injury risk. *Athl. Ther. Today* **2008**, *13*, 39–44. [CrossRef]

applied sciences

MDPI

Article

Relationships between Body Build and Knee Joint Flexor and Extensor Torque of Polish First-Division Soccer Players

Jadwiga Pietraszewska [1], Artur Struzik [2,*], Anna Burdukiewicz [1], Aleksandra Stachoń [1] and Bogdan Pietraszewski [3]

1 Department of Physical Anthropology, University School of Physical Education, Paderewskiego 35 Avenue, 51-612 Wrocław, Poland; jadwiga.pietraszewska@awf.wroc.pl (J.P.); anna.burdukiewicz@awf.wroc.pl (A.B.); aleksandra.stachon@awf.wroc.pl (A.S.)
2 Department of Team Sport Games, University School of Physical Education, Mickiewicza 58 Street, 51-684 Wrocław, Poland
3 Department of Biomechanics, University School of Physical Education, Mickiewicza 58 Street, 51-684 Wrocław, Poland; bogdan.pietraszewski@awf.wroc.pl
* Correspondence: artur.struzik@awf.wroc.pl; Tel.: +48-71-347-35-61

Received: 18 December 2019; Accepted: 20 January 2020; Published: 22 January 2020

check for updates

Featured Application: Body height, body mass and thigh and calf girths are the anthropometric variables that best describe the values of the torque of the knee flexors and extensors that can be achieved by soccer players. Based on these anthropometric variables, the coach can initially assess players' strength abilities, which can help plan individual training programmes.

Abstract: The aim of the study is to identify the relationships between anthropometric variables and knee joint extensor peak torque, knee joint flexor peak torque, and conventional hamstring-to-quadriceps ratio in Polish first-division soccer players. The study examined 37 soccer players aged 19 to 30 years (body mass: 76.8 ± 7.2 kg, body height 1.82 ± 0.06 m). Muscle torques of the knee joint were measured under isometric conditions using a Biodex 4 Pro dynamometer. The anthropometric variables such as body part lengths, breadths, and girths and skinfold thickness were measured. The strongest relationships of knee joint extensors were observed with body mass and variables describing skeleton size and lower-limb muscles. Regarding knee flexor torque, a significant relationship was found only with body mass. However, no significant relationships were observed between the conventional hamstring-to-quadriceps ratio and the anthropometric variables studied. The regression analysis results identified body height, body mass, and thigh and calf girth as the features most associated with knee joint torque. However, anthropometric measurements do not provide full information about the torque proportions of antagonist muscle groups, which is very important for injury prevention. Therefore, measurements using special biomechanical equipment are also necessary for the comprehensive analyses and control of the effects of sports training.

Keywords: anthropometric variables; hamstring-to-quadriceps ratio; H/Q ratio; injury prevention; isometric; lower limb; measurement acquisitions and techniques; strength and conditioning; sports training; strength abilities

1. Introduction

Measurements of muscle torque taken in isometric and isokinetic conditions have often been used for sports training [1–5]. Coaches use biomechanical measurement tools to analyse the sports skill level of athletes, improve movement techniques, or adjust tactics against the opponent [6,7].

Biomechanical analysis provides a fast and reliable evaluation of torque values, helps detect muscle strength imbalance, and screens for lower extremity injury [8,9]. Numerous studies have focused on the evaluation of the value of muscle torque in relation to lower limbs, which are particularly involved during sports movements. Appropriate force and proportion between antagonistic muscle groups are essential for optimal or maximum performance of various movements, which is also true in team sport games [10–14]. The dependence of force on the transverse cross-sectional area of muscles is known, but it should be noted that force is also affected by other factors related to the efficiency of the nervous system (that controls muscle function) and the properties of the muscles themselves [15].

Soccer is a sport in which the requirements for effective play are multi-factorial. The intervals and repetitive nature of the efforts during the match mean that the competitors should demonstrate adequate speed, endurance, strength, and coordination levels. In general, the players do not reach the highest levels of all these motor abilities. However, they must demonstrate a sufficiently high level of preparation in all aspects of motor abilities, with the dominance of some of them (depending on their playing position). Interindividual differences can also be observed in the anthropometric characteristics of soccer players [16]. In soccer, the lower limbs are particularly involved when kicking the ball, striking at a goal, and running at different velocities and with directional changes [17–20]. Soccer training is aimed, among other things, at increasing muscle mass and, consequently, muscle force. The adequate level of force of the knee extensors and flexors allows the player to achieve stronger strikes, jump higher, and sprint faster over short distances [21–23].

In addition to the absolute values of the generated torque, the correct ratio of the torque of extensors to that of flexors is also important. The commonly used conventional hamstring-to-quadriceps ratio (H/Q ratio) represents the ratio of concentric hamstring peak torque during lower limb flexion to concentric quadriceps peak torque during lower limb extension [24]. The functional hamstring-to-quadriceps ratio, in contrast, is a measure of the isokinetic eccentric peak torque of the hamstrings relative to the isokinetic concentric peak torque of the quadriceps during lower limb extension at equivalent angular velocities [25]. An adequately high value of these indexes is likely to effectively prevent injuries of the lower limbs (including those that occur without contact) due to effective eccentric activity to slow down or stop the movement [26].

Measurements of muscle torque under isometric conditions allow for the determination of the maximum values of force generated by a given muscle group to evaluate the potential of the player, the training programme, and muscle imbalance. It should also be noted that the amount of force generated depends on many factors, including physiological and biochemical factors. The aspect of morphological determinants is also important. Previous studies have investigated the relationships of muscle torque with body mass, body mass composition, and somatotype [27–31]. However, the results of such studies have been inconclusive. Norsuriani and Ooi [31] demonstrated positive relationships between muscle torque in the knee joint and body mass, fat-free mass, and body fat. Similar results were presented by Pietraszewski et al. [27]. Lewandowska et al. [28] emphasized positive relationships of muscle torque in the lower limb joints with characteristics of muscularity and skeletal mass (mesomorphy), negative relationships with body slenderness (ectomorphy) and lack of relationships with the fat component (endomorphy). Kim et al. [29] found that men with higher skeletal muscle mass are characterized by higher isokinetic muscle torque values in the lower limb joints. Furthermore, to our knowledge, no previous studies have examined the relationships between the conventional H/Q ratio and anthropometric variables. It should be pointed out that anthropometric measurements are the basic method used to assess the level of development of many morphological features. They can be performed anytime and anywhere, using mobile equipment. Measurements of body girths and skinfolds provide additional indirect information about changes in tissue components, which are very important in the context of functional capabilities of a player. Understanding the relationships between muscle torque and the anthropometric variables would allow coaches to easily make an initial assessment of the players' strength abilities and, at the same time, control the training effects.

Muscle torque is the product of the force F and the length of the lever arm r. The lever arm r is understood as the distance from the axis of rotation of the lever (joint) along a line perpendicular to the direction of the force F. Therefore, it can be expected that the torque of the flexors and extensors of the knee joint are related to the characteristics of muscle development (girths of the body parts) and the length of the individual lower limb parts. These features are related to the overall body size but may also show interpersonal differentiation due to the modifying effect of various environmental factors, including sports training. Due to the specificity of the training, the knowledge gained from the research of players practising various sports cannot be used uncritically. Substantial values of anthropometric variables (e.g., girths of body parts) may result from both muscularity and body fat. However, due to the characteristics of the research group (soccer players) resulting from the specificity of the motor activities performed by such athletes and the usually low body fat content in lower limbs, the relationships should be expected between muscle torque and variables indirectly describing muscularity of the lower limbs. It is important to examine morphofunctional relations in a specific sport.

The aim of the study is to identify the relationships between anthropometric variables and the values of knee joint extensor peak torque, knee joint flexor peak torque, and conventional hamstring-to-quadriceps ratio in a group of Polish first-division soccer players. Furthermore, the authors intended to identify the anthropometric variables that allow for the best possible estimation of the abovementioned biomechanical variables, which would allow coaches to monitor more frequently the effect of training loads used at different stages of the annual training cycle. Knowledge of the abovementioned biomechanical variables is also important during the choice of training loads to reduce the injury risk of athletes.

2. Materials and Methods

The study examined 37 soccer players aged 19 to 30 years. They were players playing in the Polish Ekstraklasa league. The study included only the players who had actively participated in training and league matches since the beginning of the season and had not been injured. They were previously qualified for the examinations by a sports physician. Their mean training experience was 14 years (8 to 20 years). The mean body mass (±SD) of the subjects was 76.8 ± 7.2 kg, and the body height was 1.82 ± 0.06 m. The tests were conducted in the morning, before the training sessions. On the day before the tests, the players did not play any league matches.

The study was carried out in the Biomechanical Analysis Laboratory and Scientific Research Laboratory of the Faculty of Sport (both with PN-EN ISO 9001:2009 certification) at the University School of Physical Education in Wrocław, Poland. All subjects gave their informed consent for inclusion before they participated in the study. The study was conducted in accordance with the Declaration of Helsinki, and the protocol was approved by the Senate's Research Bioethics Commission of the University School of Physical Education in Wrocław, Poland. Measurements were performed according to International Standards for Anthropometric Assessment (ISAK) [32].

The following anthropometric variables were measured: sitting height (B-vs), lower extremity length (B-tro), shank length (B-ti), biacromial diameter (a-a), biiliocristal diameter (ic-ic), humerus breadth (cl-cm), femur breath (epl-epm), chest girth, waist girth, hip girth, relaxed and flexed biceps girth, thigh girth, and calf girth. Skinfold thickness was measured at the subscapular, abdominal, suprailiac, triceps, and calf locations to assess body fat. Body height, lengths, and breadths were measured to the nearest 0.1 cm with the use of the GPM Anthropological Instruments (Siber Hegner Machinery Ltd., Zürich, Switzerland). Skinfold thickness was measured with a Tanner/Whitehouse skinfold calliper (Holtain LTD, Crosswell, Crymych, Pembs, UK) with a 0.2 mm graduation. Body mass was measured with an electronic weighing scale with an accuracy of 0.1 kg. Body mass index (BMI) was also calculated.

The peak muscle torque under isometric conditions was measured to assess the strength of the lower limbs. Peak muscle torque for extensors and flexors of the knee joint was measured at 75° and

30°, respectively; 0° at the knee joint was considered to be a full extension. The measurements were performed separately for the right and left lower limbs. The results of measurements of the dominant limb were used for analyses. Biodex System 4 Pro (Biodex Medical Systems, Inc., Shirley, NY, USA) was used for the torque measurements. The conventional hamstring-to-quadriceps ratio (H/Q ratio) was calculated using the following equation:

$$\text{H/Q ratio} = (T_f\,/\,T_e) \cdot 100\%, \tag{1}$$

where T_f denotes the concentric peak torque value of knee joint flexors of the dominant lower limb and T_e is the concentric peak torque value of extensors of the same joint and limb. The conventional H/Q ratio is possible to calculate during the measurement of peak torque only under isometric conditions [33]. Furthermore, the muscle torques of extensors and flexors were added together, and the muscle torque for the knee joint relative to body mass was calculated:

$$(T_e + T_f)\,/\,\text{BM}. \tag{2}$$

To supplement the information about the general strength abilities of the soccer players studied, back and handgrip strengths were measured using TAKEI dynamometers (Takei Scientific Instruments Co., Ltd., Niigata City, Japan).

Cluster analysis (*k*-means) was used in the statistical analysis to distinguish two groups significantly differing in the magnitude of muscle torque. The intergroup differentiation of anthropometric variables in these clusters was evaluated by means of Student's *t*-test for independent samples. Pearson's *r* correlation coefficient was used to evaluate the relationships of anthropometric variables with muscle torque.

3. Results

The soccer players studied were characterized by small and moderate intragroup variability in the anthropometric variables (Table 1). The mass-to-height ratios were correct and were confirmed by the BMI values in the entire group (23.3 ± 1.5 kg/m^2).

The obtained values of muscle torque showed a clear interindividual differentiation. To create relatively homogeneous groups in terms of the values of muscle torque generated in the knee extensors and flexors, the *k*-mean cluster analysis was used to identify two subgroups significantly differing in the magnitude of these variables: Group 1—weaker group and Group 2—stronger group. With the adopted criterion of division, the groups differed significantly in terms of the generated muscle torque and in terms of the sum of muscle torque relative to body mass (Table 2). All of the above values were significantly higher in Group 2 (stronger). Higher values were also found in most of the anthropometric variables in this group of players. However, significant intergroup differences were recorded only for body mass, BMI; chest, thigh and calf girths; and humerus and femur breadths, which are a measure of skeleton size. The soccer players from Group 2 also showed significantly higher values in the remaining strength tests, although these variables were not included in the grouping variables. Both back strength and handgrip strength were greater in the athletes from the second cluster. The mean conventional H/Q ratios were 47.0% (Group 1) and 45.6% (Group 2).

The values of correlation coefficients between muscle torque of the knee extensors and flexors and anthropometric variables were quite low, yet statistically significant in some cases (Table 3). The strongest relationships were observed for knee joint extensors with body mass, BMI, and variables describing skeleton size and lower limb muscles. In the case of the torque of knee flexors, a significant relationship was found only with body mass and BMI. However, no significant relationships were observed between the conventional H/Q ratio and the anthropometric variables studied.

Table 1. Statistical characterization of the variables analysed for the entire group of soccer players studied ($n = 37$).

Variables	Mean	Minimum	Maximum	SD
Body height (cm)	181.6	170.0	192.5	5.5
Body mass (kg)	76.8	65.0	94.5	7.2
BMI (kg/m^2)	23.3	18.6	26.0	1.5
b-vs (cm)	95.2	89.5	104.3	3.3
b-tro (cm)	95.6	88.0	105.2	4.0
b-ti (cm)	48.5	38.5	54.0	2.7
a-a (cm)	41.8	38.0	46.5	1.8
ic-ic (cm)	28.6	25.4	32.5	1.8
cl-cm (cm)	7.1	6.4	7.8	0.3
epl-epm (cm)	10.1	9.2	11.1	0.5
Chest girth (cm)	88.4	80.0	100.0	4.9
Waist girth (cm)	79.5	70.0	89.0	4.6
Relaxed biceps girth (cm)	30.3	26.0	35.0	2.4
Flexed biceps girth (cm)	33.4	29.5	37.5	2.3
Hip girth (cm)	98.5	90.0	106.0	3.5
Thigh girth (cm)	58.5	50.4	65.0	3.0
Calf girth (cm)	37.9	33.2	41.0	1.8
Subscapular skinfold (mm)	8.2	5.8	12.0	1.4
Triceps skinfold (mm)	4.3	2.8	7.0	1.0
Suprailiac skinfold (mm)	6.8	4.6	12.0	1.7
Abdominal skinfold (mm)	7.0	4.0	12.0	2.2
Calf skinfold (mm)	3.7	2.8	5.4	0.7
Right hand grip strength (kG)	47.5	36.0	62.0	6.3
Left hand grip strength (kG)	46.2	38.0	58.0	5.2
Back strength (kG)	127.0	91.5	169.5	18.7
T_e (Nm)	416.8	252.7	596.7	86.4
T_f (Nm)	191.7	106.0	294.4	46.3
H/Q ratio (%)	46.1	32.1	61.0	6.5
$(T_e + T_f)$ / BM (Nm/kg)	7.9	5.5	12.1	1.5

SD—standard deviation; BMI—body mass index; b-vs—sitting height; b-tro—lower extremity length; b-ti—shank length; a-a—biacromial diameter; ic-ic—biiliocristal diameter; cl-cm—humerus breadth; epl-epm—femur breath; T_e—peak torque value of knee joint extensors; T_f—peak torque value of knee joint flexors; H/Q ratio—conventional hamstrings-to-quadriceps ratio; $(T_e + T_f)$ / BM—sum of knee muscle torque relative to body mass.

Table 2. Statistical characteristics of the variables studied in separate clusters.

Variables	Group 1 (weaker) $n = 14$	Group 2 (stronger) $n = 23$	t	p
Body height (cm)	180.5 ± 5.0	182.2 ± 5.8	−0.95	0.349
Body mass (kg)	72.8 ± 5.1	79.2 ± 7.3	−2.91	0.006 *
BMI (kg/m^2)	22.4 ± 1.5	23.8 ± 1.3	−3.13	0.004 *
b-vs (cm)	94.7 ± 2.8	95.6 ± 3.5	−0.86	0.395
b-tro (cm)	95.6 ± 4.0	95.7 ± 4.0	−0.06	0.951
b-ti (cm)	48.4 ± 2.1	48.5 ± 3.1	−0.07	0.945
a-a (cm)	41.3 ± 1.8	42.1 ± 1.9	−1.33	0.191
ic-ic (cm)	28.4 ± 1.5	28.8 ± 2.0	−0.77	0.447
cl-cm (cm)	7.0 ± 0.3	7.2 ± 0.3	−2.60	0.014 *
epl-epm (cm)	9.9 ± 0.3	10.2 ± 0.5	−1.52	0.037 *
Chest girth (cm)	86.6 ± 5.2	89.6 ± 4.4	−1.83	0.076
Waist girth (cm)	78.0 ± 4.0	80.4 ± 4.9	−1.55	0.131
Relaxed biceps girth (cm)	29.8 ± 1.6	30.7 ± 2.7	−1.15	0.260
Flexed biceps girth (cm)	33.1 ± 1.8	33.6 ± 2.6	−0.67	0.505
Hip girth (cm)	97.3 ± 2.7	99.3 ± 3.8	−1.74	0.091
Thigh girth (cm)	57.1 ± 2.8	59.3 ± 2.8	−2.29	0.028 *
Calf girth (cm)	36.9 ± 2.0	38.5 ± 1.4	−2.83	0.008 *
Subscapular skinfold (mm)	7.5 ± 1.1	8.5 ± 1.5	−2.15	0.039 *
Triceps skinfold (mm)	4.4 ± 1.3	4.2 ± 0.9	0.36	0.719
Suprailiac skinfold (mm)	6.1 ± 1.1	7.2 ± 1.9	−1.97	0.057
Abdominal skinfold (mm)	6.4 ± 1.8	7.4 ± 2.3	−1.29	0.206
Calf skinfold (mm)	3.5 ± 0.5	3.9 ± 0.7	−1.95	0.060
Right hand grip strength (kG)	44.4 ± 5.5	49.4 ± 6.1	−2.53	0.016 *
Left hand grip strength (kG)	43.6 ± 4.6	47.8 ± 4.9	−2.56	0.015 *
Back strength (kG)	112.4 ± 11.9	135.9 ± 16.5	−4.62	0.000 *
T_e (Nm)	325.6 ± 37.2	472.2 ± 53.9	−8.93	0.000 *
T_f (Nm)	152.3 ± 26.1	215.6 ± 39.0	−5.37	0.000 *
H/Q ratio	47.0 ± 7.4	45.6 ± 6.0	0.61	0.546
$(T_e + T_f)$ / BM (Nm/kg)	6.6 ± 1.3	8.7 ± 0.8	−5.60	0.000 *

*—significant differences between Groups I and II group at the level of $p < 0.05$; p—probability of type I error; t—t-test value; BMI—body mass index; b-vs—sitting height; b-tro—lower extremity length; b-ti—shank length; a-a—biacromial diameter; ic-ic—biiliocristal diameter; cl-cm—humerus breadth; epl-epm—femur breath; T_e—peak torque value of knee joint extensors; T_f—peak torque value of knee joint flexors; H/Q ratio—conventional hamstrings-to-quadriceps ratio; $(T_e + T_f)$ / BM—sum of knee muscle torques relative to body mass.

Table 3. The correlation coefficients between the muscle torque of extensors (T_e) and flexors (T_f) of the knee joint and anthropometric variables for the entire group of soccer players (n = 37).

Anthropometric Variables	T_e	T_f	Anthropometric Variables	T_e	T_f
Body height	0.21	0.12	Flexed biceps girth	0.10	0.06
Body mass	0.46 *	0.34 *	Hip girth	0.26	0.13
BMI	0.45 *	0.37 *	Thigh girth	0.26	0.20
b-vs	0.21	0.18	Calf girth	0.40 *	0.31
b-tro	0.14	−0.01	Subscapular skinfold	0.34 *	0.14
b-ti	0.07	−0.09	Triceps skinfold	0.12	0.12
a-a	0.16	−0.02	Suprailiac skinfold	0.36 *	0.23
ic-ic	0.16	0.05	Abdominal skinfold	0.19	0.14
cl-cm	0.34 *	0.21	Calf skinfold	0.38 *	0.29
epl-epm	0.43 *	0.22	Right handgrip strength	0.33 *	0.27
Chest girth	0.31	0.13	Left handgrip strength	0.35 *	0.26
Waist girth	0.27	0.08	Back strength	0.56 *	0.25
Relaxed biceps girth	0.27	0.17			

*—statistically significant at $p < 0.05$; BMI—body mass index; b-vs—sitting height; b-tro—lower extremity length; b-ti—shank length; a-a—biacromial diameter; ic-ic—biiliocristal diameter; cl-cm—humerus breadth; epl-epm—femur breath.

Multiple regression analysis was performed, taking into account body height and anthropometric variables that were most significantly correlated with the examined muscle torque. A significant effect of anthropometric variables on the muscle torque measured in the knee joint was demonstrated. A regression model with four anthropometric variables estimating the biomechanical values was statistically significant, and all predictors explained 38% of the torque variation in the knee extensors and 25% of the torque variation in the knee flexors. Three of the four predictors in the model had a significant effect on the torque of the knee extensors and flexors. The results of the regression analysis are presented in Tables 4 and 5. The estimation error for the extensor Equation (3) was 72.3, whereas for the flexor Equation (4), this value was 42.5.

$$T_e = 1943.845 - (10.052 \cdot \text{body height}) - (25.424 \cdot \text{thigh girth}) + (13.383 \cdot \text{calf girth}) + (16.652 \cdot \text{body mass}). \quad (3)$$

$$T_f = 1019.404 - (5204 \cdot \text{body height}) - (11{,}878 \cdot \text{thigh girth}) + (5715 \cdot \text{calf girth}) + (7756 \cdot \text{body mass}). \quad (4)$$

Table 4. Results of regression analysis for knee joint extensor torque.

Independent Variables	Dependent Variable: Peak Torque Value of Knee Joint Extensors $R = 0.62$, $R^2 = 0.38$, Adj. $R^2 = 0.30$ F (4.32) = 4.86, $p < 0.01$ Standard Error of the Estimate: 72.3			
	b	Standard Error of b	t	p
Intercept coefficient *	1943.9	880.1	2.2	0.035
Body height *	−10.1	4.3	−2.4	0.025
Thigh girth *	−25.4	9.4	−2.7	0.011
Calf girth	13.4	9.8	1.4	0.180
Body mass *	16.7	5.1	3.3	0.003

*—statistically significant at $p < 0.05$.

Table 5. Results of regression analysis for knee joint flexor torque.

Independent Variables	Dependent Variable: Peak Torque Value of Knee Joint Flexors $R = 0.50$, $R^2 = 0.25$, Adj. $R^2 = 0.16$, F (4.32) = 2.71, $p < 0.05$ Standard Error of the Estimate: 42.5			
	b	Standard Error of b	t	p
Intercept coefficient	1019.4	516.7	2.0	0.057
Body height *	−5.2	2.5	−2.1	0.047
Thigh girth *	−11.9	5.5	−2.2	0.038
Calf girth	5.7	5.7	1.0	0.326
Body mass *	7.8	3.0	2.6	0.015

*—statistically significant at $p < 0.05$.

4. Discussion

The body build of the soccer players studied showed a typical profile of basic anthropometric variables typical for athletes participating in this sport. Interpersonal differences resulted from the different playing positions [34,35]. The mean body height of the soccer players studied (1.82 ± 0.06 m) was similar to the mean body height of the players of most of the best teams in the world and slightly exceeds the mean body height of young men in Poland (1.79 ± 0.07 m) [36]. Therefore, the results obtained in the study are consistent with the information provided in the literature on the body morphology of athletes. Norton and Olds [32] indicated that the mean body height of soccer players (1.79 ± 0.06 m) and its variations are very close to those of the general population. This finding was also confirmed by the research presented by Reilly and Doran [17]. However, these authors pointed to differences depending on sports skill level and nationality. The extreme values in their ranking were obtained from the players from Hong Kong (1.73 ± 0.06 m) and Norway (1.81 ± 0.05 m). The BMI of soccer players in the present study also showed values typical of soccer players [17]. The athletes tested were characterized by good musculature of all body parts, as evidenced by the values of chest and lower and upper limb girths. These features are similar to those of players on other soccer teams [34,35]. The results of our research indicated particularly muscular thighs, which enable players to generate high muscle torque. This ability is also very important when jumping, kicking, rotating, and changing the velocity of movement [14,17,19]. The harmonious development of limb and torso muscles contributes to maintaining player balance on a slippery field and allows for easier control of the ball [17]. Particularly valuable research, whose results can help athletes achieve top-level performance, was carried out among the players of the top teams. Therefore, it seems very valuable that we had an opportunity to conduct research on the Polish first-division soccer players.

Analysis of the muscle torque of knee extensors and flexors shows that human muscles are shaped to a large extent by the gravitational field. This finding is reflected in the significant muscle torque of the knee joint extensors (416.8 ± 86.4 Nm), which act against the gravitational force. The torque values of the knee flexors are approximately two times lower (191.7 ± 46.3 Nm). The mean values of isometric muscle torque in the knee joint obtained for the players in our study are much higher than the mean values obtained in Polish striker soccer players (315.3 ± 53.3 Nm for extensors and 167.7 ± 26.0 Nm for flexors) [37] and higher than those in lower-league soccer players (328.0 ± 69.7 Nm for extensors) [38]. The recorded large values of isometric muscle torque measured in our study may result from substantial training loads to which players were exposed. They trained four times a week in two sessions. In the morning, the athletes had a soccer training session (about 2.5 hours). In the afternoon, they exercised in the gym for about 1.5 hours. The exercises were selected individually. The high sports skill level of the players studied is also of great importance to the results obtained. Previous studies [1,39] indicated differences in the magnitude of muscle torque depending on the sports skill level. Measurements of muscle torque under isokinetic conditions are now also popular. However, these measurements have limitations [33,40]. Furthermore, the aim of the study was to determine the maximum strength abilities of the players, rather than the strength at a certain angular velocity of movement in the knee joint.

The relationship between the muscle torque of extensors and flexors of the knee joint is important from the standpoint of maintaining knee stability during some movement activities. For example, damage to the hamstrings can occur during the kicking of the ball with the lower limb. This damage may occur when, during knee extension, a group of antagonistic muscles does not ensure the effective eccentric deceleration of the movement [8,41,42]. It is assumed that conventional H/Q ratio values exceeding 60% can effectively prevent injuries and damage to the anterior cruciate ligament (ACL) and hamstring strains [43–48]. According to Kim and Hong [44], soccer players with a conventional H/Q ratio of more than 60% are less likely to suffer non-contact injuries to their lower limbs. In this study, the conventional H/Q ratio was 46.1% for all the soccer players studied, which is lower than the desired value and indicates unfavourable proportions of muscle torque in these muscle groups. This result suggests that coaches should pay more attention to the improvement of the strength level of the knee flexors during training. No injuries were reported the last time among the players studied. However, it can be presumed that this was influenced by motor control, which means mutual adjustment of the musculoskeletal and nervous systems [49]. In order to control movement, the central nervous system must integrate multiple sensory information (both external and proprioceptive) and trigger the signals necessary for muscle recruitment to achieve the intended goal. The performance of the movement consists in the planning of the motor strategy and its economic and flawless execution. Disturbed motor control has a negative effect on the quality of movement and increases the risk of injury. It should be expected that in elite players, sport training had led to the improved motor control. Disproportionate torque of agonist and antagonist muscle groups has been observed in male soccer players [27,37,48]. However, there are groups of soccer players for which the value of the conventional H/Q ratio remains within the recommended standard [45]. Cometti et al. [39] demonstrated that sports skill level in soccer players increases with higher H/Q ratios.

The lack of significant relationships between the conventional H/Q ratio and anthropometric variables indicates that low values of the conventional H/Q ratio in the group of soccer players (i.e., not entirely correct proportion of torque of antagonistic muscle groups) cannot be observed based on the variables describing body structure. Therefore, muscle imbalance cannot be diagnosed without special biomechanical measurements because the body build will not suggest such an abnormality. It should also be noted that high H/Q ratios may exist in the presence of relative weakness in both the hamstring and quadriceps muscle groups [50], which can also explain the lack of the above relationships. Another limitation may be that the peak values of extensors and flexors occur at different knee flexion angles [26,33].

Dividing the soccer players studied into two clusters allowed for the assessment of body build differences between the players, who significantly differed in terms of muscle torque generated within the knee joint. The athletes from both clusters did not show any differences in body height or its components, which is attributable to the genetic determinants of this feature [51]. Furthermore, the differences observed in body mass and girths that describe the size of the skeleton and muscles indicate relationships between these features and strength abilities. This is also confirmed by significant correlation coefficients for some anthropometric features (body mass, humerus breadth, femur breadth, calf girth, subscapular skinfold, suprailiac skinfold, and calf skinfold) and isometric muscle torque of the knee joint extensors and flexors. As in other studies, positive relationships were obtained between muscle torque and body mass, breadths and girths that describe the size of the skeleton and muscles [27–31]. However, contrary to previous reports, positive relationships between muscle torque and skinfolds were also found [28].

The regression analysis was used to identify the variables that can be indirectly used to estimate torque values for knee joint extensors and flexors generated by the player. The results of this analysis identify the body height, body mass, and thigh and calf girth as the features that best describe these variables. Therefore, based on these anthropometric variables, the coach may initially assess the strength abilities of soccer players, and these assessments can lead to the adjustment of training programmes to the individual predispositions of the players.

The limitations of our study are incomplete information about the applied training loads (resulting, e.g., from club changes of the players studied), the sex of the players (slightly different results can be expected in female groups, which are characterized by higher body fat), and the type of sport being practised (it is not advisable to directly refer to the results of people practising other sports, e.g., individual sports). On the other hand, all of the players studied belonged to the national soccer elite, which allows for the assumption that their training was carried out professionally, translating into a high sports skill level.

5. Conclusions

The results allow for the conclusion that the muscle torque of the knee joint extensors and flexors in soccer players is significantly related to some anthropometric variables. Due to the less complicated nature of anthropometric apparatuses and the high mobility of this type of equipment, the prediction of muscle torque in the knee joint based on anthropometry seems to be useful for screening soccer players for preliminary estimation of the abovementioned biomechanical variables. However, anthropometric methods do not provide full information about the mutual relationships between the muscle torque of the knee joint extensors and flexors, which is very important from the standpoint of injury prevention. For this reason, measurements using special biomechanical equipment are also necessary for the comprehensive analyses and control of the effects of sports training.

Author Contributions: Conceptualization, J.P. and B.P.; methodology, J.P. and B.P.; software, J.P. and B.P.; validation, J.P., A.S. (Artur Struzik), A.B., A.S (Aleksandra Stachoń). and B.P.; formal analysis, J.P., A.S. (Artur Struzik) and B.P.; investigation, J.P., A.B., A.S. (Aleksandra Stachoń) and B.P.; resources, J.P., A.B., A.S. (Aleksandra Stachoń) and B.P.; data curation, J.P. and A.S. (Artur Struzik); writing—original draft preparation, J.P., A.S. (Artur Struzik) and B.P.; writing—review and editing, J.P., A.S. (Artur Struzik) and B.P.; visualization, J.P. and A.S. (Artur Struzik); supervision, J.P.; project administration, J.P.; and funding acquisition, B.P. All authors have read and agreed to the published version of the manuscript.

Funding: This research received no external funding.

Acknowledgments: The authors thank the study participants for their effort, time devotion, and collaboration during the study.

Conflicts of Interest: The authors declare no conflict of interest. The funders had no role in the design of the study; in the collection, analyses, or interpretation of data; in the writing of the manuscript; or in the decision to publish the results.

References

1. Wit, A.; Eliasz, J.; Gajewski, J.; Janiak, J.; Jaszczuk, J.; Trzaskoma, Z. Maximal isometric muscle torque assessment in elite athletes. *Acta Bioeng. Biomech.* **2002**, *4*, 591–592.
2. Ergün, M.; Işlegen, C.; Taşkiran, E. A cross-sectional analysis of sagittal knee laxity and isokinetic muscle strength in soccer players. *Int. J. Sports Med.* **2004**, *25*, 594–598. [CrossRef] [PubMed]
3. Zakas, A. Bilateral isokinetic peak torque of quadriceps and hamstring muscles in professional soccer players with dominance on one or both two sides. *J. Sports Med. Phys. Fitness* **2006**, *46*, 28–35. [PubMed]
4. Ellenbecker, T.S.; Roetert, E.P.; Sueyoshi, T.; Riewald, S. A descriptive profile of age-specific knee extension flexion strength in elite junior tennis players. *Br. J. Sports Med.* **2007**, *41*, 728–732. [CrossRef]
5. Ruas, C.V.; Minozzo, F.; Pinto, M.D.; Brown, L.E.; Pinto, R.S. Lower-extremity strength ratios of professional soccer players according to field position. *J. Strength Cond. Res.* **2015**, *29*, 1220–1226. [CrossRef]
6. Lech, G.; Chwała, W.; Ambroży, T.; Sterkowicz, S. Muscle torque and its relation to technique, tactics, sports level and age group in judo contestants. *J. Hum. Kinet.* **2015**, *45*, 167–175. [CrossRef]
7. Sterkowicz, S.; Lech, G.; Sterkowicz-Przybycień, K.; Chwała, W.; Ambroży, T.; Pałka, T. Relationship of maximal isometric torque produced in flexors and extensors rate to technique by judo athletes. *Acta Bioeng. Biomech.* **2018**, *20*, 65–71.
8. Croisier, J.L.; Ganteaume, S.; Binet, J.; Genty, M.; Ferret, J.M. Strength imbalances and prevention of hamstring injury in professional soccer players: A prospective study. *Am. J. Sports Med.* **2008**, *36*, 1469–1475. [CrossRef]
9. Ruas, V.C.; McManus, T.R.; Bentes, M.C.; Costa, B.P. Acute effects of proprioceptive neuromuscular facilitation on peak torque and muscle imbalance. *J. Funct. Morphol. Kinesiol.* **2018**, *3*, 63. [CrossRef]
10. Bamaç, B.; Çolak, T.; Özbek, A.; Çolak, S.; Cinel, Y.; Yenigün, Ö. Isokinetic performance in elite volleyball and basketball players. *Kinesiology* **2008**, *40*, 183–189.
11. Hadzic, V.; Sattler, T.; Markovic, G.; Veselko, M.; Dervisevic, E. The isokinetic strength profile of quadriceps and hamstrings in elite volleyball players. *Isokinet. Exerc. Sci.* **2010**, *18*, 31–37. [CrossRef]
12. Teixeira, J.; Carvalho, P.; Moreira, C.; Carneiro, A.; Santos, R. Muscle strength assessment of knee flexors and extensors. Comparative study between basketball, football, handball and volleyball athletes. *Int. J. Sports Sci.* **2015**, *5*, 192–200.
13. Jones, P.A.; Thomas, C.; Dos'Santos, T.; McMahon, J.J.; Graham-Smith, P. The role of eccentric strength in 180° turns in female soccer players. *Sports* **2017**, *5*, 42. [CrossRef] [PubMed]
14. Struzik, A.; Pietraszewski, B. Relationships between hamstrings-to-quadriceps ratio and variables describing countermovement and drop jumps. *Appl. Bionics Biomech.* **2019**, *2019*, 4505481. [CrossRef] [PubMed]

15. Zatsiorsky, V.M.; Kraemer, W.J. *Science and Practice of Strength Training*, 2nd ed.; Human Kinetics: Champaign, IL, USA, 2006; pp. 17–46.
16. Reilly, T.; Bangsbo, J.; Franks, A. Anthropometric and physiological predispositions for elite soccer. *J. Sports Sci.* **2000**, *18*, 669–683. [CrossRef] [PubMed]
17. Reilly, T.; Doran, D. Fitness assessment. In *Science and Soccer*; Reilly, T., Williams, A.M., Eds.; Routledge: London, UK; New York, NY, USA, 2003; pp. 21–46.
18. Pietraszewski, B.; Siemieński, A.; Bober, T.; Struzik, A.; Rutkowska-Kucharska, A.; Nosal, J.; Rokita, A. Lower extremity power in female soccer athletes: A pre-season and in-season comparison. *Acta Bioeng. Biomech.* **2015**, *17*, 129–135.
19. Bona, C.C.; Filho, H.T.; Izquierdo, M.; Ferraz, R.M.P.; Marques, M.C. Peak torque and muscle balance in the knees of young U-15 and U-17 soccer athletes playing various tactical positions. *J. Sports Med. Phys. Fitness* **2017**, *57*, 923–929.
20. Zago, M.; Sforza, C.; Dolci, C.; Tarabini, M.; Galli, M. Use of machine learning and wearable sensors to predict energetics and kinematics of cutting maneuvers. *Sensors* **2019**, *19*, 3094. [CrossRef]
21. Newman, M.A.; Tarpenning, K.M.; Marino, F.E. Relationships between isokinetic knee strength, single-sprint performance, and repeated-sprint ability in football players. *J. Strength Cond. Res.* **2004**, *18*, 867–872.
22. Stølen, T.; Chamari, K.; Castagna, C.; Wisløff, U. Physiology of soccer: An update. *Sports Med.* **2005**, *35*, 501–536. [CrossRef]
23. Cerrah, A.O.; Gungor, E.O.; Soylu, A.R.; Ertan, H.; Lees, A.; Bayrak, C. Muscular activation patterns during the soccer in-step kick. *Isokinet. Exerc. Sci.* **2011**, *19*, 181–190. [CrossRef]
24. Dvir, Z.; Eger, G.; Halperin, N.; Shklar, A. Thigh muscle activity and anterior cruciate ligament insufficiency. *Clin. Biomech.* **1989**, *4*, 87–91. [CrossRef]
25. Aagaard, P.; Simonsen, E.B.; Magnusson, S.P.; Larsson, B.; Dyhre-Poulsen, P. A new concept for isokinetic hamstring: Quadriceps muscle strength ratio. *Am. J. Sports Med.* **1998**, *26*, 231–237. [CrossRef] [PubMed]
26. Struzik, A.; Siemieński, A.; Bober, T.; Pietraszewski, B. Ratios of torques of antagonist muscle groups in female soccer players. *Acta Bioeng. Biomech.* **2018**, *20*, 153–158.
27. Pietraszewski, B.; Zawadzki, J.; Pietraszewska, J.; Burdukiewicz, A. Body composition and muscle torques of lower limbs. *Biol. Sport* **1997**, *14*, 104–107.
28. Lewandowska, J.; Buśko, K.; Pastuszak, A.; Boguszewska, K. Somatotype variables related to muscle torque and power in judoists. *J. Hum. Kinet.* **2011**, *30*, 21–28. [CrossRef]
29. Kim, S.E.; Hong, J.; Cha, J.Y.; Park, J.M.; Eun, D.; Yoo, J.; Jee, Y.S. Relative appendicular skeletal muscle mass is associated with isokinetic muscle strength and balance in healthy collegiate men. *J. Sports Sci.* **2016**, *34*, 2114–2120. [CrossRef]
30. Camic, C.L.; Housh, T.J.; Mielke, M.; Zuniga, J.; Hendrix, C.; Johnson, G.O.; Housh, D.J.; Schmidt, R.J. Validity of fat-free weight equations for predicting isokinetic peak torque in young wrestlers. *Int. J. Sport Sci. Health* **2018**, *5*, 69–78.
31. Norsuriani, S.; Ooi, F. Bone health status, isokinetic muscular strength and power, and body composition of Malay adolescent female silat and taekwondo practitioners. *Int. J. Public Health Clin. Sci.* **2018**, *5*, 244–262.
32. Norton, K.; Olds, T. *Anthropometrica: A Textbook of Body Measurement for Sports and Health Courses*; UNSW Press: Sydney, Australia, 2002.
33. Cozette, M.; Leprêtre, P.-M.; Doyle, C.; Weissland, T. Isokinetic strength ratios: Conventional methods, current limits and perspectives. *Front. Physiol.* **2019**, *10*, 567. [CrossRef]
34. Matković, B.R.; Mišigoj-Duraković, M.; Matković, B.; Janković, S.; Ružić, L.; Leko, G.; Kondrič, M. Morphological differences of elite Croatian soccer players according to the team position. *Coll. Antropol.* **2003**, *27*, 167–174. [PubMed]
35. Hazir, T. Physical characteristics and somatotype of soccer players according to playing level and position. *J. Hum. Kinet.* **2010**, *26*, 83–95. [CrossRef]
36. Kulaga, Z.; Litwin, M.; Tkaczyk, M.; Różdżyńska, A.; Barwicka, K.; Grajda, A.; Świąder, A.; Gurzkowska, B.; Napieralska, E.; Pan, H. The height-, weight-, and BMI-for-age of Polish school-aged children and adolescents relative to international and local growth references. *BMC Public Health* **2010**, *10*, 109. [CrossRef] [PubMed]
37. Buśko, K.; Górski, M.; Nikolaidis, P.T.; Mazur-Różycka, J.; Łach, P.; Staniak, Z.; Gajewski, J. Leg strength and power in Polish striker soccer players. *Acta Bioeng. Biomech.* **2018**, *20*, 109–116.

38. Váczi, M.; Tollár, J.; Meszler, B.; Juhász, I.; Karsai, I. Short-term high intensity plyometric training program improves strength, power and agility in male soccer players. *J. Hum. Kinet.* **2013**, *36*, 17–26. [CrossRef]
39. Cometti, G.; Maffiuletti, N.A.; Pousson, M.; Chatard, J.C.; Maffulli, N. Isokinetic strength and anaerobic power of elite, subelite and amateur French soccer players. *Int. J. Sports Med.* **2001**, *22*, 45–51. [CrossRef]
40. Zawadzki, J.; Bober, T.; Siemieński, A. Validity analysis of the biodex system 3 dynamometer under static and isokinetic conditions. *Acta Bioeng. Biomech.* **2010**, *12*, 25–32.
41. Struzik, A.; Pietraszewski, B.; Bober, T. The biodex system used in the evaluation of the proportion of muscle torque and hamstring muscle injury risk. *Pol. J. Sports Med.* **2015**, *1*, 11–17.
42. Chen, Y.; Li, J.X.; Hong, Y.; Wang, L. Plantar stress-related injuries in male basketball players: Variations on plantar loads during different maximum-effort maneuvers. *Biomed. Res. Int.* **2018**, *2018*, 4523849. [CrossRef]
43. Hewett, T.E.; Myer, G.D.; Zazulak, B.T. Hamstrings to quadriceps peak torque ratios diverge between sexes with increasing isokinetic angular velocity. *J. Sci. Med. Sport* **2008**, *11*, 452–459. [CrossRef]
44. Kim, D.; Hong, J. Hamstring to quadriceps strength ratio and noncontact leg injuries: A prospective study during one season. *Isokinet. Exerc. Sci.* **2011**, *19*, 1–6. [CrossRef]
45. Cheung, R.T.; Smith, A.W.; Wong del, P. H:Q ratios and bilateral leg strength in college field and court sports players. *J. Hum. Kinet.* **2012**, *33*, 63–71. [CrossRef] [PubMed]
46. Dervišević, E.; Hadžić, V. Quadriceps and hamstrings strength in team sports: Basketball, football and volleyball. *Isokinet. Exerc. Sci.* **2012**, *20*, 293–300. [CrossRef]
47. Orchard, J.W.; Driscoll, T.; Seward, H.; Orchard, J.J. Relationship between interchange usage and risk of hamstring injuries in the Australian Football League. *J. Sci. Med. Sport* **2012**, *15*, 201–206. [CrossRef] [PubMed]
48. Daneshjoo, A.; Rahnama, N.; Mokhtar, A.H.; Yusof, A. Bilateral and unilateral asymmetries of isokinetic strength and flexibility in male young professional soccer players. *J. Hum. Kinet.* **2013**, *36*, 45–53. [CrossRef]
49. Latash, M. *Fundamentals of Motor Control*; Academic Press: Amsterdam, The Netherlands, 2012.
50. Wilkerson, G.B.; Colston, M.A.; Short, N.I.; Neal, K.L.; Hoewischer, P.E.; Pixley, J.J. Neuromuscular changes in female collegiate athletes resulting from a plyometric jump-training program. *J. Athl. Train.* **2004**, *39*, 17–23.
51. Silventoinen, K.; Kaprio, J.; Lahelma, E.; Koskenvuo, M. Relative effect of genetic and environmental factors on body height: Differences across birth cohorts among Finnish men and women. *Am. J. Public Health* **2000**, *90*, 627–630.

applied sciences

MDPI

Article

Effect of Landing Posture on Jump Height Calculated from Flight Time

Daichi Yamashita [1,*], Munenori Murata [1,2] and Yuki Inaba [1]

1 Japan Institute of Sports Sciences, 3-15-1, Nishigaoka, Kita-ku, Tokyo 115-0056, Japan; mmurata@nifs-k.ac.jp (M.M.); yuki.inaba@jpnsport.go.jp (Y.I.)

2 National Institute of Fitness and Sports in Kanoya; 1 Shiromizucho, Kanoya, Kagoshima 891-2393, Japan

* Correspondence: daichi.yamashita@jpnsport.go.jp

Received: 31 December 2019; Accepted: 20 January 2020; Published: 22 January 2020

check for updates

Abstract: Flight time is widely used to calculate jump height because of its simple and inexpensive application. However, this method is known to give different results than the calculation from vertical velocity at takeoff. The purpose of this study is to quantify the effect of postural changes between takeoff and landing on the jump height from flight time. Twenty-seven participants performed three vertical jumps with arm swing. Three-dimensional coordinates of anatomical landmarks and the ground reaction force were analyzed. Two methods of calculating jump height were used: (1) the vertical velocity of the whole-body center of mass (COM_{wb}) at takeoff and (2) flight time. The jump height from flight time was overestimated by 0.025 m compared to the jump height from the takeoff velocity ($p < 0.05$) due to the lower COM_{wb} height at landing by −0.053 m ($p < 0.05$). The postural changes in foot, shank, and arm segments mainly contributed to decreasing the COM_{wb} height (−0.025, −0.014, and −0.017 m, respectively). The flight time method is reliable and had low intra-participant variability, but it cannot be recommended for a vertical jump when comparing with others (such as at tryouts) because of the potential "cheating" effect of differences in landing posture.

Keywords: flight time; vertical jump; center of mass; landing

---

1. Introduction

Jumping ability is regarded as one of the most important aspects of many sports. Vertical jump measurement is a method to assess lower limb power [1], strength [2], and neuromuscular status [3]. Therefore, the vertical jump test has been used to assess the impact of training [2] and to select high-level players at tryouts in many sports such as American football [4] and basketball [5].

The force platform is one of the most widely used methods of vertical jump measurement and is considered the gold standard for determining the mechanical outputs of jumping [6]. Force platforms are used to measure the ground reaction force (GRF) and derive the velocity of the whole-body center of mass (COM_{wb}) using the impulse–momentum relationship. However, they are costly for sports teams and strength coaches, so their use is limited mainly to university laboratories and research institutes.

Recently, the calculation of jump height from flight time using a contact mat, a photoelectric cell, and a smartphone that utilizes a high-speed camera application has become increasingly popular due to its low cost and straightforward assessment methods. In this method, jump height is calculated using a uniform acceleration equation. The equation justifies the method only if the height of the COM_{wb} is the same at takeoff and landing. It has been reported, however, that the method overestimates the countermovement jump without arm swing (CMJ) height by 2% [7], 3–4% [8], 8% [8], and 11% [9] compared to the method using vertical velocity at takeoff from a force platform. These results suggest that the height of the COM_{wb} at landing is lower than that at takeoff, making the flight time longer. Consequently, the jump height from flight time is overestimated.

One potential determining factor for this difference, suggested by the previous studies, is that participants landed with their lower limbs partially bent, resulting in an inflated flight time [8,9]. Kibele [7] showed that knee and ankle joints were more flexed, and the COM_{wb} height was lower at landing than at takeoff. Also, a different arm posture at takeoff and landing seems to affect the difference in the COM_{wb} height when arm swing is permitted [10]. Previous studies reported that the COM_{wb} height at takeoff in the vertical jump with arm swing (VJ) was 0.024 m [11] and 0.034 m [12] higher than that in CMJ.

The flight time method of calculating jump height is widely used by laboratories and sports teams, even though many researchers have acknowledged the postural differences at takeoff and landing. However, there have been few studies which have aimed to understand the sources of error in jump height from flight time. In order to fully understand the sources of error, it is helpful to quantify the relationship between the postural difference and the difference in the COM_{wb} height. This is because the height of the center of mass of a system is given by a mathematical formulation: the mass-weighted average of the heights of the segments. Therefore, the purpose of this study is to quantify the effect of postural changes in each segment on the COM_{wb} height difference between takeoff and landing. We hypothesized that lower limb bending and arm movement are the primary factors that affect the overestimation of jump height from flight time. Understanding the sources of error in jump height from flight time would be useful for better instruction to reduce systematic bias and interpersonal variability when using the simple and low-cost method of vertical jump measurement.

2. Materials and Methods

2.1. Participants

Twenty-four males and three females (age: 19 to 42 years; height: 1.77 ± 0.11 m; mass: 75.3 ± 11.9 kg) participated in this experiment. They provided written informed consent to undergo the experimental procedures, which were conducted in accordance with the Declaration of Helsinki and were approved by the ethics committee of the Japan Institute of Sports Sciences (H29-0065).

2.2. Instrumentation

Three-dimensional coordinates of the anatomical landmarks were acquired using a 3D optical motion capture system with ten cameras (500 Hz; Vicon, Oxford, UK). Forty-seven reflective markers were placed on each participant's body—the same as in the previous study [13]. All kinematic data was filtered and interpolated using a Woltring quintic spline [14]. To choose the optimal cut-off frequency of 4.6–7 Hz, a residual analysis was performed [15]. Participants wore their athletic shoes. GRF data was obtained at 1000 Hz using two force platforms (0.9 m × 0.6 m, type 9287B; Kistler, Winterthur, Switzerland).

2.3. Procedures

Participants performed three maximal VJs after warm-up and familiarization. They were instructed to stand upright and motionless for 1 s then began the movement of the jump. They were required not to bend their lower limbs before landing. Two or more experimenters watched each trial, and if they noted that the requirement was not met (i.e., leg tucking), the trial was repeated.

2.4. Data Reduction

Two methods of calculating jump height were used: (1) the vertical velocity of the COM_{wb} at takeoff, and (2) flight time. The vertical GRFs (F_{ver}) were integrated by trapezoid rule integration to estimate the vertical velocity [6]. The vertical velocity at takeoff (V_{to}) was calculated using the following equation:

$$V_{to} = \frac{1}{m_{wb}} \int_{t_{st}}^{t_{to}} (F_{ver} - m_{wb}g)\mathrm{d}t \tag{1}$$

where m_{wb}, F_{ver}, g, t_{st}, and t_{to} represent the body mass, vertical GRF, gravitational acceleration (9.806 m/s^2 [16]), the time of the start of the initial jumping motion, and the time of its termination at takeoff, respectively. The body mass was calculated by averaging F_{ver} over the 0.3 s quiet stance [7] and dividing by gravitational acceleration. We confirmed that the coefficient of variance (CV) of F_{ver} during the quiet phase in each trial was low (less than 1%). The start of the motion was identified as the first F_{ver} detected to deviate above or below body weight by 1%. To eliminate the influence of inter-participant variance in body weight, takeoff and landing times were defined as the first intersection of F_{ver} with 1% of body weight (7.4 ± 1.2 N, range 5.0 to 9.9 N). The jump height from V_{to} (H_v) was calculated using the following equation:

$$H_v = \frac{1}{2g} V_{to}^2. \tag{2}$$

H_v was used in this study as the criterion for comparison. Jump height from flight time (H_t) was calculated using the following equation:

$$H_t = \frac{1}{8} g t_{flight}^2 \tag{3}$$

where t_{flight} represents the flight time (see Appendix A).

The COM_{wb} position was calculated as the weighted sum of a 15-segment model (i.e., head, upper trunk, lower trunk, upper arms, forearms, hands, thighs, shanks, and feet) based on body-segment parameters [17]. To compare the difference in the whole-body posture between takeoff and landing, we used a seven-segment model of the head, arm, upper trunk, lower trunk, thigh, shank, and foot (Figure 1). The positions of the arm, thigh, shank, and foot segments were the average of the right and left side.

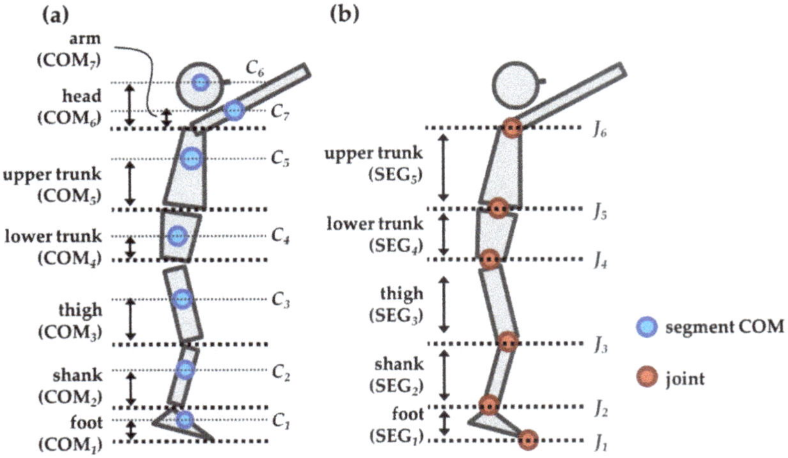

Figure 1. The definition of (**a**) vertical component of the segment center of mass length (COM_k) and (**b**) vertical component of the segment length (SEG_k).

Once an object is projected into the air, the COM of the system must follow a parabolic trajectory, and the trajectory cannot be altered in the air until landing. When the position of a segment moves relative to the COM_{wb}, it affects the other segments' positions relative to the COM_{wb} to keep the COM_{wb} trajectory constant. As a result, the difference in a segment posture influences the COM_{wb} height at landing. To understand the effects of the postural difference between takeoff and landing on the COM_{wb} height, we quantified the contributions of the changes in the vertical component of each segment on the COM_{wb} height. When one segment changes its posture, it affects (1) the segment COM height and (2) the COM height of all segments above it. We defined the vertical component of the segment COM length (COM_k) by the following equations:

$$COM_k = h_{C_k/J_k} \ (k = 1 \text{ to } 6) \tag{4a}$$

$$COM_k = h_{C_k/J_6} \ (k = 7) \tag{4b}$$

where k represents the segment number (see Figure 1) and h_{C_k/J_k} represents the height from the lower edge point (joint) of the segment to the segment COM (Figure 1a). COM_7 (i.e., the arm segment) was defined relative to the proximal joint (the suprasternal notch). In the same way, we defined the vertical component of the segment length (SEG_k) by the following equation:

$$SEG_k = h_{J_{k+1}/J_k} \tag{5}$$

where h_{J_{k+1}/J_k} represents the height from the lower edge point (joint) of the segment to the proximal joint (Figure 1b). Then, we calculated the contributions for all seven segments ($CONT_k$) using the following equations:

$$CONT_k = \frac{m_k}{m_{wb}} \Delta h_{C_k/J_k} + \sum_{i=k+1}^{7} \frac{m_i}{m_{wb}} \Delta h_{J_{k+1}/J_k} \ (k = 1 \text{ to } 5) \tag{6a}$$

$$CONT_k = \frac{m_k}{m_{wb}} \Delta h_{C_k/J_6} \qquad (k = 6 \text{ and } 7) \tag{6b}$$

where m and Δ represent the segment mass and the difference in a variable between takeoff and landing, respectively.

When a lower COM_{wb} at landing is observed, the difference makes the flight time longer, meaning that the jump height from flight time is overestimated. To understand the influence of the difference in the COM_{wb} height on jump height overestimation (ΔH), we created a contour color map using the following equations:

$$\Delta H = H_t - H_v \tag{7}$$

$$\Delta H = \frac{1}{8} g \left(\sqrt{\frac{2H_v}{g}} + \sqrt{\frac{2(H_v + \Delta COM_{wb})}{g}} \right)^2 - H_v \tag{8}$$

where ΔCOM_{wb} represents the difference in the COM_{wb} height. The term in brackets on the right side of Equation (8) is the flight time (see Appendix B). All numerical calculations were performed using MATLAB 2018b (The MathWorks, Inc., Natick, MA, USA).

2.5. Statistical Analysis

The three jumps performed with each device were averaged to provide a representative value for each variable. Means and standard deviations (SDs) were calculated after verifying the normality of distributions using Kolmogorov–Smirnov statistics. Paired-sample t-tests were used to compare the mean differences between methods and between time phases (takeoff and landing). One-sample t-tests were used to examine $CONT_k$ against zero. The magnitude of the difference was also assessed using Cohen's d, where $d > 0.8$ is a large effect, $0.5 \leq d \leq 0.8$ is a moderate effect, $0.2 \leq d \leq 0.5$ is a small effect, and $d < 0.2$ is a trivial effect [18]. The intra-participant reliability of the variables of the three jumps was examined by the intraclass correlation coefficient, one-way random-effects model ($ICC_{1,1}$). Acceptable reliability was defined as an ICC > 0.70 [19]. The analysis of the fixed bias with its upper and lower limits of agreement (LOA) between the jump heights for all 81 trials obtained from the two calculations was performed by using a Bland–Altman plot [20]. Heteroscedasticity of error (proportional bias) was defined as a coefficient of determination (r^2) > 0.1 [21]. Statistical significance was determined by a probability level of $p < 0.05$. All calculations were performed using IBM SPSS Statistics version 19 (IBM Co., Chicago, IL, USA).

3. Results

H_t was significantly higher than H_v (0.421 ± 0.081 and 0.396 ± 0.074 m, respectively, $p < 0.001$, $d = 1.046$). The mean fixed bias (with 95% LOA) between H_t and H_v was 0.025 m (with range −0.028 to 0.079 m) (Figure 2a). The further analysis of the Bland–Altman plot (Figure 3) revealed very low r^2 values ($r^2 = 0.068$), meaning outcomes estimated from H_t had no proportional bias to overestimate or underestimate jump performance. Acceptable intra-participant reliabilities were observed for both H_t and H_v ($ICC_{1,1}$ = 0.964 and 0.979, respectively).

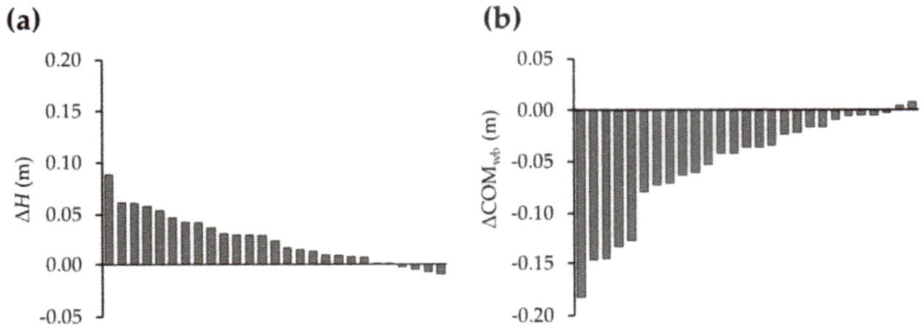

Figure 2. (**a**) The difference in jump height (ΔH) and (**b**) the difference in whole-body COM height (ΔCOM_{wb}) for each participant. Each bar represents a participant. They are arranged in descending (**a**) and ascending (**b**) order.

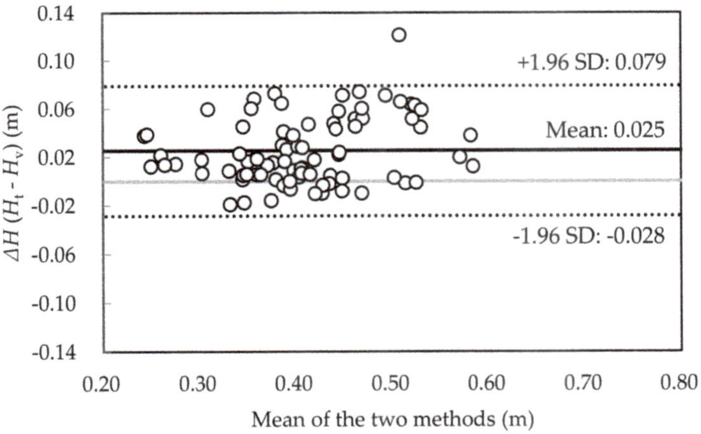

Figure 3. Bland–Altman plot for the jump height from the vertical velocity at takeoff (H_v) and the jump height from flight time (H_t). The central line represents the absolute average difference between the methods, and the upper and the lower lines represent ± 1.96 standard deviation (SD).

The COM_{wb} was significantly lower at landing than at takeoff (1.087 ± 0.100 m and 1.140 ± 0.071, respectively, $p < 0.001$, $d = 1.17$) (Table 1). Acceptable intra-participant reliabilities were observed for the COM_{wb} at both takeoff and landing ($ICC_{1,1}$ = 0.991 and 0.967, respectively). Inter-participant variability ranged from −0.182 to 0.008 m (Figure 2b). COM_{arm}, COM_{shank}, and COM_{foot} showed lower values at landing compared to at takeoff ($p < 0.05$, large effect size) (Table 1). SEG_{shank} and SEG_{foot} showed lower values at landing ($p < 0.05$, large effect size) compared to at takeoff (Table 2). The intra-participant reliabilities of those variables were acceptable ($ICC_{1,1} > 0.7$). $CONT_{arm}$, $CONT_{shank}$, and $CONT_{foot}$ showed lower values compared to zero ($p < 0.05$, moderate to large effect size) (Table 3 and Figure 4).

Table 1. The vertical component of the segment center of mass (COM_k) at takeoff and landing.

Segment	Takeoff (m)	Landing (m)	Difference (m)	Effect Size (*d*)
Whole body	1.140 ± 0.071	1.087 ± 0.100 *	–0.053	1.17
Head	0.143 ± 0.024	0.143 ± 0.028	0.000	0.02
Arm	0.109 ± 0.133	–0.067 ± 0.187 *	–0.175	1.33
Upper Trunk	0.183 ± 0.021	0.184 ± 0.019	0.001	0.13
Lower Trunk	0.086 ± 0.010	0.086 ± 0.010	0.000	0.00
Thigh	0.225 ± 0.013	0.227 ± 0.015	0.002	0.25
Shank	0.237 ± 0.018	0.229 ± 0.021 *	–0.009	1.64
Foot	0.129 ± 0.009	0.108 ± 0.031 *	–0.021	0.97

* $p < 0.05$.

Table 2. The vertical component of the segment length (SEG_k) at takeoff and landing.

Segment	Takeoff (m)	Landing (m)	Difference (m)	Effect Size (*d*)
Upper Trunk	0.321 ± 0.035	0.323 ± 0.032	0.002	0.13
Lower Trunk	0.220 ± 0.026	0.220 ± 0.028	0.000	0.00
Thigh	0.427 ± 0.026	0.430 ± 0.003	0.003	0.25
Shank	0.400 ± 0.032	0.385 ± 0.035 *	–0.015	1.64
Foot	0.199 ± 0.011	0.173 ± 0.037 *	–0.025	0.99

* $p < 0.05$.

Table 3. The contribution of the difference in the vertical component of the segment length (SEG_k) to the difference in the whole-body center of mass (COM_{wb}) height ($CONT_k$).

Segment	Contribution (m)	Effect Size (*d*)
Head	0.000 ± 0.001	0.023
Arm	–0.017 ± 0.013 *	1.259
Upper Trunk	–0.001 ± 0.005	0.119
Lower Trunk	0.000 ± 0.006	0.009
Thigh	0.003 ± 0.011	0.244
Shank	–0.014 ± 0.009 *	1.446
Foot	–0.025 ± 0.032 *	0.781

* $p < 0.05$.

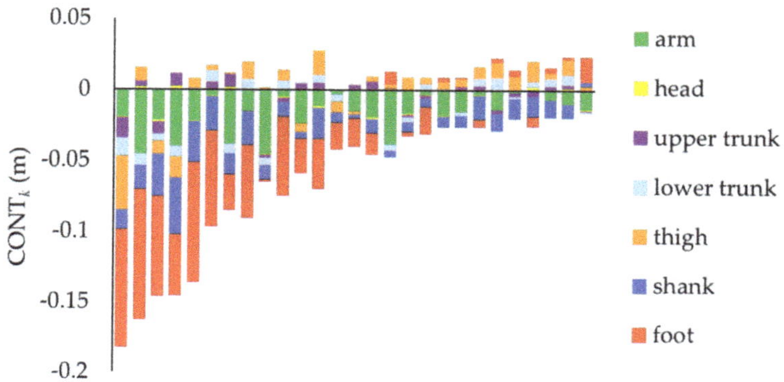

Figure 4. The contribution of the difference in the vertical component of the segment length (SEG_k) to the difference in the whole-body center of mass (COM_{wb}) height (contributions for all seven segments, $CONT_k$). The sum of the differences in all of the segments is the difference in the COM_{wb} height. Each bar shows the result for a participant; the values are arranged in ascending order.

From Equation (8), the difference in jump height (ΔH) was influenced by ΔCOM_{wb} and jump height (H_v). The contour map (Figure 5) showed that the jump height did not greatly affect the overestimation of jump height.

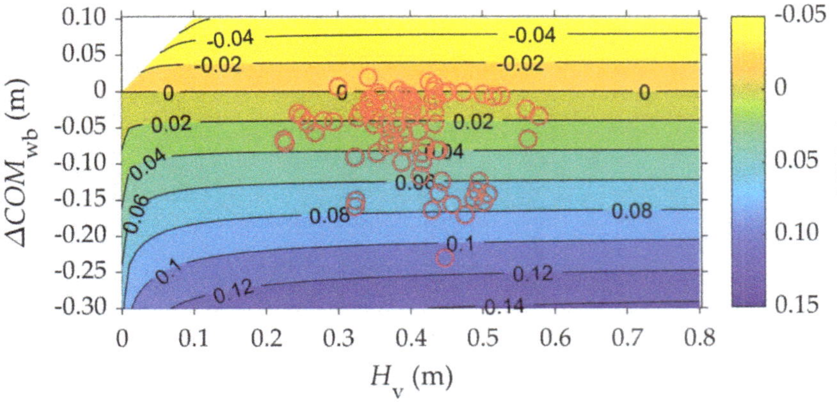

Figure 5. An illustration of the influence of the difference in the COM_{wb} height (ΔCOM_{wb}) and jump height (H_v) on jump height overestimation (ΔH) (Equation (8)). The top-left triangular area in white shows there are no real roots because ΔCOM_{wb} cannot be greater than H_v. The red circles represent each experimental data point.

4. Discussion

The purpose of the present study was to quantify the effect of postural changes between takeoff and landing on jump height overestimation. The jump height from flight time was 0.025 m (6.4%) higher than the jump height calculated from velocity. We confirmed that the current result was reasonable compared to the previous studies, which showed 2–11% overestimation of CMJ height [7–9].

The difference in the vertical components of the foot and shank segment lengths were the main contributions to the difference in the COM_{wb} height. Also, the inter-participant differences were large (range −0.092 to 0.018 m for foot and −0.040 to 0.003 m for the shank, see Figure 4). Therefore, the observed lower COM_{wb} height at landing was mostly due to lower ankle dorsiflexion. In some previous studies, experimenters instructed each participant not to flex their knees [22] and hips [23] at landing. In other studies, experimenters instructed participants to land in a similarly extended position at takeoff [24,25]. From these previous studies, it is suggested that ankle dorsiflexion at landing was considered a less serious effect on the jump height, whereas our results indicate that it is the most critical motion. In these studies, each trial was also watched and judged by the experimenter subjectively to ensure that the instructions had been followed. To reduce the difference in the COM_{wb} height, it is recommended that an experimenter instruct participants about the landing technique for the jump tests, primarily focusing on foot and shank segments, such as "landing with toes pointing downwards" [26]. However, such instruction seems to be inappropriate. One reason for this is that the posture at landing is essential because high impacts cause lower joint injuries [27]. Preparatory flexing at the hip, knee, and ankle is an effective strategy to reduce the impact of landing [28]. Moreover, a previous study revealed that extra attention increased the impact of landing [29]. It may be difficult to control the posture at landing in detail without increasing the risk of injury and excess stress.

It is unlikely—but not impossible—that the control of the upper body affects the lower limb bending. Because the total momentum and total angular momentum of a system both remain constant unless acted upon by an external influence, when a segment moves relative to the COM_{wb}, the other segments have to move to compensate. Therefore, there is a possibility that the upper trunk movement affected the shank and foot postures. It is notable that the causal relationships between these postural effects are unknown.

In this study, the difference in the arm COM height relative to the suprasternal notch also affected the difference in the COM_{wb} height between takeoff and landing, though there was a large inter-participant variability (from −0.046 m to 0.008 m). Although no previous study has reported the effect of arm posture on flight time overestimation, some studies reported that an arm swing contributes to increased the COM_{wb} height [11]. The height of the arm COM in 23 out of 27 participants was above the proximal joint at takeoff, and in 14 of the 23 participants, it was below the proximal joint at landing. When VJ is performed, experimenters do not instruct the participants regarding the height of arm movement before takeoff, because they want to evaluate jump performance using arm swing as much as possible, comparing it to jumps without arm swing. On the other hand, it might be possible to control the height of the arm movement at landing through instruction, such as "arms above the shoulder at landing." No studies justify that the height of the COM_{wb} should be the same at takeoff and landing. At least, the current instruction focusing on lower limb posture cannot prevent the potential "cheating" that can be accomplished by lowering arms as much as possible at landing. Previous studies have shown that arm swing improved jump height [30], but the improvement might be somewhat overestimated when the jump height was calculated from flight time. Therefore, the flight time method cannot be recommended for a vertical jump with arm swing, especially when compared with others, such as at tryouts.

From the contour curve (Figure 5), jump height did not greatly affect the overestimation. For example, if the jump height is 0.20 m and the COM_{wb} height is 0.04 m lower at landing compared to takeoff, then the jump height from flight time is overestimated by 0.0195 m. if the jump height changes to 0.60 m, the overestimation from the same difference in the COM_{wb} height is 0.0198 m. The relationship between the difference in the COM_{wb} height and the overestimation of jump height is in a ratio of almost two to one.

Many studies have considered force platforms as the "gold standard" to evaluate jump height [22,31,32], but this confuses the instrumentation with the calculation method. We can calculate jump height by two methods using force platforms: (1) vertical velocity at takeoff, and (2) the time in the air [8,9]. To clarify the validity and reliability of the simple methodology to calculate jump height from flight time, force platforms are considered the gold standard to calculate flight time because the method contained is valid under certain conditions, as described above. The jump height from vertical velocity at takeoff is the true gold standard for jump height measurement.

Calculating jump height from flight time is still useful for coaches who want to measure changes in an individual resulting from their training program because of its low cost, simplicity, and ease of implementation. Recently, many commercial devices are have been developed to measure jump height from flight time, such as an iPhone app [33] and inertial measurement unit [24]. Other methods have also been in development, such as linear position transducers, but these showed overestimation by 7.0 cm compared to the jump height from flight time [34]. In this study, we confirmed that the flight time method has high intra-participant reliability and no proportional bias, though there is a fixed bias. Researchers and coaches are usually interested in comparing jump height before and after training. If the same device is used for both pre- and post-tests, it is useful.

5. Conclusions

In conclusion, we found that jump height from flight time is overestimated compared to the jump height from takeoff velocity as a result of the lower limb and arm postures at landing. Understanding the sources of error in jump height from flight time can be used to develop better instruction to reduce the systematic error.

Author Contributions: Conceptualization, D.Y., M.M. and Y.I.; methodology, D.Y., M.M. and Y.I.; software, D.Y., M.M. and Y.I.; validation, D.Y., M.M. and Y.I.; formal analysis, D.Y., M.M. and Y.I.; investigation, D.Y., M.M. and Y.I.; writing—original draft preparation, D.Y., M.M. and Y.I.; visualization, D.Y.; funding acquisition, D.Y., M.M. and Y.I. All authors have read and agreed to the published version of the manuscript.

Funding: This work was supported by the research grant of the Japanese Society of Biomechanics.

Acknowledgments: The authors would like to thank the executive committee members of "KEIHIROBA", a conference of the Japanese Society of Biomechanics, for their helpful comments.

Conflicts of Interest: The authors declare no conflict of interest.

Appendix A. Calculation of Jump Height from Flight Time

It is noted that the assumption for this calculation is that the height of the COM_{wb} is the same at takeoff and landing of the jump (H_v is equal to H_t). Once an object is projected into the air, the COM_{wb} must follow a parabolic trajectory, and the trajectory cannot be altered in the air until landing because only the gravitational acceleration is applied to it. Therefore, the vertical velocity of the COM_{wb} is calculated as

$$V(t) = V_0 - gt \tag{A1}$$

where $V(t)$ represents the vertical velocity, V_0 represents the initial velocity, and t represents the time of travel. As $V(t)$ becomes zero at the highest point during flight phase the time from the takeoff to the highest point (t_{top}) is expressed as

$$V_{to} - gt_{up} = 0 \tag{A2a}$$

$$V_{to} = gt_{up} \tag{A2b}$$

t_{up} should be half of the flight time, with the peak of the jump happening at exactly the midpoint of the flight time, expressed as follows:

$$V_{to} = \frac{1}{2}gt_{flight} \tag{A3}$$

Substituting Equation (A3) for Equation (2), we obtain Equation (3) as follows:

$$H_t = H_v = \frac{1}{2g}\left(\frac{1}{2}gt_{flight}\right)^2 \tag{A4}$$

$$H_t = \frac{1}{8}gt_{flight}{}^2 \tag{A5}$$

Appendix B. Calculation of the Flight Time from the Vertical Displacement of the COM_{wb}

The vertical displacements of the COM_{wb} travelling from takeoff to the highest point (vertical velocity becomes zero) and from the highest point (vertical velocity is zero) to landing are both expressed as

$$h(t) = \frac{1}{2}gt^2 \tag{A6}$$

where $h(t)$ represents the vertical displacement. During the time from the takeoff to the highest point (t_{up}), $h(t)$ is equal to H_v, and Equation (A6) gives

$$H_v = \frac{1}{2}gt_{up}{}^2 \tag{A7}$$

$$t_{up} = \sqrt{\frac{2H_v}{g}} \tag{A8}$$

On the other hand, during the time from the highest point to landing (t_{down}), $h(t)$ is the sum of H_v and ΔCOM_{wb}, and Equation (A6) gives

$$H_v + \Delta COM_{wb} = \frac{1}{2}gt_{down}{}^2 \tag{A9}$$

$$t_{\text{down}} = \sqrt{\frac{2(H_v + \Delta\text{COM}_{\text{wb}})}{g}} \tag{A10}$$

The flight time (t_{flight}) is the sum of t_{up} and t_{down}. Therefore, t_{flight} is expressed as

$$t_{\text{flight}} = \sqrt{\frac{2H_v}{g}} + \sqrt{\frac{2(H_v + \Delta\text{COM}_{\text{wb}})}{g}} \tag{A11}$$

References

1. Ache-Dias, J.; Dal Pupo, J.; Gheller, R.G.; Kulkamp, W.; Moro, A.R. Power Output Prediction From Jump Height and Body Mass Does Not Appropriately Categorize or Rank Athletes. *J. Strength Cond. Res.* **2016**, *30*, 818–824. [CrossRef]
2. Young, W.B.; Bilby, G.E. The Effect of Voluntary Effort to Influence Speed of Contraction on Strength, Muscular Power, and Hypertrophy Development. *J. Strength Cond. Res.* **1993**, *7*, 172–178. [CrossRef]
3. Claudino, J.G.; Cronin, J.; Mezencio, B.; McMaster, D.T.; McGuigan, M.; Tricoli, V.; Amadio, A.C.; Serrao, J.C. The countermovement jump to monitor neuromuscular status: A meta-analysis. *J. Sci. Med. Sport* **2016**, *20*, 397–402. [CrossRef]
4. Sierer, S.P.; Battaglini, C.L.; Mihalik, J.P.; Shields, E.W.; Tomasini, N.T. The National Football League Combine: Performance differences between drafted and nondrafted players entering the 2004 and 2005 drafts. *J. Strength Cond. Res.* **2008**, *22*, 6–12. [CrossRef]
5. Teramoto, M.; Cross, C.L.; Rieger, R.H.; Maak, T.G.; Willick, S.E. Predictive Validity of National Basketball Association Draft Combine on Future Performance. *J. Strength Cond. Res.* **2017**. [CrossRef]
6. Lake, J.; Mundy, P.; Comfort, P.; McMahon, J.J.; Suchomel, T.J.; Carden, P. Concurrent Validity of a Portable Force Plate Using Vertical Jump Force-Time Characteristics. *J. Appl. Biomech.* **2018**, *34*, 410–413. [CrossRef]
7. Kibele, A. Possibilities and Limitations in the Biomechanical Analysis of Countermovement Jumps: A Methodological Study. *J. Appl. Biomech.* **1998**, *14*, 105. [CrossRef]
8. Moir, G.L. Three Different Methods of Calculating Vertical Jump Height from Force Platform Data in Men and Women. *Meas. Phys. Educ. Exerc. Sci.* **2008**, *12*, 207–218. [CrossRef]
9. Aragón, L.F. Evaluation of four vertical jump tests: Methodology, reliability, validity, and accuracy. *Meas. Phys. Educ. Exerc. Sci.* **2000**, *4*, 215–228. [CrossRef]
10. Dowling, J.J.; Vamos, L. Identification of kinetic and temporal factors related to vertical jump performance. *J. Appl. Biomech.* **1993**, *9*, 95–110. [CrossRef]
11. Lees, A.; Vanrenterghem, J.; De Clercq, D. Understanding how an arm swing enhances performance in the vertical jump. *J. Biomech.* **2004**, *37*, 1929–1940. [CrossRef]
12. Hara, M.; Shibayama, A.; Takeshita, D.; Fukashiro, S. The effect of arm swing on lower extremities in vertical jumping. *J. Biomech.* **2006**, *39*, 2503–2511. [CrossRef]
13. Suzuki, Y.; Ae, M.; Takenaka, S.; Fujii, N. Comparison of support leg kinetics between side-step and cross-step cutting techniques. *Sport. Biomech.* **2014**, *13*, 144–153. [CrossRef]
14. Woltring, H.J. A Fortran package for generalized, cross-validatory spline smoothing and differentiation. *Adv. Eng. Softw.* **1986**, *8*, 104–113. [CrossRef]
15. Winter, D.A. *Biomechanics and Motor Control of Human Movement*, 4th ed.; John Wiley & Sons, Inc: Hoboken, NJ, USA, 2009.
16. Street, G.; McMillan, S.; Board, W.; Rasmussen, M.; Heneghan, J.M. Sources of error in determining countermovement jump height with the impulse method. *J. Appl. Biomech.* **2001**, *17*, 43–54. [CrossRef]
17. Ae, M.; Tang, H.; Yokoi, T. Estimation of inertia properties of the body segments in Japanese athletes. *Biomechanisms* **1992**, *11*, 23–33. [CrossRef]
18. Cohen, J. *Statistical Power Analysis for the Behavioral Sciences*; Routledge: Abington, UK, 2013.
19. Baumgartner, T.A.; Chung, H. Confidence Limits for Intraclass Reliability Coefficients. *Meas. Phys. Educ. Exerc. Sci.* **2001**, *5*, 179–188. [CrossRef]
20. Bland, J.M.; Altman, D. Statistical methods for assessing agreement between two methods of clinical measurement. *Lancet* **1986**, *327*, 307–310. [CrossRef]

21. García-Ramos, A.; Pérez-Castilla, A.; Martín, F. Reliability and concurrent validity of the Velowin optoelectronic system to measure movement velocity during the free-weight back squat. *Int. J. Sports Sci. Coach.* **2018**, *13*, 737–742. [CrossRef]
22. Rogers, S.A.; Hassmén, P.; Hunter, A.; Alcock, A.; Crewe, S.T.; Strauts, J.A.; Gilleard, W.L.; Weissensteiner, J.R. The Validity and Reliability of the MyJump2 Application to Assess Vertical Jumps in Trained Junior Athletes. *Meas. Phys. Educ. Exerc. Sci.* **2018**, 1–9. [CrossRef]
23. Rodriguez-Rosell, D.; Mora-Custodio, R.; Franco-Marquez, F.; Yanez-Garcia, J.M.; Gonzalez-Badillo, J.J. Traditional vs. Sport-Specific Vertical Jump Tests: Reliability, Validity, and Relationship With the Legs Strength and Sprint Performance in Adult and Teen Soccer and Basketball Players. *J. Strength Cond. Res.* **2017**, *31*, 196–206. [CrossRef]
24. Casartelli, N.; Müller, R.; Maffiuletti, N.A. Validity and reliability of the Myotest accelerometric system for the assessment of vertical jump height. *J. Strength Cond. Res.* **2010**, *24*, 3186–3193. [CrossRef]
25. Glatthorn, J.F.; Gouge, S.; Nussbaumer, S.; Stauffacher, S.; Impellizzeri, F.M.; Maffiuletti, N.A. Validity and reliability of Optojump photoelectric cells for estimating vertical jump height. *J. Strength Cond. Res.* **2011**, *25*, 556–560. [CrossRef]
26. Kenny, I.C.; Cairealláin, A.Ó.; Comyns, T.M. Validation of an electronic jump mat to assess stretch-shortening cycle function. *J. Strength Cond. Res.* **2012**, *26*, 1601–1608. [CrossRef]
27. McKay, G.D.; Goldie, P.A.; Payne, W.R.; Oakes, B.W. Ankle injuries in basketball: Injury rate and risk factors. *Br. J. Sports Med.* **2001**, *35*, 103–108. [CrossRef]
28. Iida, Y.; Kanehisa, H.; Inaba, Y.; Nakazawa, K. Activity modulations of trunk and lower limb muscles during impact-absorbing landing. *J. Electromyogr. Kinesiol.* **2011**, *21*, 602–609. [CrossRef]
29. Shinya, M.; Wada, O.; Yamada, M.; Ichihashi, N.; Oda, S. The effect of choice reaction task on impact of single-leg landing. *Gait Posture* **2011**, *34*, 55–59. [CrossRef]
30. Young, W.B.; MacDonald, C.; Flowers, M.A. Validity of double-and single-leg vertical jumps as tests of leg extensor muscle function. *J. Strength Cond. Res.* **2001**, *15*, 6–11.
31. Lesinski, M.; Muehlbauer, T.; Granacher, U. Concurrent validity of the Gyko inertial sensor system for the assessment of vertical jump height in female sub-elite youth soccer players. *BMC Sports Sci. Med. Rehabil.* **2016**, *8*, 35. [CrossRef]
32. McMahon, J.J.; Jones, P.A.; Comfort, P. A Correction Equation for Jump Height Measured Using the Just Jump System. *Int. J. Sports Physiol. Perform.* **2016**, *11*, 555–557. [CrossRef]
33. Balsalobre-Fernandez, C.; Glaister, M.; Lockey, R.A. The validity and reliability of an iPhone app for measuring vertical jump performance. *J. Sports Sci.* **2015**, *33*, 1574–1579. [CrossRef]
34. O'Donnell, S.; Tavares, F.; McMaster, D.; Chambers, S.; Driller, M. The validity and reliability of the GymAware linear position transducer for measuring counter-movement jump performance in female athletes. *Meas. Phys. Educ. Exerc. Sci.* **2018**, *22*, 101–107. [CrossRef]

applied sciences

MDPI

Article

Electromyographic Evaluation of Specific Elastic Band Exercises Targeting Neck and Shoulder Muscle Activation

Ying Gao [1,2], Lars A. Kristensen [3], Thomas S. Grøndberg [3], Mike Murray [3], Gisela Sjøgaard [3] and Karen Søgaard [3,4,*]

1 Department of Sports Science, College of Education, Zhejiang University, Hangzhou 310028, China; yigao@zju.edu.cn
2 Faculty of Sport and Health Sciences, University of Jyväskylä, 40700 Jyväskylä, Finland
3 Department of Sports Science and Clinical Biomechanics, University of Southern Denmark, DK-5230 Odense, Denmark; larsakristensen1@gmail.com (L.A.K.); thomas-stig@hotmail.com (T.S.G.); mike_ster@hotmail.com (M.M.); gsjogaard@health.sdu.dk (G.S.)
4 Department of Clinical Research, University of Southern Denmark, DK-5230 Odense, Denmark
* Correspondence: ksogaard@health.sdu.dk; Tel.: +45-6550-4409

Received: 12 December 2019; Accepted: 16 January 2020; Published: 21 January 2020

check for updates

Featured Application: This electromyographic evaluation of elastic band exercises documents that the type and intensity of exercise played a role in the activity in three muscle groups of the neck and shoulder. However, importantly, shrugs and reverse flyes are simple to perform and also highly effective training exercises as they induced high muscle activity of both shoulder and neck muscles at 12RM. The practical implications of these findings are that these simple exercises can be preferred for training programs targeting neck and shoulder muscle either for increased general strength or as prevention for work related pain and disorders.

Abstract: Background: Specific strength training at a high intensity is effective in reducing work related neck/shoulder pain. However, it remains to be documented as to which exercises most specifically target neck and shoulder muscles at high activation level while using simple equipment as e.g., elastic bands. We hypothezised that selected exercises would specifically target the respective muscles, as follows: (1) shrugs and reverse flyes: the upper trapezius muscle, (2) cervical extension and lateral flexion: the upper neck extensor muscle, and (3) cervical flexion and rotation: the sternocleidomastoideus muscle. Methods: Eleven healthy males (25.9 ± 1.4 years, BMI 24.3 ± 1.4) with no neck/shoulder pain (VAS = 0) performed the six exercises with elastic bands at 12RM (repetition maximum) and 20RM in a randomized order. Electromyography was bilaterally recorded from the three muscles and it was normalized to maximal voluntary activation (%MVE). Exercises that evoke more than 60%MVE were considered as high intensity activation. Results: High muscle activation level was attained during 12RM in the upper trapezius muscle during shrugs (100.3 ± 29.8%MVE) and reverse flyes (91.6 ± 32.8%MVE) and in the upper neck extensor muscle during cervical extension (67.6 ± 29.8%MVE) and shrugs (61.9 ± 16.8%MVE). In the sternocleidomastoideus muscle, the highest activity was recorded during cervical flexion (51.7 ± 16.4%MVE) but it did not exceed 60%MVE. The overall activity was ~10% higher during 12RM when compared to 20RM. Conclusion: The simple exercises shrugs and reverse flyes resulted in high intensity activation of both the upper trapezius and neck extensors, while no exercises activated sternocleidomastoideus at high intensity.

Keywords: physical training intensity; therapeutic exercises; work related neck pain; musculoskeletal pain

1. Introduction

Neck and shoulder pain are frequent in the general population and within several occupational groups that are characterized by monotonous work tasks that require the stability of the head and neck in constrained postures [1–6]. In often male dominated jobs, helmets add to the weight of the head increasing the demands on the neck muscles and a high prevalence of neck pain is reported among e.g., construction workers, helicopter, and fighter pilots [7–9]. Physical exercise—in particular specific high intensity strength training at the job—has been shown to effectively reduce musculoskeletal pain in the neck and shoulders [10–13].

In clinical practice, at work sites, and in home-based rehabilitation the use of conventional resistance training equipment, such as mascines or free weight, may not be a possibility, as it requires high financial cost and advanced facilities. Thus, it is important to identify more feasible alternatives that are easy to perform at the workplace and still effectively target muscles in the neck and shoulder regions. One alternative is the use of elastic resistance bands that has recently gained interest, because they are simple to handle, low in cost, and easy to transport. However, with elastic bands, it might specifically for the neck extensor muscles be difficult to add enough external load as resistance to achieve the requested high activity level. To attain this, a special headband has been invented for more specific, but also more complex, elastic band neck exercises. These complex exercises have been applied in training interventions of helicopter pilots [8]. However, the attendance rate was low, which was possibly due to the more complex head band exercises; and, in that study, it was not investigated whether the more complex exercises actually resulted in a higher activation for the specific muscles when compared to the more simple neck exercises, like shrugs and reverse flyes [8,14]. The aim of the present study was to assess whether simple exercises as compared to the more complex exercises may suffice for sufficient activation of the neck and shoulder muscles.

Regarding the level of sufficient exercise intensity in strength training programs, this is often defined as the percentage of maximal voluntary force exerted [15]. In practice, one repetition maximum (RM) weight is often used as reference for the maximal weight that can be lifted in an exercise for a given number of repetitions. Activation levels of >60% of maximal voluntary contraction force (MVC) are generally recommended to attain a physiological effect of resistance training [15]. A higher force per contraction can be performed the lower the number of repetitions per set, e.g., at 1RM the performed force is close to 100% MVC, decreasing to around 60% MVC when performing 12RM—depending highly on training status. In well trained persons 60% MVC might be repeated more times, e.g., corresponding to 20RM.

To access the activation level, electromyography (EMG) is commonly used to measure the activity of the specific muscles that are involved in an exercise [16,17]; the bilateral upper trapezius muscle (mUTR), upper neck extensor muscles (mUNE), and sternocleidomastoideus muscle (mSCM) were the muscle groups included here. In the prevention and treatment of neck and shoulder pain such EMG measurements can ensure that a given mode and intensity of training will activate the target muscles to the sufficient level [18]. In addition, EMG measures allow for exploratory analysis, such as the development of muscle fatigue during multiple repetitions that might be an indicator of the long-term effect on hypertrophy, as well as the reduced perception of pain. The myoelectrical manifestations used as an indication of muscle fatigue is an increase in the amplitude and a shift in the power spectrum towards lower frequencies [19–21]. Previous studies with advanced exercises have presented important knowledge for selecting the most adequate exercises to target specific health effects [16,17]. However, studies using simple exercises have similarly shown an effect on both pain relieve and increase in strength [2,22]. The present study will expand on this by using simple (shrugs, SH, and reverse flyes, RF), as well as more advanced (cervical extension, CE and cervical flexion, CF), and quite difficult (cervical lateral flexion, CL and cervical rotation, CR) neck/shoulder exercises and compare the activation in the muscles specifically targeted with each exercise at 20RM and 12RM.

The main hypothesis was that specific training exercises were requested to maximize the activation of the individual neck and shoulder muscles: with mUTR activation being the highest during SH and

RF, mUNE during CE and CL, and mSCM during CF and CR. Further, it was hypothesized that the intensity of the exercises significantly affected muscle activity: 12RM, resulting in higher activity than 20RM during the specific training exercises. Finally, we hypothesized that the muscle activity in the concentric phase was higher than in the eccentric phase and fatigue development was present in each set of an exercise being shown as an increase in EMG amplitude and a decrease in EMG frequency in the targeted muscle groups.

2. Materials and Methods

2.1. Study Design and Subjects

This study was conducted over three separate days that were interspaced with at least three days with no exercises. On the first day participants were familiarized with the elastic band training exercises and practiced the proper technique under supervision. On the second day, individual loadings were determined for each training exercise during low intensity, high repetition sets (20RM), and during high intensity, moderate repetition sets (12RM). All of the exercises were performed while using concentric and eeccentric muscle contractions in a controlled manner, with slow to moderate lifting velocity corresponding to the execution in earlier intervention studies [16]. Subsequently, the actual duration of concentric and eccentric phases were measured and shown to have an equal duaration each of minimum three seconds. The length and color code of elastic bands were adjusted to reduce or increase the resistance to match 12RM and 20RM for each participant. The task was repeated until the participant was exhausted and/or not able to continue the exercise according to the basics of procedures in the literature to define the individual 12RM and 20RM loads for each exercise [15]. The participant was deemed not able to continue the exercise if making compensating adjustment in body posture or not obeying contraction time and range of movement. The exercise was repeated until the loading corresponding to 20RM and 12RM was determined. On the third day, the determined loadings for each participant were applied and all EMG measurements were performed, while the participant performed all exercises at the intensity of 20RM sets, followed by 12RM sets. The order of exercises was randomized and each exercise was separated by 3-minutes rest intervals to avoid a gradient of fatigue through the test protocol. At the end, isometric maximal voluntary contractions (MVC) were performed for the normalization of the EMG data. For each targeted muscle, three MVCs with a duration of 5 s and 1 minute rest between consecutive trials were performed to induce the maximal muscle activity in a seated position and according to standardized procedures described in details in a recent study [23]. In short, maximal activity in mUTR was obtained during resisted bilateral shoulder elevation, with resistance being placed above the acromions. Maximal activity for mUNE was obtained during a resisted cervical extension and for mSCM during a resisted cervical flexion with resistance being placed on the external occipital protuberance and above the eyebrows, respectively.

The subjects participating were a group of 11 healthy males (age 25.9 ± 1.4 years, height 183.6 ± 5.0 cm, body mass 82.1 ± 6.0 kg, BMI 24.3 ± 1.4), with no pain in neck or shoulders (VAS = 0). The inclusion of only males was justified by the aviation personell—who used these exercises in a previous RCT study—all being males [8]. All of the subjects were informed regarding the benefits and risks of the investigation prior to signing the institutionally approved informed consent document to participate in the study. The local Ethics Committee of Southern Denmark (S-20120121) approved the procedures in the study.

2.2. Procedures

2.2.1. Electromyography (EMG)

The activity of the neck and shoulder muscles was measured through a bipolar surface EMG configuration (Neuroline 720-01-K, Medicotest A/S, Ølstykke, Denmark). The electrodes (Ag/AgC1, Ambu Blue Sensor, N-00-S/25, Denmark) were positioned bilaterally, in accordance with the SENIAM recommendations and earlier studies [24–27]) with an inter-electrode distance of 20 mm. The electrodes were placed (1) above the upper trapezius muscle (mUTR), 20% medially from the centerpoint between the lateral edge of the acromion and seventh cervical vertebra, (2) above the upper neck extensor (mUNE) muscle, on the most prominent part of the muscle at the level of the fourth cervical vertebra, and (3) above the sternocleidomastoideus muscle (mSCM), over the main muscle belly at the 1/3 distance from the sternal notch to the mastoid process (Figure 1). Before affixing the electrodes, skin was shaved and cleaned with scrubbing gel (Acqua gel, Meditec, Parma, Italy) and 70% alcohol, to effectively lower the impedance to less than 10 kΩ.

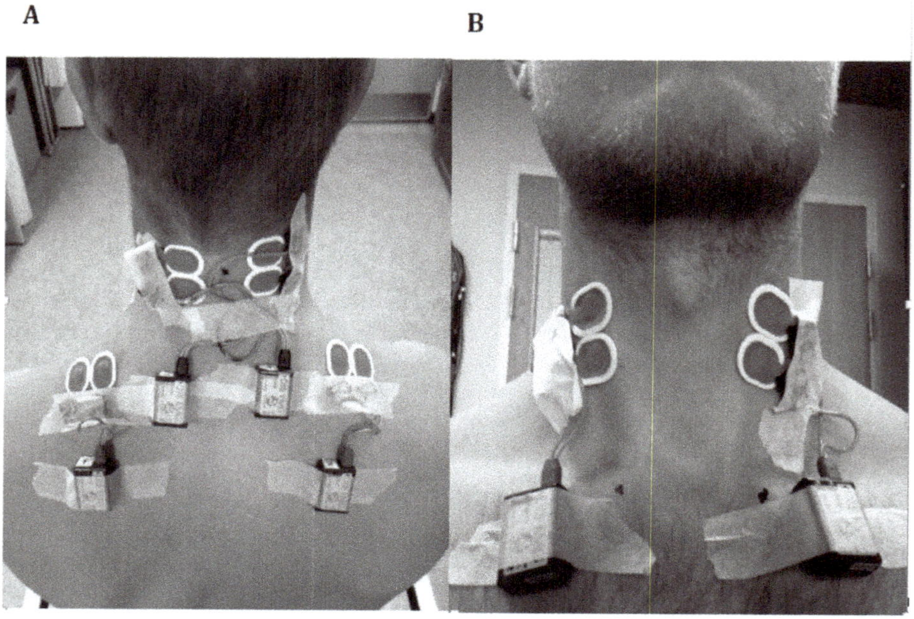

Figure 1. Bilateral electrode placement above the upper trapezius muscle (mUTR) and upper neck extensors muscles (mUNE) (**A**) and sternocleidomastoideus muscles (mSCM) (**B**).

The EMG signal was pre-amplified with a gain of 400 and then transferred through an EMG probe (Direct Transmission System (DTS), Noraxon Inc, Scottsdale, AZ, USA) to a belt-receiver (TELEMyo ™ 2400T, DTS, Noraxon Inc, Scottsdale, AZ, USA) with a bandwidth of 10–500 Hz and a common mode rejection ratio better than 100 dB. The signal was sampled at 1000 Hz while using a 16-bit A/D converter (DAQCard Al 16XE 50, National Instruments, Austin, TX, USA) and stored on a computer via the laboratory interface (CED 1401, Spike2 software, Cambridge Electronic Devices, Cambridge, UK).

2.2.2. Elastic Band Training Exercises

Six specifically tailored elastic band training exercises were performed targeting the neck and shoulder muscles. The choice of exercises was evidence-based and designed by an interdisciplinary team, including sports exercise training specialists, physical therapists, medical doctors, and chiropractors. The exercises were performed with a head harness (Neck Flex ®, Leesburg, VA, USA) while using different color-coded elastic resistance bands (Thera-Band®, The Hygenic Corporation, Akron, OH, USA). The individual exercises are briefly presented below, as a detailed description can be found in an earlier publication [14] and video material is available online: Neck exercises.

In short the exercises are the following:

(1) Shrugs (SH) were performed in standing position with the elastic band fixated under the feet and arms placed along the side of the body and shoulders relaxed. The head was held in an anatomically neutral position with eyes looking straight forward. During the exercise, shoulders were elevated as high as possible toward the ears against resistance and lowered again.
(2) Reverse Flyes (RF) were performed seated with the elastic band fixated under the feet and with a straight back. Participants were instructed to position their head in an anatomically neutral position, lean the trunk forward (~20–30°), and place both arms pointing towards the floor. Elbows were kept in a static and slightly flexed position (~5°). During the exercise, both of the arms were raised toward a horizontal level against resistance and lowered again.
(3) Cervical Extension (CE) was performed seated. The participants were instructed to keep a straight back, position their head in an anatomically neutral position, and lean the trunk forward (~20–30°). Arms were held strait with the hands placed underneath the knees. An elastic band was stretched between the hands and front of the head harness. The exercise was performed with a low cervical spine flexion, followed by a low cervical spine extension (against resistance).
(4) Cervical Flexion (CF) was performed in the same way as during cervical extension, but the elastic band was stretched between a door anchor and the back of the head harness. During the exercise, the participants performed a low cervical spine flexion (against resistance), followed by a low cervical spine extension.
(5) Cervical Lateral flexion (CL) was performed standing erect with the head in an anatomically neutral position. One hand was placed horizontally against a wall and an elastic band was stretched between the hand and side of the head harness. The exercise was performed with a low lateral spine flexion against resistance, followed by a low lateral spine extension. The exercise was performed for the right and left side, respectively.
(6) Cervical rotation (CR) was performed seated with a straight back and trunk leaned forward (~20°). The head was held in an anatomically neutral position and rotated approximately 45° to either the right or left side. An elastic band was stretched between the head harness and a door anchor. Keeping a static upper body, the hips were flexed and the body flexed (against resistance), followed by an extension. The exercise was performed to the right and left side.

2.3. Outcomes

Analyses of the EMG signals were the outcome data. All data analysis was performed while using a custom-made script in Matlab (MathWorks Inc. USA) and Hedera 2.0 (University of Southern Denmark, Denmark). The peak EMG amplitude was calculated by root mean square (RMS) in a 500 ms moving window. For each repetition during exercises, the peak RMS was normalized to the peak RMS that was obtained during the MVC of each muscle as percentage of maximal voluntary EMG signal (%MVE) in standardized postures [23]. The 3rd, 4th, and 5th repetition of each exercise were averaged and used as the muscle activation for the exercises. The EMG activity in the concentric and eccentric phases of the muscle contraction was analyzed for each muscle during the exercise, resulting in the highest activation. The concentric and eccentric phases of each exercise were marked as an event in a separate channel in the Spike2 software, during each recording, As each repetition was performed with a 3 s concentric and 3 s eccentric phase, the midpoint was used to divide these phases. Additionally, the mean power frequency (MPF) was calculated in the last 1000 ms of the concentric phase of each contraction for the muscle groups during the exercise that represents the highest activation during the contraction.

2.4. Statistical Analysis

All of the data are presented as mean ± SD. The Q-Q plot and the Shapiro-Wilk test with skewness were used to test for normal distribution of data. Differences between left and right muscles during the symmetrical exercises SH, RF, CE, and CF, or between ipsilateral and contralateral muscles during CL and CR performed towards left and right side were examined while using paired-t test. In the case of no difference, the mean value of the two sides was used. If significant differences were found, the side specific values are reported and the highest used to represent the muscle activation level for that exercise. Regarding the main hypotheses of the muscle activity (%MVE) for each muscle group, one way repeated measures ANOVA was used to compare the different exercises (SH, RF, CE, CF, CL, and CR). If the assumption of Sphericity was violated, the Greenhouse-Geisser correction was used. Planned contrasts were used to localize the differences between the exercise, resulting in the highest activation of specific muscle group and the other exercises. Paired t-test was used for testing possible differences between the two exercise intensities. Furthermore, paired t-test was used to compare muscle activity in the concentric phase and eccentric phase of each muscle group. The fatigue development was tested by paired t-test to compare the third repetition and the last one repetition in normalized EMG amplitude (%MVE) and EMG frequency (MPF). A substantial addition in muscle activation level of at least 20% was requested in order to recommend an advanced, instead of a simple, exercise. This level is well above the SD of approx. 10% in mean level of trapezius activation both found within groups and variability of repeated EMG recordings during high intensity contractions, see, for instance, mean activation levels for trapezius in [17]. A power analysis showed that eight subjects would be sufficient to reveal a difference in muscle activation level of 20%. Within the recruiting period, we managed to recruit 11, who all conducted the full protocol. All of the results were analyzed using SPSS 16.0 statistical software (SPSS Inc, Chicago, IL, USA). A probability level of $p < 0.05$ (two-tailed) was considered to be statistically significant.

3. Results

Table 1 presents the muscle activation levels for the six exercises at two intensities. There were no differences between left and right side of muscles during the symmetrical exercises and no difference between ipsilateral and contralateral when the exercises CL and CR were performed to right versus left side. Therefore, the data were averaged over left and right muscles or ipsilateral and contralateral muscles, except during CL for mUNE and mSCM. This is due to the higher muscle activity during CL in ipsilateral than contralateral for mUNE (30.2% ± 13.6% vs. 12.8% ± 6.8%, $p < 0.001$, for 20RM, and 45.5% ± 14.6% vs. 17.2% ± 7.1%, $p < 0.001$, for 12RM) and for mSCM (25.6% ± 7.7% vs. 11.6% ± 6.3%, $p = 0.001$, for 20RM, and 41.5% ± 14.6% vs. 15.3% ± 8.4%, $p = 0.001$, for 12RM). Therefore, for CL the higher value of %MVE in ipsilateral of mUNE and mSCM were used as the results (Table 1).

Table 1. Mean muscle activity (%MVE) for representative muscle groups shown for each exercise at 12RM and 20RM intensities. Values are presented as mean ± SD. Values > 60% MVE are marked in bold. Italic values are unilateral data, see text.

Muscle	Intensity	Exercise					
		SH	RF	CE	CF	CL	CR
mUTR	20RM	**78.8 ± 24.0**	**75.9 ± 30.7**	3.4 ± 1.8	1.9 ± 1.3	8.1 ± 4.6	3.7 ± 2.3
	12RM	**100.3 ± 29.8**	**91.6 ± 32.8**	4.8 ± 2.5	2.4 ± 2.0	9.1 ± 4.1	4.3 ± 2.8
mUNE	20RM	45.7 ± 16.7	39.1 ± 22.3	50.9 ± 25.8	7.6 ± 3.2	*30.2 ± 13.6*	10.4 ± 4.6
	12RM	**61.9 ± 16.8**	55.4 ± 16.3	**67.6 ± 29.8**	8.5 ± 3.8	*45.5 ± 14.6*	12.0 ± 4.7
mSCM	20RM	16.7 ± 13.5	14.2 ± 8.9	15.1 ± 11.3	44.8 ± 16.6	*25.6 ± 7.7*	28.8 ± 9.9
	12RM	26.2 ± 16.5	22.9 ± 14.8	10.1 ± 6.3	51.7 ± 16.4	*41.5 ± 14.6*	42.5 ± 11.6

The repeated measures ANOVAs revealed significantly different muscle activation between the types of exercise for: mUTR ($F(1.099, 10.987) = 97.622$, $p < 0.001$), mUNE ($F(2.359, 23.59) = 29.038$, $p < 0.001$), and mSCM ($F(5, 50) = 18.045$, $p < 0.001$), respectively. Figure 2 presents the contrast of muscle activity during the exercise, which induced the highest activation with other exercises. For mUTR, SH induced highest activity, and this was significantly higher than all of the remaining exercises, except RF. For mUNE, CE induced the highest activity and this was higher than during CF, CL and CR, but not significantly different from SH, and RF. For mSCM, CF induced the highest activity and this was higher than in all the remaining exercises. Furthermore, when compared with 20RM, the mean muscle activity of all exercises was significantly higher (relative increase 34%) during 12RM. For the specific muscle groups the relative increase were 26% in mUTR ($p < 0.001$), 41% in mUNE ($p < 0.001$), and 35% in mSCM ($p < 0.001$). (Figure 3).

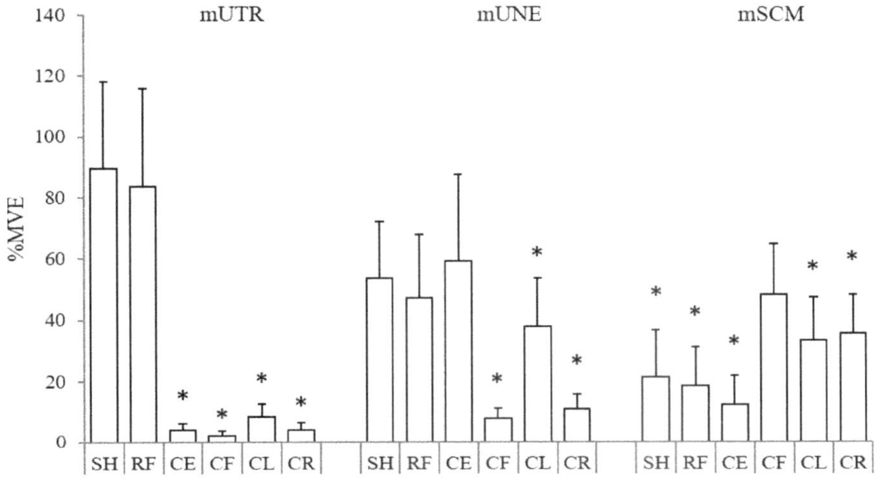

Figure 2. The muscle activity of the upper trapezius muscle (mUTR), upper neck extensors muscles (mUNE) and sternocleidomastoideus muscles (mSCM) were compared in six different types of exercises. The highest muscle activation (%MVE) was found during SH for mUTR, during CE for mUNE and during CF for mSCM. * marks exercises with significantly lower activation than the exercise inducing the highest muscle activation.

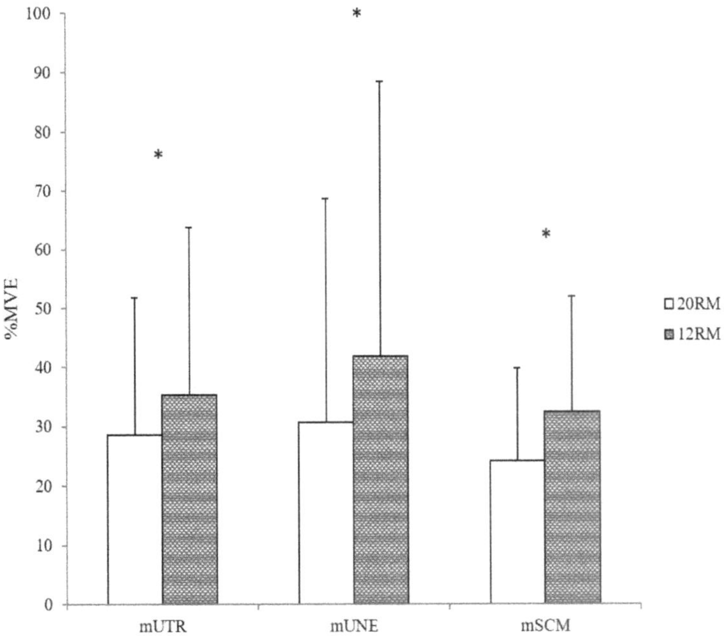

Figure 3. Mean muscle activity (%MVE) for all exercises in the upper trapezius muscle (mUTR), upper neck extensors muscles (mUNE), and sternocleidomastoideus muscles mSCM during intensities of 12RM and 20RM. * indicates significant difference between training intensities.

Analysis while considering activity separately in the concentric and eccentric contraction modes showed that the muscle activity of mUTR, mUNE, and mSCM in the concentric phase when compared to the eccentric phase was significantly increased by relative 75% for SH ($p < 0.001$), 32% for CE ($p = 0.009$), and 22% for CF ($p = 0.002$) (Figure 4).

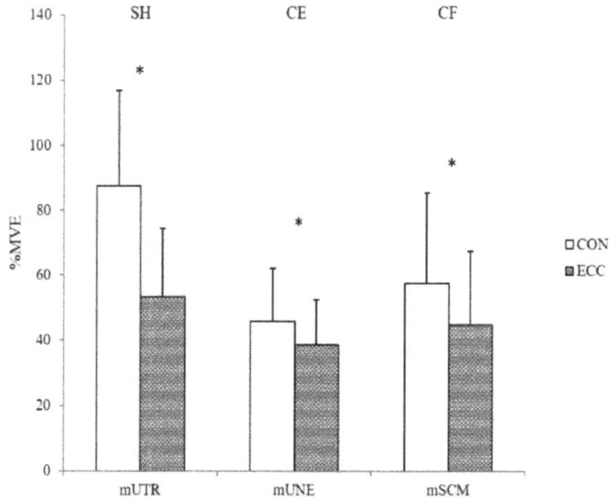

Figure 4. The muscle activity (%MVE) of the upper trapezius muscle (mUTR), upper neck extensors muscles (mUNE), and sternocleidomastoideus muscles (mSCM) in the concentric phase (CON) and eccentric phase (ECC) during shugs (SH), cervical extension (CE), and cervical flexion (CF), respectively. * Indicates significant difference between CON and ECC within exercise.

Analysis of fatigue development showed that, in each set of an exercise, a significant increase occurred in %MVE and a significant decrease occurred in EMG frequency (MPF) during SH, CE, and CF for each muscle groups of mUTR, mUNE, and mSCM, respectively (Table 2).

Table 2. The fatigue development of mean muscle activity (%MVE) and mean power frequency (MPF) for the upper trapezius muscle (mUTR), upper neck extensors muscles (mUNE), and sternocleidomastoideus muscles mSCM at intensities of 12 and 20RM. #Values are presented as mean ± SD for the exercises shrugs (SH), cervical extension (CE), and cervical extension (CF) that resulted in the highest activation of each muscle.

Muscle	Exercise	Fatigue Development	20RM			12RM		
			Rep 3.	Rep 19.	*p*-Values	Rep 3.	Rep 11.	*p*-Values
mUTR	SH	%MVC	81.2 ± 27.1	95.6 ± 25.6	0.001	94.3 ± 31.4	114.1 ± 30.2	<0.001
		MPF	78.5 ± 5.9	69.8 ± 8.2	<0.001	80.3 ± 12.0	71.6 ± 10.2	<0.001
mUNE	CE	%MVC	50.7 ± 26.2	67.5 ± 33.7	0.006	64.5 ± 28.7	82.4 ± 33.7	0.010
		MPF	66.0 ± 7.6	57.2 ± 8.9	0.004	63.6 ± 9.4	57.6 ± 10.6	0.005
mSCM	CF	%MVC	43.9 ± 17.4	53.5 ± 20.7	0.016	51.4 ± 17.1	65.0 ± 21.5	0.027
		MPF	104 ± 14.4	95.2 ± 14.4	0.005	104.0 ± 13.4	99.1 ± 12.7	0.012

4. Discussion

4.1. Main Finding

Specific activation patterns were identified in the three neck/shoulder muscles that were studied, in accordance with the hypothesis exercise type. Interestingly, the simple exercises, SH and RF, elicited muscle activities of >60% MVE in mUNE as well as in mUTR. The more advanced exercise, CE, showed numerically slightly higher—but not significantly—activation of mUNE, and CF produced the highest activity in mSCM, but this did not attain >60% MVE neither during 20RM nor 12RM and can therefore, not be recommend in case of mSCM insufficiencies. The two muscle groups most often attracting work related pain are mUTR and mUNE [1,3–5]. It is therefore interesting that both muscles can be intensively activated by two simple exercises (SH and RF), as this might potentially counteract the work related pain. The CL and CR, which are the most difficult exercises to perform correctly, did not elicit significantly higher percentages of muscle activation than any of the other exercises and can therefore be disregarded, unless for extra training variation. Thereby, our main hypothesis was rejected.

Earlier studies have reported significantly higher activation levels of mUTR during SH than RF [17,28], but this was not the case in the present study. The similar activation levels of mUTR during SH and RF may have great clinical importance due to the difference in force resistance that is required for performing the exercises. When performing the SH exercise for all subjects, a combination of three high resistance elastic bands were needed, whereas only one low resistance elastic band was used during RF. Thus, the SH exercise requires significant grip strength and core stability, which might be challenging for some populations. The mUTRs role as a scapula up- and outward rotator in combination with the significant external moment arm during the RF exercise may explain the high muscle activation in spite of low external resistance. Therefore, the RF exercise might, for some populations, be a more appropriate choice of exercise as compared to the SH.

The activity of mUNE was not significantly higher in CE compared to SH and RF. To our knowledge this is the first study to demonstrate high activity of neck muscles during SH and RF. This observation implies that the use of sophisticated exercises might not be needed as simple exercises effectively can activate both neck and shoulder muscles. This finding is supported by an earlier finding on high (>60% MVE) muscle activity of m. splenius capitis during lateral raise, which is similar to the RF that was used in the present study [29]. As splenius capitis is a part of the upper neck extensor muscle group, it will contribute to the gross activity measured as mUNE in the present study. Based on functional anatomy, the cervical spine will be forced in a protracted position during SH and RF, when performed with an external loading [30–32]. The neck muscles must be activated in order to stabilize and keep a neutral positio. This essential role of neck stability during SH, RF, and lateral raise may thus be a possible explanation for the high muscle activity in neck extensor muscle group. In this regard, it must be noted that the stabilizing effect arrives from static muscle contractions and the subsequent physiological effect might not be the same as from dynamic muscle contractions during the CE exercise [33–35]. Studies using wire electrodes to separate activity in the deep and superficial neck extensor muscles have pointed out the importance of activity specifically in the deep extensor muscles for patients with chronic neck pain [36]. Since surface EMG records from all of the underlying muscles, we cannot rule out that the same activity in SH and CE represent different contribution from the neck extensor muscles that may be of importance for specific patient groups. However, this distinction might not be equally important for the purpose of general strengthening of the neck extensors to relieve work related pain.

4.2. Contraction Intensities

In general, it is assumed that muscle activation increases with increased external load. Our hypothesis regarding a higher intensity during 12RM as compared 20RM was confirmed. We found that the EMG activity increased by 9% MVE from an overall mean of 28% MVE at 20RM to 37% MVE at 12RM. This corresponds to the estimated increase from 60% of 1RM at 20RM to the 70% of 1RM at 12RM. However, this general increase consisted of largely different increases for each of the targeted muscles. The muscles have a rather different function in each exercise and a high degree of co-contraction occurring in the neck region. Only for the mUTR would the 20RM evoke intensities >60% MVE that is normally considered to be most efficient for a training effect. From a pain reduction perspective, the high intensive exercises have been shown to be most effective and, therefore, the 12RM might be the best gain for the work time invested into exercises at the workplace [11,18].

4.3. Muscle Activation Patterns

The exercises included a concentric and an eccentric phase and EMG recordings clearly demonstrated that the highest activation occurred in the concentric phase, which is in concert with general physiological findings of the recruitment pattern explained by the force-velocity relationship [37,38]. Elastic band training exercises has been shown to compare well to activation patterns seen during traditional dumbbell exercises [29] and this underpins that elastic band exercise training might well substitute training with more advanced equipment. EMG recording also demonstrated that, during each set of contractions, significant muscle fatigue developed in all recorded muscle groups. In spite of the difference in activation level between the 12RM and 20RM in the first contraction of the sets then at the last contraction of the 12RM and 20RM both attained levels >60%MVE, which may potentially evoke an increase in muscle capacity [15]. Hereby, the present hypotheses regarding the contraction mode and fatigue EMG patterns were confirmed.

4.4. Study Strenghts and Limitations

We consider this study as a first step in the investigation of recommendable training and rehabilitation exercises for the neck/shoulder muscle based on elastic bands. The strengths of the study is the direct measures using EMG to assess the activation of representative neck muscle groups and instead of subjective assessment. Additionally, the design of the study using randomization of the exercises is a strength, since this ensures that the fatigue development within each type of exercise is not biased by its timewise performance.

The limitations of present study relate to the population only including a small group of males similar to majority of employees in the job groups specifically challenged with high exposures of the neck muscles due to helmet wearing or high accelerations (fighter pilots, helicopter pilots, construction workers, etc.). Additionally, the population consist of healthy subjects without serious neck/shoulder disorders. Therefore, extrapolation of our results to patients with neck/shoulder injury should be undertaken with caution. We cannot conclude that patients suffering from neck/shoulder pain will show similar muscle activation patterns when performing the exercises that are studied here.

Furthermore, EMG activity during the dynamic exercises was normalized to measurements of EMG activity during static MVCs to quantify the level of muscle activation. Given the inherent methodological limitations that are associated with surface EMG, only a rough estimate of the absolute level of muscle activation can be inferred while using this method. Fine wire electrodes for some of the deeper layers of neck extensor muscles could have further detailed the contribution of activity from each of the neck extensors. Finally, more muscles and contraction intensities could be studied to give an even more complete activation pattern for optimal recommendations of training exercises.

4.5. Future Perspectives

While considering the limitations above, future studies should include participants with work-related neck pain in a randomized controlled trial preferably designed with three groups: (A) training only using the simple exercises, (B) training with a variety of exercises, including the advanced and difficult exercises, and (C) a control group. Importantly the training volume, progression, undulation, and duration should be similar for groups A and B, as we have exemplified in previous studies [9–11,13]. The primary outcome should be neck pain testing the hypothesis that groups A and B would have equally positive effects. Additionally, secondary outcomes should be included, like muscle strength and adherence to the exercises; the latter in particular to test the hypothesis implied in the present study that adherence for group A would be larger that for group B.

5. Conclusions

The simple exercises (SH and RF) were highly effective, as they induced high muscle activation level of both shoulder and neck muscles and induced fatigue within the set of exercises. The more advanced neck exercises (CE and CF) and quite difficult exercises (CL and CR) did not, in general, add significantly higher muscle activation levels. The practical implications of these findings are that these simple exercises can be preferred for self-administered training programs—e.g., at the worksite—targeting neck and shoulder muscle either for increased general strength or as prevention for work related pain and disorders.

Author Contributions: Responsible for the conceptualization of the study were, M.M., G.S. and K.S.; the methodology, software, and validation was taken care of by M.M., L.A.K. and T.S.G.; formal analysis, M.M., L.A.K. and T.S.G.; investigation and experimental work, M.M., L.A.K. and T.S.G.; resources, G.S., K.S. and Y.G.; data curation and statistics, Y.G.; writing—original draft preparation, Y.G.; writing—review and editing, all authors.; visualization, Y.G., L.A.K. and T.S.G.; supervision, G.S. and K.S.; project administration, K.S. All authors have read and agreed to the published version of the manuscript.

Funding: This research received no external funding.

Conflicts of Interest: The authors declare no conflict of interest.

References

1. Sjøgaard, G.; Lundberg, U.; Kadefors, R. The role of muscle activity and mental load in the development of pain and degenerative processes at the muscle cell level during computer work. *Eur. J. Appl. Physiol.* **2000**, *83*, 99–105. [CrossRef]
2. Hagberg, M.; Harms-Ringdahl, K.; Nisell, R.; Hjelm, E.W. Rehabilitation of neck-shoulder pain in women industrial workers: A randomized trial comparing isometric shoulder endurance training with isometric shoulder strength training. *Arch. Phys. Med. Rehabil.* **2000**, *81*, 1051–1058. [CrossRef]
3. Unge, J.; Ohlsson, K.; Nordander, C.; Hansson, G.A.; Skerfving, S.; Balogh, I. Differences in physical workload, psychosocial factors and musculoskeletal disorders between two groups of female hospital cleaners with two diverse organizational models. *Int. Arch. Occup. Environ. Health* **2007**, *81*, 209–220. [CrossRef]
4. Van den Heuvel, S.G.; van der Beek, A.J.; Blatter, B.M.; Bongers, P.M. Do work-related physical factors predict neck and upper limb symptoms in office workers? *Int. Arch. Occup. Environ. Health* **2006**, *79*, 585–592. [CrossRef]
5. Van den Oord, M.H.; De Loose, V.; Meeuwsen, T.; Sluiter, J.K.; Frings-Dresen, M.H. Neck pain in military helicopter pilots: Prevalence and associated factors. *Mil. Med.* **2010**, *175*, 55–60. [CrossRef]
6. Adam, J. *Results of NVG-Induced Neck Strain Questionnaire Study in CH-146 Griffon Aircrew*; DRDC Toronto, Defence R&D Canada: Toronto, ON, Canada, 2004.
7. Lange, B.; Toft, P.; Myburgh, C.; Sjøgaard, G. Effect of targeted strength, endurance, and coordination exercise on neck and shoulder pain among fighter pilots: A randomized-controlled trial. *Clin. J. Pain* **2013**, *29*, 50–59. [CrossRef]
8. Murray, M.; Lange, B.; Nornberg, B.R.; Sogaard, K.; Sjogaard, G. Self-administered physical exercise training as treatment of neck and shoulder pain among military helicopter pilots and crew: A randomized controlled trial. *BMC Musculoskelet. Disord.* **2017**, *18*, 147. [CrossRef]
9. Gram, B.; Andersen, C.; Zebis, M.K.; Bredahl, T.; Pedersen, M.T.; Mortensen, O.S.; Jensen, R.H.; Andersen, L.L.; Sjøgaard, G. Effect of training supervision on effectiveness of strength training for reducing neck/shoulder pain and headache in office workers: Cluster randomized controlled trial. *BioMed Res. Int.* **2014**, *2014*, 693013. [CrossRef]
10. Blangsted, A.K.; Sogaard, K.; Hansen, E.A.; Hannerz, H.; Sjogaard, G. One-year randomized controlled trial with different physical-activity programs to reduce musculoskeletal symptoms in the neck and shoulders among office workers. *Scand. J. Work. Environ. Health* **2008**, *34*, 55–65. [CrossRef]
11. Andersen, L.L.; Kjaer, M.; Sogaard, K.; Hansen, L.; Kryger, A.I.; Sjogaard, G. Effect of two contrasting types of physical exercise on chronic neck muscle pain. *Arthritis Rheum.* **2008**, *59*, 84–91. [CrossRef]
12. Zebis, M.K.; Andersen, L.L.; Pedersen, M.T.; Mortensen, P.; Andersen, C.H.; Pedersen, M.M.; Boysen, M.; Roessler, K.K.; Hannerz, H.; Mortensen, O.S.; et al. Implementation of neck/shoulder exercises for pain relief among industrial workers: A randomized controlled trial. *BMC Musculoskelet. Disord.* **2011**, *12*, 205. [CrossRef]

13. Dalager, T.; Bredahl, T.G.; Pedersen, M.T.; Boyle, E.; Andersen, L.L.; Sjøgaard, G. Does training frequency and supervision affect compliance, performance and muscular health? A cluster randomized controlled trial. *Man. Ther.* **2015**, *20*, 657–665. [CrossRef] [PubMed]
14. Murray, M.; Lange, B.; Nornberg, B.R.; Sogaard, K.; Sjogaard, G. Specific exercise training for reducing neck and shoulder pain among military helicopter pilots and crew members: A randomized controlled trial protocol. *BMC Musculoskelet. Disord.* **2015**, *16*, 198. [CrossRef] [PubMed]
15. Ratamess, N.A.; Alvar, B.A.; Evetoch, T.K.; Housh, T.J.; Kibler, W.B.; Kraemer, W.J.; Triplett, N.T. American College of Sports Medicine position stand. Progression models in resistance training for healthy adults. *Med. Sci. Sports Exerc.* **2009**, *41*, 687–708. [CrossRef]
16. Andersen, C.H.; Zebis, M.K.; Saervoll, C.; Sundstrup, E.; Jakobsen, M.D.; Sjogaard, G.; Andersen, L.L. Scapular muscle activity from selected strengthening exercises performed at low and high intensities. *J. Strength Cond. Res.* **2012**, *26*, 2408–2416. [CrossRef]
17. Andersen, L.L.; Kjaer, M.; Andersen, C.H.; Hansen, P.B.; Zebis, M.K.; Hansen, K.; Sjogaard, G. Muscle activation during selected strength exercises in women with chronic neck muscle pain. *Phys. Ther.* **2008**, *88*, 703–711. [CrossRef]
18. Chen, X.; Coombes, B.K.; Sjogaard, G.; Jun, D.; O'Leary, S.; Johnston, V. Workplace-Based Interventions for Neck Pain in Office Workers: Systematic Review and Meta-Analysis. *Phys. Ther.* **2018**, *98*, 40–62. [CrossRef]
19. Beck, T.W.; Stock, M.S.; Defreitas, J.M. Shifts in EMG spectral power during fatiguing dynamic contractions. *Muscle Nerve* **2014**, *50*, 95–102. [CrossRef]
20. Farina, D.; Merletti, R.; Enoka, R.M. The extraction of neural strategies from the surface EMG: An update. *J. Appl. Physiol.* **2014**, *117*, 1215–1230. [CrossRef]
21. Enoka, R.M.; Duchateau, J. Muscle fatigue: What, why and how it influences muscle function. *J. Physiol.* **2008**, *586*, 11–23. [CrossRef]
22. Ris, I.; Sogaard, K.; Gram, B.; Agerbo, K.; Boyle, E.; Juul-Kristensen, B. Does a combination of physical training, specific exercises and pain education improve health-related quality of life in patients with chronic neck pain? A randomised control trial with a 4-month follow up. *Man. Ther.* **2016**, *26*, 132–140. [CrossRef] [PubMed]
23. Murray, M.; Lange, B.; Chreiteh, S.S.; Olsen, H.B.; Nornberg, B.R.; Boyle, E.; Sogaard, K.; Sjogaard, G. Neck and shoulder muscle activity and posture among helicopter pilots and crew-members during military helicopter flight. *J. Electromyogr. Kinesiol.* **2016**, *27*, 10–17. [CrossRef] [PubMed]
24. Holtermann, A.; Roeleveld, K.; Mork, P.J.; Gronlund, C.; Karlsson, J.S.; Andersen, L.L.; Olsen, H.B.; Zebis, M.K.; Sjogaard, G.; Sogaard, K. Selective activation of neuromuscular compartments within the human trapezius muscle. *J. Electromyogr. Kinesiol.* **2009**, *19*, 896–902. [CrossRef] [PubMed]
25. Gosselin, G.; Rassoulian, H.; Brown, I. Effects of neck extensor muscles fatigue on balance. *Clin. Biomech.* **2004**, *19*, 473–479. [CrossRef] [PubMed]
26. Juul-Kristensen, B.; Kadefors, R.; Hansen, K.; Bystrom, P.; Sandsjo, L.; Sjogaard, G. Clinical signs and physical function in neck and upper extremities among elderly female computer users: The NEW study. *Eur. J. Appl. Physiol.* **2006**, *96*, 136–145. [CrossRef]
27. Falla, D.; Dall'Alba, P.; Rainoldi, A.; Merletti, R.; Jull, G. Location of innervation zones of sternocleidomastoid and scalene muscles-a basis for clinical and research electromyography applications. *Clin. Neurophysiol.* **2002**, *113*, 57–63. [CrossRef]
28. Jakobsen, M.D.; Sundstrup, E.; Andersen, C.H.; Zebis, M.K.; Mortensen, P.; Andersen, L.L. Evaluation of muscle activity during a standardized shoulder resistance training bout in novice individuals. *J. Strength Cond. Res.* **2012**, *26*, 2515–2522. [CrossRef]
29. Andersen, L.L.; Andersen, C.H.; Mortensen, O.S.; Poulsen, O.M.; Bjornlund, I.B.; Zebis, M.K. Muscle activation and perceived loading during rehabilitation exercises: Comparison of dumbbells and elastic resistance. *Phys. Ther.* **2010**, *90*, 538–549. [CrossRef]
30. Rivard, J.; Unsleber, C.; Schomacher, J.; Erlenwein, J.; Petzke, F.; Falla, D. Activation of the semispinalis cervicis and splenius capitis with cervical pulley exercises. *Musculoskelet. Sci. Pract.* **2017**, *30*, 56–63. [CrossRef]
31. Schomacher, J.; Erlenwein, J.; Dieterich, A.; Petzke, F.; Falla, D. Can neck exercises enhance the activation of the semispinalis cervicis relative to the splenius capitis at specific spinal levels? *Man. Ther.* **2015**, *20*, 694–702. [CrossRef]

32. Gabriel, D.A.; Matsumoto, J.Y.; Davis, D.H.; Currier, B.L.; An, K.N. Multidirectional neck strength and electromyographic activity for normal controls. *Clin. Biomech.* **2004**, *19*, 653–658. [CrossRef]
33. Folland, J.P.; Williams, A.G. The adaptations to strength training: Morphological and neurological contributions to increased strength. *Sports Med.* **2007**, *37*, 145–168. [CrossRef]
34. Higbie, E.J.; Cureton, K.J.; Warren, G.L., 3rd; Prior, B.M. Effects of concentric and eccentric training on muscle strength, cross-sectional area, and neural activation. *J. Appl. Physiol.* **1996**, *81*, 2173–2181. [CrossRef]
35. Vikne, H.; Refsnes, P.E.; Ekmark, M.; Medbo, J.I.; Gundersen, V.; Gundersen, K. Muscular performance after concentric and eccentric exercise in trained men. *Med. Sci. Sports Exerc.* **2006**, *38*, 1770–1781. [CrossRef]
36. Jull, G.; Falla, D. Does increased superficial neck flexor activity in the craniocervical flexion test reflect reduced deep flexor activity in people with neck pain? *Man. Ther.* **2016**, *25*, 43–47. [CrossRef]
37. Aura, O.; Komi, P.V. Mechanical efficiency of pure positive and pure negative work with special reference to the work intensity. *Int. J. Sports Med.* **1986**, *7*, 44–49. [CrossRef]
38. Komi, P.V.; Kaneko, M.; Aura, O. EMG activity of the leg extensor muscles with special reference to mechanical efficiency in concentric and eccentric exercise. *Int. J. Sports Med.* **1987**, *8* (Suppl. 1), 22–29. [CrossRef]

applied sciences

MDPI

Article

Configurable 3D Rowing Model Renders Realistic Forces on a Simulator for Indoor Training

Ekin Basalp [1,*], Patrick Bachmann [1], Nicolas Gerig [2], Georg Rauter [1,2] and Peter Wolf [1]

[1] Sensory-Motor Systems Lab, Department of Health Sciences and Technology, Institute of Robotics and Intelligent Systems, ETH Zurich, 8092 Zurich, Switzerland; patricba@student.ethz.ch (P.B.); georg.rauter@unibas.ch (G.R.); peter.wolf@hest.ethz.ch (P.W.)

[2] BIROMED-Lab, Department of Biomedical Engineering, University of Basel, 4123 Basel, Switzerland; nicolas.gerig@unibas.ch

* Correspondence: basalp.ekin@hest.ethz.ch

Received: 31 December 2019; Accepted: 17 January 2020; Published: 21 January 2020

check for updates

Abstract: In rowing, rowers need outdoor and indoor training to develop a proficient technique. Although numerous indoor rowing machines have been proposed, none of the devices can realistically render the haptic, visual, and auditory characteristics of an actual rowing scenario. In our laboratory, we developed a simulator to support rowing training indoors. However, rendered forces with the initial rowing model, which was based on a simplified fluid dynamic model that approximated the drag/lift forces, were not perceived realistic enough for indoor training by expert rowers. Therefore, we implemented a new model for the blade–water interaction forces, which incorporates the three-dimensional rotation of the oar and continuously adjusts drag/lift coefficients. Ten expert rowers were asked to evaluate both models for various rowing aspects. In addition, the effect of individualization of model parameters on the perceived realism of rowing forces was elaborated. Based on the answers of the experts, we concluded that the new model rendered realistically resistive forces and ensured a smooth transition of forces within a rowing cycle. Additionally, we found that individualization of parameters significantly improved the perceived realism of the simulator. Equipped with a configurable rowing model, our simulator provides a realistic indoor training platform for rowers.

Keywords: 3D force modeling; rowing biomechanics; robot-assisted training; individualized training; virtual reality simulator; training of experts; sports engineering; rowing simulator; tendon based parallel robot; transversal vibration control

---

1. Introduction

In rowing races, success is determined by the minimum time to complete a pre-defined course track of commonly 2000 m [1]. Thus, mathematically, the main performance metric for competitive rowers is the average boat speed to finish the course [2]. Attaining a faster boat speed requires not only a physical effort but also an appropriate boat setup and a skillful technique to transfer the propulsion forces into speed [3–5]. Therefore, fitness and technique trainings are crucial for competitive rowers [5].

Outdoor rowing training may not always be feasible since performance of rowers can be affected by unpleasant weather conditions [6,7]. Indoor training conducted with rowing machines offers a valuable alternative to outdoor practice [8]. Over the past decades, various rowing machines have been developed to yield a realistic rowing training [9]. Overall, the developed machines can be grouped in four categories depending on the technical aspects: indoor rowing tanks, ergometers, ergometer simulators, and virtual reality simulators [10].

For indoor-tank rowing, two large unit water-tanks are used next to a fixed rowing boat [11]. Since the oars directly interact with the real water, the resistance at the oar blade is naturally present. However, water tanks require specially constructed large spaces, which might be challenging to maintain.

Ergometers provide a useful strength training opportunity indoors. On an ergometer, a rower interacts with a handle to simulate the resistance. Rendered resistance is mainly dependent on the pulling speed of the handle. Ergometers generate the desired resistance forces through friction-based interaction of a flywheel with air or water [12]. Since ergometers built with a stationary footrest were found to increase inertial forces and joint injuries compared to the on-water rowing [13], moving footrest systems were also proposed for ergometers such as Rowperfect® Indoor Sculler [14] and Concept2® Dynamic [15]. Thus, the boat's acceleration and deceleration at each stroke can be simulated on such ergometers.

Ergometers do not render the essential kinematic and dynamic characteristics of an oar movement. Thus, advanced ergometer simulators that incorporate the actual handling of a rowing oar with three degrees of freedom were developed [16–18]. In ergometer simulators, rowers manipulate one (sweep) [16] or two shortened oars (sculling) [17] as in the case of actual rowing boats. The rolling motion of an actual boat can also be simulated on ergometer simulators. However, rowing simulators do not incorporate actual water dynamics to render the resistive forces. Resistance is rendered with an air fan and disc brake system, and the resultant force perceived by the rower is mainly dependent on the speed of the oar. Additionally, none of the ergometer simulators provide any visual or auditory displays to bring the realism of the training closer to the on-water rowing experience. Thus, virtual reality simulators were developed to augment the realism of indoor rowing training [19–22].

The virtual rowing training platform SPRINT was developed at Scuola Superiore S.Anna, Pisa, Italy [19,20]. The SPRINT platform included two shortened oars with three degrees of freedom (DoF) each, a fixed foot stretcher, and a sliding seat. Visual rendering of the rowing scenario was coupled to the output of a mathematical model that combined the movement of boat, oars, and a rower. Rendering of hydrostatic vertical forces and oar weight was realized by a mass–spring system. The vertical force model only depended on the vertical rotation of the oar, i.e., the effect of longitudinal rotation of the oar was neglected. Rendering of the horizontal forces in the virtual water was based on a modified form of a dissipation mechanism used in Concept2® ergometers. Due to this mechanism, haptic rendering of the horizontal forces did not capture the fluid-dynamics characteristics of the oar blade–water interaction. Therefore, the haptic rendering of the overall forces in the SPRINT platform was not based on the actual kinematic and dynamic features of a rowing movement.

In our laboratory, we also developed a rowing simulator to support realistic training for sculling and sweep rowing indoors [23]. Thanks to the real-time control and dedicated hardware setup, our rowing simulator has been used for training of rowers [22] and motor learning related research such as augmented feedback designs [24,25] and robotic training strategies [26,27].

The haptic rendering of the rowing forces incorporated the effect of the 3-DoF rotation of an actual rowing oar and of the boat dynamics [28]. The propulsion forces were calculated according to an empirical model [29], which approximated the drag and lift forces acting on the blade [21]. Although the model was previously shown to be suitable for technique training [22], expert rowers suggested that the realism of the rendered resistive forces on the simulator could be improved. Therefore, in this study, we proposed a new three-dimensional force model that incorporates 3-DoF rotation of the oar and accounts for a smooth blade transition between air and water.

The new force model was implemented on our rowing simulator and assessed by ten expert rowers in a training experiment. In the first part of the experiment, rowers were asked to qualitatively assess both the existing oar-blade model [21] and the new model. We hypothesized that the new blade force model would be perceived as more realistic than the existing model. In the second part of the experiment, the benefit of individualizing the force model on expert rowers' acceptance of the rowing simulator as a training platform was investigated. We hypothesized that the individualization of model parameters would result in higher realism of perceived rowing forces of the simulator.

2. Materials and Methods

2.1. Participants

Ten expert rowers (five females, five males; age = 18–27 years, mean age = 21.8 years; experience of rowing = 5–15 years, mean experience = 7–8 years) were recruited for the experiment. Participants were either students at ETH Zurich and/or affiliated with a rowing club in Zurich or neighboring cities. We set the inclusion criteria for the 'expertness' in rowing as (i) intensive training with port-side sweep rowing for at least 3 years; and (ii) participation in at least one national and one international regatta as a port-side rower.

The current study was approved by the ETH Zurich Ethics Commission (EK 2017-N-75). Each participant was invited for a 2-h session conducted on a single day. Expert participants were verbally instructed about the goal, experimental protocol, assessment by means of a questionnaire, and the potential risks that might occur. In addition, participants were informed that they could withdraw from the study at any time without stating a reason and without facing any consequences. All participants signed a consent for participation and were reimbursed for their commitment with a gift (maximum value of CHF 20).

2.2. Rowing Setup

In the study, our custom-made rowing simulator was used (Figure 1) [21]. The rowing simulator consisted of an instrumented, shortened, single scull boat (Stämpfli Racing Boats AG, Zürich, Switzerland). The boat was placed on a platform in the middle of three 4.4 m × 3.3 m sized screens, i.e., one screen in front of the stern and one screen to each the left and right side of the boat [30].

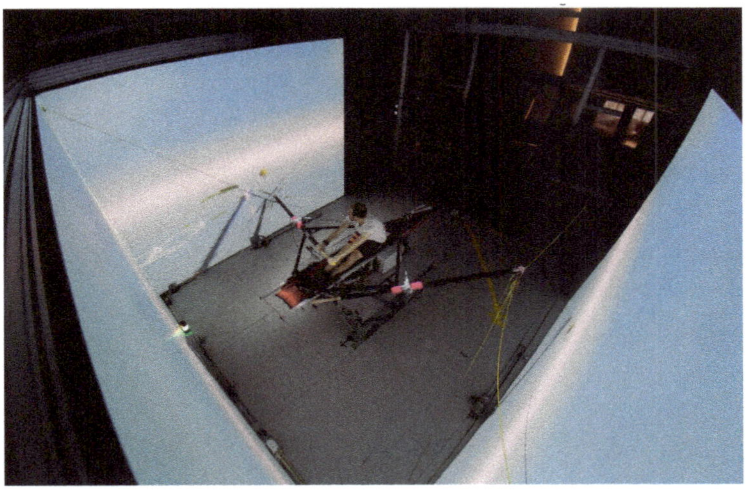

Figure 1. Rowing simulator setup with a shortened single scull boat and three screens surrounding the boat. The person rowing in the boat is the first author of the paper.

A visual scenario representing a plain lake environment along with a line of buoys defining the course was displayed with three projectors (Projection Design F3+) on the large screens. The visual scenario was programmed in Unity (Unity Technologies ApS, CA, USA) and synchronized to the kinematic input from the participant with a minimum update rate of 30 fps.

The auditory rendering of the rowing scenario was developed in C++. Two speakers (DELL A525 Zylux Multimedia Computer Speaker System), which were placed behind the bow of boat, were used to deliver the task inherent, virtual water–oar interaction sounds with an update rate of 30 Hz.

The haptic rendering of the rowing simulator was realized by a tendon-based parallel robot [31]. The robot consisted of five ropes attached to the end-effector, i.e., shortened oar, in parallel, and five actuation units that were placed on a fixed frame [21]. The ropes were guided from the respective motors through fixed deflection units to control the motion of the oar for rowing application. At the other end of the oar, the handle of the oar was held by the user with both hands.

In the setup, the shortened oar was longitudinally fixed inside the oarlock with a clamp and bearing, which restrained the end-effector to move on a spherical surface. Therefore, the user could manipulate the oar in the three remaining rotational degrees of freedom. These three rotational oar movements were the horizontal oar angle θ, the vertical oar angle δ, and the longitudinal oar angle φ. Although the robot rendered forces in all three directions at the end-effector, the longitudinal forces were not transferred to the user due to the mechanical constraints of the oarlock. Rendered forces at the end-effector were haptically displayed to the user as torques along the horizontal oar angle θ and the vertical oar angle δ. The longitudinal oar angle φ was only controlled by the user.

The hardware configuration and rowing model application on our simulator was set up for portside sweep rowing. In sweep rowing, an even number of rowers are seated in the boat. Each rower holds a single oar at the oar handle with both hands. The haptic rendering of the rowing forces for this study was calculated for the portside and mirrored onto the bow side in the Matlab/Simulink® model to simulate a pair on the rowing simulator.

2.3. *3D Modeling of Rowing Forces*

2.3.1. Sub-Phases of a Rowing Stroke

Rowing is a cyclic movement that is executed with the coordination of a complete set of body limbs to control an oar motion to propel the boat [32]. One rowing cycle, i.e., stroke, can be partitioned into four sub-phases, which features specific kinematic and kinetic characteristics. A cycle begins with the catch, followed by drive, release, and the recovery [1]. In the catch, the rower rapidly lifts the oar handle to insert the blade into the water and prepares for application of propulsive forces. During drive, the rower pulls the oar handle towards his/her body by a well-coordinated movement of legs, trunk, and arms, while the seat is moved towards the bow and the oar blade is moved through water with power. In the release, the oar handle is swiftly pushed down to remove the blade from the water and oar is feathered to reduce the air resistance [33]. In the recovery, when the blade is fully in the air, the rower flexes trunk and legs while extending the arms towards the stern of the boat to prepare for the next cycle [32].

2.3.2. Simulator Coordinates for the Rowing Models

To define the motion of the oar and resulting rowing forces, four different coordinate systems were defined. A boat-fixed coordinate system, i.e., $\chi_{Boat} = \{B_x, B_y, B_z\}$, was defined on the boat's center of gravity such that B_y was directed towards the stern and B_z upwards (Figure 2). An oar-fixed coordinate, i.e., $\chi_{Oar} = \{O_x, O_y, O_z\}$, was defined on the tip of the oar blade and was rotated with the blade movement. A special configuration of χ_{Oar} was defined as the initial oar coordinate system, i.e., $\chi_{Oar}^{initial} = \{O_x^i, O_y^i, O_z^i\}$, when the portside oar was in the initial configuration ($\theta = 0°, \delta = 0°, \varphi = 0°$) and perpendicular to the boat. Vertical (δ), longitudinal (φ), and horizontal (θ) rotations of the oar were the mathematically positive angles, i.e., defined in a counterclockwise (CCW) direction around O_x^i, O_y^i, O_z^i axes, respectively. To calculate the forces acting against the oar blade resulting from the relative blade velocity, a stream coordinate system, i.e., $\chi_{Stream} = \{S_x, S_y, S_z\}$, was defined on the blade such that $S_y = -B_y$ and $S_z = B_z$.

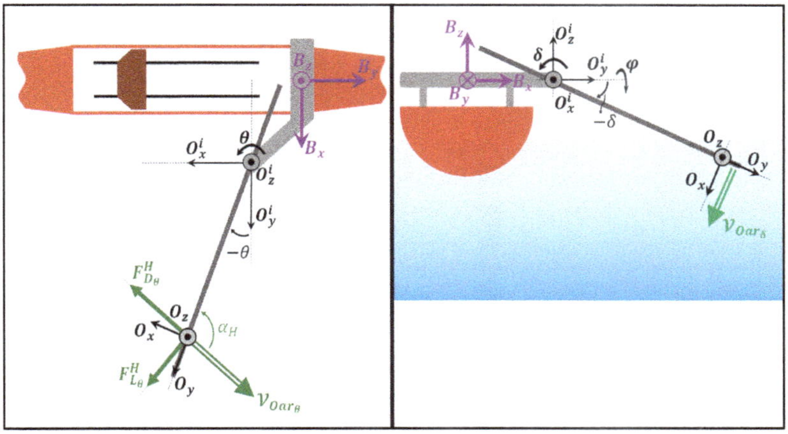

Figure 2. Coordinate systems and forces acting on the blade in the BIFC (binary blade immersion, fixed drag/lift coefficients) model. Forces due to horizontal oar movement are shown in the left pane (top view). the vertical force on the blade is shown in the right pane (back view of the boat).

Calculated rowing forces at the portside (right) oar blade were mirrored onto the bow side (left) blade, simulating identical rowers for a pair boat. As a result, the rotation of the simulated boat around the B_z-axis was assumed to be 0°, and the linear displacement of the boat was constrained to be along the B_y-axis for simplicity.

2.3.3. Description and Implementation of Base (1D) Models of Rowing for Blade–Water Interaction

Both rowing models presented in this paper were originally based on the modeling of drag and lift forces acting on a blade immersed into the water, which was provided by [34] as follows:

$$F_{Drag} = \frac{1}{2} C_D \rho A \left(\vec{V}_{o/w}\right)^2 \tag{1}$$

$$F_{Lift} = \frac{1}{2} C_L \rho A \left(\vec{V}_{o/w}\right)^2 \tag{2}$$

where $\rho = 1000 \frac{\text{kg}}{\text{m}^3}$ is the density of the water; $\vec{V}_{o/w}$ is the relative blade velocity between blade and water; A is the projected area of the blade (calculated perpendicular to the front of the blade); C_D and C_L are the angle of attack dependent dimensionless drag and lift force coefficients, respectively.

Both equations were based on an experiment in which the blade was inserted into the water tank and held fixed against a one dimensional (1D) incoming flow [34]. Thus, the relationship between the forces and the velocity of the blade was only considered on the horizontal plane.

An alternative version of the 1D rowing model described by Equations (1) and (2) was implemented for our rowing simulator by [30]. In the simplified version of the model, drag and lift forces acting on the oar blade in the horizontal plane were approximated as follows:

$$F^H_{Drag} = C_D(\alpha_H) \left(\vec{V}_{Oar_\theta}\right)^2 \tag{3}$$

$$F^H_{Lift} = C_L(\alpha_H) \left(\vec{V}_{Oar_\theta}\right)^2 \tag{4}$$

where

$$C_D(\alpha_H) = 2 C_L^{max} sin^2 \alpha_H \tag{5}$$

$$C_L(\alpha_H) = C_L^{max} sin2\alpha_H \tag{6}$$

and the angle of attack α_H was defined between the longitudinal axis of the oar and $\vec{V}_{Oar_\theta}$, i.e., the oar blade velocity that depends on the 1-DoF horizontal rotation of the oar (θ). C_L^{max} was calculated from the maximum value of lift coefficient C_L whose values were presented based on α_H in [34].

In principle, the force model presented in [34] with Equations (1) and (2), and the interaction forces calculated with Equations (3)–(6) are based on the assumption that the magnitude of drag and lift forces depend quadratically on the blade velocity. To our knowledge, Equations (3) and (4) were initially mentioned for drag and lift oar blade forces in [29] based on the authors' interpretation of the results from the experiments and force model presented in [35]. In [35], the authors empirically found an approximation of the drag C_D and lift C_L coefficients that depend on the angle of attack (Equations (5) and (6)), which were further considered for the modeling of rowing in [29,36], and calculation of the efficiency of rowing oars [37].

In [30], the oar blade was assumed to be squared (vertical with respect to water surface, i.e., $\varphi = 0°$), and the resultant force acting on the blade, i.e., F_θ, was only dependent on the horizontal (θ) movement of the oar.

$$F_\theta = \sqrt{\left(F_{Drag}^H\right)^2 + \left(F_{Lift}^H\right)^2} = 2C_L^{max}\left(\vec{V}_{Oar_\theta}\right)^2 sin\alpha_H \tag{7}$$

However, for a real rowing stroke, the resultant forces on a blade do not only depend on the horizontal rotation of the oar θ but also on the vertical δ and longitudinal rotation φ. Thus, two rowing models that incorporate three rotations of the oar are presented in the following sub-sections.

2.3.4. 3D rowing Model with Binary Blade Immersion and Fixed Drag/Lift Coefficients (BIFC)

In 2010, the 1D rowing model implemented by [30] was upgraded by [21] to a 3D rowing model, which incorporates the 3-DoF rotation of the oar (Figure 2). In [21], Equation (7) that was used for the θ angle-dependent forces was also applied to calculate the δ angle-dependent forces on the oar blade, i.e., F_δ, due to the vertical rotation (δ):

$$F_\delta = 2C_L^{max}\left(\vec{V}_{Oar_\delta}\right)^2 \tag{8}$$

where $V_{Oar_\delta} = l_{Oar}\dot{\delta}$, and l_{Oar} is the length of the oar.

The underlying assumption in Equation (8) was that the oar blade moved in the water with a feathered orientation (parallel with respect to the water surface, i.e., $\varphi = 90°$). No additional angle of attack was employed since the influence of an incoming stream in this direction was omitted. F_θ and F_δ were combined with the longitudinal oar angle (φ) to calculate the horizontal (F_{Oar}^H) and vertical (F_{Oar}^V) rowing forces based on three rotational angles of the oar blade as follows:

$$F_{Oar}^H = F_\theta cos^2\varphi + F_\delta \frac{sin2\varphi}{2} \tag{9}$$

$$F_{Oar}^V = F_\theta \frac{sin2\varphi}{2} + F_\delta sin^2\varphi \tag{10}$$

from which the overall oar blade force (F_{Oar}) could be calculated as

$$F_{Oar} = \sqrt{\left(F_{Oar}^H\right)^2 + \left(F_{Oar}^V\right)^2} = F_\theta cos\varphi + F_\delta sin\varphi. \tag{11}$$

Finally, to calculate the F_θ and F_δ, the value of the constant C_L^{max} was chosen as $42.25 \frac{\mathrm{N}}{(\mathrm{m/s})^2}$ for a big blade area (A) of 0.13 m^2 based on the calculations in [29,37].

2.3.5. 3D Rowing Model with Linear Blade Immersion and Adjusted Drag/Lift Coefficients (LIAC)

In 2019, we developed a new 3D rowing model (linear blade immersion, adjusted drag/lift coefficients or LIAC), which was essentially based on Equations (1) and (2). In the model, F_{Drag} was defined in the opposite direction of the oar blade velocity while F_{Lift} was perpendicular to drag. During a continuous movement of the oar blade in the water, the magnitude and the direction of F_{Drag} and F_{Lift} change based on the values of drag (C_D) and lift (C_L) force coefficients, respectively. Furthermore, C_D and C_L were dependent on the instantaneous angle of attack. For the horizontal plane ($\mathcal{P}_{xy}$), we used a look-up table based on [34], which delivered the values of C_D and C_L depending on the calculated instantaneous angle of attack. Extending the model to the vertical plane ($\mathcal{P}_{yz}$), we employed a separate instantaneous angle of attack (Figure 3).

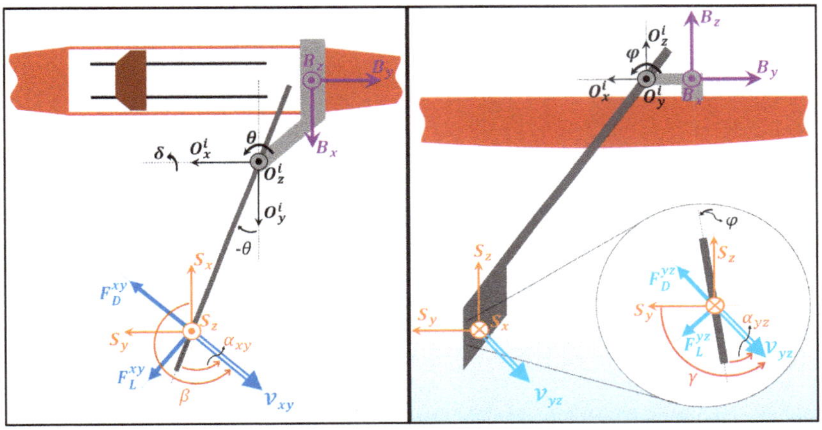

Figure 3. Coordinate systems and forces acting on the oar blade in the LIAC (linear blade immersion, adjusted drag/lift coefficients) model. Blade forces on the horizontal plane are shown in the left pane (top view). Blade forces on the vertical plane are shown in the right pane (side view of the boat). In the right pane, the enlarged circle shows the cross-section of the blade.

As a first step, the relative oar-blade velocity vector was defined in the boat-fixed coordinate system (χ_{Boat}) as $\vec{V}_{xyz} = \begin{bmatrix} V_{Oar_x} & V_{Oar_y} & V_{Oar_z} \end{bmatrix}^T$. To calculate the angles of attack for the oar blade motion in the horizontal ($\mathcal{P}_{xy}$) and vertical ($\mathcal{P}_{yz}$) planes, the projection of $\vec{V}_{xyz}$ vector onto these planes were described as $\vec{V}_{xy} = \begin{bmatrix} V_{Oar_x} & V_{Oar_y} \end{bmatrix}^T$ for the horizontal and $\vec{V}_{yz} = \begin{bmatrix} V_{Oar_y} & V_{Oar_z} \end{bmatrix}^T$ for the vertical planes.

According to the model presented in [34], the angle of attack used for the horizontal plane ($\mathcal{P}_{xy}$) was defined between the velocity of the oar blade ($\vec{V}_{xy}$) and the longitudinal (transverse) axis of the oar blade, yielding α_{xy}. Following a similar approach, the angle of attack used for the vertical plane ($\mathcal{P}_{yz}$) was defined between the $\vec{V}_{yz}$ component of the oar blade velocity and the cross section (vertical axis) of the blade, resulting in α_{yz}.

For the implementation of α_{xy} and α_{yz} calculations in a Matlab/Simulink® model, the following relationships were used (Figure 3):

$$\alpha_{xy} = \beta - \pi - \theta \tag{12}$$

$$\alpha_{yz} = \gamma - \frac{\pi}{2} + \varphi, \tag{13}$$

where β is an angle that is defined between $\vec{V}_{xy}$ and S_x in the horizontal plane (CCW around S_z), and γ is the angle between $\vec{V}_{yz}$ and the S_y (CCW around S_x).

Subsequently, drag and lift coefficients for both the horizontal and vertical forces were defined as $C_D^{xy} = f_D(\alpha_{xy})$, $C_L^{xy} = f_L(\alpha_{xy})$, $C_D^{yz} = f_D(\alpha_{yz})$, and $C_L^{yz} = f_L(\alpha_{yz})$, where the functions f_D and f_L were implemented as a look-up table based on the presented experimental values [34].

In the final step, the projected areas of the oar blade in the horizontal and vertical planes were calculated as follows:

$$A^{xy} = A_b p_{imm} cos\varphi \tag{14}$$

$$A^{yz} = A_b p_{imm} \tag{15}$$

where $A_b = 0.13$ m^2 was chosen for a big blade used in sweep rowing. p_{imm} is the proportion of the oar blade in the water, which was calculated as

$$p_{imm} = \begin{cases} 0 & \text{for } \delta_{Surface} < \delta \\ \left| \frac{\delta - \delta_{Surface}}{\delta_{InWater} - \delta_{Surface}} \right| & \text{for } \delta_{InWater} < \delta \leq \delta_{Surface} \\ 1 & \text{for } \delta \leq \delta_{InWater} \end{cases} \tag{16}$$

where $\delta_{Surface}$ and $\delta_{InWater}$ are the vertical oar angles at which the tip of the blade touches the water surface and the blade is fully immersed in the water, respectively. Placement of Equation (16) in Equations (14) and (15) allowed for a smooth transition of forces between the drive and recovery.

For the calculation of A^{xy}, the projected area was assumed to be dependent only on the longitudinal oar angle φ since horizontal rotation of the oar θ was included in the calculation of horizontal angle of attack α_{xy}, which would update the magnitude of drag and lift coefficients according to the fluid dynamic model [34]. Moreover, for the calculation of A^{yz}, we assumed that the projected area would only depend on the immersion ratio of the blade since the vertical angle of attack α_{yz} was dependent on the longitudinal oar angle φ. In addition, a further trigonometric adjustment of the A^{yz} based on the horizontal angle θ resulted in low forces in the vertical plane during the pilot tests conducted on the simulator, which were not preferred by the experts. Therefore, we did not include any dependency of other oar angles to estimate A^{yz}.

Based on the previous calculations, the drag and lift forces of the model (LIAC) that resulted from the horizontal, vertical, and longitudinal rotation of the oar blade were calculated as follows:

$$F_{Drag}^{xy} = \frac{1}{2} C_D^{xy} \rho A^{xy} \left(\vec{V}_{xy}\right)^2 \tag{17}$$

$$F_{Lift}^{xy} = \frac{1}{2} C_L^{xy} \rho A^{xy} \left(\vec{V}_{xy}\right)^2 \tag{18}$$

$$F_{Drag}^{yz} = \frac{1}{2} C_D^{yz} \rho A^{yz} \left(\vec{V}_{yz}\right)^2 \tag{19}$$

$$F_{Lift}^{yz} = \frac{1}{2} C_L^{yz} \rho A^{yz} \left(\vec{V}_{yz}\right)^2. \tag{20}$$

To calculate the emerging horizontal ($F_H^{\chi_B}$) and vertical ($F_V^{\chi_B}$) plane forces acting on the oar blade in the boat-fixed coordinate system (χ_{Boat}), the drag and lift forces resulting from Equations (17)–(20) were transformed in the horizontal and vertical planes:

$$\vec{F}_H^{\chi_B} = {}^{Boat}\mathcal{R}_{Stream}(-\pi)\, {}^{Stream}\mathcal{R}_{xy}(\beta) \begin{bmatrix} F_{Drag}^{xy} & F_{Lift}^{xy} & 0 \end{bmatrix}^T \tag{21}$$

$$\vec{F}_V^{\chi_B} = {}^{Boat}\mathcal{R}_{Stream}(-\pi)\, {}^{Stream}\mathcal{R}_{yz}(\gamma) \begin{bmatrix} 0 & F_{Drag}^{yz} & F_{Lift}^{yz} \end{bmatrix}^T \tag{22}$$

where ${}^{Boat}\mathcal{R}_{Stream}(\theta_{rot} = -\pi)$ was used to define the χ_{Stream} in the boat-fixed coordinate system (χ_{Boat}), and ${}^{Stream}\mathcal{R}_{xy}(\beta)$ and ${}^{Stream}\mathcal{R}_{yz}(\gamma)$ were the transformation matrices in the horizontal and vertical planes, respectively.

According to the model presented by [34], the resulting plane forces of $\overrightarrow{F}_H^{X_B}$ and $\overrightarrow{F}_V^{X_B}$ define two distinct 2D models in the horizontal ($\mathcal{P}_{xy}$) and vertical ($\mathcal{P}_{yz}$) planes, respectively. In order to generate a 3D force (F_{Oar}) from two 2D models ($\overrightarrow{F}_H^{X_B}$ and $\overrightarrow{F}_V^{X_B}$), we employed a combination factor, which would satisfy the following constraints:

$$F_{Oar} = f\left(V_{xyz}^2\right)C_1 f\left(V_{xy}^2\right) + C_2 f\left(V_{xy}^2\right), \text{ where } C_1,\ C_2 \leq 1 \text{ and } C_1 + C_2 = 1. \tag{23}$$

C_1 and C_2 are the combination factors for the horizontal and vertical planes, respectively. The underlying assumption for the relationship in Equation (23) was that the blade forces calculated in each plane, i.e., $f\left(V_{xy}^2\right)$ and $f\left(V_{yz}^2\right)$, were dependent on the square of blade velocity that was projected onto the horizontal and vertical planes, respectively. Thus, to satisfy the relationship in Equation (23), C_1 and C_2 were defined as follows:

$$C_1 = \frac{V_{xy}^2}{V_{xy}^2 + V_{yz}^2} \tag{24}$$

$$C_2 = \frac{V_{yz}^2}{V_{xy}^2 + V_{yz}^2}. \tag{25}$$

Therefore, the 3D force resulting from Equations (23)–(25) was defined as

$$\overrightarrow{F}_{Oar} = C_1 \overrightarrow{F}_H^{X_B} + C_2 \overrightarrow{F}_V^{X_B}. \tag{26}$$

2.3.6. Rendering of Forces on the Rowing Simulator

To calculate the forces, rowing models were implemented in Matlab/Simulink® (r2013b, The MathWorks, Inc., Natick, MA, USA). To render the rowing forces at the end-effector, the actuators driving each rope were simultaneously controlled by the Matlab/Simulink® model running on an xPC-target at 1 kHz. The implemented rowing models required all oar angles (θ, δ, φ), the seat position x_s, and the seat acceleration $\ddot{x}_s$ as inputs to calculate the forces. Longitudinal oar angle φ and both of the seat parameters were measured with respective sensors in the instrumented boat. To retrieve the values of horizontal θ and vertical δ oar angles, the end-effector's position was initially calculated from the forward kinematics of the robot, i.e., change of rope lengths. Subsequently, a coordinate transformation was applied to the end-effector's position to calculate the horizontal θ and vertical δ oar angles. The longitudinal oar angle φ was measured with the potentiometers on the oar.

2.3.7. Control of Rope Tension Forces during Recovery

In the presented rowing models, the resistance forces resulting from air friction during recovery were not explicitly modeled. Instead, the resistance was provided with the residual forces acting at the oarlock, which results from tensioning the ropes of the tendon based parallel robot to control the translational motion of the end-effector, i.e., shortened oar.

A technical challenge regarding the use of a tendon based parallel robot is that rendering of high rowing forces at the end-effector due to rapid oar movements can result in unwanted transversal vibrations on the ropes and resonance on the robot. Since the vibrations mainly occur when the ropes are flexed, the stiffness control of the robot plays an important role in the haptic rendering of the rowing models on the simulator. Therefore, the natural (eigen) frequency of the system needs to be determined to understand the vibrational characteristics and assure the stiffness of the robot.

Assuming that the ropes are modeled as taut strings in our robot, the wave equation that governs the natural transversal vibration of the rope can be described as follows [38]:

$$\omega_n = \frac{n\pi}{l_i}\sqrt{\frac{\tau_i}{\zeta}} \tag{27}$$

where ω_n^i is the eigen frequency of the *i*th rope, *n* is *n*th frequency, l_i is the length of the *i*th rope, τ_i is the tension force on the *i*th rope, and ζ is the mass per unit rope length.

According to Equation (27), the natural frequency for transversal vibration of the rope is proportional to the square root of the rope tension and inversely dependent on the length of the rope. Confirming the relationship presented in Equation (27), three out of five ropes of the tendon based parallel robot are lengthened during catch-to-drive transition in the simulator, which reduces ω_n and thus results in increased transversal vibrations on the ropes. However, the vibration on the ropes can be suppressed if the eigenfrequency ω_n is increased. Although increasing the tension on the ropes would yield a stiffer tendon-based parallel robot, increased stiffness would also yield higher friction forces at the oarlock, which may cause unrealistically high resistive forces during recovery. Therefore, we could not use fixed, high tension forces on the ropes during whole rowing cycle to counteract the vibrations resulting from high rowing forces during catch and drive.

In the previous studies on the rowing simulator, the vibration problem could be suppressed by setting the minimum rope tension force (τ_{rope}^{min}) to a fixed value of 50 N , since the rendered forces at the end-effector were low due to the fact that body–arm rowing was used as the rowing task [24–27]. However, this minimum rope tension of 50 N yielded neither realistic recovery forces nor eliminated the transversal vibration issue for the rapid, full-body rowing movement. Therefore, we implemented a position-dependent τ_{rope}^{min} adjustment in the LIAC model. Three different thresholds for τ_{rope}^{min} were defined dependent on the vertical movement of the oar (δ) (Figure 4): τ_{rec}^{min} for the rope tension during recovery, $\tau_{transit}^{min}$ during catch and release, and τ_{drive}^{min} during drive.

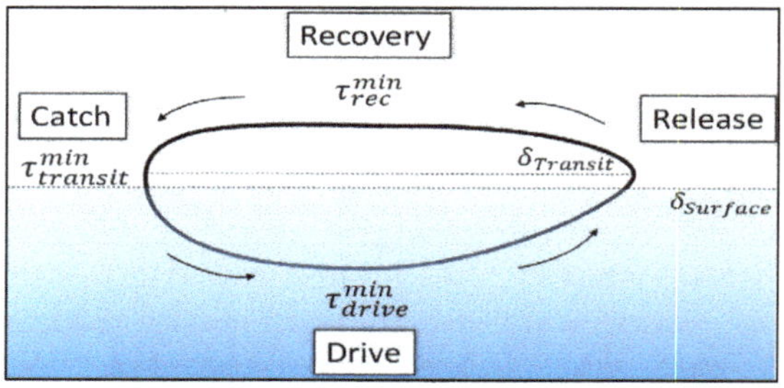

Figure 4. Illustration of the vertical oar rotation-dependent rope tension adjustment for the tendon based parallel robot. A typical rowing cycle is illustrated with the thick elliptic line. Four arrows show the direction of the oar blade within the cycle.

The linear adjustment of the rope tension forces were defined as follows:

$$\tau_{rope}^{min} = \begin{cases} \tau_{rec}^{min} & \text{for } \delta_{Transit} < \delta \\ \tau_{rec}^{min} + \left(\tau_{transit}^{min} - \tau_{rec}^{min}\right)\left(\frac{\delta_{Transit}-\delta}{\delta_{Transit}-\delta_{Surface}}\right) & \text{for } \delta_{Surface} < \delta \leq \delta_{Transit} \\ \tau_{transit}^{min} + p_{imm}\left(\tau_{drive}^{min} - \tau_{transit}^{min}\right) & \text{for } \delta \leq \delta_{Surface} \end{cases} \quad (28)$$

where $\delta_{Transit}$ is an empirically set value (1°) for the vertical oar angle above the simulated water surface. In Equation (28), initially set values for the parameters were $\tau_{transit}^{min} = 50$ N, $\tau_{drive}^{min} = 60$ N.

2.3.8. Configurable Parameters of LIAC for an Individualized Rowing Experience on the Simulator

A pilot study with three expert rowers, who were not included in main study, was conducted to assess and calibrate the newly developed LIAC model. The goal was to adjust the LIAC parameters, which would yield realistic forces for all sub-phases of the cycle and a corresponding boat velocity. Thus, we chose four parameters as follows that modulated the calculation of the forces: drag and lift adjustment coefficient (G_{dl}) for drive, curvature of the rowing blade (φ_{bias}) for catch and release, minimum rope tension force (τ_{rec}^{min}) for recovery, and boat drag coefficient (C_{Bdrag}) for the boat velocity.

For drive, three expert rowers reported that the resulting 3D model forces were greater than the forces felt during actual rowing. Thus, to calibrate the forces in the horizontal and vertical plane, we employed an additional gain, i.e., G_{dl}, for the adjustment of drag and lift coefficients:

$$C_{D'}^{xy} = G_{dl} f_D\left(\alpha_{xy}\right) \tag{29}$$

$$C_{L'}^{xy} = G_{dl} f_L\left(\alpha_{xy}\right) \tag{30}$$

$$C_{D'}^{yz} = G_{dl} f_D\left(\alpha_{yz}\right) \tag{31}$$

$$C_{L'}^{yz} = G_{dl} f_L\left(\alpha_{yz}\right) \tag{32}$$

where D' and L' subscripts define the adjusted drag and lift coefficients, which were replaced in Equations (17)–(20). In the end of the pilot test, G_{dl} was set to 0.35 by taking the average of the preferred gains reported by three expert rowers.

Another critique of the pilot test rowers was that the vertical forces during the air–water transition were low. Thus, we increased the vertical forces in the catch and release by introducing a bias angle, i.e., φ_{bias}, which was assumed to account for the lateral curvature of the rowing blade. Originally, the oar blade presented in the LIAC model was designed as a flat plate. Thus, introducing φ_{bias} yielded:

$$\varphi' = \varphi_{bias} + \varphi_{act} \tag{33}$$

where $\varphi_{act} = \varphi$ is the actual longitudinal rotation of the oar blade and φ' is the adjusted oar blade rotation used in the Matlab/Simulink® implementation. The initial value of φ_{bias} was set to 2°.

The minimum rope tension force in the tendon based parallel robot was adjusted for the rowing forces during recovery (τ_{rec}^{min}). Since τ_{rope}^{min} was set to a fixed value of 50 N for the overall cycle in the BIFC model, the initial value of τ_{rec}^{min} in the LIAC model was also set to 50 N for the recovery ($p_{imm} = 0$). All three rowers judged that the recovery force rendered by LIAC was unrealistically resistive; thus, an average value of 35 N was set for a smooth manipulation of the oar in the recovery.

Finally, pilot study rowers were asked to assess how well the perceived boat velocity matched the applied forces on the simulator. Based on the implementation of the rowing model, the boat velocity $\dot{y}_{boat}$, was dependent on the driving forces applied by the rowers on the oar blades ($F_{propulsion}$), inertia forces resulting from the movement of the rowers and the oar blades ($F_{inertia}$), and drag forces acting on the boat ($F_{BoatDrag}$).

$$m_{boat}\ddot{y}_{boat} = F_{Net} = F_{propulsion} - F_{inertia} - F_{BoatDrag} \tag{34}$$

where m_{boat} is the combined mass of rowers and the boat, and $\ddot{y}_{boat}$ is the acceleration of the boat in B_y-axis [28]. In Equation (34), $F_{BoatDrag}$ is calculated as in [29] as follows:

$$F_{BoatDrag} = C_{Bdrag}\left(\dot{y}_{boat}\right)^2 \tag{35}$$

According to Equation (34) and Equation (35), a modulation of C_{Bdrag} affected the acceleration and velocity of the boat, which also altered the calculated drag and lift forces from Equations (17)–(20).

The initial value of C_{Bdrag} in the LIAC model was set to 4.94 $\frac{\text{N}}{(\text{m/s})^2}$ based on the calculations for a pair type of boat in [30]. Pilot study rowers judged that the perceived boat acceleration and velocity (from the visual display as well as resulting forces on the blade) felt realistic for the given initial value. Although C_{Bdrag} was not further modulated in the pilot test but depended on the type and size of the boat [29], C_{Bdrag} was implemented in the model to be configurable by the rowers.

In the pilot test, we observed that the three rowers individually preferred slightly different values for the presented parameters, yielding a variety of rowing model settings. Indeed, rowers may differ in many aspects, e.g., anthropometric measures, technique, strength, accustomed water conditions (river, lake, or sea), preferred boat rigging, etc. Therefore, we decided to present the previously chosen model parameters, i.e., G_{dl}, φ_{bias}, τ_{rec}^{min}, and C_{Bdrag}, also to the main study expert rowers as a means of adjusting the rowing model to their individual preferences. Initial values of all the model parameters for the main study were determined by taking the average of preferred values of pilot study rowers.

2.4. Experimental Protocol

The study consisted of two parts. In the first part, five training blocks were conducted (Figure 5). Each of the blocks served to focus on a different aspect of the rowing movement. In each block, both the BIFC and LIAC models were presented twice in a randomized manner, yielding four trials within the block. The order of model presentation was based on Latin square randomization. The order of models presented within a training block was different in each block for a participant. Participants were not explicitly informed about the number of different models or the order of the model presentation. They were only instructed to focus on the evaluation of the movement aspect in each of the trials within the training block. Each trial with the presented model lasted for two minutes unless the participant decided to finish the trial and rate the model on the questionnaire.

Trainings	Training 1					Training 2					Training 3					Training 4					Training 5					Training 6	
Trial	1	2	3	4		1	2	3	4		1	2	3	4		1	2	3	4		1	2	3	4		Conf.	Assm.
Model	BIFC	LIAC	LIAC	BIFC	Break	BIFC	LIAC	BIFC	LIAC	Break	LIAC	LIAC	BIFC	BIFC	Break	LIAC	BIFC	BIFC	LIAC	Break	BIFC	BIFC	LIAC	LIAC	Break	LIAC	
Questions	Q1 & Q2					Q3 & Q4					Q5 & Q6					Q7 & Q8					Q9 & Q10					Q1 to Q10	
Time [min]	2	2	2	2	2	2	2	2	2	2	2	2	2	2	2	2	2	2	2	2	2	2	2	2	5	5	5

Figure 5. Experimental protocol. Training 1 to 5 were run in the first part of the experiment. Training 6 was run in the second part. Conf. and Assm. stand for Configuration and Assessment, respectively.

Prior to the Training-1, participants were not provided with an additional familiarization trial with any of the models. After the completion of each training block (four trials), a two-minute break was given. The first part of the experiment ended with the completion of Training-5.

In the second part of the experiment, i.e., Training-6, the participant was only provided with the LIAC model. The participants were given a minute to get familiar (again) with the model. Thereafter, experts were free to configure the parameters that would adjust the relevant rowing forces within four to five minutes. Configuration of the LIAC model resulted in the 'individualized' model, i.e., IND. After the individual configuration, participants were given one more minute to row with the final set of configuration. Finally, participants were asked to re-evaluate each of the questions in the previously presented questionnaire. The final evaluation typically took five minutes.

2.5. Task

The goal of the experiment was to qualitatively assess the realism of the presented rowing models for different aspects of the rowing movement. Experts were asked to mainly perform a full-body rowing cycle by coordinating their body limbs. They could also try other types of movements, e.g., half-slide, body-arm, only-arms rowing, if they wanted. They were free to decide to row with a feathered, i.e., horizontal oar blade with respect to water surface, or a squared blade in the recovery. In Training-1 and Training-2, the rowers could also pause at a certain posture and insert the squared blade into virtual water or tap the feathered blade onto the virtual water surface.

Experts were instructed to consider the experiment on the rowing simulator as if it was a technical training on the water, i.e., rowing with low to medium stroke rates and 40–70% of maximum force application. To capture the realism of the rowing forces with both compared models, participants were not explicitly constrained to any stroke rate. However, due to our instruction, they were observed to row with a range between 12 and 28 strokes per minute.

In the first part of the study, expert rowers were asked to focus on specific aspects of the rowing movement. To familiarize the expert rowers with the simulator and compare the presented force models, rather secondary/inessential aspects of a rowing stroke were considered in the first two trainings. In Training-1, the focus was on the horizontal movement of the blade above the water surface. Rowers deliberately slid the feathered blade on the virtual water surface and tried braking with the squared blade, which is a typical move to stop the boat during water outings. In Training-2, vertical movement of the oar during air to water transition was considered. In this training, rowers tried to distinguish the vertical forces between the squared blade insertion into water and the feathered blade tapping onto water surface. In addition, they were requested to purposely square the oar blade near to the water surface to feel the vertical reaction forces from the water surface.

In Training-3, the focus was on the forces during drive and recovery. In Training-4, the focus was given to the oar blade forces during the air–water transition, i.e., catch and release. In Training-5, the rowers were asked to rate the overall perception of the rowing forces considering the whole cycle. Finally, they were asked to rate the realism of the flow (speed) of the visual scenario with respect to their stroke rate and the force they applied. Completion of every trial was followed by filling out a questionnaire.

In the second part of the study, i.e., Training-6, participants were asked which particular forces they would have liked to adjust for their individual preferences. Upon participants' requests, the study instructor set the respective parameters from the graphical user interface to update the relevant forces of the active rowing model. Participants were asked to give their verbal comments to fine tune the force model according to their rowing preferences. The configurable parameters were G_{dl}, φ_{bias}, τ_{rec}^{min}, and C_{Bdrag}. Once the participants were satisfied with the final individualized parameter settings, they were asked to assess the realism again.

2.6. Questionnaire

Two questionnaires with the same questions were prepared for both parts of the experiment. The questionnaires were used for the qualitative comparison of the rowing models not only to each other but also to the real rowing scenario on the water.

The questionnaires consisted of one page for each training block. On top of every page, the sub-goal of the training block and an 11-point Likert-scale was presented. On the Likert-scale, '0 = Not realistic at all', '5 = Moderate (Neutral)', and '10 = Feels like on the water' statements were inserted below the respective values. In the middle of the page, two questions were stated for every training aspect. A written instruction and an illustration were present to help participants understand the questions. The designs of both questionnaires were the same except the trial counts within the blocks. For the first part, i.e., Training-1 to 5, Questionnaire-1 included four boxes (one for each trial) for the rating input for two questions at the bottom of the page. For the second part, Questionnaire-2 included two boxes for two questions: an upper box for the mean value of LIAC that was calculated by the study instructor after the first part, and a lower box for the new evaluation rating input.

Both questionnaires were prepared in Microsoft Word. The questionnaires were displayed on a monitor, which was placed to the front-left side of the participant. Participants were also provided a keypad input device, which was placed on the left side of the boat, to rate the respective aspect of the rowing movement. Participants entered numerical values from zero to ten using the keypad to fill the question boxes for each trial. Participants were left alone in the simulator, and the study instructor waited outside to avoid affecting the experts by any means during the model evaluation.

2.7. Questionnaire Analysis (Statistics)

For the statistical analysis of questionnaire evaluation, Matlab® 2017a and RStudio (Integrated Development Environment for R, version 1.1.463, RStudio Team, 2018) software were used. Separate analyses were conducted for each part of the study since the goal in each analysis was different.

For the first part, to check if the compared base rowing models (BIFC, LIAC) significantly differed in terms of realism based on the qualitative evaluation with the Questionnaire-1, the following linear mixed effect (*lme*) model was used.

$$Q_i \sim BaseModel + (1 \mid Expert) \qquad (lme\text{-}1)$$

where Q_i is the dependent variable that represents the questions in the questionnaire, and $i = 1, \ldots, 10$ is the index of each question. *BaseModel* is the categorically independent variable with two levels defined as BIFC and LIAC. Since the models were presented twice in the training blocks, both rating values were used for each rowing model for every Q_i in the *lme-1*. *Expert* is defined as the random factor since ratings in both models belonged to one participant.

A separate analysis was conducted for the effect of individualization in the second part of the study. To examine if the individual adjustment of forces yielded a significant improvement in the realism ratings in Questionnaire-2 for the rowing on the simulator, *lme-2* was constructed as follows.

$$Q_i \sim Configuration + (1 \mid Expert) \qquad (lme\text{-}2)$$

where $Configuration$ is the categorical independent variable with two levels: the configurable base model, i.e., LIAC, and the individually configured model, i.e., IND. Since the configured model IND was rated once with Questionnaire-2 (after the individual adjustment in Training-6), the mean of two ratings for the LIAC model was retrieved from the first part of the study (Questionnaire-1) in *lme-2*.

For both lme models, *"lmerTest"* package was used to check the significance by retrieving the p-values. To check the main effect of *BaseModel* in *lme-1* and $Configuration$ in *lme-2*, the *"anova"* method was used. Normality of the residuals from both lme models were visually inspected using Q–Q plots. Resultant p-values below 0.05 were regarded to show significance. The p-values between 0.05 and 0.1 were considered as 'trending towards significance.

3. Results

For every participant, individual initial and final values of the LIAC model are listed in Table 1.

Table 1. Pre- and post-individualization values of the LIAC model parameters in Training 6.

Experts	Individually Configured Parameters of the LIAC Model							
	Gain for Drag and Lift Coefficient (G_{dl})		Minimum Rope Forces in Recovery (τ_{rec}^{min}) (N)		Gain for Boat Drag Coefficient (C_{Bdrag}) ($\frac{N}{(m/s)^2}$)		Oar Blade Curvature (φ_{bias}) (°)	
	Initial Value	Final Value	Initial Value	Final Value	Initial Value	Final Value	Initial Value	Final Value
E1	0.35	0.32	35	32	4.94	2.47	2	2
E2	0.35	0.38	35	35	4.94	4.94	2	0
E3	0.35	0.40	35	30	4.94	4.94	2	0
E4	0.35	0.37	35	30	4.94	6.18	2	0
E5	0.35	0.32	35	30	4.94	5.43	2	0
E6	0.35	0.32	35	30	4.94	4.94	2	0
E7	0.35	0.40	35	32	4.94	4.94	2	0
E8	0.35	0.35	35	30	4.94	5.19	2	0
E9	0.35	0.35	35	32	4.94	6.18	2	2
E10	0.35	0.35	35	32	4.94	4.45	2	0

3.1. Comparison of Base Rowing Models (BIFC vs. LIAC)

3.1.1. Aspect 1 and 2: Interaction between Oar Blade and Water Surface

For the horizontal and vertical blade movements above the water surface, the *lme-1* did not show any significant main effect of the *BaseModel* except the feathered blade sliding on the water surface (Table 2). For Question 1, LIAC was rated as more realistic than BIFC ($F_{BM}(1,\ 30) = 26.81$; $p < 0.001$) with a slightly above-medium value ($\mu^{Q1}_{LIAC} = 5.95$).

Table 2. Statistical results from lme-1 and lme-2 models for each question in the questionnaires.

Aspect of Rowing	Quest. Index	*lme-1:* *BaseModel* (BIFC vs. LIAC) Mean Rating of Models	Main Effect of Group and *p*-Value	*lme-2:* *Configuration* (LIAC vs. IND) Mean Rating of Models	Main Effect of Group and *p*-Value
A1	Q1	$\mu_{BIFC} = 4.05$ $\mu_{LIAC} = 5.95$	$F_{BM}(1,\ 30) = 26.81$ $p < 0.001$	$\mu_{LIAC} = 5.95$ $\mu_{IND} = 6.60$	$F_C(1,\ 10) = 6.01$ $p = 0.03$
	Q2	$\mu_{BIFC} = 6.15$ $\mu_{LIAC} = 5.85$	$F_{BM}(1,\ 30) = 0.57$ $p = 0.455$	$\mu_{LIAC} = 5.85$ $\mu_{IND} = 6.80$	$F_C(1,\ 10) = 5.56$ $p = 0.041$
A2	Q3	$\mu_{BIFC} = 6.00$ $\mu_{LIAC} = 5.75$	$F_{BM}(1,\ 30) = 0.23$ $p = 0.633$	$\mu_{LIAC} = 5.75$ $\mu_{IND} = 6.30$	$F_C(1,\ 10) = 4.86$ $p = 0.052.$
	Q4	$\mu_{BIFC} = 5.30$ $\mu_{LIAC} = 5.65$	$F_{BM}(1,\ 30) = 0.50$ $p = 0.484$	$\mu_{LIAC} = 5.65$ $\mu_{IND} = 6.40$	$F_C(1,\ 10) = 3.19$ $p = 0.104$
A3	Q5	$\mu_{BIFC} = 4.85$ $\mu_{LIAC} = 7.25$	$F_{BM}(1,\ 30) = 39.36$ $p < 0.001$	$\mu_{LIAC} = 7.25$ $\mu_{IND} = 8.20$	$F_C(1,\ 10) = 6.13$ $p = 0.032$
	Q6	$\mu_{BIFC} = 3.85$ $\mu_{LIAC} = 6.25$	$F_{BM}(1,\ 30) = 30.37$ $p < 0.001$	$\mu_{LIAC} = 6.25$ $\mu_{IND} = 8.00$	$F_C(1,\ 10) = 54.44$ $p < 0.001$
A4	Q7	$\mu_{BIFC} = 5.10$ $\mu_{LIAC} = 7.15$	$F_{BM}(1,\ 30) = 24.86$ $p < 0.001$	$\mu_{LIAC} = 7.15$ $\mu_{IND} = 8.00$	$F_C(1,\ 10) = 7.21$ $p = 0.023$
	Q8	$\mu_{BIFC} = 5.50$ $\mu_{LIAC} = 6.65$	$F_{BM}(1,\ 30) = 6.72$ $p = 0.015$	$\mu_{LIAC} = 6.65$ $\mu_{IND} = 7.80$	$F_C(1,\ 10) = 6.95$ $p = 0.025$
A5	Q9	$\mu_{BIFC} = 4.65$ $\mu_{LIAC} = 6.70$	$F_{BM}(1,\ 30) = 39.12$ $p < 0.001$	$\mu_{LIAC} = 6.70$ $\mu_{IND} = 7.90$	$F_C(1,\ 10) = 16.74$ $p = 0.002$
	Q10	$\mu_{BIFC} = 5.35$ $\mu_{LIAC} = 6.05$	$F_{BM}(1,\ 30) = 3.91$ $p = 0.057$	$\mu_{LIAC} = 6.05$ $\mu_{IND} = 6.40$	$F_C(1,\ 10) = 1.16$ $p = 0.306$

3.1.2. Aspect 3 and 4: Interaction of Oar Blade and Virtual Water during Sub-Phases of the Rowing Cycle

For all four sub-phases of the rowing cycle, *lme-1* revealed a significant main effect of the *BaseModel* (Question 5, 6, 7, and 8, Table 2). The LIAC was rated more realistic than BIFC for the drive (Q5: $F_{BM}(1,\ 30) = 39.36$; $p < 0.001$), recovery (Q6: $F_{BM}(1,\ 30) = 30.37$; $p < 0.001$), catch (Q7: $F_{BM}(1,\ 30) = 24.86$; $p < 0.001$), and release (Q8: $F_{BM}(1,\ 30) = 6.72$; $p = 0.015$). For all the sub-phases of the cycle, rating of LIAC was above medium. However, BIFC was rated below medium for drive and recovery and medium for the catch and release.

3.1.3. Aspect 5: Overall Realism of the Rowing Forces and Synchronized Flow of Visual Scenario

The *lme-1* revealed a significant main effect of *BaseModel* for the overall realism of the rowing forces and a trending main effect for the flow of visual scenario (Q10: $F_{BM}(1,\ 30) = 3.91$; $p = 0.057$). For Question 9, LIAC was rated more realistic than BIFC (Q9: $F_{BM}(1,\ 30) = 39.12$; $p < 0.001$) that was rated below medium. For Question 10, both models were rated with a slightly above-medium value (Table 2).

3.2. Effect of Individualized Configuration (LIAC vs. Individually Configured model (IND))

3.2.1. Aspect 1 and 2: Interaction between Oar Blade and Water Surface

The *lme-2* showed a main effect of *Configuration* for the horizontal movements of the oar blade above the water surface (Q1: $F_C(1, 10) = 6.01$; $p = 0.034$ and Q2: $F_C(1, 10) = 5.56$; $p = 0.041$) (Table 2). The realism rating value for the forces of oar blade sliding (Q1) increased by 11% compared to the LIAC model ($\mu_{IND}^{Q1} = 6.60$). Rating of the braking forces increased by 16%. Regarding the water surface interaction forces resulting from the vertical movement of the blade above surface, there was a trend for a main effect of the *Configuration* for the Question 3 (Q3: $F_C(1, 10) = 4.86$; $p = 0.052$). The rating value for the Question 3 increased by 10%. The *Configuration* did not yield any significance for the Question 4 although the rating value increased by 13%.

3.2.2. Aspect 3 and 4: Interaction of Oar Blade and Virtual Water during Sub-Phases of the Rowing Cycle

The *lme-2* showed a significant main effect of the *Configuration* for all four sub-phases of the rowing stroke (for Question 5, 6, 7, and 8, Table 2). Compared to base LIAC model, the configured LIAC model, i.e., IND, was rated more realistic for the drive (Q5: $F_C(1, 10) = 6.13$; $p = 0.032$), recovery (Q6: $F_C(1, 10) = 54.44$; $p < 0.001$), catch (Q7: $F_C(1, 10) = 7.21$; $p = 0.023$), and release (Q8: $F_C(1, 10) = 6.95$; $p = 0.025$). The rating values of the drive, recovery, catch, and release were increased by 13%, 28%, 12%, and 17%, respectively.

3.2.3. Aspect 5: Overall Realism of the Rowing Forces and Synchronized Flow of Visual Scenario

The *lme-2* revealed a main effect of *Configuration* only for the perception of overall rowing forces (Q9: $F_C(1, 10) = 16.74$; $p = 0.002$) (Table 2). The rating of Question 9 increased by 18% after individual configuration ($\mu_{IND}^{Q9} = 7.90$). Tuning of the rowing LIAC model did not significantly improve the perception of the visual scenario (Table 2), and the rating value increased by 6%.

4. Discussion

Robotic simulators provide an important environment for training of surgeons [39], pilots [40], motor learning of complex sports [41,42], and training of athletes for the performance improvement in real-life sports [22]. To benefit from simulator training in real-life tasks, simulators are required to realistically render the task aspects [43,44].

In this study, we propose a novel 3D rowing model to render the interaction forces between the oar blade and water for a rowing simulator. The implemented model was assessed by expert rowers in terms of the realism of the rendered rowing forces. The assessment of the realism was based on the subjective perception of the experts about how realistic the rowing forces were rendered. At first, two distinct 3D models were compared in numerous aspects of the rowing movement. We hypothesized that the new 3D rowing model would render the forces on the simulator more realistically than the previously implemented 3D model [21]. Secondly, we elaborated whether individualizing parameters of the new rowing model further increases the perceived realism of the interaction forces.

The consensus from the experts about the rowing simulator was that it provided a more realistic indoor training platform than any other rowing devices in the market due to incorporation of not only visual and auditory rendering of rowing aspects but also a realistic task force rendering with 3-DoF oar manipulation. Results from both parts of the study along with the impact of model parameters on the perceived realism are discussed in the following sub-sections.

4.1. Comparison of Two 3D Rowing Models (BIFC vs. LIAC)

Presented 3D rowing models that calculated drag and lift forces acting on the oar blade mainly differed in two parameters, i.e., use of angle of attack and the blade area in the water. In comparison to the BIFC model, the LIAC model not only incorporated the effect of angle of attack in horizontal plane

(α_{xy}) but also in the vertical plane (α_{yz}). In the LIAC model, the resultant drag and lift coefficients in horizontal and vertical planes were continuously adjusted according to the instantaneous angle of attack. On the contrary, a pre-determined value of C_L^{max} was used in the BIFC model for the forces in both planes and the vertical forces were not dependent on the angle of attack due to modeling assumptions. In addition, the BIFC model did not incorporate the effect of the oar-blade area in the water. However, forces from the LIAC model depended on the proportion of the blade area in the water.

We also investigated the effect of rope tension forces (τ_{rec}^{min}) on the rendering of recovery forces. In real rowing, rowers push the oar handle with a minimal force to move the oar from release to catch position during recovery. Thus, to provide a realistic feeling on the simulator during recovery, the rendered forces on the oar blade should be the lowest compared to other sub-phases of the rowing cycle. Therefore, we initially set the τ_{rec}^{min} value for the LIAC model to 35 N based on pilot study and compared to the τ_{rec}^{min} value of the BIFC model that was kept at 50 N .

4.1.1. Rowing Cycle and Sub-Phases

As hypothesized, the newly proposed 3D rowing LIAC model was rated as significantly more realistic than BIFC in drive (Q5), recovery (Q6), catch (Q7), release (Q8), and overall rowing cycle (Q9).

For drive, expert rowers mentioned that the rendered forces with LIAC were realistically resistive. The development of the force during early drive and the reduction of the force in the late drive was also judged as realistic. Experts also reported a smoother force increase and decrease in the catch and release, respectively. Inclusion of parameter p_{imm} in the blade area calculation accounted for a smooth transition of the oar blade during catch and release, which plays an important role in the exerted propulsion forces, continuity of the oar motion and elimination of unwanted decelerating blade drag forces [45].

On the other hand, the resistance provided by the BIFC model was reported to be low during the drive. It could be argued that a selection of a higher C_L^{max} value might have yielded a more realistically resistive force in drive. However, the force calculation in the BIFC model did not incorporate a parameter for the 'immersed blade area in the water' as in the case of LIAC. Thus, due to rapidly increased forces, the binary approach in the force modeling might have introduced a discrepancy during blade transition between air and water in the catch and release.

For recovery, superiority of the LIAC model over BIFC is not surprising because in the pilot tests, initially selected values of rope tension for LIAC ($\tau_{rec}^{min} = 50$ N) were also rated to be unrealistically high by three pilot experts. Nevertheless, we wanted to inspect and compare the overall ratings from main study expert rowers; thus, τ_{rec}^{min} forces in the BIFC model were maintained at 50 N . If we had lowered the τ_{rec}^{min} tension forces for the BIFC model, the observed significance regarding the realism of recovery forces of LIAC over BIFC would not have been found in our opinion.

4.1.2. Auxiliary Aspects of Rowing Movement

During actual rowing outings, rowers usually may need to stop the boat. Thus, rowers immerse the squared blade into the water to apply a braking motion. In the simulator, expert rowers were also asked to simulate a similar scenario, in which they would immerse the squared blade into water to decelerate the boat (Question 2). In this aspect, the perceived horizontal braking forces were not perceived differently for BIFC and LIAC. Some rowers described that braking forces were higher than they expected, which also decelerated the simulated boat more compared to an actual boat. This observation may be explained by the fact that effects of the Froude number and the turbulence were omitted in both models. In real blade–water interaction, the immersed area of the oar blade in the water may reduce the drag and lift force coefficients due to a change of the Froude number [46]. In addition, the turbulence was also shown to reduce the drag and lift coefficients in a computational study [47]. Thus, both effects should be implemented to yield more realistic interaction forces.

Another simplification was the geometrical modeling of the oar blade. In both rowing models BIFC and LIAC, the oar blade was initially modeled as a flat rectangular plate, i.e., the longitudinal and lateral curvature of an original rowing blade [34] design was omitted. Based on the feedback from pilot study rowers regarding the vertical forces in the catch and release, φ_{bias} was implemented in LIAC to account for a simplified form of a lateral curvature of the blade. However, due to the simplified blade geometry, the contact area between the blade and water surface increased. Thus, rendered horizontal and vertical forces above the water surface were generally perceived as higher for the simulator than a real rowing scenario.

In real rowing, due to rolling of the boat to one side, rowers may sometimes unintentionally touch and slide the blade on the water surface when moving the oar from release to catch with a feathered blade. In the simulator, participants were asked to make a deliberate attempt to slide the feathered blades to simulate this scenario (Question 1). Rowers reported that the BIFC model rendered horizontal forces that decelerated the boat when the blade was slid parallel to the water surface (feathered oar). According to the mathematical modeling of BIFC, the ideal forces along the B_y-axis should be zero. However, the lack of curved blade design in addition to the binary modeling of the oar immersion, i.e., either fully in or fully out of the water, might have facilitated the blade immersion during the feathered position, which may have induced the reported braking forces.

Although LIAC was rated more realistic than BIFC for the horizontal forces resulting from feathered blade sliding, the mean rating was only above average ($\mu^{Q1}_{LIAC} = 5.95$). Braking forces were also reported for LIAC; however, the magnitude was perceived as lower in comparison to BIFC. The reason for the reported forces for LIAC was the introduction of φ_{bias}, whose initial value was set to 2° to simulate a rotated blade instead of a (laterally) flat one. Thus, especially a rapid movement towards the catch while sliding the blade on the water surface resulted small resistive forces.

For both rowing models, the lack of curved blade design also resulted in excessive vertical contact forces. Naturally, the more parallel a blade is oriented with respect to surface, the greater the vertical forces become acting on the blade due to the water contact. In Question 3 and 4, experts were asked to rate the realism of the vertical forces resulting from water surface contact for a range of oar blade rotation ($0° \leq \varphi \leq 90°$). For BIFC, experts judged that immersion or tapping with all φ orientations rendered higher vertical forces than real rowing. This observation may be attributed to the binary modeling of the oar immersion for the calculation of rowing forces. For LIAC, the squared blade insertion ($\varphi = 0°$) into water was reported to render realistically low vertical forces. However, tapping with the feathered blade ($\varphi = 90°$) rendered much higher vertical forces than the real rowing.

To encounter high vertical forces and ensure the safety of the users in the simulator, rendered vertical forces on the simulator were limited to 200 N in B_z-axis, which was mainly considered for the general rowing movement with the squared blade orientation. However, the selected force limit was perceived too high for rendering the vertical forces during feathered blade tapping onto the water surface. Thus, the vertical force limit should be lowered in the future implementation.

4.1.3. Visual Scenario

Although the velocity model proposed in [28] was not further modified for the first part of the study, the resultant velocity of the simulated boat might have differed due to the design of distinct force models. In the simulator, the visual scenario shown on the screens was updated according to the boat speed ($\dot{y}_{boat}$) that was calculated through Equation (34). The result from Question 10 showed that the LIAC model was only marginally better than BIFC for the realism of the visual scenario flow. However, ratings for both models were only above medium. Discussions with the experts revealed that the design of the visual scenario itself could be improved for a more realistic virtual reality immersion, e.g., placement of the virtual buoys, rendering of the regatta lane, etc. Thus, although the goal was to judge how well the visual scenario updated with respect to the performed stroke rate and applied force, the majority of the experts were distracted by certain features in the visual scene such as the visual rendering of blade and buoys.

4.2. Effect of Individualization on the Perceived Realism of a Rowing Simulator (LIAC vs. IND)

In the second part of the study, experts were presented with a pool of model parameters, which can be individually configured to attain the most preferred rowing scenario on the simulator. The configurable parameters were chosen as G_{dl}, φ_{bias}, τ_{rec}^{min}, and C_{Bdrag}.

Adjustment of G_{dl} was mainly proposed for the magnitude of the drag/lift forces in drive, catch and release; however, it also influenced the forces resulting from blade and water surface interaction. Seven out of ten rowers preferred a different value of G_{dl} (range: 0.32–0.40, Table 1) after testing the simulator with a variety of G_{dl} settings. The modulation of G_{dl} resulted in an analogous effect as the damper setting on an ergometer. Four out of five male rowers preferred a higher setting while three females settled on a lower setting than the initial value of 0.35.

The bias factor φ_{bias} was originally introduced to modulate the vertical forces during catch and release. Change of φ_{bias} affected the value of the projected blade area for both the horizontal and vertical plane forces. In Training-6, eight out of ten rowers did not prefer the previously introduced φ_{bias} for rendering higher vertical forces in the catch and release (Table 1). φ_{bias} was set back to 0° for these participants. Thus, while rowing with a squared blade, decreased value of φ_{bias} resulted in lower vertical forces.

Configuration of τ_{rec}^{min} was offered for the resistive forces during recovery. Overall, nine experts decided to configure the rope tension forces (τ_{rec}^{min}) to a lower value than the initial setting (initial: 35 N , range after configuration: 30–35 N, Table 1).

Finally, modification of C_{Bdrag} was proposed to alter the boat velocity and thus, the resultant drag forces. Originally, the parametrization of C_{Bdrag} was suggested during pilot test to account for different rowing boat types since the rowers may be more familiar with training on a certain type of boat. However, during the main study, some rowers criticized the decelerating drag forces acting on the blade at the water entry and exit. The decelerating drag forces were caused by the stream velocity (negative of calculated boat velocity), which acted in the $+B_y$-direction. Therefore, C_{Bdrag} was the only relevant configurable parameter available to us, which could influence the magnitude of decelerating drag forces through boat velocity. For example, increment of C_{Bdrag} resulted in a reduced boat velocity (Equation (34)), which reduced the braking forces in $+B_y$-direction at the water entry and exit.

4.2.1. Rowing Cycle and Sub-Phases

The individualization of the model parameters significantly improved the perceived realism of forces in drive, recovery, catch, release, and overall rowing cycle.

For drive, individualization of G_{dl} was the key factor for the increased realism of resistive forces in the water. All participants except one were content with the magnitude of resistance, the peak force, and the change of forces in the water ($\mu_{IND}^{Q5} = 8.20$). One participant reported that he felt constrained by the simulator when he tried to apply high propelling forces. This observation can be attributed to the fact that the rendering of the horizontal forces was also bounded with a safety limit of 450 N in the B_y-axis. For the experiment, rowers were instructed to row with 40–70% of maximum force application and within low to medium stroke rates to assess the general capability of the simulator. Since the participants were not explicitly informed about the safety limit, their expected force peak was not rendered for the rowers who applied higher propelling forces. Therefore, the simulator will be tested in future for a higher force rendering to account for higher propelling forces by the rowers.

For recovery, individualization of rope tension forces significantly improved the perceived realism of forces ($\mu_{IND}^{Q6} = 8.00$). Although a setting of 30 N rendered reduced recovery forces, some rowers preferred to experience the slight resistance, referring to the real rowing in a windy weather. Few rowers actually preferred to set even lower values than 30 N for τ_{rec}^{min} stating that the residual resistance could still be perceived. However, due to design criteria of a tendon based parallel robot, the ropes must always be under tension to control the end-effector movement. Additionally, the selected value for τ_{rec}^{min} should ensure a smooth transition from τ_{rec}^{min} to τ_{drive}^{min} without causing disruptive tension–force jumps on the ropes. Thus, for control and safety reasons the lowest limit of τ_{rec}^{min} was set to 30 N . Nevertheless, the selected value was based on the initial trials on the simulator, and further experiments may yield a safe and less resistive τ_{rec}^{min} tension forces on the ropes.

Overall, individualization of φ_{bias}, G_{dl}, and C_{Bdrag} led to significant improvement in the ratings of catch and release ($\mu_{IND}^{Q7} = 8.00$, $\mu_{IND}^{Q8} = 7.80$). A final setting of $\varphi_{bias} = 0°$ was mainly reported to have rendered more realistic vertical forces during blade transition. In addition, four rowers who preferred a higher value of C_{Bdrag} judged that the increment encountered the previously criticized high decelerating drag forces at the initial and final blade contact with the water. However, other rowers who preferred the same or lower C_{Bdrag} values notified us that lower braking forces would be preferred. If the rendering of high braking forces at the water entry and exit had been reported in the pilot study, implementation of Froude number and turbulence effects to reduce the drag/lift coefficients might have been a more convenient approach than adjusting C_{Bdrag}.

4.2.2. Auxiliary Aspects of Rowing Movement

Overall, the individualization of the G_{dl}, φ_{bias}, and C_{Bdrag} significantly improved the perceived realism of the horizontal forces resulting from blade contact with the water. However, due to previously presented blade modeling simplifications and safety limits, the increment on the rating of vertical contact forces were insignificant.

Setting φ_{bias} back to 0°, as preferred by the majority of experts, led to a significant increase in the realism perception of the horizontal forces when sliding the blade (Question 1). Although experts were more consistent with the rating values for the configured model IND, on average the realism was only above average ($\mu_{IND}^{Q1} = 6.60$). This result may be explained with the lack of longitudinal curvature in the modeling of the blade surface. In a real rowing case, inserting a feathered blade into the water requires a deliberate attempt to push the oar handle upwards due to the curved design of the blade. However, due to the flat-plate modeling of the oar blade, rowers were still able to insert the feathered blade into the simulated water, which produced slight decelerating drag forces.

Significant improvement in the realism of the braking forces (in Question 2) may be attributed to the rowers who settled on a lower value for G_{dl} and/or C_{Bdrag} during individualization, which reduced the high braking forces encountered in the first part of the study. Nonetheless, the average rating for the realism of braking forces were only above average ($\mu_{IND}^{Q2} = 6.80$). The result can be attributed to the rowers, who preferred higher values for the both drag coefficients. Thus, the previously reported issue regarding the excessive braking forces was repeated.

For the vertical forces acting on the blade due to contact with water surface, individualization of G_{dl} only resulted in a marginally significant increment ($\mu_{IND}^{Q3} = 6.30$). The majority of the rowers, who preferred $\varphi_{bias} = 0°$, mentioned that the minor reduction of the vertical forces were more realistic compared to $\varphi_{bias} = 2°$. However, for the tapping with feathered blade onto the water surface, previously reported unrealistically high vertical force rendering was stated again in Question 3 and 4. Repetition of the issue was an expected outcome due to the lack of realistic blade modeling and used safety limit of 200 N . in B_z-axis, which were not among the configurable parameters.

4.2.3. Visual Scenario

During model individualization in Training-6, rowers were instructed to focus on the configuration of the rowing forces. When the participants settled on their preferred rowing forces on the simulator, they were asked to evaluate the realism of the simulated boat speed from the visual scenario displayed on the screens.

Results showed that although a slight improvement was observed in the rating, force model configuration failed to significantly improve the perception regarding the flow of visual scenario ($\mu_{IND}^{Q9} = 6.40$). Based on the rowers' feedback, the design of the visual scenario was the main factor that might have distracted the experts. Rowers especially stated that the virtual buoy placement in the visual scenario was perceived narrower than a normal regatta lane, which prevented them from comparing the simulator to the real life rowing. A few rowers also mentioned that they did not have adequate experience on a pair (2-) boat. These rowers reported that they were accustomed to rowing in bigger boats such as four (4-) and eight (8+), which are faster than a pair (2-) according to regatta results [48]. Thus, the resultant simulated boat velocity might have been slower for these rowers.

4.3. Practical Implications

To experience a realistic rowing training scenario in a simulator, expert rowers demand well-designed and synchronized rendering of visual, auditory, and haptic aspects of the task. In our simulator, forces rendered by the implemented rowing model and the synchronous rendering of visual and auditory displays were generally found to be realistic by the expert rowers. However, simplifications during modeling and technical limitations of the hardware setup might have led to perception of unnatural forces or unrealistic auditory and visual features, which may have confounded the overall realism of rowing scenario in the simulator. Therefore, in cases where residual technical problems cannot be readily resolved due to practical or safety reasons, individualization of the model parameters might be an appropriate approach to maintain or increase the perceived realism of the simulator.

Indeed, in our study, individualization of the model parameters was observed to yield a significant improvement in the acceptance of training on the simulator. Thus, expert rowers were also asked which other parameters of the rowing model and features to enhance the indoor training they would prefer to have on the simulator.

The majority of the experts proposed the parametrization of the inboard and outboard length of the oar, which could be virtually configured based on the rigging preferences of the rower. In real rowing, the preferred inboard and outboard oar lengths are usually found after a trial and testing process in the trainings. The required length of the oar also needs to be adjusted for the weather conditions. Thus, our rowing simulator can provide an easy and fast rigging within a training session.

In addition, expert rowers were interested in visualizing the errors related to the kinematic and dynamic outcomes of their rowing strokes in the simulator. Although commercial products, which can provide propulsion forces and oar kinematics while rowing on a boat, are available, these devices are not yet capable of providing feedback about the resulting errors, i.e., difference between desired and actual performance [49,50]. Therefore, as previously studied with beginner rowers, three modalities rendered on the rowing simulator could also be used to present feedback to the expert rowers for performance enhancement [26,51,52].

Some expert rowers were also interested in the visual display of an 'opponent boat', against which the rower in the simulator can train or race to simulate a regatta scenario. According to this idea, the 'opponent boat' could be set to propel with a desired speed setting, and the rower in the simulator could train to match the set goals for the training. Therefore, realistic rendering of the rowing forces can further be employed to augment the simulator to train professional athletes [53].

Finally, thanks to the achieved haptic realism of a sweep rowing application, potential sources of injuries associated with the asymmetries unique to the sweep rowing can be systematically assessed on our virtual reality simulator [54–56].

4.4. Limitations

Although rolling motion is important in real rowing for training the balance of the boat, the actuation of rolling angle on the rowing simulator was not included for safety reasons and controlling issues of the tendon based parallel robot. As reported by the participants, fixation of the boat for rolling motion helped to focus more on the relevant features and forces of rowing motion, although on the other hand, it lessened the overall haptic realism of the simulator compared to a training with an actual rowing boat on water. This technical drawback may be counteracted by introducing augmented reality headsets that can combine real life view of the rowing simulator and virtual rendering of boat rolling on the simulated water.

Another technical limitation was the use of fixed oarlock height for all the participants. The oarlock height was not adjustable due to the attachment of the end-effector (shortened oar) with the tendon based parallel robot. To operate the simulator, the oar and the boat had to be kept at a pre-defined position for the calibration of rope tension forces in the tendon based parallel robot. Since changing the end-effector position (by changing oarlock height) affected the calibration process of the robot, the oarlock was not adjusted. Therefore, some participants could not row with their most preferred rigging setting, which might also reflect the overall rating of the force models.

Due to the scope of the study, assessment of the realism of the rendered rowing forces were based on the subjective perception of the expert rowers. Therefore, future studies are required to objectively assess the realism by comparing the rendered rowing forces on the simulator to the on-water force measurements.

5. Conclusions

In this study, we presented a new rowing model that computed the blade–water interaction forces based on three rotational movements of the oar and implemented the model in our rowing simulator. We asked expert rowers to qualitatively assess the realism of the resulting forces for various rowing aspects on our rowing simulator. Based on the evaluation of experts, we found that the newly proposed 3D rowing model, which incorporated continuously adjusted drag/lift coefficients and immersed area of the blade, was highly realistic for the sub-phases and overall rowing cycle. We also showed that when the rowing model parameters were configured according to individual preferences of expert rowers, the perceived realism significantly increased.

The implementation of the proposed 3D rowing model for the haptics and the audiovisual rendering of the rowing scenario on the virtual reality simulator provide a realistic indoor training opportunity to rowers with different expertise levels. Thanks to the parametric design of the model and the real-time operation of the simulator, rowers and coaches can choose and immediately test the inquired settings to increase the rowing performance. The pool of configurable parameters can easily be extended to include the boat and oar rigging, which may further help rowers to find their most preferred rowing configuration in a boat. Therefore, our rowing simulator does not only provide an indoor rowing training facility but also a platform, which assists rowers and coaches to find the most optimal rowing settings.

Author Contributions: E.B. contributed to the rowing model development, vibration control of the parallel robot, experimental protocol design, questionnaire preparation, conducting of the experiments, data acquisition, and statistical analysis and prepared the manuscript. P.B. conducted his MSc. semester thesis, in which he contributed to the rowing model development and vibration control of the parallel robot. N.G. and G.R. contributed to the rowing model conceptualization, solution of technical problems encountered during model development, and they revised the manuscript. P.W. participated in the experimental protocol design and questionnaire preparation, contributed to discussion of the statistical analysis results, and revised the manuscript. All authors have read and agreed to the published version of the manuscript.

Funding: This research received no external funding.

Acknowledgments: We would like to thank the rowers who took part in testing the models during the development phase and pilot study. We also express our gratitude to Michael Herold-Nadig for his technical support and safety testing of the rowing simulator.

Conflicts of Interest: The authors declare no conflict of interest.

References

1. Smith, R.M.; Loschner, C. Biomechanics feedback for rowing. *J. Sports Sci.* **2002**, *20*, 783–791. [CrossRef] [PubMed]
2. Warmenhoven, J.; Cobley, S.; Draper, C.; Smith, R. Over 50 Years of Researching Force Profiles in Rowing: What Do We Know? *Sports Med.* **2018**, *48*, 2703–2714. [CrossRef] [PubMed]
3. Hawkins, D. A new instrumentation system for training rowers. *J. Biomech.* **2000**, *33*, 241–245. [CrossRef]
4. Hofmijster, M.J.; Landman, E.H.; Smith, R.M.; Van Soest, A.J. Effect of stroke rate on the distribution of net mechanical power in rowing. *J. Sports Sci.* **2007**, *25*, 403–411. [CrossRef]

5. Nolte, V. *Rowing Faster*, 2nd ed.; Human Kinetics: Champaign, IL, USA, 2011.
6. Schmidt, R.A.; Wrisberg, C.A. *Motor Learning and Performance: A Situation-Based Learning Approach*; Human kinetics: Champaign, IL, USA, 2008.
7. Kleshnev, V.; Nolte, V. Learning from racing. In *Rowing Faster*; Human Kinetics: Champaign, IL, USA, 2011.
8. Soper, C.; Hume, P.A. Towards an ideal rowing technique for performance. *Sports Med.* **2004**, *34*, 825–848. [CrossRef]
9. A Brief History of Rowing Machines. Available online: www.allrowers.com/history-of-rowing-machines.html (accessed on 21 December 2019).
10. Stettler, D. Improvement of oar water interactions in the M3-rowing simulator. Bachelor's Thesis, Swiss Federal Institute of Technology (ETH) Zurich, Zurich, Switzerland, June 2014.
11. Indoor Rowing Tanks. Available online: www.durhamboat.com/rowing-tanks/ (accessed on 22 December 2019).
12. Air vs. Magnetic Rowing Machine: What Is the Difference? Available online: www.rowingmachineking.com/air-vs-magnetic-rowing-machine-difference/ (accessed on 22 December 2019).
13. Kleshnev, V. Comparison of on-water rowing with its simulation on Concept2 and Rowperfect machines. In Proceedings of the ISBS-Conference Proceedings Archive, Beijing, China, 22–27 August 2005.
14. Indoor Sculler. Available online: www.rowperfect.com.au/ (accessed on 22 December 2019).
15. Dynamic Indoor Rower for Athletes & Teams. Available online: www.concept2.com/indoor-rowers/dynamic (accessed on 22 December 2019).
16. Rowing Innovations. Available online: www.rowinginnovations.com/swingulator-sweep-trainer/ (accessed on 22 December 2019).
17. Coffey Corporation. Available online: coffeycorporation.com/SimulatOar.php (accessed on 22 December 2019).
18. Biorower. Available online: www.biorower.com/ (accessed on 22 December 2019).
19. Filippeschi, A.; Ruffaldi, E. Boat dynamics and force rendering models for the sprint system. *IEEE Trans. Hum. Mach. Syst.* **2013**, *43*, 631–642. [CrossRef]
20. Ruffaldi, E.; Filippeschi, A. Structuring a virtual environment for sport training: A case study on rowing technique. *Robot. Auton. Syst.* **2013**, *61*, 390–397. [CrossRef]
21. Rauter, G.; von Zitzewitz, J.; Duschau-Wicke, A.; Vallery, H.; Riener, R. A tendon-based parallel robot applied to motor learning in sports. In Proceedings of the 2010 3rd IEEE RAS & EMBS International Conference on Biomedical Robotics and Biomechatronics, Tokyo, Japan, 26–29 September 2010; pp. 82–87.
22. Rauter, G.; Sigrist, R.; Koch, C.; Crivelli, F.; van Raai, M.; Riener, R.; Wolf, P. Transfer of complex skill learning from virtual to real rowing. *PLoS ONE* **2013**, *8*, e82145. [CrossRef]
23. M3 Rowing Simulator. Available online: www.rowing.ethz.ch/ (accessed on 22 December 2019).
24. Basalp, E.; Gerig, N.; Marchal-Crespo, L.; Sigrist, R.; Riener, R.; Wolf, P. Visual augmentation of spatiotemporal errors in a rowing task. In Proceedings of the Human Movement and Technology: Book of abstracts-11th Joint Conference on Motor Control & Learning, Biomechanics & Training, Darmstadt, Germany, 28–30 September 2016.
25. Gerig, N.; Basalp, E.; Sigrist, R.; Riener, R.; Wolf, P. Visual error amplification showed no benefit for non-naïve subjects in trunk-arm rowing. *Curr. Issues Sport Sci. (CISS)* **2019**, *3*. [CrossRef]
26. Rauter, G.; Gerig, N.; Sigrist, R.; Riener, R.; Wolf, P. When a robot teaches humans: Automated feedback selection accelerates motor learning. *Sci. Robot.* **2019**, *4*, eaav1560. [CrossRef]
27. Basalp, E.; Marchal-Crespo, L.; Rauter, G.; Riener, R.; Wolf, P. Rowing simulator modulates water density to foster motor learning. *Front. Robot. AI* **2019**, *6*, 74. [CrossRef]
28. Rauter, G. Enhancing Robot-Assisted Motor Learning by a Virtual Trainer. Ph.D. Thesis, Swiss Federal Institute of Technology (ETH) Zurich, Zurich, Switzerland, 2013.
29. Cabrera, D.; Ruina, A.; Kleshnev, V. A simple 1+ dimensional model of rowing mimics observed forces and motions. *Hum. Mov. Sci.* **2006**, *25*, 192–220. [CrossRef] [PubMed]
30. von Zitzewitz, J.; Wolf, P.; Novaković, V.; Wellner, M.; Rauter, G.; Brunschweiler, A.; Riener, R. Real-time rowing simulator with multimodal feedback. *Sports Technol.* **2008**, *1*, 257–266. [CrossRef]
31. Zitzewitz, J.v.; Rauter, G.; Steiner, R.; Brunschweiler, A.; Riener, R. A versatile wire robot concept as a haptic interface for sport simulation. In Proceedings of the 2009 IEEE International Conference on Robotics and Automation, Kobe, Japan, 12–17 May 2009; pp. 313–318.
32. Dawson, R.; Lockwood, R.; Wilson, J.; Freeman, G. The rowing cycle: Sources of variance and invariance in ergometer and on-the-water performance. *J. Mot. Behav.* **1998**, *30*, 33–43. [CrossRef] [PubMed]

33. Tessendorf, B.; Gravenhorst, F.; Arnrich, B.; Tröster, G. An imu-based sensor network to continuously monitor rowing technique on the water. In Proceedings of the 2011 Seventh International Conference on Intelligent Sensors, Sensor Networks and Information Processing, Adelaide, Australia, 6–9 December 2011; pp. 253–258.
34. Caplan, N.; Gardner, T.N. A fluid dynamic investigation of the Big Blade and Macon oar blade designs in rowing propulsion. *J. Sports Sci.* **2007**, *25*, 643–650. [CrossRef]
35. Wang, Z.J.; Birch, J.M.; Dickinson, M.H. Unsteady forces and flows in low Reynolds number hovering flight: Two-dimensional computations vs robotic wing experiments. *J. Exp. Biol.* **2004**, *207*, 449–460. [CrossRef]
36. On Rowing. Available online: home.hccnet.nl/m.holst/RoeiWeb.html (accessed on 14 December 2019).
37. Cabrera, D.; Ruina, A. Propulsive efficiency of rowing oars. *J. Appl. Biomech.* **2006**.
38. Diao, X.; Ma, O. Vibration analysis of cable-driven parallel manipulators. *Multibody Syst. Dyn.* **2009**, *21*, 347–360. [CrossRef]
39. Enayati, N.; Okamura, A.M.; Mariani, A.; Pellegrini, E.; Coad, M.M.; Ferrigno, G.; De Momi, E. Robotic assistance-as-needed for enhanced visuomotor learning in surgical robotics training: An experimental study. In Proceedings of the 2018 IEEE International Conference on Robotics and Automation (ICRA), Brisbane, Australia, 21–25 May 2018; pp. 6631–6636.
40. De Winter, J.C.; Dodou, D.; Mulder, M. Training effectiveness of whole body flight simulator motion: A comprehensive meta-analysis. *Int. J. Aviat. Psychol.* **2012**, *22*, 164–183. [CrossRef]
41. Marchal-Crespo, L.; van Raai, M.; Rauter, G.; Wolf, P.; Riener, R. The effect of haptic guidance and visual feedback on learning a complex tennis task. *Exp. Brain Res.* **2013**, *231*, 277–291. [CrossRef]
42. Rauter, G.; Sigrist, R.; Riener, R.; Wolf, P. Learning of temporal and spatial movement aspects: A comparison of four types of haptic control and concurrent visual feedback. *IEEE Trans. Haptics* **2015**, *8*, 421–433. [CrossRef] [PubMed]
43. Gopher, D. Skill training in multimodal virtual environments. *Work* **2012**, *41*, 2284–2287. [CrossRef] [PubMed]
44. Miles, H.C.; Pop, S.R.; Watt, S.J.; Lawrence, G.P.; John, N.W. A review of virtual environments for training in ball sports. *Comput. Graph.* **2012**, *36*, 714–726. [CrossRef]
45. Rowing Technique for Coaches—Catch, Drive, Release and Recovery. Available online: http://rowingact.org.au/former%20website/SDO/TECHNIQUE_1.html (accessed on 19 December 2019).
46. Macrossan, M.N.; Kamphorst, M. *Computational Study of the Froude Number Effects on the Flow around a Rowing Blade*; Department of Mechanical Engineering Report, University of Queensland: Queensland, Australia, October 2009.
47. Coppel, A.L. A Computational Fluid Dynamic Investigation of Rowing Oar Blades. Ph.D. Thesis, University of Birmingham, Edgbaston, Birmingham, UK, 2010.
48. 2019 World Rowing Championships. Available online: http://www.worldrowing.com/events/2019-world-rowing-championships/schedule-results (accessed on 19 December 2019).
49. OARS - smartOar. Available online: https://www.smartoar.com/oars/ (accessed on 20 December 2019).
50. Biorow Sensors and Electronics. Available online: http://biorow.com/products/biorow_sensors/ (accessed on 19 December 2019).
51. Sigrist, R.; Rauter, G.; Riener, R.; Wolf, P. Terminal feedback outperforms concurrent visual, auditory, and haptic feedback in learning a complex rowing-type task. *J. Mot. Behav.* **2013**, *45*, 455–472. [CrossRef]
52. Sigrist, R.; Rauter, G.; Marchal-Crespo, L.; Riener, R.; Wolf, P. Sonification and haptic feedback in addition to visual feedback enhances complex motor task learning. *Exp. Brain Res.* **2015**, *233*, 909–925. [CrossRef]
53. Wellner, M.; Sigrist, R.; von Zitzewitz, J.; Wolf, P.; Riener, R. Does a virtual audience influence rowing? *Proc. Inst. Mech. Eng. P J. Sports Eng. Technol.* **2010**, *224*, 117–128. [CrossRef]
54. Readi, N.; Rosso, V.; Rainoldi, A.; Vieira, T. Do sweep rowers symmetrically activate their low back muscles during indoor rowing? *Scand. J. Med. Sci. Sports* **2015**, *25*, e339–e352. [CrossRef]
55. Fohanno, V.; Nordez, A.; Smith, R.; Colloud, F. Asymmetry in elite rowers: Effect of ergometer design and stroke rate. *Sports Biomech.* **2015**, *14*, 310–322. [CrossRef]
56. Parkin, S.; Nowicky, A.V.; Rutherford, O.M.; McGregor, A.H. Do oarsmen have asymmetries in the strength of their back and leg muscles? *J. Sports Sci.* **2001**, *19*, 521–526. [CrossRef]

applied sciences

MDPI

Article

Immediate Effect of Whole-Body Vibration on Skin Temperature and Lower-Limb Blood Flow in Older Adults with Type 2 Diabetes: Pilot Study

Kennedy Freitas Pereira Alves [1], Ana Paula de Lima Ferreira [1], Luana Caroline de Oliveira Parente [1], François Talles Medeiros Rodrigues [1], Thais Vitorino Marques [1], Gabriel Barreto Antonino [1], Luiz Carlos de Melo [1], Débora Wanderley Villela [1], Marcelo Renato Guerino [1], Wagner Souza Leite [1], Shirley Lima Campos [1], Larissa Coutinho de Lucena [2], Redha Taiar [3] and Maria das Graças Rodrigues de Araújo [1,*]

1 Department of Physical Therapy, Universidade Federal de Pernambuco (UFPE), Recife-PE 50670-901, Brazil kennedyfpa@hotmail.com (K.F.P.A.); apllima@yahoo.com.br (A.P.d.L.F.); luanaparente00@gmail.com (L.C.d.O.P.); francoismedeirosfisiot@gmail.com (F.T.M.R.); thaisvmarques@hotmail.com (T.V.M.); gabrielbarreto@live.com (G.B.A.); luiz-carlos-melo@hotmail.com (L.C.d.M.); deborawanderley84@hotmail.com (D.W.V.); marceloguerino@hotmail.com (M.R.G.); wagnerszleite@gmail.com (W.S.L.); shirleylcampos@uol.com.br (S.L.C.)

2 Faculdade Nova Esperança (FACENE), João Pessoa-PB 58067-698, Brazil; larissacoutinho@gmail.com

3 GRESPI, Université de Reims Champagne-Ardenne, 51100 Reims, France; rtaiar@chu-reims.fr

* Correspondence: mgra@ufpe.br

Received: 20 December 2019; Accepted: 11 January 2020; Published: 19 January 2020

check for updates

Abstract: The purpose of this study was to evaluate the response of a single whole-body vibration (WBV) training session to peripheral skin temperature and peripheral blood flow of older adults with type 2 diabetes. A double-blind, controlled clinical trial was conducted following the Consolidated Standards of Reporting Trials (CONSORT) guidelines. A single session of WBV (24 Hz; amplitude 4 mm; vibration time 45 s, with a series of eight repetitions with recovery between repetitions of 30 s; total time of 10 min) or sham vibration on the Kikos P204 Vibrating Platform was employed. To assess skin temperature, the FLIR E40bxs thermographic camera and the ultrasonic vascular Doppler for flow velocity were used. Evaluation occurred before and after a WBV or sham intervention. The sample consisted of three men and 17 women. In the WBV group, there was a decrease in the temperature from 29.7 °C (±1.83) to 26.6 °C (±2.27), with $p = 0.01$. Temperature following sham decreased from 28.6 °C (±1.84) to 26.3 °C (±2.49), with $p = 0.01$. Regarding blood flow, there was a decrease in the analyzed arteries, especially the left posterior tibial artery, where there was a statistically significant flow reduction from 27.1 m/s (±25.36) to 20.5 m/s (±19.66), post WBV ($p = 0.01$). In the sham group, an increased flow velocity was observed for all the arteries analyzed, except for the left dorsal artery. Immediately following a full-body vibration session, peripheral skin temperature and lower-limb blood flow tend to decrease in diabetic patients. However, from the design of study developed, we cannot infer the maintenance of this effect in the medium and long term.

Keywords: type 2 diabetes mellitus; blood flow velocity; skin temperature; vibration

1. Introduction

Diabetes mellitus (DM) is one of the largest global health emergencies of the 21st century and, each year, more people develop and live with this condition. In addition to the 415 million adults diagnosed with diabetes, there are 318 million adults with glucose intolerance, resulting in a high risk of aggravating the condition and improving future disease development [1]. In Brazil, the number increased from 4.5 million in 2000 to 11.3 million in 2013, with an estimated range of 19.2 million by 2035 [2].

DM is a metabolic disorder of multiple etiologies characterized by hyperglycemia, which may be caused by defects in insulin secretion and/or action, involving specific pathological causes, such as the destruction of pancreatic β-cells responsible for the production of this hormone. This condition generates numerous side effects, such as the consequent disturbance mainly of carbohydrates, as well as proteins and fats, which leads to complications in various organs and may be accompanied by dyslipidemia, systemic arterial hypertension (SAH), endothelial dysfunction, obesity, smoking, and physical inactivity. Other macrovascular complications are either myocardial infarction and stroke, whereas microvascular complications include retinopathies and nephropathies [3–6].

The classification of the Brazilian Diabetes Society (BDS) includes four types defined according to specific disabilities or processes. Type 2 DM has an unknown etiology and is linked to relative insulin deficiency when it has a state of resistance combined with failure to secrete insulin, requiring the use of synthetic insulin to achieve glycemic control. It covers about 90% of diabetes cases in the population and usually manifests in adults aged over 45 with overweight or central fat deposition, physical inactivity, and family history of type 2DM [6,7].

Endothelial dysfunction is an important early factor in the development of macrovascular and microvascular complications of diabetes. Vascular endothelium is a dynamic tissue lining the blood vessels and the heart, and its performance is essential for homeostasis. The endothelium has several functions like maintenance of vascular tone, intravascular homeostasis, participation in the inflammatory process, and maintenance of blood flow. Reduction in peripheral blood flow and, consequently, local cooling of the skin are relevant markers of endothelial dysfunction that can be used as predecessors of cardiovascular disease. In older adults with type 2 diabetes, vascularization of the lower limbs is compromised [7–9]. The basal skin temperature in patients with type 2 diabetes is significantly lower than in patients without diabetes [10]. and this fact can be explained by the dilation of arteriovenous anastomoses and reduced blood supply in the capillaries due to autonomic neuropathy [11].

Many studies showed reduced blood glucose, insulin resistance, dyslipidemia, hypertension, and improved vascular function after regular physical activity, which are recognized as an important tool in diabetes control. Evidence of vasodilation after aerobic exercise and resistance training in diabetic individuals suggests increased vascular diameter and glucose uptake in the musculoskeletal system, which contributes to blood supply and lower-limb irrigation [12–15].

Reduced exercise tolerance is an important feature in patient with type 2 diabetes, who often have muscle weakness, making exercise more challenging. Whole-body vibration (WBV) is an alternative therapy which consists of a training program that exposes the whole body to mechanical vibrations. It was shown to be a safe alternative intervention to stimulate balance, and to improve muscle strength, body composition, and endurance in sedentary individuals who cannot perform conventional exercises such as walking and functional training. Protocols vary according to number of sessions per week, number of weeks, and frequency used in WBV in hertz (Hz) and amplitude (mm) [16–19].

However, there are few reports in the literature about the immediate effect of WBV on skin temperature and blood flow in the lower limbs of older adults with type 2 diabetes. Thus, the aim of this study was to evaluate the response of a single whole-body vibration training session to peripheral blood circulation and peripheral skin temperature in older adults with type 2 diabetes.

2. Methods

This is a pilot, randomized, double-blind, controlled clinical study that followed the guidelines set out in the Consolidated Standards of Reporting Trials (CONSORT) conducted with patients diagnosed with type 2 diabetes mellitus.

The research was carried out in the Laboratório de Cinesioterapia e Recursos Terapêuticos Manuais (LACIRTEM) of the Physiotherapy Department of the Universidade Federal de Pernambuco (UFPE). This study was approved by the Research Ethics Committee of the CCS/UFPE (No.3.219.332) and registered at www.clinicaltrials.gov (NCT04207853), respecting the ethical aspects of Resolution 466/12 of the National Health Council and the Declaration of Helsinki. All participants signed the free and informed consent form (FICF) after being informed of the objectives, risks, and benefits of the research.

The study was disseminated in various ways to attract volunteers (social networks, leaflets, lectures, scientific events, etc.). Volunteers were instructed to contact the research team to receive information and study guidance. When the volunteer met the research criteria, they were instructed to attend the day and time scheduled for exams to confirm the disease, while wearing appropriate clothes for the study.

The study included 20 clinically classified type 2 diabetic men (three) and women (17) who were aged 60 to 80 years, functionally independent, without severe foot deformities requiring the use of therapeutic footwear, without orthopedic deficiencies, without indication of deep vein thrombosis, without the use of walking aids, and with cognitive ability to respond and perform the exercises assessed by the Mini-Mental State Examination (MMSE), using 19 points as the cutoff point for illiteracy [20].

Volunteers who performed another physical activity and those who presented peaks of hypertension during treatment or who reported discomfort during the session that prevented the completion of training were discontinued.

Randomization was performed by a researcher not involved in any research stage, using the tool offered by www.randomization.com, and the order of participants was placed in numbered, sealed, and opaque envelopes, which were only opened at the time of the survey intervention. Thus, participants were randomly divided into two groups: whole-body vibration training (WBV), simulated vibration training (sham), and a single treatment session for WBV and sham. All participants underwent an initial and final evaluation.

Initially, anthropometric data were collected: gender, age, body mass index (BMI), vital signs (heart rate (HR), blood pressure (BP), and peripheral oxygen saturation (SpO_2), capillary glucose, and clinical history at the time of evaluation. Subsequently, the volunteers were assessed for peripheral circulation through blood flow and peripheral skin temperature on both limbs and then instructed to perform training on the vibrating platform. At the end of the intervention, blood flow, peripheral skin temperature, vital signs, and capillary blood glucose were re-evaluated.

2.1. Vital Signs

Heart rate (HR) measured in beats per minute (BPM) and peripheral oxygen saturation (SpO_2) expressed as a percentage (%) were evaluated using a Contec model CMS50M digital pulse oximeter (Contec Medical System Co., Ltd., Qinhuangdao, Hebei Province, China) placed on the second left toe with pressure. Blood pressure (BP) was measured in millimeters of mercury (mmHg) using a PREMIUM aneroid sphygmomanometer and a Rappaport Premium adult stethoscope (Accumed-Produtos Medico-Hospitalares Ltda, Rio de Janeiro-RJ, Brazil) using the left arm supported during measurement, with the patient sitting comfortably after 15 min of rest at baseline and immediately after the intervention.

2.2. Capillary Blood Glucose

Capillary blood glucose expressed in milligrams per deciliter (mg/dL) was verified by an Accu-Chek Active GU blood glucose monitor (Roche Diabetes Care Ltda, São Paulo-SP, Brazil), with the participant sitting comfortably with their arms resting on their legs before and at the end of each intervention to monitor and control the patient's glycemic index.

2.3. Anthropometric Data

Body composition was evaluated using a digital electronic anthropometric scale with a retractable ruler model MIC 200PPA (Micheletti Industria e Comercio de Equipamentos EIRELI, São Paulo-SP, Brazil). Body mass index (BMI) was defined as weight in kilograms divided by the square of height in meters (kg/m^2).

2.4. Blood Flow Velocity

The evaluation of blood flow velocity expressed in meters per second (m/s) was performed by ultrasonic vascular Doppler BV-620VP (Shenzhen Bestman Instarument Co., Ltd., Shenzhen, Guangdong, China), with sine wave emission. Prior to evaluation, a wide trichotomy of the examination site was performed, and then the area was coated with a water-soluble gel to promote better contact between the transducer and the skin. The transducer was positioned at a 45° angle to the blood vessels and was analyzed for flow. For the posterior tibial artery and the dorsal artery of the foot, recordings of the high-frequency transducer (8 Hz) were made, using the distal third of the calf and the instep as a reference. Confirmation of the most appropriate site for the evaluation of each blood vessel was performed through the acoustic signals generated by the equipment and, at this point, a marking was made to facilitate re-evaluation after the intervention. Three measurements were taken at 30-s intervals, and the final result was the average of the values obtained in the three collections. The exam was repeated after the intervention via which the patient was submitted, respecting the markings of the previous collection.

2.5. Skin Temperature

Skin temperature was evaluated using an FLIR E40bx thermographic camera (FLIR Systems Inc., Wilsonville, OR, USA) with a thermal sensitivity of 0.05 °C and a resolution of 160 × 120 pixels. The camera was positioned at a distance of 1 m from the assessed region, at an angle of 90° to the camera lens. Participants were instructed regardingthe preparatory measures for the thermographic exam, according to the recommendations of Priego Quesada (2017). Prior to the exam, the volunteers rested for 10 min in a room with a temperature between 18 and 23 °C and relative humidity between 40% and 70%, monitored with the help of athermo-hygrometer (KT-908). Reflected temperature was measured by the "reflective method" described in ISO 18434-1: 2008 for thermographic camera calibration [21]. We noted that the human body emissivity is assumed to be constant (about 0.98) [22] to ensure that the influence of the environment reflection is negligible.

Image analyses were performed according to the Glamorgan (2008) protocol for the anterior and posterior leg and dorsal region and sole of the feet, using the average temperature of a polygon, which followed the contour of the studied areas [23].

Images were collected before and after the intervention. Subsequently, the images obtained were organized, stored, and analyzed in the FLIR Tools software (FLIR Systems Inc., Wilsonville, OR, USA) provided by the camera manufacturer (Figure 1).

2.6. Whole-Body Vibration

The vibratory platform used was the Kikos P204–110v (Kikos Fitness, São Paulo-SP, Brazil), whose vibration direction is of the lateral oscillatory type. The vibrating plate features non-slip material on its surface for better foot attachment at the time of training. The established vibration frequency was 24 Hz. The 4-mm displacement peak was maintained throughout the program and was determined by the width of the volunteers' position on the platform. The intervention time was 45 s, with eight series being performed and, for each, the recovery time was maintained at 30 s for all intervals, totaling 600 s of intervention.

The subjects, barefoot, adopted a 40° squat position, verified with a universal goniometer The calcans were supported on a silicone insole, consisting of a particular type of semi-rigid sole used to correct the distribution of body weight and biomechanical changes found in the foot. Only the forefoot and midfoot were in direct contact with the platform in order to reduce the transmission of vibrations to the head. This posture was used for the both groups.

The simulated vibration "sham" was performed with the platform disconnected. A sound device was connected producing a noise similar to the connected platform for a time equivalent to the treatment protocol because the vibratory stimulus could not be visibly distinguished. Participants who experienced false vibration had no contact with those who performed the protocol effectively.

On the day of intervention, participants were instructed not to do intense exercises, requiring at least one week apart from the last moderate/high-intensity physical activity. In addition, all participants were instructed to wear comfortable and appropriate exercise clothing.

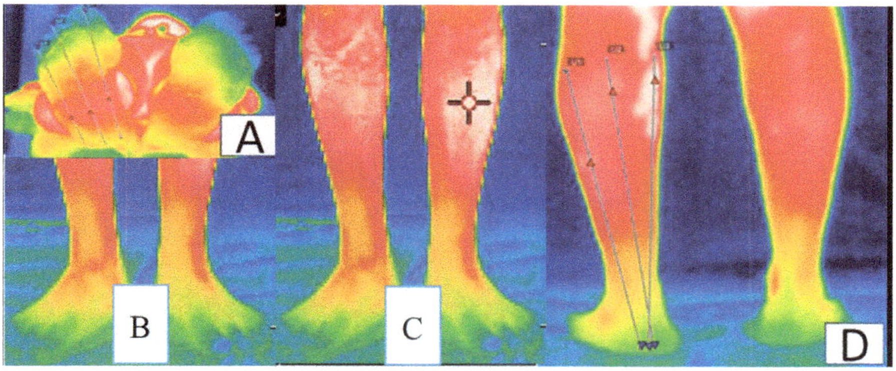

Figure 1. Regions of interest for infrared thermographic imaging: (**A**) sole of the feet; (**B**) back of the feet; (**C**) anterior leg; (**D**) back of the legs.

2.7. Data Analysis

Statistical analysis was performed using the statistical software SPSS version 23.0 (IMB Software Group, Chicago, IL, USA). To test the hypothesis of normality of each quantitative variable, the Shapiro–Wilk adherence test was used, and then the behavior of the quantitative variables was evaluated using the Mann–Whitney and Wilcoxon comparison test for the inter- and intragroup analysis, respectively.

The effect size was calculated by Cohen's d. An effect size of 0.2 was considered a small effect, an effect size of 0.5 was considered a moderate effect, and an effect size of 0.8 was considered a large effect. Descriptive statistical analysis was performed using the mean, standard deviation (SD), and percentage for qualitative variables. All tests were assigned a significance level of 95% ($p < 0.05$).

3. Results

Each group consisted of 10 participants, from WBV (two men and eight women) and sham (one man and nine women), as shown in Figure 2. For all variables, the groups were considered statistically comparable at baseline, as shown in Table 1).

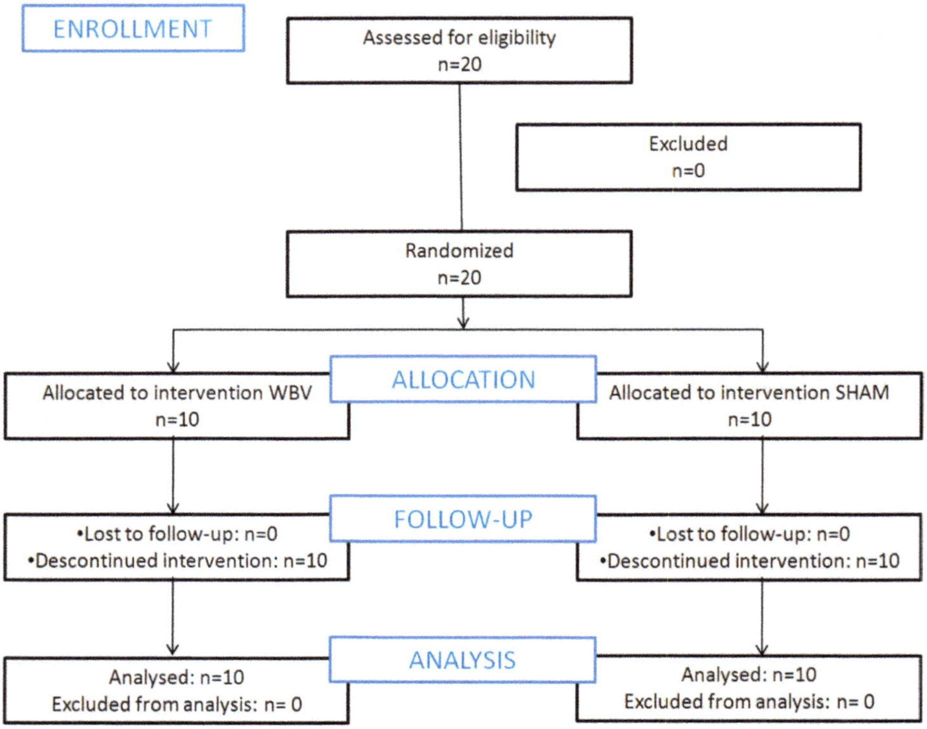

Figure 2. Flow chart which describes the recruitment, randomization, and allocation of patients.

Table 1. Baseline characteristics of patients in the whole-body vibration (WBV) and sham groups.

Variables	WBV G1 (n=10)	Sham G2 (n=10)	*p*-Value
Age (years)	64.4 (±3.99)	65.2 (±4.54)	0.83
Diagnostic Time (years)	7.29 (±4.53)	9.13 (±6.81)	0.55
Sex	10 (100%)	10 (100%)	
Male	2 (20%)	1 (10%)	0.54
Female	8 (80%)	9 (90%)	
Smoker			
Yes	3 (30%)	1 (10%)	
No	7 (70%)	9 (90%)	0.4
Former smoker	2 (28.6%)	3 (33.3%)	
Passive smoker	-	-	
Anthropometric data			
BMI	30.4 (±4)	27.9 (±3.9)	0.18

Table 1. *Cont.*

Variables	WBV G1 (n=10)	Sham G2 (n=10)	*p*-Value
Schooling			
Illiterate	-	-	
Incomplete elementary	1 (10%)	1 (10%)	
Complete elementary	2 (20%)	2 (20%)	
Incomplete medium	-	1 (10%)	0.42
Medium complete	2 (20%)	4 (40%)	
Incomplete higher	2 (20%)	1 (10%)	
Graduated	3 (30%)	-	
Postgraduate	-	1 (10%)	
Marital status			
Single	3 (30%)	-	
Married	4 (40%)	4 (40%)	0.24
Divorced	2 (20%)	3 (30%)	
Widowed	1 (10%)	3 (30%)	
Family income			
<1 MW	3 (30%)	3 (30%)	
1–3 MW	4 (40%)	4 (40%)	1
4–6 MW	2 (20%)	2 (20%)	
>7 MW	1 (10%)	1 (10%)	
HPAQ	8.49 (±1.19)	7.9 (±1.37)	0.32
Blood pressure			
SBP	120 (±9.43)	116 (±6.99)	0.3
DBP	77 (±8.23)	78 (±9.66)	0.81
Capillary blood glucose	133.6 (±36.75)	135.4 (±35.2)	0.91
SpO_2	98.3 (±0.82)	95.3 (4.8)	0.08
Diabetes knowledge	20.85 (±11.27)	21.42 (±9.54)	0.9

Data are expressed as mean (±standard deviation) or absolute frequency (relative frequency). Abbreviations: body mass index, BMI; minimum wage, MW; SpO_2, peripheral oxygen saturation; QAFH, usual physical activity questionnaire; BMI, body mass index; SBP, systolic blood pressure; DBP, diastolic blood pressure.

Only the plantar region of the right foot suffered a statistically significant reduction in skin temperature in both groups. In the WBV group, there was a reduction from 29.7 °C (±1.83) to 26.6 °C (±2.27), $p = 0.01$. In the sham group, the reduction was from 28.6 °C (±1.84) to 26.3 °C (±2.49), $p = 0.01$. There was no statistically significant difference between the groups. For all results, the effect size was considered low. Thermographic data are expressed in Table 2.

Table 2. Mean skin temperature analysis in °C (and standard deviation) before and after intervention (intra- and intergroup) and effect size.

Variables	WBV Before	WBV Post	p^+	Sham Before	Sham Post	p^+	Effect Size	*p*-Value
Plant right foot	29.7 (±1.83)	26.6 (±2.27)	0.01 *	28.6 (±1.84)	26.3 (±2.49)	0.01 *	0.12	0.78
Left foot plant	26.9 (±2.33)	26.7 (±1.89)	0.84	28.7 (±2)	26.3 (±2.21)	0.05	0.18	0.67
Back right foot	29.6 (±1.64)	29 (±1.7)	0.33	30.2 (±1.93)	29.2 (±1.23)	0.07	−0.16	0.77
Back left foot	27.9 (±1.85)	29.3 (±1.06)	0.06	30.2 (±1.93)	29.7 (±1.25)	0.30	−0.32	0.45
Anterior right leg	30.1 (±1.59)	30.1 (±1.59)	0.52	30.3 (±1.06)	30.3 (±1.06)	0.48	−0.19	0.74
Anterior left leg	30.1 (±1.52)	30.1 (±1.52)	0.21	30.4 (±1.26)	30.4 (±1.26)	0.36	−0.24	0.64
Posterior right leg	30.4 (±0.97)	30.3 (±1.06)	0.80	30.4 (±.97)	30.4 (±0.84)	1	0.12	0.82
Posterior left leg	30.6 (±0.97)	30.3 (±1.34)	0.54	30.5 (±1.08)	30.6 (±0.84)	0.76	−0.32	0.56

Data are expressed as mean (± standard deviation). Abbreviations: WBV, whole-body vibration; sham, dummy vibration. p^+, Wilcoxon; *p*-value, Mann–Whitney; * statistically significant result.

There was a reduction in blood flow in the WBV group after the intervention, but there was only a statistically significant reductionin the left tibial artery, from 27.1 m/s (±25.36) to 20.5 m/s (±19.66), $p = 0.01$. In the sham group, there was only a reduction in flow velocity in the left dorsal artery and an increase in all other evaluated arteries. However, none of them had a statistically significant difference, just as there was no statistically significant difference between the groups in this variable. Alow effect size was also observed for this variable. The complete analysis of blood flow velocity is detailed in Table 3.

Table 3. Analysis of mean blood flow velocity in m/s (and standard deviation) before and after intervention (intra and intergroup) and effect size.

Variables	WBV			Sham			Effect Size	*p*-Value
	Before	Post	p^+	Before	Post	p^+		
Right dorsal artery	29.3 (±25.79)	23.4 (±24.61)	0.31	27.9 (±29.14)	29.3 (±25.79)	0.81	−0.23	0.61
Left dorsal artery	24.6 (±21.65)	20.3 (±20.25)	0.37	26.8 (±23.58)	23.8 (±21.21)	0.13	−0.17	0.71
Right tibial artery	23.1 (±25.92)	21.4 (±23.68)	0.58	21.3 (±24.76)	23.2 (±26)	0.06	−0.07	0.87
Left tibial Artery	27.1 (±25.36)	20.5 (±19.66)	0.01 *	24 (±26.24)	27.2 (±25.42)	0.69	−0.26	0.52

Abbreviations: WBV, whole-body vibration; sham, dummy vibration. p^+ Wilcoxon; *p*-value, Mann–Whitney; * statistically significant result.

4. Discussion

In the present study, in both groups (WBV and sham), there was a reduction in skin temperature and blood flow velocity, corroborating previous studies. Fujiwara et al. (2000) evaluated 61 diabetic participants and 16 healthy subjects, and they reported that basal skin temperature in diabetic patients was significantly lower than in healthy patients, in addition to lower temperature variation and recovery after intervention. These findings agree with the present study, where there was little variation in some of the evaluated points and, in some of them, the temperature remained constant [10]. According to Uccioli et al. (1994), this fact can be explained by the dilation of arteriovenous anastomoses and reduced blood supply in the capillaries due to autonomic neuropathy [11].

The study by Kistler et al. (1998) inferred that capillary blood flow and blood pressure are lower in autonomic diabetic neuropathy because the anatomical walls are larger, directing flow and damaging peripheral sympathetic nerves. This promotes a slow recovery of skin temperature, justifying the smallest temperature variation in the present study [24].

Manimmanakorn et al. (2017) performed a study with a vibratory platform with 36 diabetic participants, of which 17 were trained on the platform for 12 weeks and 19 participants maintained their daily activities only. Reductions in peak systolic velocity and diastolic pressure were noticed in the WBV group, but without significant difference from the control group. In the present study, although no peak systolic velocity and diastolic pressure were evaluated, we observed a reduction in which there was a reduction in blood flow velocity in the evaluated arteries. This finding can be explained by the possible vibration-induced dilation of the smaller blood vessels present in the muscles, reducing peripheral vascular resistance by inhibiting endothelin, an important vasoconstrictor [25].

Saudoudo et al. (2013) evaluated 40 diabetics, 20 of whom underwent 12 weeks of whole-body vibration training, and reported increased blood flow velocity after 12 weeks of intervention, with a significant difference compared with the control group. This divergence in the results found in relation to the present study is possibly due to the accomplishment of a training program over a long period of time, enhancing the effects of vibration and promoting long-term results, which may not be the case in the first session of the study training [8].

Bagavathiappan et al. (2010) analyzed 112 patients diagnosed with type 2 diabetes through thermographic images of the plantar region, correlating the presence of diabetic neuropathy and the cutaneous temperature of these individuals. Higher values of mean and global foot temperature were found in patients with neuropathy, in addition to a positive correlation with temperature in the hallux regions, bilaterally. This finding may be justified by neuropathic feet being warm, with palpable pulses and distended veins, indicating increased blood flow, although slow, and possible congestion in this region [26].

Due to the neurovascular lesions triggered by hyperglycemia, as well as the response to lesions or infections, the vascular response of these individuals is slower, making them more susceptible to ulcer development and possible long-term necrosis and amputation. Thus, due to the increase in skin temperature after the use of WBV, it is plausible to think that the use of this equipment may be beneficial for the vascular maintenance of these individuals, and thermographic monitoring is an ally in the maintenance of these patients' lesions [26]. The findings cited are discordant with the present study.

However, Moreira-Marconi et al. (2019) evaluated 19 healthy subjects and reported a reduction in skin temperature in most participants, regardless of the frequency used in WBV (0 Hz, 30 Hz, and 50 Hz). This reduction can be explained by the fast and constant isometric muscle action that promotes work and increases the body's metabolic rate, increasing intramuscular temperature, modifying the thermal balance, activating the feedback-mediated heat loss mechanism. This change leads to redirection of blood flow from inactive to active areas. Then, during vibration, blood flow redistributes to the periphery, allowing heat exchange with the external environment, cooling the skin, and causing blood to flow to the lower-temperature muscle regions, thus preventing hyperthermia and maintaining homeostasis [27].

In addition, a larger blood supply is needed in the recruited musculature, requiring greater blood supply, initially diverting from adjacent peripheral areas. Under these conditions, a vasoconstrictor response induced by the mechanical action of vibration occurs, promoting a reduction in peripheral perfusion, and reducing the local skin temperature [26]. The above findings of the study corroborate the results found in the present study.

WBV training is an exercise program that does not increase excessive physical effort to the skeletal muscle system compared to conventional exercise. Mastering this training with WBV can help with choosing the most appropriate training program for older adults with type 2 diabetes.

5. Conclusions

Immediately following a full-body vibration session, peripheral skin temperature and lower-limb blood flow tend to be reduced in older adults with type 2 diabetes. However, from the design of study developed, we cannot infer the maintenance of this effect in the medium and long term.

6. Study Limitations

As this clinical trial was a pilot with a small number of participants, this study can serve as a basis for sample size estimation in future studies, taking into account the trends pointed out in the results presented. No condition in the metabolic control of diabetes was considered because it was a single-session study, given that glycated hemoglobin is the average glycaemia in the last 3–4 months. An increase in the sample size and a higher male presence (balanced gender repartition) in a future study should be considered to infer more robust results.

Author Contributions: Conceptualization, L.C.d.O.P., K.F.P.A. and M.d.G.R.d.A.; data curation, D.W.V. and A.P.d.L.F.; formal analysis, W.S.L. and S.L.C.; funding acquisition, M.d.G.R.d.A. and M.R.G.; Investigation, F.T.M.R., G.B.A., T.V.M. and L.C.d.O.P.; methodology, L.C.d.M., L.C.d.O.P., K.F.P.A., F.T.M.R., T.V.M., L.C.L, A.P.d.L.F., D.W.V., W.S.L. and M.d.G.R.d.A.; project administration, M.R.G., L.C.d.L. and M.d.G.R.d.A.; writing-original draft, L.C.d.O.P., K.F.P.A., G.B.A. and M.d.G.R.d.A.; writing-review and editing, K.F.P.A., S.L.C., R.T. and M.d.G.R.d.A. All authors have read and agreed to the published version of the manuscript.

Funding: This research was funded by Fundação de Amparo à Ciência e Tecnologia do Estado de Pernambuco (FACEPE) through financial support APQ No. 0337-13.

Conflicts of Interest: The authors declare that there is no conflict of interests regarding the publication of this paper.

References

1. International Diabetes Federation IDF. IDF Atlas, Seventhedition, Brussels, Belgium. 2015. Available online: www.diabetesatlas.org (accessed on 10 November 2019).
2. Cecilio, H.P.M.; Arruda, G.O.; Teston, E.F.; Santos, A.L.; Marcon, S.S. Comportamentos e comorbidades associados às complicações microvasculares do diabetes. *Acta Paul. Enferm.* **2015**, *28*. [CrossRef]
3. Brasil. Ministério da Saúde. Secretaria de Atenção à Saúde. Departamento de Atenção Básica. *Estratégias Para o Cuidado da Pessoa com Doença Crônica: Diabetes Mellitus*; Ministério da Saúde, Secretaria de Atenção à Saúde. Departamento de Atenção Básica: Brasília, Brazil, 2013.
4. Del Pozo-Cruz, J.; Alfonso-Rosa, R.M.; Ugia, J.L.; McVeigh, J.G.; Del Pozo-Cruz, B.; Sañudo, B. A Primary Care–Based Randomized Controlled Trial of 12-Week Whole-Body Vibration for Balance Improvement in Type 2 Diabetes Mellitus. *Arch. Phys. Med. Rehabil.* **2013**, *94*, 2112–2118. [CrossRef] [PubMed]
5. Irvine, C.; Taylor, N.F. Progressive resistance exercise improves glycaemic control in people with type 2 diabetes mellitus: A systematic review. *Aust. J. Physiother.* **2009**, *55*, 237–246. [CrossRef]
6. Sivanandam, S.; Anburajan, M.; Venkatraman, B.; Menaka, M.; Sharath, D. Medical thermography: A diagnostic approach for type 2 diabetes based on non-contact infrared thermal imaging. *Endocrine* **2012**, *42*, 343–351. [CrossRef]
7. Sociedade Brasileira de Diabetes. Diretrizes da Sociedade Brasileira de Diabetes (2017-2018). São Paulo: Editora Clannad. 2017. Available online: http://www.diabetes.org.br/profissionais/images/2017/diretrizes/diretrizes-sbd-2017-2018.pdf (accessed on 10 November 2019).
8. Sañudo, B.; Alfonso-Rosa, R.; De Pozo-Cruz, B.; De Pozo-Cruz, J.; Galiano, D.; E Figueroa, A. O treinamento de vibração do corpo inteiro melhora o fluxo sanguíneo das pernas e a adiposidade em pacientes com diabetes mellitus tipo 2. *Eur. J. Appl. Physiol.* **2013**, *113*, 2245–2252. [CrossRef]
9. Silber, H.A.; Lima, J.A.; Bluemke, D.A.; Astor, B.C.; Gupta, S.N.; Foo, T.K.; Ouyang, P. Arterial reactivity in lower extremities is progressively reduced as cardiovascular risk factors increase: Comparison with upper extremities using magnetic resonance imaging. *J. Am. Coll. Cardiol.* **2007**, *49*, 939–945. [CrossRef]
10. Fujiwara, Y.; Inukai, T.; Aso, Y.; Takemura, Y. Thermographic measurement of skin temperature recovery time of extremities in patients with type 2 diabetes mellitus. *Exp. Clin. Endocrinol. Diabetes* **2000**, *108*, 463–469. [CrossRef]
11. Uccioli, L.; Mormile, F.; Monticone, G.; Mennuni, G.; Durola, L.; Menzinger, G.; Russo, F. Autonomic neuropathy influences great toe blood pressure. *Diabetes Care* **1994**, *17*, 284–287. [CrossRef]
12. Sato, Y.; Nagasaki, M.; Kubota, M.; Uno, T.; Nakai, N. Clinical aspects of physical exercise for diabetes/metabolic syndrome. *Diabetes Res. Clin. Pract.* **2007**, *77* (Suppl. S1), S87–S91. [CrossRef]
13. Chudyk, A.; Petrella, R.J. Effects of exercise on cardiovascular risk factors in type 2 diabetes: A meta-analysis. *Diabetes Care* **2011**, *34*, 1228–1237. [CrossRef]
14. Okada, S.; Hiuge, A.; Makino, H.; Nagumo, A.; Takaki, H.; Konishi, H.; Goto, Y.; Yoshimasa, Y.; Miyamoto, Y. Effect of exercise intervention on endothelial function and incidence of cardiovascular disease in patients with type 2 diabetes. *J. Atheroscler. Thromb.* **2010**, *17*, 828–833. [CrossRef] [PubMed]
15. Bloor, C.M. Angiogenesis during exercise and training. *Angiogenesis* **2005**, *8*, 263–271. [CrossRef] [PubMed]
16. Behboudi, L.; Azarbayjani, M.A.; Aghaalinejad, H.; Salavati, M. Effects of aerobic exercise and whole body vibration onglycaemia control in type 2 diabetic males. *Asian J. Sports Med.* **2011**, *2*, 83–90. [CrossRef]
17. Regensteiner, J.G. Type 2 diabetes mellitus and cardiovascular exercise performance. *Rev. Endocr. Metab. Disord.* **2004**, *5*, 269–276. [CrossRef] [PubMed]
18. Egger, A.; Niederseer, D.; Diem, G.; Finkenzeller, T.; Ledl-Kurkowski, E.; Forstner, R.; Pirich, C.; Patsch, W.; Weitgasser, R.; Niebauer, J. Different types of resistance training in patients with type 2 diabetes mellitus: Effects on glycemic control, muscle mass and strength. *Eur. J. Prev. Cardiol.* **2013**. [CrossRef] [PubMed]
19. Machado, A.; Garcia Lupez, D.; Gonzalez Gallego, J.; Garstschea, N. Whole body vibration training increases muscle strength and mass in older women: A randomize controlled trial. *Scand. J. Med. Sci. Sports* **2010**, *20*, 200–207. [CrossRef] [PubMed]
20. Lourenço, R.A.; Veras, R.P. Mini-Exame do Estado Mental: Característicaspsicométricas em idosos ambulatoriais. *Revista de Saúde Pública* **2006**, *40*, 712–719. [CrossRef]

21. Quesada, J.I.P.; Kunzler, M.R.; Carpes, F.P. Methodological aspects of infrared thermography in human assessment. In *Application of Infrared Thermography in Sports Science. Biological and Medical Physics, Biomedical Engineering*; Priego Quesada, J.I., Ed.; Springer International Publishing: Cham, Switzerland, 2017; pp. 49–67. [CrossRef]
22. Polidori, G.; Renard, Y.; Lorimier, S.; Pron, H.; Derruau, S.; Taiar, R. Medical Infrared Thermography assistance in the surgical treatment of axillary Hidradenitis Suppurativa: A case report. *Int. J. Surg. Case Rep.* **2017**, *34*, 56–59. [CrossRef]
23. Ammer, K. The Glamorgan Protocol for recording and evaluation of thermal images of the human body. *Thermol. Int.* **2008**, *18*, 125–144.
24. Kistler, A.; Mariauzouls, C.; Von-Berlepsch, K. Fingertp temperature as an indicator for sympathetic responses. *Int. J. Psychophysiol.* **1998**, *29*, 35–41. [CrossRef]
25. Manimmanakorn, N.; Manimmanakorn, A.; Phuttharak, W.; Hamlin, M.J. Effects of whole body vibration on glycemic indices and peripheral blood flow in type II diabetic patients. *Malays. J. Med. Sci.* **2017**, *24*, 55–63. [CrossRef] [PubMed]
26. Bagavathiappan, S.; Philip, J.; Jayakumar, T.; Raj, B.; Rao, P.N.; Varalakshmi, M.; Mohan, V. Correlation between plantar foot temperature and diabetic neuropathy: A case study by using an infrared thermal imaging technique. *J. Diabetes Sci. Technol.* **2010**, *4*, 1386–1392. [CrossRef] [PubMed]
27. Moreira-Marconi, E.; Moura-Fernandes, M.C.; Lopes-Souza, P.; Teixeira-Silva, Y.; Reis-Silva, A.; Marchon, R.M. Evaluation of the temperature of posterior lower limbs skin during the whole body vibration measured by infrared thermography: Cross-sectional study analysis using linear mixed effect model. *PLoS ONE* **2019**, *14*, e0212512. [CrossRef] [PubMed]

applied sciences

MDPI

Article

Inter-Segmental Coordination during a Unilateral 180° Jump in Elite Rugby Players: Implications for Prospective Identification of Injuries

Kellen T. Krajewski [1], Carla McCabe [2], Aaron M. Sinnott [1], Gavin L. Moir [3], Hugh S. Lamont [4], Susan Brown [5] and Chris Connaboy [1,*]

1 Neuromuscular Research Laboratory, Rehabilitation & Health Sciences, University of Pittsburgh, Pittsburgh, PA 15260, USA; ktk20@pitt.edu (K.T.K.); ams626@pitt.edu (A.M.S.)
2 School of Sport, Ulster University, Coleman BT370QB, UK; c.mccabe@ulster.ac.uk
3 Exercise Science Department, East Stroudsburg University, East Stroudsburg, PA 18301, USA; gmoir@po-box.esu.edu
4 Department of Kinesiology, Coastal Carolina University, Conway, SC 29528, USA; hlamont@coastal.edu
5 School of Applied Sciences, Edinburgh Napier University, Edinburgh EH11 4DY, UK; Su.Brown@napier.ac.uk
* Correspondence: connaboy@pitt.edu

Received: 6 December 2019; Accepted: 27 December 2019; Published: 7 January 2020

check for updates

Abstract: Musculoskeletal injuries often occur during the execution of dynamic sporting tasks that involve rotation. The prescription of appropriate prevention strategies of musculoskeletal injury relies on assessments to identify risk, but current assessment tools focus on uniplanar movements. The purpose of this paper is to demonstrate the utility of the unilateral 180° jump as a potential assessment tool for injury risk in the lower body by (1) providing descriptive kinematics of the knee, thigh, and pelvis (2) conducting inter-segmental coordination analysis, and (3) comparing the knee kinematics between the dominant and non-dominant limb (NDL) during the loading (LOP) and landing phase (LAP). Elite rugby players completed one session, performing five 180° unilateral jumps on each limb while collecting kinematic data. Independent *t*-tests were used to compare peak angles of DL and NDL. Continuous Relative Phase (CRP) plots were constructed for thorax and pelvis in the transverse plane. At the loading phase, the non-dominant limb had greater peak knee abduction (ABD) ($p = 0.01$). At the landing phase, the dominant limb had greater peak knee adduction (ADD) ($p = 0.05$). At the landing phase, the non-dominant limb had greater peak knee ABD ($p = 0.01$). CRP plots indicate participants can utilize a thorax-led, pelvis-led, or synchronized rotational method. Bilateral asymmetries were observed, indicated by significant differences in the bilateral landing phase peak ADD/ABD, which is of particular interest considering all participants were healthy. Therefore, additional research is needed to determine thresholds for injury risk during rotational tasks.

Keywords: functional; movement; evaluation; assessment; screen

1. Introduction

Athletes in sports that involve jumping, pivoting, forceful change of direction, and contact are at a higher risk of lower-body injury [1–5]. Non-contact anterior cruciate ligament (ACL) tears are one of the most common lower-body injuries and have been largely attributed to the torsional forces associated with cutting and rotational tasks [4–6]. Rugby incorporates all of the aforementioned characteristics and it has been reported that lower limb joint/ligament injuries are the most common location and type of injury [7] occurring in both contact and non-contact situations [8]. Knee injuries result in players missing the most days from training, with ACL injuries accounting for the greatest proportion [9].

Clinicians and practitioners attempt to pre-emptively identify modifiable risk factors and address through strength, flexibility, and neuromuscular training. The identification of risk factors requires the utilization of a physical assessment tool with valid and reliable prognostic value for sporting tasks.

Pre-participation screenings are designed to determine potential intrinsic injury risk factors by identifying characteristics of the musculoskeletal system that may predispose an athlete to injury or identify incomplete recovery from a previous injury [10]. Traditionally, screening methods require a battery of tests, including static and dynamic protocols [10–16]. Static tests focus on measurements of joint range of motion, muscle strength, and muscle flexibility. The limitations of static tests are their applicability to dynamic situations, yielding limited meaningful information in the context of a dynamic sporting environment [17]. Previous research has demonstrated that scores obtained during static balance tests do not reflect scores obtained during dynamic balance tests, further indicating their limitations as injury predictors in a sporting context [16,18]. The recognition of these limitations led to the emergence of more functionally relevant or task-specific screening tests; where the functional aspect refers to the adoption and inclusion of assessments that more closely replicate activities of daily living and dynamic sports movement [19].

The purpose of functional and dynamic testing is to evaluate movements similar to sports actions that require the muscles to co-activate in integrated patterns to control a multi-joint movement [20]. A failure of the muscles to activate in a coordinated manner in the control of a movement often results in the development of a compensatory but predictively variable strategy to complete the task, which, after a period of time, can lead to injury and pain around the affected joint [11,17,21]. Functional testing is intended to enable the tester to identify 'weak links', suggesting an uncontrolled movement system (joint), within the series of linked joints [17]. Identification of these movement dysfunctions may serve as a predictor of injury and can be addressed by the appropriate professionals prior to the emergence of symptoms within a sporting context. Specifically, Mottram and Comerford [17] stated that the ability to identify uncontrolled movement strategies is imperative to enable risk management strategies to be developed, which can mitigate the propensity for (re)injury through increased localized strength and/or enhanced motor control.

Several mechanical dysfunctions have been associated with functional/dynamic assessments as injury predictors. Multiple studies have suggested that uncontrolled knee motion in the frontal plane when landing is a good predictor of ACL injury [22–24]. Additionally, compensatory strategies have been demonstrated in jump landings following injury [25,26]. The neuromuscular control associated with this form of dysfunctional, maladaptive movement pattern has been suggested as a potentially modifiable risk factor [27]. The utilization of unilateral tasks further enhances the assessment through the identification of limb asymmetry when compared to bilateral assessments [15,16,18]. However, research regarding cutting tasks (change of direction) suggests that large frontal plane knee excursions are necessary in order to complete the task and may not be indicative of uncontrolled motion [28–31]. Therefore, drop jumping landing tasks may not be appropriate for assessing injury risk during rotational sporting tasks. Designing an appropriate task to identify the predictors of rotary injury risk during dynamic sports-specific activity must include several components: high-velocity loading force, a certain degree of motor control complexity to complete the task, bilateral assessment, and multi-planar movement. A novel unilateral 180° jump task could potentially fulfill all the aforementioned requirements and better mimics the dynamic multi-planar nature of sports that involve jumping, pivoting, and forceful changes in direction (i.e., rugby, soccer, basketball, American football). Additionally, as athletes performing rotational movements are the target population for this assessment, using elite-level athletes, such as rugby players, will provide greater external validity of findings.

Considering the unilateral 180° jump as a potential physical assessment tool for rotational injury prediction in the lower body first requires a demonstration of its utility. Therefore, the purpose of this investigation is to analyze the unilateral 180° jump in elite-level rugby players by (1) providing descriptive kinematic of the knee, thigh, and pelvis, (2) conducting inter-segmental coordination analysis between the thorax and pelvis, and (3) comparing the knee kinematics between the dominant and non-dominant limb during the loading and landing phase. It is hypothesized that the greater rotational component will demonstrate greater frontal plane kinematics (greater peak abduction/adduction angles) of the knee during landing to compensate for the high torsional energy transfer through the kinetic chain. Similarly, greater peak frontal plane knee kinematics will be associated with a shoulder led transverse thorax–pelvis relative phase angle. More simply, we believe that greater knee adduction angles will be observed in individuals who utilize a shoulder led rotation pattern.

2. Methods and Materials

2.1. Participants

Fourteen male, elite rugby players from a National Union Academy (age: 20.0 ± 1.9 years; height: 184.5 ± 7.2 cm; mass 94.7 ± 12.0 kg) volunteered their participation. The inclusion criteria required that all participants were free of any current injury and had no history of serious lower limb injuries (injuries requiring surgery). To ascertain technique limb dominance, participants were asked which leg they preferred to kick a ball with. All participants indicated that their preferred kicking leg was their right leg. As a result, the right leg will be referred to as the dominant leg and the left as the non-dominant leg. All participants were 18 years of age or older. The study adhered to the guidelines set by the Declaration of Helsinki, ethical approval was obtained from the local University Ethics Committee (Edinburgh Napier University) prior to the investigation, and all participants provided written informed consent.

2.2. Procedures

Participants were requested to refrain from strenuous activity at least 24 h before attending the testing session. All participants were given verbal instructions indicating how to complete the jumping task. These included to initiate a unilateral stance, perform a countermovement action followed by a 180° jump as high as possible (maximal effort) landing on the same leg. Participants were asked to jump in a clockwise direction from their right leg and counterclockwise from their left leg. Participants were instructed not to use any arm movements during the jumps by placing "their hands on their hips". Following a demonstration of the jump, all participants practiced during an individualized warm-up to familiarize themselves with the protocol. Subsequent to the warm-up, each participant rested for 3 min before performing five single-leg 180° jumps either with their dominant or non-dominant leg (with a balanced pause between each jump), followed by five additional jumps with the other leg. The order of the starting jump leg was randomized to avoid learning effects.

2.3. Motion Analysis

A three-dimensional analysis was carried out using a 12-camera high speed (240 Hz) motion capture system (ProReflex, Qualisys AB., Gothenburg, Sweden). All participants wore fitted clothing and were barefoot to permit the accurate attachment of 25 retro-reflective markers (19 mm diameter) on the following anatomical landmarks (on both left and right sides): head of the first and fifth metatarsal bones, lateral and medial malleolus, posterior calcaneus, lateral and medial femoral epicondyles, anterior superior iliac spine (ASIS), posterior superior iliac spines (PSIS), acromion process, 4th lumbar vertebra, 10th thoracic vertebra, 7th cervical vertebra, sternum jugular notch, and xiphoid process. To reduce skin movement artifact error, clusters of 4 retro-reflective markers fixed to lightweight rigid plastic plates (Qualisys AB, Gothenburg, Sweden) were attached to both thighs and shanks to track the leg movement during the unilateral 180° jump, and a cluster of 4 markers placed on the skin around T7 was used to track thorax motion. Following the marker placement procedure, a static calibration file in the anatomical position was collected for each participant for the purpose of 3D model building and data generation.

2.4. Data Analysis

Marker trajectories were smoothed with a 6 Hz fourth-order low-pass Butterworth filter and kinematic data processed using Visual 3D^{TM} software (C-Motion Inc., Rockville, MD, USA). A power spectrum density calculation was conducted and 99% of the power was contained in the first 6 Hz, and, therefore, a 6 Hz cutoff frequency was chosen. All kinematic variables were quantified within the loading and landing phases of the unilateral 180° jump. The loading phase was defined as the instant the knee commenced flexion (following a stationary single-leg stance) until the instant of maximum knee flexion of the same leg. The landing phase was defined as the instant the landing foot made contact with the ground (determined from kinematic data when the metatarsal markers stopped their downward trajectory) until maximum knee flexion was attained.

Individual segment pose and segment coordinate systems (SCS) for each participant was generated with Visual 3D^{TM} from the static calibration trial using a multi-anatomical landmark optimized method [32] and utilizing the markers as follows: The pelvis was defined using the X-Y plane passing through the ASIS and PSIS, the origin of the SCS at the midpoint between the 2 ASIS markers, which allowed the segment X axis to be determined by the vector between the origin and the right ASIS marker. The SCS Z axis was determined by the axis perpendicular to the X-Y plane in the vertical direction, and, lastly, the SCS Y axis calculated as the cross product of the X and Z axes. All remaining segments were defined using the principle of creating a frontal plane using medial and lateral proximal and distal markers and then determining the SCS Z (vertical) axis as the unit vector directed from the distal segment end to the proximal segment end. The SCS Y axis was then perpendicular to the frontal plane and Z axis, and, finally, the X axis determined by the application of the right-hand rule. The origin of each SCS was located at the proximal end for each segment.

Knee joint angles were calculated according to the International Society of Biomechanics recommendations as the shank segment relative to the thigh segment resolved into the proximal segment SCS, using the Cardan sequence XYZ, where movement in the X plane denotes flexion (−)/extension (+), Y plane abduction (−)/adduction (+), and Z plane axial internal (+)/external (−) rotation [33]. Knee alignment in the Y plane was defined as zero when the long axes of the thigh and shank were aligned. Axial rotation of the thigh segment (Z plane) was calculated relative to the pelvis segment [33].

For both adduction/abduction and axial rotation, the non-dominant leg was multiplied by negative one to allow comparison between limbs. Pelvis motion was assessed in terms of flexion/extension, adduction/abduction, and axial rotation, which was computed as the angle of the pelvic segment relative to the fixed laboratory/global coordinate system (GCS) using Cardan sequence XYZ and the right-hand thumb rule [33]. Given the dynamic nature of the activity and magnitude of axial rotation involved in this activity, the sequence XYZ was selected instead of ZYX as described by Baker [34]. All variables were averaged across the five trials per leg for each participant; these means were then used to calculate group means.

The continuous relative phase (CRP) was calculated as a representative measure of inter-segmental coordination between the thorax and pelvis segmental rotations about the vertical (z) axis. To calculate the CRP between two segments, the phase-angle (Φ) from the phase-plane portraits of each segment were calculated. Phase-portraits were constructed with angular displacement (θ) on the horizontal (x) axis and the first derivative, angular velocity (ω) on the vertical (y) axis. Prior to calculating Φ, the phase-portrait values were normalized to the minimum and maximum values found in each axis using the protocol outlined by Li et al., [35], and to 101 data points. The normalization procedures minimize the influence of different segmental movement amplitudes [35] and allow comparisons of jumps with different temporal structures. The Φ was defined as the angle between the right horizontal and a line drawn to a specific data point (θ_i,ω_i) from the origin (0,0). The CRP was calculated as the difference between the thorax and pelvis segment angles at each of the 101 data points. Ensemble curves were produced for the individual CRP profiles by determining the mean CRP value at each of the 101 data points from the 5 jump trials. The variability in the CRP was displayed as the standard deviation of the 5 trials at each data point. The CRP provides a means to interpret both the coordination between the relative segments and its variability. This information can give insight into the relative stability (change in variability) of a pattern of movement over time, helping to identify which, if any, coordination patterns are important and common across multiple individuals. Measuring the relative phase between limb/segment movements (oscillations) has been regularly employed to examine the organization of a system at a synergistic level, as phase differences reflect the fundamental cooperation and competition evident within a movement system. The tendency of the synchronization of the reversal points of frequency-locked coupled oscillators is to adopt either an in-phase (0°) or anti-phase (180°) relationship with the movements initiating and/or terminating simultaneously [36]. According to Swinnen et al., [37] synchronization of the reversal points can be interpreted as intermittent loci of control, where reversal points act as anchors for the organization of the system. In contrast, asynchronous phase differences (i.e., 90°, 270°, etc.) are more difficult to maintain, requiring effort and considerable practice [38].

2.5. Statistical Analysis

All statistical analyses were performed using the Statistical Package for Social Sciences 14.0 (SPSS Inc., Chicago, IL, USA, 2004). The Kolmogorov–Smirnov test was used to determine the normality of the data distribution for each variable. Measures of central tendency and distribution of the data were reported as means and sample standard deviations. Paired sample *t*-tests were used to determine if any statistically significant differences existed between the mean values of the dominant and non-dominant legs for peak knee angle (flexion/extension and adduction/abduction), time to reach peak knee angle, axial rotation of the femur with respect to the pelvis and axial rotation of the pelvis with respect to the GCS in both the loading and landing phases following a 180° unilateral jump. Significance was accepted at $p \leq 0.05$ for all statistical tests. To measure the magnitude of the difference between the dominant and non-dominant legs relative to the variability, effect size (d) calculations were performed across all variables. Interpretation of the data was based on Cohen's (1992) guidelines, whereby effect sizes greater than 0.2 and less than 0.5 are considered small, greater than 0.5 and less than 0.8 are moderate, and greater than 0.8 are large. The intra-class correlation coefficient (ICC) was calculated to determine the reliability of each variable across the repeated trials. The standard error of measurement (SEM) was also calculated to assess the test's reliability, for example, a larger SEM indicates a lower test reliability and was calculated as follows:

$$SEm = SD * \left(\sqrt{1} - ICC\right)$$

3. Results

The results are presented with respect to the knee joint motion, thigh motion, pelvis motion, and inter-segmental coordination employed during the unilateral 180° jump. The ICC values (see Table 1) show a large range (ICC's Dominant Limb: Pelvis 0.03–0.83; Knee 0.34–0.92; Non-Dominant Limb: Pelvis 0.13–0.74; Knee 0.32–0.89) of test–retest reliability scores across the variables of interest, with a tendency for the dominant limb to demonstrate greater reliability across a greater number of variables.

3.1. Knee Joint Motion

During both the loading and landing phases, peak knee abduction was found to differ significantly between the legs, with the knee of the non-dominant leg abducting more than the dominant leg (see Table 2). Peak knee adduction between the legs differed significantly in the landing phase, with the knee of the dominant leg adducting more than the knee of the non-dominant leg. Peak knee adduction and peak knee abduction occurred within the early and late phases, respectively, during both the loading and landing phases. All other knee variables did not differ between the dominant and non-dominant leg within the loading and landing phases.

Table 1. Reliability measures.

Phase	Variable	Non-Dominant Limb		Dominant Limb	
		ICC [95% CI]	SEM	ICC [95% CI] (Dominant)	SEM
Loading	Peak Flexion (degs)	0.83 [0.68–0.93]	1.56	0.83 [0.68–0.94]	1.55
	Peak Adduction (degs)	0.89 [0.77–0.96]	0.63	0.83 [0.67–0.93]	1.28
	Peak Abduction (degs)	0.87 [0.74–0.95]	0.92	0.90 [0.80–0.96]	0.58
	Time Peak Add (%SC)	-	-	-	-
	Time Peak Abd (%SC)	-	-	-	-
Landing	Peak Extension (degs)	0.70 [0.47–0.87]	1.44	0.78 [0.59–0.91]	1.03
	Peak Flexion (degs)	0.73 [0.52–0.89]	2.35	0.77 [0.58–0.91]	1.96
	Peak Adduction (degs)	0.86 [0.73–0.95]	0.66	0.79 [0.61–0.92]	1.26
	Peak Abduction (degs)	0.87 [0.75–0.95]	0.81	0.92 [0.83–0.97]	0.44
	Time Peak Add (%SC)	-	-	-	-
	Time Peak Abd (%SC)	-	-	-	-

ICC 0.5–0.75 = moderate reliability, ICC 0.75–0.90 = good reliability, & ICC > 0.90 = excellent reliability. Larger SEM indicates less reliability.

Table 2. Knee motion for both loading and landing phases. Abd—Abduction; Add—Adduction.

Phase	Variable	Non-Dominant	Dominant	*p*-Value	Mean Diff [95% CI]	Effect Size (d)
Loading	Peak Flexion (degs)	−58.2 ± 9.2	−59.7 ± 9.1	0.67	1.5 [15.2, −12.2]	0.16
	Peak Adduction (degs)	−0.3 ± 5.7	4.6 ± 7.5	0.06	5.0 [14.2, −4.3]	0.75
	Peak Abduction (degs)	−8.2 ± 7.1	−0.8 ± 5.8	0.01 *	7.3 [17.0, −2.3]	1.13
	Time Peak Add (%SC)	22 ± 25	30 ± 24	0.39	7.9 [42.3, −26.4]	0.32
	Time Peak Abd (%SC)	77 ± 23	67 ± 29	0.33	9.9 [47.9, −27.9]	0.38
Landing	Peak Extension (degs)	−13.8 ± 4.7	−11.3 ± 4.7	0.16	2.5 [9.2, −4.1]	0.53
	Peak Flexion (degs)	−47.8 ± 8.7	−46.1 ± 8.5	0.61	1.7 [14.2, −10.9]	0.20
	Peak Adduction (degs)	2.8 ± 4.7	7.0 ± 6.0	0.05 *	4.2 [11.7, −3.3]	0.79
	Peak Abduction (degs)	−8.7 ± 6.2	−2.2 ± 5.5	0.01 *	6.5 [15.0, −2.0]	1.11
	Time Peak Add (%SC)	16 ± 18	23 ± 24	0.46	6.1 [36.9, −24.8]	0.29
	Time Peak Abd (%SC)	72 ± 20	62 ± 31	0.33	9.9 [47.3, −27.5]	0.39

Mean ± Standard Deviation. * Significant at $p \leq 0.05$. Cohen's d calculated for effect size (0.2–0.5 are considered small, 0.5–0.8 considered moderate, and >0.8 considered a large effect).

3.2. Thigh Motion

There was no difference in axial rotation of the thigh between the dominant and non-dominant leg during both the loading ($p = 0.89$, d = 0.04) and landing phases ($p = 0.77$, d = 0.14). During the loading phase thigh rotation for the dominant leg was 2.2 ± 5.3 degrees and the non-dominant leg was 2.4 ± 5.0 degrees. During the landing phase, thigh rotation for the dominant leg was 6.1 ± 4.5 degrees and the non-dominant leg was 5.6 ± 4.1 degrees. In preparation for executing and following the single-leg 180° jump, the thigh segments of the dominant and non-dominant legs internally rotated.

3.3. Pelvis Motion

During both the loading and landing phases, pelvis motion was not significantly different between dominant and non-dominant legs (Table 3). In preparation for taking-off, the pelvis was found to extend (sagittal plane), adduct (frontal plane), and externally rotate (transverse plane). On landing, the pelvis was found to extend, abduct, and internally rotate.

Table 3. Pelvis alignment for both loading and landing phases. Flex—Flexion; Ext—Extension; Add—Adduction; Abd—Abduction; Rot—Rotation.

Phase	Variable (degs)	Left	Right	*p* Value	Mean Diff [95% CI]	Effect Size (d)
Loading	Flex/Ext	−12.0 ± 7.5	−9.2 ± 5.5	0.27	2.8 [12.1, −9.3]	0.43
	Add/Abd	−3.6 ± 3.1	−4.3 ± 2.4	0.46	0.7 [4.3, −3.6]	0.26
	Axial Rot.	4.4 ± 6.0	0.8 ± 4.6	0.09	3.6 [11.1, −7.5]	0.68
Landing	Flex/Ext	−1.2 ± 3.8	−1.1 ± 3.1	0.95	0.1 [6.5, −6.4]	0.03
	Add/Abd	−7.8 ± 4.1	−7.0 ± 3.3	0.60	0.8 [6.7, −5.9]	0.22
	Axial Rot.	−18.0 ± 9.0	−18.7 ± 8.0	0.82	0.7 [12.9, −12.2]	0.08

Mean ± Standard Deviation. Cohen's d calculated for effect size (0.2–0.5 are considered small, 0.5–0.8 considered moderate and >0.8 considered a large effect).

3.4. Inter-Segmental Coordination

The three exemplar CRP plots presented (Figure 1) denote differences in pelvis/thorax segmental coordination for both the dominant and non-dominant take-off legs within the 180° jump. As shown in Figure 1, all participants tended to remain in-phase with minimal variability during the loading/take-off phase of the jump irrespective of the take-off leg. All three participants remain in-phase with low variability throughout the unilateral 180° jump when taking off from the non-dominant leg. Participant C employs a similar pattern of inter-segmental coordination (and variability) irrespective of the take-off leg. Participants A and B show a tendency to move out-of-phase and demonstrate high levels of CRP variability throughout the landing phase of the unilateral 180° jump.

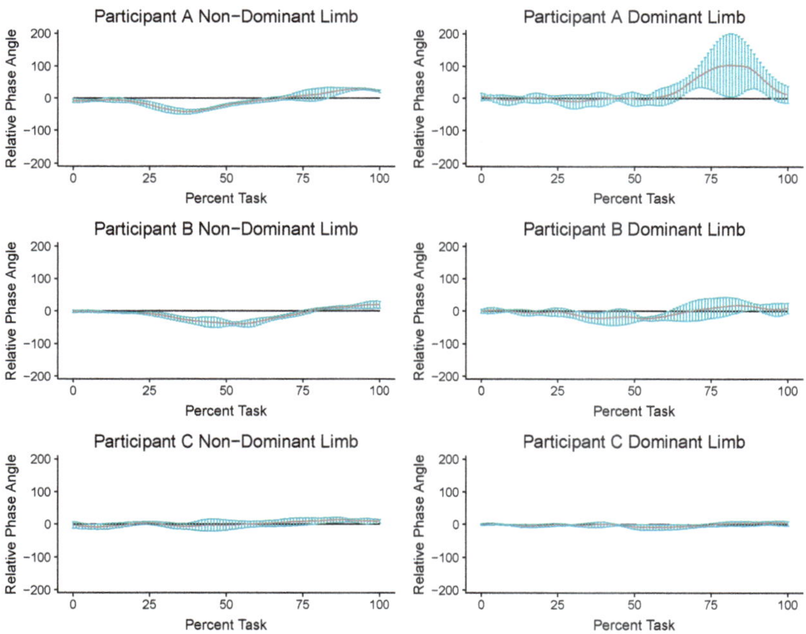

Figure 1. Exemplar continuous relative phase plots.

The solid red line represents the mean relative phase angle of all five jumps with a phase angle of 0 representing completely in phase or both the thorax and pelvis rotating together at the same rate. A phase angle > 0 (positive angle) indicates that the thorax is rotating ahead or leading the pelvis. A phase angle < 0 (negative angle) indicates the pelvis is rotating ahead or leading the thorax. The loading phase is represented by the first ~20%. The landing phase is represented by the last ~50%.

Vertical error bars indicate the standard deviation at points of the CRP plot. Therefore, larger vertical bars indicate greater movement variability jump to jump, while smaller bars represent less variability.

In phase is represented by a relative phase angle of 0. Out of phase (thorax-led) is represented by a positive relative phase angle.

4. Discussion

The principal aim of this study was to demonstrate the utility of the unilateral 180° jump as a potential physical assessment tool by providing descriptive kinematic data. All participants demonstrated similar knee and thigh motion and pelvis motion. During the loading phase there were no significant differences between limbs, with all participants executing internal rotation of the thigh while extending, adducting, and externally rotating the pelvis. These movements portray a "counter movement" pattern opposite to the desired action of rotation, by loading the system up similar to the downward movement (knee flexion) preceding forceful knee extension one would observe in a vertical jump. This rotational "counter movement" pattern may be performed in an attempt to load the associated musculature and utilize strain energy to facilitate the stretch-shortening cycle. Previous research has indicated hip flexion with hip internal rotation has a greater correlation with large knee valgus moments [29], therefore the observed external rotation of the hips in healthy athletes may be a subconscious attempt to reduce potentially deleterious forces at the knee.

During the landing phase there were no significant differences between limbs for thigh rotation ($p = 0.77$, d = 0.14) or pelvis movement (Flex/Ext [$p = 0.95$, d = 0.03], Add/Abd [$p = 0.6$, d = 0.22], Axial Rotation [$p = 0.82$, d = 0.08]). For all participants, the thigh internally rotated and the pelvis extended, abducted and internally rotated. For cutting tasks with up to a 110° change in direction, hip internal rotation in conjunction with hip abduction was associated with larger knee valgus moments [30,31]. The larger the degree of rotation the greater the rotational forces [30], indicating a necessity to execute the task and not necessarily a greater risk of injury. Since the participants of this study were healthy elite-level athletes, the pelvis moving in line with the rotation of the jump may have been a strategy to reduce torsional forces experienced at the knee and ankle by dissipating energy at a link in the kinetic chain with a greater range/number of degrees of freedom (DOF).

There was a significant difference in the frontal plane knee motion during the landing phase between limbs (Add [$p = 0.05$, d = 0.79], Abd [$p = 0.01$, d = 1.11]). Greater knee adduction was observed in the dominant limb (~4.2° greater), whereas greater knee abduction was observed in the non-dominant limb (~6.5° greater). Moreover, it should be noted that these specific findings had some of the strongest ICC values and lowest SEM values (see Table 1). This finding indicates that a bilateral asymmetry is observable even in healthy athletic populations. Due to the cross-sectional nature of this study, it is impossible to indicate which strategy is more beneficial. Uncontrolled knee movement during landing has been associated with greater ACL injury risk in uniplanar drop jump tasks [23] but rotational tasks require greater knee frontal plane excursions to complete [29–31]. Therefore, prospective evidence is needed to indicate which motion contributes to a greater risk of injury. Furthermore, one motion is most likely not good or bad, but stresses different soft tissue structures more greatly (for example, abduction: ACL, adduction: MCL). Likewise, how the system (body) moves as a whole is more important for understanding the potential injury risk.

Visual qualitative assessment of CRP plots indicated that there was little variation between participants during the loading/take-off phase (refer to the first 20% in Figure 1). Moreover, the relative phase angle between the shoulders and the pelvis for the majority of the participants was very close to 0°. A phase angle at or close to 0° indicates that the shoulders and pelvis were in-phase or moving together. Coordinated coupling of the upper and lower body segments during the loading phase may be a result of the population tested. Healthy athletes presumably have better motor control and coordinate their torsos in a synergistic fashion with their lower extremities to improve mechanical efficiency by reducing counter-productive movement. Furthermore, this may demonstrate torso control by the participants to reduce torsional forces experienced at the knee that has been observed in cutting tasks when torso lean/excursion is excessive [28]. Likewise, the amelioration of the torsional force and modulation of body segments may help maintain the center of gravity within the base of support so that the individual successfully executes the task (does not fall over). Prospective research should be conducted on unhealthy or non-athletic populations to determine if uncontrolled torso movement can be observed during the loading phase of a unilateral 180° jump task. Investigating 'less athletic' or impaired populations will further elucidate motor control capabilities, differentiating key differences in task execution compared to elite-level athletes when performing this specific task.

Whereas the loading/take-off phase exhibited relative uniformity between participants, the greatest amount of variation was observed during the landing phase of the 180° jump (refer to last 50% of Figure 1). CRP plots displayed three common landing strategies amongst participants: thorax (shoulder)-led movement, pelvis-led, or simultaneous in-phase landing. Although these three strategies were performed, it should be noted that each individual only utilized one of them and with varying degrees of phase coupling. In some cases, individuals were close to a relative phase angle of 180°, which represents an anti-phase movement (similar to arms swinging in separate directions at the same rate during normal gait). Furthermore, the phase coupling between the thorax and pelvis was not constant, meaning the rates at which the two segments rotated in conjunction with one another were at different rates throughout the second half of the unilateral 180° jump task, suggesting a dissociation between the segments during the landing portion of the task.

CRP best describes the passage of energy through a system [39,40], in this case, the human body. Each participant utilized a slightly different strategy to dissipate the high torsional energy in order to execute the task and land successfully. The fact that many strategies with different degrees of variation were observed demonstrates the principle of equifinality of movement solutions due to the vast number of DOF [41,42]. Although there are multitudes of strategies to complete the task, it stands to reason a shoulder-led strategy may be the most deleterious for structures around the knee. If the shoulders lead with the pelvis following, upon landing when a high load rate force is transmitted up the kinetic chain the shoulders will have little range left for further excursion to enable energy dissipation. Instead, it could likely result in a resultant force transmitted back down to attempt to be dissipated by the pelvis movement. This could result in greater hip flexion and hip internal rotation, which has been linked to greater knee valgus moments [29–31]. A pelvis-led strategy may enable better dissipation of excessive energy because it can pass up through the entire kinetic chain. However, because this study is cross-sectional this theory is conjecture on the part of the authors. Similar to research involving drop jumps [23,43,44], future research could yield information regarding which strategy portends greater injury risk and what relative phase angles indicate problematic upper torso and lower body coupling during rotational sporting tasks.

Further evidence to support this approach can be observed in the test–retest reliability (ICC) scores (see Table 1). Large variability can be observed across the ICC scores, and can be interpreted as being indicative of the presence of variability in specific movements, thus, enabling the production of more constrained, and reliable movements in key structures (depending on the movement strategy employed). There are some limitations to this study, mainly, the small sample size (n = 14). However, the use of elite-level athletes (rugby players) not only strengthens the external validity of this investigation (the application for athletes), the smaller sample size is representative of the smaller percentage of the population that achieved elite athlete level status (professional). Furthermore, for significant findings the effect sizes were moderate to large (>0.79) (see Table 2) and the reliability measures (ICC [0.79–0.92] and SEM [0.44–1.26]) for those that were significant were moderate to strong (see Table 1). Considering the knee is a major focal point of injury risk assessment, these aforementioned values support the unilateral 180° jump as a potential rotary injury risk assessment tool.

5. Practical Application

The unilateral 180° jump provides a plethora of biomechanical data. Firstly, the unilateral 180° jump task can be performed in a confined space making it a more feasible alternative to performing cutting tasks when assessing lower body rotary risk. Requiring the participant to jump as high as possible could provide kinetic data and a simultaneous metric on vertical jump performance. Although jump heights will not be as large in magnitude as a traditional countermovement jump, frequently, jumps are made off a single leg and unbalanced (not having time to properly align and set the body) during play, thus, the 180° jump task replicates scenarios experienced in sport. Furthermore, the requirement of maximal jump height forces the participant to utilize muscle contraction velocities that are similar to game conditions, unlike other previously mentioned assessments [2,11,12,18,45]. Likewise, landing from a maximal vertical height while rotating will exert a multi-planar loading force (vertical and torsional) typical to athletes in sports such as soccer, rugby, football, or basketball. Moreover, landing unilaterally while rotating more closely mimics one of the most common mechanisms of ACL injury as compared to landing from a uniplanar jump task [4].

Future studies should investigate the unilateral 180° jump with pathological and non-pathological populations prospectively (over the course of an athletic season) to associate movement patterns with injury predictors. Due to the multiple variables that can be collected at once with the unilateral 180° jump and the ability to visually assess the movement, the task has the potential to be a valid and efficient objective prognostic physical assessment tool that may provide greater sensitivity for identifying rotational injury risk.

Author Contributions: C.C. conceived the concept of this investigation and supervised the project which was carried out by C.M. Data was reduced and analyzed by S.B. K.T.K. wrote the manuscript and created the tables/figures with the support A.M.S., G.L.M. and H.S.L. All authors have read and agreed to the published version of the manuscript.

Funding: This research received no external funding.

Conflicts of Interest: The authors declare no conflict of interest.

References

1. Daniel, D.M.; Stone, M.L.; Dobson, B.E.; Fithian, D.C.; Rossman, D.J.; Kaufman, K.R. Fate of the ACL-injured patient. A prospective outcome study. *Am. J. Sports Med.* **1994**, *22*, 632–644. [CrossRef] [PubMed]
2. Gokeler, A.; Hof, A.L.; Arnold, M.P.; Dijkstra, P.U.; Postema, K.; Otten, E. Abnormal landing strategies after ACL reconstruction. *Scand. J. Med. Sci. Sports* **2010**, *20*, e12–e19. [CrossRef] [PubMed]
3. McLean, D.A. Analysis of the physical demands of international rugby union. *J. Sports Sci.* **1992**, *10*, 285–296. [CrossRef]
4. Boden, B.P.; Dean, G.S.; Feagin, J.A., Jr.; Garrett, W.E., Jr. Mechanisms of anterior cruciate ligament injury. *Orthopedics* **2000**, *23*, 573–578. [CrossRef]
5. Cochrane, J.L.; Lloyd, D.G.; Buttfield, A.; Seward, H.; McGivern, J. Characteristics of anterior cruciate ligament injuries in Australian football. *J. Sci. Med. Sport* **2007**, *10*, 96–104. [CrossRef]
6. McNair, P.; Marshall, R.N.; Matheson, J.A. Important features associated with acute anterior cruciate ligament injuru. *N. Z. Med. J.* **1990**, *103*, 537–539.
7. Fuller, C.; Clarke, L.; Molloy, M.G. Risk of injury associate with rugby union played on artificial turf. *J. Sports Sci.* **2010**, *28*, 563–570. [CrossRef]
8. Brooks, J.; Fuller, C.W.; Kemp, S.P.; Reddin, D.B. Epidemiology of injuries in English professional rugby union: Part 1 match injuries. *Br. J. Sports Med.* **2005**, *39*, 757–766. [CrossRef]
9. Dallalana, R.J.; Brooks, J.H.; Kemp, S.P.; Williams, A.M. The epidemiology of knee injuries in English professional rugby union. *Am. J. Sports Med.* **2007**, *35*, 818–830. [CrossRef]
10. Dennis, R.; Finch, C.F.; Elliott, B.C.; Farhart, P.J. The reliability of musculoskeletal screening tests used in cricket. *Phys. Ther. Sport* **2008**, *9*, 25–33. [CrossRef]
11. Chorba, R.; Chorba, D.J.; Bouillion, L.E.; Overmyer, C.A.; Landis, J.A. Use of a functional movement screening tool to determine injury risk in female collegiate athletes. *Nat. Am. J. Sports Phys. Ther.* **2010**, *5*, 47–54.
12. Cook, G. Weak links: Screening an athlete's movement patterns for weak links can boost your rehab and training efforts. *Train. Cond.* **2002**, *12*, 29–37.
13. Duke, S.R.; Martin, S.E.; Gaul, C.A. Preseason functional movement screen predicts risk of time-loss injury in experienced male rugby union athletes. *J. Strength Cond. Res.* **2017**, *31*, 2740–2747. [CrossRef]

14. Gutowski, A.; Rosene, J.M. Preseason performance testing battery for men's lacrosse. *Strength Cond. J.* **2011**, *33*, 16–22. [CrossRef]
15. Maulder, P.; Cronin, J. Horizontal and vertical jump assessment: Reliability, symmetry, discriminative and predictive ability. *Phys. Ther. Sport* **2005**, *6*, 74–82. [CrossRef]
16. Ross, S.E.; Guskiewicz, K.M. Examination of static and dynamic postural stability in individuals with functionally stable and unstable ankles. *Clin. J. Sport Med.* **2004**, *14*, 332–338. [CrossRef]
17. Mottram, S.; Comerford, M. A new perspective on risk assessment. *Phys. Ther. Sport* **2008**, *9*, 40–51. [CrossRef]
18. Hrysomallis, C.; Mclaughlin, P.; Goodman, C. Relationship between static and dynamic balance tests among elite Australian footballers. *J. Sci. Med. Sport* **2006**, *9*, 288–291. [CrossRef]
19. Arnason, A.; Sigurdsson, S.B.; Gudmundsson, A.; Holme, I.; Engebretsen, L.; Bahr, R. Risk factors for injuries in football. *Am. J. Sports Med.* **2004**, *32*, 5S–16S. [CrossRef]
20. Davids, K.; Glazier, P.; Araujo, D.; Bartlett, R. Movement systems as dynamical systems: The functional role of variability and its implications for sports medicine. *Sports Med.* **2003**, *33*, 245–260. [CrossRef]
21. Elphinston, J.; Hardman, S.L. Effect of an integrated functional stability program on injury rates in an international netball squad. *J. Sci. Med. Sport* **2006**, *9*, 169–176. [CrossRef]
22. Chappell, J.; Yu, B.; Kirkendall, D.T.; Garrett, W.E. A comparison of knee kinetics between male and female recreational athletes in stop-jump tasks. *Am. J. Sports Med.* **2002**, *30*, 261–267. [CrossRef]
23. Hewett, T.E.; Myer, G.D.; Ford, K.R.; Heidt, R.S., Jr.; Colosimo, A.J.; McLean, S.G.; van den Bogert, A.J.; Paterno, M.V.; Succop, P. Biomechanical measures of neuromuscular control and valgus loading of the knee predict anterior cruciate ligament injury risk in female athletes: A prospective study. *Am. J. Sports Med.* **2005**, *33*, 492–501. [CrossRef]
24. Mclean, S.; Walker, K.; Ford, K.R.; Myer, G.D.; Hewett, T.E.; Van Den Bogert, A.J. Evaluation of a two dimensional analysis method as ligament injury a screening and evaluation tool for anterior cruciate ligament injury. *Br. J. Sports Med.* **2005**, *39*, 355–362. [CrossRef]
25. Ernst, G.; Saliba, E.; Diduch, D.R.; Hurwitz, S.R.; Ball, D.W. Lower extremity compensations follwoing anterior cruciate ligament reconstruction. *Phys. Ther.* **2000**, *80*, 251–260. [CrossRef]
26. Webster, K.; Gonzalez-Adrio, R.; Feller, J.A. Dynamic joint loading follwoing hamstring and patellar tendon anterior cruciate ligament reconstruction. *Knee Surg. Sports Traumatol. Arthrosc.* **2004**, *12*, 15–21. [CrossRef]
27. Herrington, L.; Comfort, P. Training for prevention of ACL injury: Incorporation of progressive landing skill challenges into a program. *Strength Cond. J.* **2013**, *35*, 59–65. [CrossRef]
28. Jamison, S.T.; Pan, X.; Chaudhari, A.M. Knee moments during run-to-cut maneuvers are associated with lateral trunk positioning. *J. Biomech.* **2012**, *45*, 1881–1885. [CrossRef]
29. McLean, S.; Huang, X.; van den Bogert, A.J. Association between lower extremity posture at contact and peak knee valgus moment during sidestepping: Implications for ACL injury. *Clin. Biomech.* **2005**, *20*, 863–870. [CrossRef]
30. Sigward, S.; Cesar, G.M.; Havens, K.L. Predictors of frontal plane knee moments during side-step cutting to 45 and 110 men and women: Implications for ACL injury. *Clin. J. Sport Med.* **2016**, *25*, 529–534.
31. Sigward, S.M.; Powers, C.M. Loading characteristics of females exhibiting excessive valgus moments during cutting. *Clin. Biomech. (Bristol, Avon)* **2007**, *22*, 827–833. [CrossRef]
32. Cappello, A.; Cappozzo, A.; La Palombara, P.F.; Lucchetti, L.; Leardini, A. Multiple anatomical landmark calibration for optimal bone pose estimation. *Hum. Mov. Sci.* **1997**, *16*, 259–274. [CrossRef]
33. Lees, A.; Barton, G.; Robinson, M. The influence of Cardan rotation sequence on angular orientation data for the lower limb in the soccer kick. *J. Sports Sci.* **2010**, *28*, 445–450. [CrossRef] [PubMed]
34. Baker, R. Pelvic angles: A mathematically rigorous definition which is consistent with a conventional clinical understanding of the terms. *Gait Posture* **2001**, *13*, 1–6. [CrossRef]
35. Li, L.; Van Den Bogert, E.C.; Caldwill GEVan Emmerik, R.E.; Hamill, J. Coordination patterns of walking and running at similar frequencies. *Hum. Mov. Sci.* **1999**, *18*, 67–85. [CrossRef]
36. Haken, H.; Kelso, J.A.; Bunz, H. A theoretical model of phase transitions in human hand movements. *Biol. Cybern.* **1985**, *51*, 347–356. [CrossRef]
37. Swinnen, S.; Walter, C.B.; Lee, T.D.; Dounskaia, N. The organisation and control of new patterns of interlimb coordination against the backdrop of pre-existing preferred coordination modes. In Proceedings of the 1st Annual Congress of the European College of Sports Science, Nice, France, 28–31 May 1996.

38. Zanone, P.; Kelso, J.A. Evolution of behavioural attractors with leaning: Non-equilibrium phase transitions. *J. Exp. Psychol. Hum. Percept. Perform.* **1992**, *18*, 403–421. [CrossRef]
39. Ruf, L.; Chéry, C.; Taylor, K.L. Validity and reliability of the load-velocity relationship to predict the one-repetition maximum in deadlift. *J. Strength Cond. Res.* **2018**, *32*, 681–689. [CrossRef]
40. Lamb, P.; Stöckl, M. On the use of continuous relative phase: Review of current approaches and outline for a new standard. *Clin. Biomech.* **2014**, *29*, 484–493. [CrossRef]
41. Bernstein, N. *The Coordination and Regulation of Movements*; Pergamon Press: London, UK, 1967.
42. Cusamano, J.; Cesari, P. Body-goal variability mapping in an aiming task. *Biol. Cybern.* **2006**, *94*, 367–379. [CrossRef]
43. Hewett, T.; Torg, J.S.; Boden, B.P. Video analysis of trunk and knee motion during non-contact anterior curciate ligament injury in female athletes: Lateral trunk and knee abduction motion are combined components of the injury mechanism. *Br. J. Sports Med.* **2009**, *43*, 417–422. [CrossRef] [PubMed]
44. Padua, D.; DiStefano, L.J.; Beutler, A.I.; de la Motte, S.J.; DiStefano, M.J.; Marshall, S.W. The landing error scoring system as a screening tool for an anterior cruciate ligament injury-prevention program in elite-youth soccer athletes. *J. Athl. Train.* **2015**, *50*, 589–595. [CrossRef] [PubMed]
45. Nagai, T.; Sell, T.C.; House, A.J.; Abt, J.P.; Lephart, S.M. Knee proprioception and strength and landing kinematics during a single-leg stop-jump task. *J. Athl. Train.* **2013**, *48*, 31–38. [CrossRef] [PubMed]

applied sciences

MDPI

Article

Do Grade II Ankle Sprains Have Chronic Effects on the Functional Ability of Ballet Dancers Performing Single-Leg Flat-Foot Stance? An Observational Cross-Sectional Study

Bruno Dino Bodini [1], Giacomo Lucenteforte [2], Pietro Serafin [1], Lorenzo Barone [2], Jacopo A. Vitale [1,*], Antonio Serafin [1], Valerio Sansone [1,3] and Francesco Negrini [1]

1 IRCCS Istituto Ortopedico Galeazzi, Via Riccardo Galeazzi 4, 20161 Milan, Italy; bodinibruno@hotmail.com (B.D.B.); pietro.knw@gmail.com (P.S.); antonio.serafin@grupposandonato.it (A.S.); valerio.sansone@unimi.it (V.S.); francesco.negrini@gmail.com (F.N.)

2 Physical Medicine and Rehabilitation Residency Program, University of Catania, Piazza Università 2, 95124 Catania, Italy; giacomo.lucenteforte@outlook.com (G.L.); lorenzob100887@gmail.com (L.B.)

3 Department of Biomedical Sciences for Health, Università degli Studi di Milano, Via Mangiagalli 31, 20133 Milan, Italy

* Correspondence: jacopo.vitale@grupposandonato.it; Tel.: +39-026-621-4980

Received: 19 November 2019; Accepted: 18 December 2019; Published: 24 December 2019

check for updates

Abstract: Ballet dancers have a higher risk than the general population of ankle sprains Ankle proprioception is of the utmost importance for executing static and dynamic positions typical of ballet dancing. Ankle sprains can create changes in functional ability that may affect ballet performance. The aim of this cross-sectional observational study is to evaluate if non-professional ballet dancers that were previously injured with a grade II ankle sprain carry a long-term stability deficit in ballet specific positions (passé, arabesque) and in single-leg flat-foot stance, thereby affecting ballet performance. We enrolled 22 amateur female ballet dancers, 11 who previously had a grade II ankle injury and 11 who had no history of ankle injury. Stabilometric data (Center of Pressure Speed and Elipse Area) were assessed with the postural electronic multisensory baropodometer in normal, arabesque, and passè positions with both open and closed eyes. Using an unpaired *t*-test, we compared healthy and pathological feet of the ankle injury group for a standard monopodalic position and two ballet-specific positions. No difference between pathological and healthy feet of non-professional ballet dancers who suffered grade II ankle injury was detected. According to the parameters considered in this study, grade II ankle sprains seem to have a favorable prognosis in the sample that we evaluated.

Keywords: stabilometry; ballet dancers; ankle injury; chronic ankle instability; balance

1. Introduction

During performance, ballet dancers have a great need for stability and balance control [1,2], along with a high level of general "flexibility", which are peculiar characteristics that improve with ballet training [1,3–5]. Ankle control is of the utmost importance for executing static and dynamic positions typical of ballet dancing [6,7]. In 2017, Vassallo et al. reported the epidemiology of dance-related injuries between 2000 and 2013, defining ankle sprains as the most common injury [8]. Later, Smith et al. and Smith et al. collected data about their incidence and prevalence [9,10]. Ankle sprains could cause chronic functional ability changes, as is common in groups of dancers and non-dancers alike [11,12], also increasing the risk of re-injury [13]. In addition, it was seen that people with functional ankle instability had a less refined kinesthesia, in particular for inversion movements [14]. However, a prospective study conducted in 1996 by Leanderson et al., based on stabilometry, demonstrated only short-term consequences after ankle sprain in ballet dancers [15]. Data about chronic ankle instability (CAI) in the general sporting population are, to date, not satisfactory to determine the presence of common aspects of CAI within individual sports [16]. However, an exploratory study conducted by Simon et al. revealed a condition of CAI in 75.9% of university dance majors [17]. Despite new wearable technologies, such as Inertial Measurament Unit IMU [18,19] and wireless electromyography (EMG) [20], having been proposed to investigate balance and postural control, stabilometry remains the most commonly used tool [21]. To date, there is a lack of studies about the chronic effects of an ankle injury in ballet dancers using stabilometry evaluation, but there is an increasing body of literature concerning this method [22,23], including not previously injured ballet dancers [24]. A study performed by Lin et al., using stabilometry to assess dancers with ankle injuries, found that non-acute ankle injuries could worsen postural stability of professional ballet-dancers during ballet-specific postures. However, they did not focus on a specific injury nor on a precise time-frame for the post-injury evaluation [25]. Another relevant aspect in the evaluation of balance control is represented by the visual system's role in proprioception, which has the potential to mask proprioceptive issues in athletes [26]. For this reason, it can be argued that in order to obtain reliable proprioceptive measures, visual occlusion is mandatory [27,28]. As ankle control is crucial for successfully executing ballet specific positions and it could be impaired in ankle sprains, a position specific testing with closed eyes is needed. Moreover, some studies showed that athletic and sport training is task specific and skills acquired in a specific task did not completely translate to the same skills in different tasks [29–31]. Recently, Thalassinos et al. discovered that there were many sport skill-specific bias about sensory inputs for spatial orientation and postural control between experienced soccer athletes, ballet dancers, and nonathletes [32]. Considering the high incidence and prevalence of ankle injuries in professional and non-professional ballet dancers (from 3% to 25.6% of all musculoskeletal ballet injuries [4,8,10,33]), and the primary role of ankle control to perform ballet-specific tasks, the aim of our study was to understand if non-professional ballet dancers, previously injured with a grade II ankle sprain, compared to dancers who did not suffer any ankle injury, carry a long-term stability deficit in ballet specific positions (passé, arabesque), thereby affecting ballet performance.

2. Materials and Methods

This was a cross-sectional observational multi-center study. We enrolled 22 subjects (N = 11 in the pathological group and N = 11 in the control group) from six different schools of classical ballet dance in Italy. The inclusion criteria were: age between 15 and 25 years old; classical ballet dancer; at least 10 years of training (average of 2–3 sessions of training a week); right footed in ballet practice; unilateral ankle injuries; clinical diagnosis of grade II inversion right ankle sprain that occurred at least 6 months before our evaluations made by an orthopedic surgeon [34]; and informed consent signed (parents' consent was collected if subjects were under 18 years old). The exclusion criteria were: professional dancer; another lower extremity injury; history of neurological or motor deficits; and any visual impairment. The control group had the same inclusion and exclusion criteria, except for the history of ankle injury. For grading the severity of the ankle injury, we used the classification proposed by Malliaropoulos et al. [34], and classified grade II ankle injury patients positive to anterior drawer test but negative to talar tilt test. We defined grade II ankle injury as a partial tear of the lateral ligament complex of the ankle without decreased motion and loss of function. All pathological subjects were treated conservatively with soft bandages applied for 2 weeks. No patient underwent a specific rehabilitation protocol. At the moment of inclusion in the study, all patients were symptom free and fully participating in dance at their pre-injury level. Before entering the study, the participants were fully informed about the study's aims and procedures, and written informed consent was obtained before testing. The study protcol was approved by the Ethics Committee of the University of Milan (approved on 12/10/15, Prot. N. 54/15) in accordance with current national and international laws and regulations governing the use of human subjects (Declaration of Helsinki II).

3. Sample Size Calculation

In order to determine the sample size of the study, we carried out an a-priori power analysis on the basis of the scientific literature. Since our main outcome was to determine if there was a difference in monopodalic stance ellipse area between healthy and previously pathological ballet dancers, we used data from the 2017 paper by de Mello et al. [24]. With a reported monopodalic ellipse area in ballet dancers of 155 ± 64 mm^2 and an expected difference of 70 mm^2 (effect size = 1.1), we calculated that we would need a sample of 22 subjects (11 for each group) to detect a significant change with a power of 0.80 and an alpha of 0.05. To calculate the sample size, we used the GPower software (Universitat Dusseldorf-Germany).

4. Tools and Procedures

The internationally approved stabilometric parameters [35] we evaluated in this paper are: (1) Ellipse area (mm^2), contains 90% of the positions sampled of the center of pressure (CoP) and represents the dispersion of the oscillations and the precision of the system; and (2) CoP speed (mm/s), the average speed related to the CoP oscillations. The acquisition of stabilometric data was performed with the postural electronic multisensory baropodometer Diasu®, equipped with Milletrix® interface, which has previously been shown to be a valid and reliable tool [36] This tool features a scanning frequency of 100 frames per second in real time, acquisition surface of 40 cm^2 per module, accuracy of +/−5%, and maximum point pressure of 150 N/cm^2. The evaluations were carried out in six different gyms with a dedicated medical room, characterized by the absence of external perturbations, especially acoustic. Prior to the test, each dancer stretched and warmed up the muscles of the lower limbs via routine self-selected exercises for 5 min. Each dancer wore comfortable and adherent clothes, and a pair of cotton socks.

During the test, an expert podiatrist and a nurse were present in the room. The baropodometer was placed at a distance of two meters from a homogeneous color wall that the subjects faced during the whole time of examination. Each evaluation session was held in the early afternoon and each subject had to be at rest from training in the previous day. The subjects were evaluated with both eyes open and closed. To ensure a reference point to be observed during acquisitions, a mark was positioned on the wall at the same height and equidistant to the patient's eyes, and at a distance of 2 m. We acquired each measurement three times with eyes open and three times with eyes closed. The first try was considered as a demonstration and we did not record it. For each foot we tested, the patients in the following conditions both with eyes opened and closed.

Monopodalic stance with the contralateral limb slightly detached from the ground (5 s length) (Figure 1).

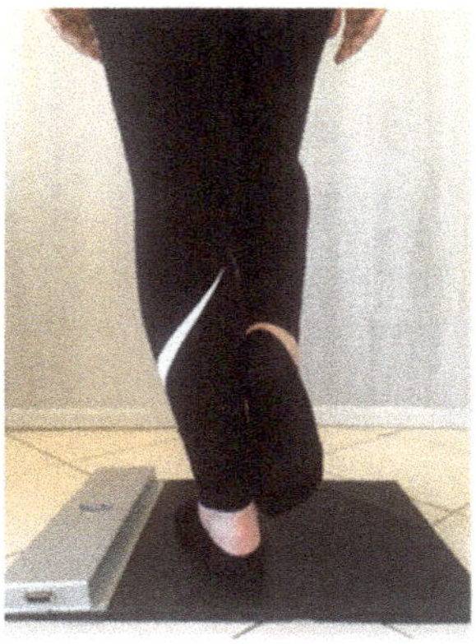

Figure 1. Monopodalic position.

Monopodalic with the limb in contact with the ground in "en dehors" position, and the contralateral in "arabesque" position, defining a 45° angle between them (5 s length) (Figure 2).

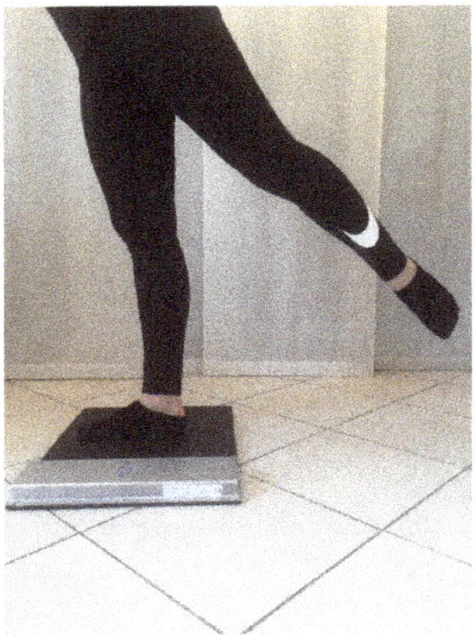

Figure 2. Arabesque.

Monopodalic with the limb in contact with the ground in "en dehors" position and the contralateral in "passé" position with a back "retiré" (5 s length) (Figure 3).

Figure 3. Passè.

5. Statistical Analysis

For each subject, parameters were determined by calculating the mean of the trials; subsequently, the grand average and the standard deviation over the subjects in each position were computed. Preliminary tests for normality (Kolmogorov-Smirnov test) and for equality of sample variances (Levene's test) provided the basis for using parametric statistics. All parameters were normally distributed. First, we checked if there were any significant statistical differences between the two groups as far as age, body mass index (BMI), years of dance practice, or age when they started practicing using an unpaired Student's *t*-test. Second, we compared stabilometric parameters, of each position, between the left and right foot in healthy subjects, and between injured and non-injured feet in the post-injury patients by the unpaired Student's *t*-test. Third, the difference between open and closed eyes in each condition, for each foot, was calculated; then, we checked if the differences in stabilometric parameters between closed and open eyes were larger in pathological vs. healthy feet using the unpaired Student's *t*-test. The level of significance was set to $p < 0.05$. Statistical analysis was performed using IBM SPSS 20.

6. Results

Demographical data of the study population are shown in Table 1. We could not find any statistical difference between the two groups for the parameters of age, BMI, years of dance practice, or age when they started practicing.

Table 1. Demographical data of participants.

	Post-Injury (N = 11)	Healthy (N = 11)	*p*-Values
Age (years)	16.36 ± 2.81	17.5 ± 3.37	>0.05
BMI (Kg/m^2)	19.85 ± 2.43	19.37 ± 2.33	>0.05
Years of Practice	11.91 ± 2.11	12.37 ± 4.07	>0.05
Age When They Started Practicing (years)	4.36 ± 1.23	5.18 ± 1.47	>0.05

Data are reported as Mean ± SD. Abbreviation: BMI, body mass index.

First, we analyzed the healthy subjects. We compared stabilometric parameters of the right and left feet for each parameter and no statistical differences for all parameters were detected (Table 2).

Table 2. Comparison between right and left foot in healthy subjects (N = 11) for stabilometric parameters acquired with open eyes.

	Right Foot	Left Foot	*p*-Values
Monopodalic Ellipse Area (mm^2)	105.91 ± 67.87	82.06 ± 47.43	>0.05
Monopodalic CoP Speed (mm/s)	13.84 ± 3.24	13.20 ± 5.34	>0.05
Arabesque Ellipse Area (mm^2)	274.40 ± 255.50	152.23 ± 87.54	>0.05
Arabesque CoP Speed (mm/s)	21.84 ± 8.46	19.65 ± 5.60	>0.05
Passè Ellipse Area (mm^2)	222.49 ± 130.06	133.26 ± 72.00	>0.05
Passè CoP Speed (mm/s)	22.22 ± 4.98	18.17 ± 5.58	>0.05

The pathological group was then analyzed. We compared stabilometric parameters between the injured feet and the healthy feet and no statistically significant differences were observed (Table 3).

Table 3. Comparisons between injured and healthy feet in pathological subjects (N = 11) for stabilometric parameters acquired with open eyes.

	Trauma	No Trauma	*p*-Values
Monopodalic Ellipse Area (mm^2)	128.25 ± 93.83	83.23 ± 49.15	>0.05
Monopodalic CoP Speed (mm/s)	14.23 ± 5.05	13.28 ± 5.55	>0.05
Arabesque Ellipse Area (mm^2)	165.17 ± 127.18	117.16 ± 59.89	>0.05
Arabesque CoP Speed (mm/s)	19.32 ± 4.38	18.54 ± 5.63	>0.05
Passè Ellipse Area (mm^2)	148.15 ± 82.43	119.66 ± 66.75	>0.05
Passè CoP Speed (mm/s)	20.75 ± 6.92	18.58 ± 5.97	>0.05

Data are reported as Mean ± SD. Abbreviation: CoP, Centre of Pressure.

We calculated the differences between open and closed eyes for all the parameters, and with reference to the pathological group, data collected with the feet that suffered a grade II ankle injury were compared with healthy contralateral limbs. As reported in Table 4, no significant differences between the sides were observed.

Lastly, we also compared right foot parameters of healthy and pathological subjects, both with open eyes (Table 5) and the difference between open and closed eyes (Table 6), and no significant differences were detected.

Table 4. Comparison between injured and healthy feet in pathological subjects (N = 11). Mean delta values (difference between open and closed eyes) of stabilometric parameters.

	Trauma	No Trauma	p-Values
Monopodalic Ellipse Area (mm^2)	321.47 ± 151.77	358.42 ± 310.67	>0.05
Monopodalic CoP Speed (mm/s)	15.77 ± 3.77	16.88 ±7.81	>0.05
Arabesque Ellipse Area (mm^2)	449.63 ± 203.33	440.82 ±378.02	>0.05
Arabesque CoP Speed (mm/s)	15.28 ± 9.60	15.29 ±9.60	>0.05
Passè Ellipse Area (mm^2)	330.93 ± 294.15	302.69 ± 216.10	>0.05
Passè CoP Speed (mm/s)	16.90 ± 9.22	12.66 ± 7.53	>0.05

Data are reported as Mean ± SD. Abbreviation: CoP, Centre of Pressure.

Table 5. Comparison between right feet of pathological and healthy subjects for stabilometric parameters acquired with open eyes.

	Right Foot Post-Injury (N = 11)	Right Foot Healthy (N = 11)	p-Values
Monopodalic Ellipse Area (mm^2)	105.91 ± 67.87	128.25 ± 93.83	>0.05
Monopodalic CoP Speed (mm/s)	13.84 ± 3.24	14.23 ± 5.05	>0.05
Arabesque Ellipse Area (mm^2)	274.40 ± 255.50	165.17 ± 127.18	>0.05
Arabesque CoP Speed (mm/s)	21.84 ± 8.46	19.32 ± 4.38	>0.05
Passè Ellipse Area (mm^2)	222.49 ± 130.06	148.15 ± 82.43	>0.05
Passè CoP Speed (mm/s)	22.22 ± 4.98	20.75 ± 6.92	>0.05

Data are reported as Mean ± SD. Abbreviation: CoP, Centre of Pressure.

Table 6. Comparison between right foot of pathological and healthy subjects. Mean delta values (difference between open and closed eyes) of stabilometric parameters.

	Right Foot POST-Injury (N = 11)	Right Foot Healthy (N = 11)	p-Values
Monopodalic Ellipse Area (mm^2)	321.47 ± 151.77	397.47 ± 252.87	>0.05
Monopodalic CoP Speed (mm/s)	15.77 ± 3.77	15.85 ± 5.29	>0.05
Arabesque Ellipse Area (mm^2)	449.63 ± 203.33	307.73 ± 252.13	>0.05
Arabesque CoP Speed (mm/s)	15.28 ± 9.60	14.64 ± 5.04	>0.05
Passè Ellipse Area (mm^2)	330.93 ± 294.15	294.46 ± 85.47	>0.05
Passè CoP Speed (mm/s)	16.90 ± 9.22	15.32 ± 7.14	>0.05

Data are reported as Mean ± SD. Abbreviation: CoP, Centre of Pressure.

7. Discussion

Our study showed that there is no ballet specific long-term balance deficit for grade II ankle injuries in ballet dancers. In order to have this specific conclusion, we checked if our sample of non-professional, but experienced, ballet dancers had any difference in balance while performing ballet specific positions on the dominant or non-dominant foot. We did not find any difference between the two sides. This finding is in contrast with a classic study performed by Leanderson et al. in 1996, where they tested professional athletes and non-athletes in the monopodalic position [15]. However, the study cited is 22 years older, so training techniques and materials such as shoes have changed, and are performed in a different, monocentric, and professional setting. This could very well explain the differences found. Moreover, the verified parameters of validity and reliability of the tools we used [36] and the power set at 80% for the sample size calculation, could be some of the reasons of the differences found between the outcomes of our study and the study published by Leanderson et al. Since the right and left foot were found to have no difference in the healthy group, we compared injured and non-injured side of our pathological group. We found out that, after six months, there was no significant difference in balance between the injured and non-injured sides. The internal (same subject) comparison is of the utmost importance, as any kind of difference in training could be an important bias [37]. In fact, using the contralateral healthy feet as a control, we were able to eliminate biases caused by potential differences in skills, body types, and other physical characteristics between athletes. Recently, a few studies pointed out that a crossover effect of injuries can exist, especially tendinitis, from one side to the other [38,39]. While it is not clear the mechanism of action of this crossover effect (e.g., central nervous system or anticipatory postural adjustments [40]), it still could be a possible confounder of our results as the pathological and healthy side could be similar because of it. Our study arrived to a different conclusion than a similar 2011 study [25]; however, we focused on a more selective group of subjects who had only grade II ankle injury. We supposed that patients with more severe ankle sprains could have long-term impairment in postural control.

One strength of our study was the use of stabilometry in a new way, comparing ballet-specific static positions for determining ankle injury effects. We found just two studies that examined ballet-specific positions; one examined the single-leg retiré position [37] and the other the passé en demi-pointe position [24]. However, neither of the studies focused on post-injury athletes. Another strength of the study was the comparison between healthy and injured feet in the same athletes that eliminate the role of possible differences in skills between different subjects, allowing us to better study the role of the previous ankle injury. Furthermore, we recruited the subjects in six different ballet schools. Different centers have different training strategies, so the results obtained are probably not linked to a peculiar type of training but to the natural history of grade II ankle injuries.

A limitation of our study is the small number of subjects involved; however, we enrolled 22 patients (11 for each group) as resulted with the sample size calculation. Our inclusion criteria were very strict to guarantee the internal validity of our study, and despite screening six different centers, we were able to find 11 patients meeting the required inclusion criteria for the pathological subjects group. Furthermore, because of our study design, we cannot exclude an important role of specific rehabilitation protocol in speeding up the recovery process, as we tested our patients long after their injury.

In the future, it could be important to study in a similar fashion different grades of ankle injuries, classified using imaging support, as well as different types of ankle injuries, such as inversion and eversion. Moreover, it could be interesting to evaluate the parameters used in our study in relevé, another ballet-specific position. One more possible future study could investigate the role of treatment (e.g., immobilization (functional or rigid) vs. no immobilization) and rehabilitation in the time to return to optimal performance as well as potential long-term effects in ankle injuries of higher clinical grading.

Our finding could encourage many dancers all over the world that suffered and will suffer ankle injuries [8]. Dancers with a grade II ankle injury should be reassured by clinicians that their functional ability as ballet dancers performing single-leg flat-foot stance will likely not be hindered in the long term.

8. Conclusions

Our findings show that grade II ankle injury does not compromise the ability of a ballet dancer, even if treated with just soft bandage applied for two weeks, even without a specific rehabilitation protocol.

Author Contributions: B.D.B., G.L., P.S., and F.N. designed the present study, checked the literature, and wrote the manuscript; J.A.V. performed the statistical analysis; L.B., J.A.V., A.S., and V.S. critically revised the manuscript; all authors gave the approval for the final version of the review submitted for publication. All authors have read and agreed to the published version of the manuscript.

Funding: This work was supported by the Italian Ministry of Health (Ricerca Corrente).

Conflicts of Interest: The authors have no conflict of interest. The funders had no role in the design of the study; in the collection, analyses, or interpretation of data; in the writing of the manuscript, or in the decision to publish the results.

References

1. Steinberg, N.; Tenenbaum, S.; Stern, M.; Zeev, A.; Siev-Ner, I. The association between menarche, intensity of training and passive joint ROM in young pre-professional female dancers: A longitudinal follow-up study. *Phys. Ther. Sport* **2018**, *32*, 59–66. [CrossRef]
2. Krityakiarana, W.; Jongkamonwiwat, N. Comparison of Balance Performance between Thai Classical Dancers and Non-Dancers. *J. Dance Med. Sci.* **2016**, *20*, 72–78. [CrossRef]
3. Tekin, D.; Agopyan, A.; Baltaci, G. Balance Training in Modern Dancers: Proprioceptive-Neuromuscular Training vs. Kinesio Taping. *Med. Probl. Perform. Art.* **2018**, *33*, 156–165. [CrossRef] [PubMed]
4. Bruyneel, A.V.; Mesure, S.; Paré, J.C.; Bertrand, M. Organization of postural equilibrium in several planes in ballet dancers. *Neurosci. Lett.* **2010**, *485*, 228–232. [CrossRef]
5. Yin, A.X.; Geminiani, E.; Quinn, B.; Owen, M.; Kinney, S.; McCrystal, T.; Stracciolini, A. The Evaluation of Strength, Flexibility, and Functional Performance in the Adolescent Ballet Dancer during Intensive Dance Training. *PM & R* **2019**, *11*, 722–730.
6. Krasnow, D.; Wilmerding, M.V.; Stecyk, S.; Wyon, M.; Koutedakis, Y. Biomechanical research in dance: A literature review. *Med. Probl. Perform. Art.* **2011**, *26*, 3–23. [PubMed]
7. da Silveira Costa, M.S.; de Sá Ferreira, A.; Felicio, L.R. Static and dynamic balance in ballet dancers: A literature review. *Fisioter. Pesqui.* **2013**, *20*, 299–305.
8. Vassallo, A.J.; Hiller, C.; Stamatakis, E.; Pappas, E. Epidemiology of Dance-Related Injuries Presenting to Emergency Departments in the United States, 2000-2013. *Med. Probl. Perform. Art.* **2017**, *32*, 170–175. [CrossRef]
9. Smith, T.O.; Davies, L.; de Medici, A.; Hakim, A.; Haddad, F.; Macgregor, A. Prevalence and profile of musculoskeletal injuries in ballet dancers: A systematic review and meta-analysis. *Phys. Ther. Sport* **2016**, *19*, 50–56. [CrossRef]
10. Smith, P.J.; Gerrie, B.J.; Varner, K.E.; McCulloch, P.C.; Lintner, D.M.; Harris, J.D. Incidence and Prevalence of Musculoskeletal Injury in Ballet: A Systematic Review. *Orthop. J. Sports Med.* **2015**, *3*. [CrossRef]
11. Lin, C.-W.; Su, F.-C.; Lin, C.-F. Influence of ankle injury on muscle activation and postural control during ballet grand plié. *J. Appl. Biomech.* **2014**, *30*, 37–49. [CrossRef] [PubMed]
12. Glencross, D.; Thornton, E. Position sense following joint injury. *J. Sports Med. Phys. Fitness* **1981**, *21*, 23–27. [PubMed]
13. Malliaropoulos, N.; Ntessalen, M.; Papacostas, E.; Longo, U.G.; Maffulli, N. Reinjury after acute lateral ankle sprains in elite track and field athletes. *Am. J. Sports Med.* **2009**, *37*, 1755–1761. [CrossRef] [PubMed]
14. Sousa, A.S.P.; Leite, J.; Costa, B.; Santos, R. Bilateral Proprioceptive Evaluation in Individuals with Unilateral Chronic Ankle Instability. *J. Athl. Train.* **2017**, *52*, 360–367. [CrossRef] [PubMed]
15. Leanderson, J.; Eriksson, E.; Nilsson, C.; Wykman, A. Proprioception in classical ballet dancers. A prospective study of the influence of an ankle sprain on proprioception in the ankle joint. *Am. J. Sports Med.* **1996**, *24*, 370–374. [CrossRef] [PubMed]
16. Attenborough, A.S.; Hiller, C.E.; Smith, R.M.; Stuelcken, M.; Greene, A.; Sinclair, P.J. Chronic ankle instability in sporting populations. *Sports Med. Auckl. NZ* **2014**, *44*, 1545–1556. [CrossRef]
17. Simon, J.; Hall, E.; Docherty, C. Prevalence of chronic ankle instability and associated symptoms in university dance majors: an exploratory study. *J. Dance Med. Sci.* **2014**, *18*, 178–184. [CrossRef]
18. Fusca, M.; Negrini, F.; Perego, P.; Magoni, L.; Molteni, F.; Andreoni, G. Validation of a Wearable IMU System for Gait Analysis: Protocol and Application to a New System. *Appl. Sci.* **2018**, *8*, 1167. [CrossRef]
19. Ghislieri, M.; Gastaldi, L.; Pastorelli, S.; Tadano, S.; Agostini, V. Wearable Inertial Sensors to Assess Standing Balance: A Systematic Review. *Sensors* **2019**, *19*. [CrossRef]
20. Lee, Y.-J.; Liang, J.N.; Wen, Y.-T. Characteristics of Postural Muscle Activity in Response to a Motor-Motor Task in Elderly. *Appl. Sci.* **2019**, *9*, 4319. [CrossRef]
21. McKeon, P.O.; Hertel, J. Systematic review of postural control and lateral ankle instability, part I: Can deficits be detected with instrumented testing. *J. Athl. Train.* **2008**, *43*, 293–304. [CrossRef] [PubMed]
22. Nardone, A.; Godi, M.; Grasso, M.; Guglielmetti, S.; Schieppati, M. Stabilometry is a predictor of gait performance in chronic hemiparetic stroke patients. *Gait Posture* **2009**, *30*, 5–10. [CrossRef] [PubMed]
23. Bruyneel, A.-V.; Humbert, A.; Bertrand, M. Comparison of balance strategies in mountain climbers during real altitude exposure between 1.500m and 3.200m: Effects of age and expertise. *Neurosci. Lett.* **2017**, *657*, 16–21. [CrossRef] [PubMed]

24. de Mello, M.C.; de Sá Ferreira, A.; Ramiro Felicio, L. Postural Control During Different Unipodal Positions in Professional Ballet Dancers. *J. Dance Med. Sci.* **2017**, *21*, 151–155. [CrossRef] [PubMed]
25. Lin, C.-W.; Lin, C.-F.; Hsue, B.-J.; Su, F.-C. A comparison of ballet dancers with different level of experience in performing single-leg stance on retiré position. *Motor Control* **2014**, *18*, 199–212. [CrossRef] [PubMed]
26. Witchalls, J.; Waddington, G.; Adams, R.; Blanch, P. Evaluation of the relative contribution of peripheral and focal vision to proprioceptive differentiation of underfoot inversion angles in young elite athletes. *Percept. Mot. Skills* **2013**, *117*, 923–934. [CrossRef]
27. Han, J.; Anson, J.; Waddington, G.; Adams, R.; Liu, Y. The Role of Ankle Proprioception for Balance Control in relation to Sports Performance and Injury. *BioMed Res. Int.* **2015**, *2015*, 842804. [CrossRef]
28. Marulli, T.A.; Harmon-Matthews, L.E.; Davis-Coen, J.H.; Willigenburg, N.W.; Hewett, T.E. Eyes-Closed Single-Limb Balance is Not Related to Hypermobility Status in Dancers. *J. Dance Med. Sci.* **2017**, *21*, 70–75. [CrossRef]
29. Casabona, A.; Leonardi, G.; Aimola, E.; La Grua, G.; Polizzi, C.M.; Cioni, M.; Valle, M.S. Specificity of foot configuration during bipedal stance in ballet dancers. *Gait Posture* **2016**, *46*, 91–97. [CrossRef]
30. Asseman, F.; Caron, O.; Crémieux, J. Is there a transfer of postural ability from specific to unspecific postures in elite gymnasts? *Neurosci. Lett.* **2004**, *358*, 83–86. [CrossRef]
31. Giboin, L.-S.; Gruber, M.; Kramer, A. Three months of slackline training elicit only task-specific improvements in balance performance. *PLoS ONE* **2018**, *13*, e0207542. [CrossRef] [PubMed]
32. Thalassinos, M.; Fotiadis, G.; Arabatzi, F.; Isableu, B.; Hatzitaki, V. Sport Skill-Specific Expertise Biases Sensory Integration for Spatial Referencing and Postural Control. *J. Mot. Behav.* **2018**, *50*, 426–435. [CrossRef] [PubMed]
33. Biernacki, J.L.; Stracciolini, A.; Fraser, J.; Micheli, L.J.; Sugimoto, D. Risk Factors for Lower-Extremity Injuries in Female Ballet Dancers: A Systematic Review. *Clin. J. Sport Med.* **2018**. [CrossRef] [PubMed]
34. Malliaropoulos, N.; Papacostas, E.; Papalada, A.; Maffulli, N. Acute lateral ankle sprains in track and field athletes: an expanded classification. *Foot Ankle Clin.* **2006**, *11*, 497–507. [CrossRef]
35. Scoppa, F.; Capra, R.; Gallamini, M.; Shiffer, R. Clinical stabilometry standardization: Basic definitions–acquisition interval–sampling frequency. *Gait Posture* **2013**, *37*, 290–292. [CrossRef]
36. Cuccia, A.M. Validity and reliability of spatio-temporal gait parameters in adolescents. *Iran. J. Pediatr.* **2013**, *23*, 610–611.
37. Lin, C.-W.; Chen, S.-J.; Su, F.-C.; Wu, H.-W.; Lin, C.-F. Differences of ballet turns (pirouette) performance between experienced and novice ballet dancers. *Res. Q. Exerc. Sport* **2014**, *85*, 330–340. [CrossRef]
38. Sansone, V.; Maiorano, E.; Applefield, R.C.; Gandola, M.; Negrini, F. Strength Reduction in Unilateral Shoulder Pain: Is the Healthy Side Really Healthy in Rotator Cuff Disease? *Am. J. Phys. Med. Rehabil.* **2019**, *98*, 382–386. [CrossRef]
39. Heales, L.J.; Lim, E.C.W.; Hodges, P.W.; Vicenzino, B. Sensory and motor deficits exist on the non-injured side of patients with unilateral tendon pain and disability–implications for central nervous system involvement: a systematic review with meta-analysis. *Br. J. Sports Med.* **2014**, *48*, 1400–1406. [CrossRef]
40. Cavallari, P.; Bolzoni, F.; Bruttini, C.; Esposti, R. The Organization and Control of Intra-Limb Anticipatory Postural Adjustments and Their Role in Movement Performance. *Front. Hum. Neurosci.* **2016**, *10*, 525. [CrossRef]

applied sciences

MDPI

Article

Effects of the Whole-Body Vibration and Auriculotherapy on the Functionality of Knee Osteoarthritis Individuals

Cristiane Ribeiro Kütter [1,2], Eloá Moreira-Marconi [2,3], Ygor Teixeira-Silva [2,4], Marcia Cristina Moura-Fernandes [2,3], Alexandre Gonçalves de Meirelles [1,2], Mario José dos Santos Pereira [2], Shyang Chang [5], José Alexandre Bachur [6], Laisa Liane Paineiras-Domingos [2,4,7], Redha Taiar [8,*], Mario Bernardo-Filho [2] and Danúbia da Cunha de Sá-Caputo [2,4,6]

1 Mestrado Profissional em Saúde, Medicina Laboratorial e Tecnologia Forense, Instituto de Biologia Roberto Alcantara Gomes, Universidade do Estado do Rio de Janeiro, Rio de Janeiro, RJ 20950-003, Brazil; crkutter@hotmail.com (C.R.K.); meirelles.ale@gmail.com (A.G.d.M.)
2 Laboratório de Vibrações Mecânicas e Práticas Integrativas e Complementares, Departamento de Biofísica e Biometria, Instituto de Biologia Roberto Alcantara Gomes e Policlínica Américo Piquet Carneiro, Universidade do Estado do Rio de Janeiro, RJ 20950-003, Brazil; eloamarconi@gmail.com (E.M.-M.); silvarogy@hotmail.com (Y.T.-S.); marciafernandesfisio@hotmail.com (M.C.M.-F.); mariojsp@gmail.com (M.J.d.S.P.); laisanit@gmail.com (L.L.P.-D.); bernardofilhom@gmail.com (M.B.-F.); dradanubia@gmail.com (D.d.S.-C.)
3 Programa de Pós-graduação em Fisiopatologia Clínica e Experimental, Faculdade de Ciências Médicas, Universidade do Estado do Rio de Janeiro, Rio de Janeiro, RJ 20.551-030, Brazil
4 Programa de Pós-graduação em Ciências Médicas, Faculdade de Ciências Médicas, Universidade do Estado do Rio de Janeiro, Rio de Janeiro, RJ 20.551-030, Brazil
5 Department of Electrical Engineering, National Tsing Hua University, Hsinchu 30013, Taiwan; shyang@ee.nthu.edu.tw
6 Departamento de Medicina e Fisioterapia, Universidade de Franca, Franca, SP 14401-426, Brazil; jabachur@hotmail.com
7 Departamento de Fisoterapia, Faculdade Bezerra de Araújo, Rio de Janeiro, RJ 23052-180, Brazil
8 GRESPI, Université de Reims, 51100 Reims, France
* Correspondence: redha.taiar@univ-reims.fr

Received: 19 October 2019; Accepted: 25 November 2019; Published: 29 November 2019

check for updates

Featured Application: The protocol proposed by this study (WBV alone or associated with TA) can be used by health professionals as a non-pharmacological intervention to improve the functionality of individuals with knee osteoarthritis.

Abstract: Knee osteoarthritis (KOA) is a degenerative disease of the knee joint. This study aims to evaluate the effects of whole-body vibration (WBV), auriculotherapy (AT), and the association of these techniques with the functionality of KOA individuals. Individuals (n = 120) were allocated an AT group (GAT), a WBV group (GWBV), an association group (GWBV + AT), and their respective controls (CGAT, CGWBMV, CGWBMV + AT). The WBV intervention was performed with 5–14 Hz in 3 min of working time with 1 min rest. The control group performed the protocol with the vibrating platform (VP) turned off. The AT intervention was performed with adhesive tapes, with seeds placed in the both ears on the Shenmen point, knee joint, and kidney. The control groups had seedless tape placed on both ears. The participants were instructed to press the adhesive tapes with the fingers three times per day (for 6 days) and to remove the adhesive tapes on the seventh day, before returning to the laboratory. The International Knee Documentation Committee (IKDC), the short physical performance battery (SPPB), and the anterior trunk flexibility (ATF) tests were applied. Acute and cumulative effects were determined. In first session (acute effect of the first session), significant improvements were observed in the groups GWBV ($p = 0.03$) and GWBV + AT ($p = 0.04$), and in the cumulative effect a significant improvement was observed in the groups GWBV ($p = 0.02$) and GWBV + AT ($p = 0.01$). Concerning the overall score of the SPPB, significant improvements were observed in the individuals of the GWBV ($p = 0.01$) and GWBV + AT ($p = 0.03$) groups (cumulative effect). No changes were found in the score for the IKDC. The WBV alone or associated with AT, besides being a safe and feasible strategy, likely produces physiological responses that improve the functionality of KOA individuals, considering the findings of the ATF and the score of the SPPB.

Keywords: mechanical vibration; traditional Chinese medicine; balance; flexibility; muscle strength

1. Introduction

Knee osteoarthritis (KOA) is characterized by the degradation and erosion of the articular cartilage, inflammation of the synovial membrane, and formation of osteophytes and ligament laxity [1,2]. These conditions could lead to a relevant level of pain, a reduction in the movement of the knees and joint stiffness with commitment of functionality. Ferreira el al. (2018), published an umbrella review, in which it is pointed out that KOA evolution can improve, remain stable, or get worse. Moreover, it is established that treatment strategies include pharmacological, non-pharmacological, surgical, and non-surgical interventions. Furthermore, it is highlighted that there are several non-pharmacological and non-surgical interventions that are safe and with low cost, which can be incorporated into self-management performed at home or in the community, and that have a relevant public health impact [3]. In this context, therapeutic interventions with less side effects for KOA individuals would be desirable, such as various modalities of exercise [4], such as whole-body vibration (WBV) exercises [3,5], physiotherapy techniques, and the procedures related to traditional Chinese medicine (TCM), such as acupuncture [6,7] and auriculotherapy (AT) [5].

One technique among the TCM procedures is AT [8,9]. In TCM, disease is considered to be caused by the imbalance of the energy, qi, in the body [10].

Similar points to acupuncture are also found around areas of the sensory organs, such as on the ears in the AT [8,11,12]. AT is a procedure in which there is a stimulation of point (acupoint) on specific areas of the inner and outer ear lobe that treat disease or illness [10].

In AT, different types of materials, including plant seeds (*Semen vaccariae*), are applied to stimulate the acupoints for therapeutic purposes. The effectiveness of AT would be associated with functional changes in the body, interconnected with TCM, neurological, and embryological theories, microsystems, and a hormonal basis [9,12–15].

In general, AT involves the stimulation of a point named the "corresponding region" related to the organic dysfunction to be treated. Other points are also chosen considering the disease. AT has demonstrated beneficial effects in various clinical disorders [12,14,16,17], such as pain and anxiety [8], pain associated with cancer [18], knee arthroscopy, hip fractures, and hip arthroplasty [19].

An important alternative to physical activity is the WBV exercises, which have been used to improve fitness performance [20,21] or as an option for the rehabilitation process [22]. WBV exercises take place when the individual has direct contact with the base of a vibrating platform (VP). In this case, there is a transmission of the mechanical vibration generated in the VP to the body of the individual [23,24].

Studies involving WBV exercises suggested important findings, such as mass gain and muscle strength, balance, walking speed, decrease of pain [22–24], improvement in flexibility [5,25,26], and a decrease in the risk of falling [27].

Different tools are used to assess the functionality of various populations [28,29], for example, The International Knee Documentation Committee (IKDC) [29]. In addition, a short physical performance battery (SPPB) is used to evaluate the physical performance of individuals and evaluate their functional capacity considering balance, gait speed, and lower limb strength [30]. The functionality can also be assessed using the anterior trunk flexibility (ATF) test [25], in which the individual is asked to measure the distance between their middle finger to floor.

Considering the relevance of the use of non-pharmacological procedures to the management of individuals with KOA, the aim of this study was to analyze the effect of WBV and AT and the association of these two techniques on the functionality of KOA individuals. Thus, the hypothesis states that WBV and AT (associated or isolated) could be a safe and feasible strategy for improving the functionality of KOA individuals.

2. Materials and Methods

This cross-section and cross-over study was approved by the Ethics Committee in Research of the Hospital Universitário Pedro Ernesto (HUPE) and the Universidade do Estado do Rio de Janeiro (UERJ), and it was registered in the clinical trial registration (CAAE-19826413.8.0000.5259 and RBR-7dfwct, respectively). All the procedures are embedded by the Declaration of Helsinki. The interventions were performed in Laboratório de Vibrações Mecânicas e Práticas Integrativas (LAVIMPI), UERJ, from 2013 to 2019.

2.1. Inclusion Criteria

Participants aged ≥40 years old of both sexes who werr outpatients of the HUPE with a diagnosis of KOA.

2.2. Exclusion Criteria

Participants with untreated hypertension, pregnancy, neurological diseases, "fear" of the movements on the VP, neurological alterations, which may hamper understanding of assessment instruments, severe or incapacitating clinical diseases, concomitant musculoskeletal disorders, prosthesis in the knee or hip, criterium of the investigator (aiming for the safety of the participant), and those individuals who refused to sign the Informed Agreement Form to participate in the study.

2.3. Sample Size

For a statistical power of 95% and a significance level of 5%, a sample size of 13 subjects in each group, according to the variable functionality, was calculated [31]. Considering a possible decrease in sample size, 222 individuals were recruited. The subjects were referred by a physician of the Orthopedics Department of the HUPE/UERJ. The individuals were asked to maintain the regular dietary habits and level of physical activity during the study period.

2.4. Participants

A total of 102 individuals were excluded due to the exclusion criteria or economic conditions or because the individual gave up. A total of 120 individuals were allocated into six groups: the AT group (GAT), the WBV group (GWBV), the group with the association of interventions (GWBV + AT), and their respective controls (CGAT, CGWBV, CGWBV + AT). In all the groups there were, at least, two men. In the GWBV and GWBV + AT, the intervention was performed twice a week and, in the GAT, once a week following a previous publication [5]. The control groups performed the same procedure, but with the side alternating vibrating platform (SAVP) turned off.

To try to optimize the study, the allocation of the individuals in each group considered the order of the moment of the arrival in the LAVIMPI. The first individual entered in the GAT, the second in the GWBV, and in sequence up to GWBV + AT, CGAT, CGWBV, and CGWBV + AT. The seventh individual and the next individuals followed a new sequence.

2.5. Whole-Body Vibration Intervention

The 5-week, twice a week protocol used a SAVP (Novaplate Fitness Evolution®, Brazil), following a previous publication [5]. In these platforms mechanical vibrations are generated. Mechanical vibrations are physical agents with oscillatory, sinusoidal, and deterministic movements around an equilibrium point [25,32,33]. The individual was seated in a chair with a bare foot on the base of the SAVP, with knee flexion and the hands resting on the knees (Figure 1). The biomechanical parameters of the mechanical vibration used in this protocol were peak-to-peak displacement (D) with 2.5, 5.0, and 7.5 mm, peak acceleration from 0.12 up to 2.95× g. and a working time of 3 min interspersed by a 1 min rest. This sequence was applied three times. The frequency varied from 5 Hz (in the first day) to 14 Hz (in the last day), increasing by 1 Hz per session. This protocol followed a previous publication [5].

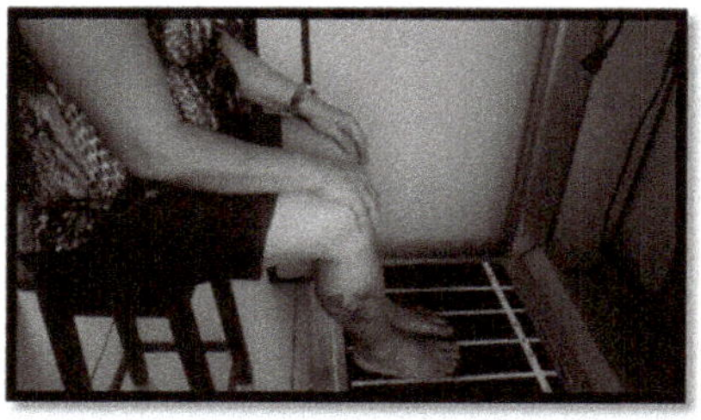

Figure 1. Position of the individual in the auxiliary chair with their feet on the SAVP.

2.6. Auricular Intervention

The auricular points were chosen according to AT principles [12,14,16,17]. The Shenmen point is an analgesic point, the Knee joint point is the corresponding point, and the kidney has a function related to the bones and lower limbs. The knee joint is located at the superior crus of the antihelix, at the same level of the superior border of the inferior crus of the antihelix. The Shenmen is located at the bifurcation of the crura of the antihelix. The kidney lies in the upper part of the cymba conchae at the superior portion of the acupuncture point of the small intestine [14,15,34].

Adhesive tapes with two mustard (*Semen vaccariae*) seeds (Figure 2A up) were placed in both ears in referred points (Figure 2A down). The control groups had the seedless (Figure 2B up) tape placed in the lobes of both ears (Figure 2B down).

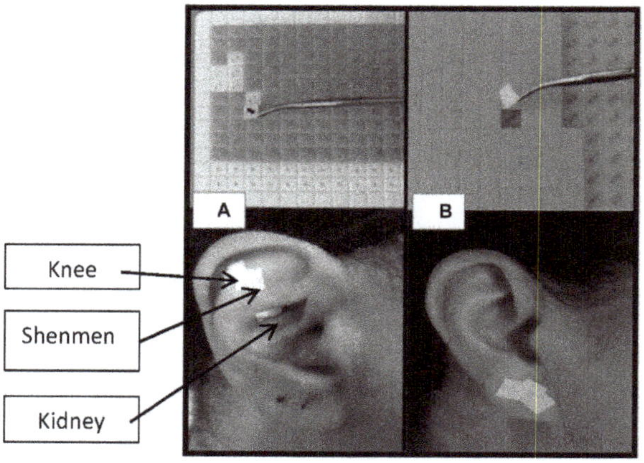

Figure 2. (**A**) Auriculotherapy (AT) points (knee, kidney, and Shenmen); (**B**) control group AT.

The participants were instructed to use the fingers to press the adhesive tapes for 10 min, three times per day (six days), and to remove the adhesive tapes on the seventh day, before returning to the laboratory. This procedure was done for five weeks. Before the placement of the adhesive tapes, local visual inspection and local asepsis with alcohol at 70% were performed. Figure 3 shows the steps of the interventions using adhesive tapes in the ears of the participants.

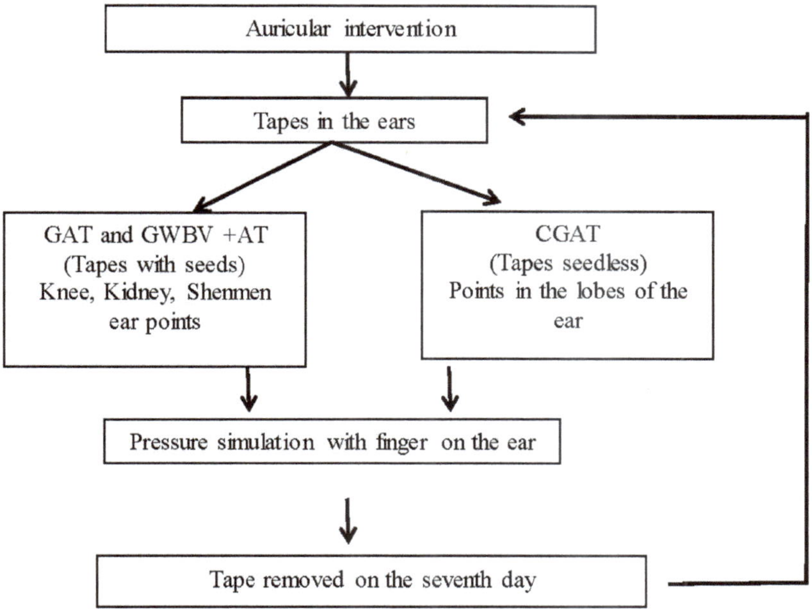

Figure 3. Steps of the interventions using adhesive tapes in the ears of the participants. group (GAT), control of the AT group (CGAT), **whole-body vibration group** (GWBV) + AT group with WBV and AT associated.

3. Primary Outcome Measures

The evaluations of anterior trunk flexion (ATF), the International Knee Documentation Committee (IKDC), and short physical performance battery (SPPB) tests were performed longitudinally, as described in the next paragraphs of the methodology.

3.1. Anterior Trunk Flexion (ATF)

The individuals of the all groups performed the ATF, as shown in Figure 4. To measure flexibility, the individual was asked to perform an anterior trunk flexion with the lower limbs extended and with the head relaxed at the maximum flexion. The difference (Δ ATF) between the measurements before and after the first session (acute effect of the first session) and before and after the last session (acute effect of the last session) were considered. The difference (Δ ATF) between the measurements before the first session and after last session (cumulative effect) were also considered [33].

3.2. International Knee Documentation Committee

The International Knee Documentation Committee created a subjective from which was validated in Brazil [35] attached as a Supplementary Document to this work. It contains 18 items/questions to verify symptoms, function, and sports activities of the knee joint. The lower patient's score was related to the lower function and the intensity of the symptoms [29]. For the score of the IKDC, "0" represents the highest level of symptoms and "10" represents the lowest level of symptoms [29]. This tool was applied before the first session and after the last session (cumulative effect).

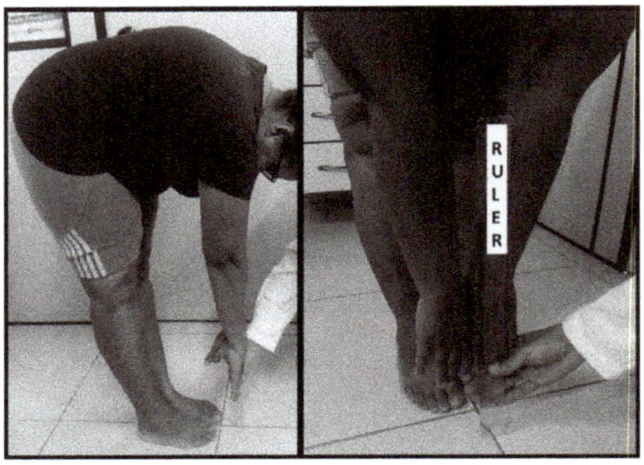

Figure 4. Individual performing anterior trunk flexion.

3.3. *Short Physical Performance Battery (SPPB)*

The SPPB was validated in Brazil and it consists of three tests [30,32]: (a) Balance test: The individual was asked to stay in an appropriated posture for 10 s in three different positions (feet together, semi-tandem, and tandem); (b) walking time test: The individual was asked to walk two times for 3 m (marked on the floor) and it was timed, and the shortest time between the two was considered; (c) five times sit-to-stand test: The individual was asked to get up and sit on the chair 5 times, with the arms crossed on the chest, as fast as possible (timed). This tool was applied before the first session and after the last session (cumulative effect).

4. Statistical Analysis

The Shapiro–Wilk test was used to verify the normality of the data. The differences between the intra-groups was analyzed using the Wilcoxon rank test. The comparisons between the findings of the intervention group and the respective control group was carried out with the Mann–Whitney test. Kruskal Wallis and student Newman–Keuls tests were used to compare the anthropometric characteristics of all the groups. The value used was $p \leq 0.05$ as the significant. The software used was BioEstat 5.3.

5. Results

The enrollment of the participants in the study is shown in the Figure 5 in the flow diagram. Individuals (n = 222) were recruited and 102 were eliminated from the work (due to exclusion criteria or economic conditions or because the individual gave up). Consequently, 120 individuals (106 women and 14 men) participated in this study. They were allocated in the AT group (GAT), the WBV group (GWBV), and the group with the association of interventions (GWBV + AT) and their respective controls (CGAT, CGWBV, and CGWBV + AT). The number of individuals that performed the evaluations (ATF, IKDC, and SPPB) is also indicated.

Statistical analyses of some anthropometric parameters (age, stature, and body mass) of the individuals in the various groups was performed. Although only three anthropometric characteristics were considered, the results in Table 1 indicate no difference among the groups ($p > 0.05$). This finding shows a homogeneity among the groups.

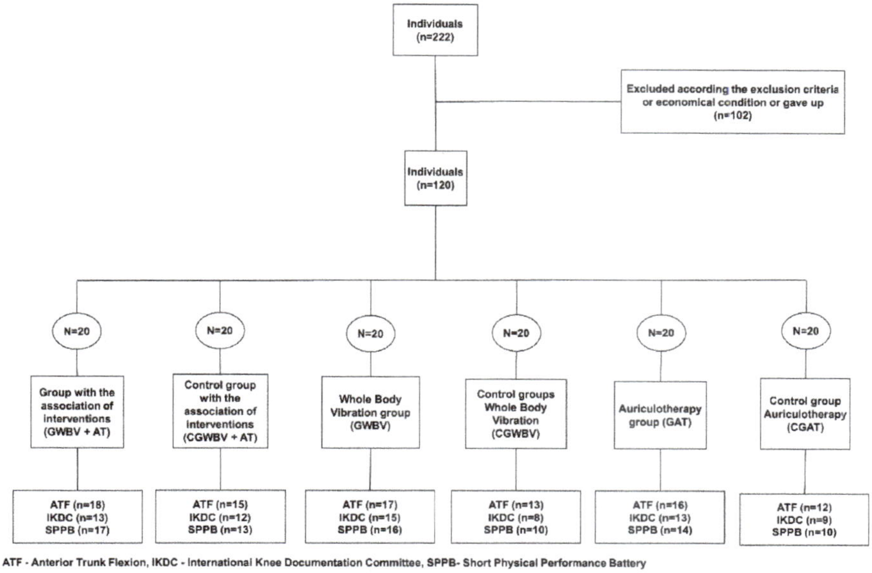

Figure 5. Enrollment of the participants in the study.

Table 1. Anthropometric characteristics of the individuals included in the work.

Groups	Age (Year) (Md ± SD)	Stature (m) (Md ± SD)	Body Mass (kg) (Md ± SD)
CGAT	63.00 ± 7.00	1.60 ± 0.08	85.67 ± 14.98
GAT	64.00 ± 7.00	1.59 ± 0.09	84.58 ± 14.90
CGWBV	66.00 ± 12.00	1.57 ± 0.07	84.50 ± 13.38
GWBV	67.00 ± 10.00	1.57 ± 0.05	84.59 ± 14.15
CGWBV + AT	65.00 ± 7.00	1.55 ± 0.04	84.96 ± 14.59
GWBV + AT	65.00 ± 8.00	1.55 ± 0.08	85.49 ± 14.36
p-value	0.852	0.310	0.999

Media (Md), standard deviation (SD), meter (m), kilogram (kg), control group auriculotherapy group (CGAT), auriculotherapy group (GAT), control whole-body vibration group (CGWBV), whole-body vibration group (GWBV), control whole-body vibration group and auriculotherapy (CGWBV + AT), whole-body vibration group and auriculotherapy (GWBV + AT), Kruskal Wallis and student Newman–Keuls tests, level of significance $p \leq 0.05$.

The Δ ATF was calculated considering the measurements before and after the first session (acute effect of the first session), and significant improvements were observed in the groups GWBV ($p = 0.03$) and GWBV + AT ($p = 0.04$) (Figure 6A).

The Δ ATF was calculated considering the measurements before and after the last session, and there were no significant changes (acute effect) (Figure 6B).

The Δ ATF was calculated considering the measurements before the first session and after the last session (cumulative effect), and significant improvements were observed in the groups GWBV ($p = 0.02$) and GWBV + AT ($p = 0.01$) (Figure 6C).

The findings concerning the IKDC (for all groups) are presented in Table 2. The analysis of the results shows no significant changes ($p > 0.05$) among all the groups and respective controls.

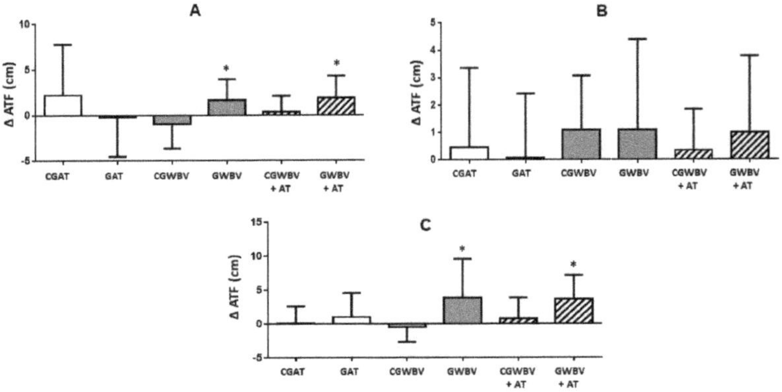

Figure 6. The findings concerning the anterior trunk flexion (ATF) test of all groups. Control auriculotherapy group (CGAT), auriculotherapy group (GAT), control whole-body vibration group (CGWBV), whole-body vibration group (GWBV), control whole-body vibration group and auriculotherapy (CGWBV + AT), whole-body vibration group and auriculotherapy (GWBV + AT). (**A**) The acute effect of the first session. GWBV ($p = 0.03$) and GWBV + AT ($p = 0.04$). (**B**) The acute effect of the last session. (**C**) The cumulative effect. GWBV ($p = 0.02$) and GWBV + AT ($p = 0.01$).

Table 2. The data concerning the International Knee Documentation Committee score.

	Control			Intervention		
	Before	**After**		**Before**	**After**	
	Md ± SD /100	Md ± SD /100	*p*-value*	Md ± SD /100	Md ± SD /100	*p*-value*
GAT	24.68 ± 9.85	27.58 ± 9.71	0.098	33.59 ± 16.84	35.91 ± 17.70	0.37
GWBV	28.87 ± 13.69	34.19 ± 11.45	0.299	31.55 ± 13.21	33.79 ± 9.07	0.43
GWBV + AT	27.07 ± 7.86	30.26 ± 8.42	0.059	31.03 ± 11.85	34.49 ± 14.67	0.14

Media, (Md) standard deviation (SD), auriculotherapy group (GAT), whole-body vibration group (GWBV), whole-body vibration group and auriculotherapy (GWBV + AT), *p*-value, level of significance $p \leq 0.05$*.

The findings concerning the overall score of SPPB involving all groups are presented in the Table 3. Significant changes ($p \leq 0.05$) were observed in the individuals of the GWBV and GWBV + AT groups (the cumulative effect). No significant changes ($p > 0.05$) were verified in the GAT and in the control groups.

Table 3. The data concerning the short physical performance battery (SPPB) overall score.

	Control			Intervention		
	Before	**After**		**Before**	**After**	
	Med ± IQR (score)	Med ± IQR (score)	*p*-value*	Med ± IQR (score)	Med ± IQR (score)	*p*-value*
GAT	7.00 ± 1.75	7.00 ± 2.00	0.529	7.00 ± 3.75	7.50 ± 2.00	0.29
GWBV	7.00 ± 2.00	7.00 ± 7.00	0.798	7.00 ± 2.00	8.00 ± 2.00	0.01*
GWBV + AT	7.00 ± 4.00	7.00 ± 3.00	0.834	7.00 ± 1.50	8.50 ± 1.75	0.03

Median (Med), interquartile range (IQR), auriculotherapy group (GAT), whole-body vibration group (GWBV), whole-body vibration group and auriculotherapy (GWBV + AT), *p*-value, level of significance $p \leq 0.05$*.

6. Discussion

As hypothesized, the current work has demonstrated that WBV alone, or in association with AT, improves the functionality of KOA individuals, considering the ATF and overall score of the SPPB.

However, these improvements were not associated with the responses in the IKDC score. A suggestion might be associated with a subjective approach of the IKDC.

Costa et al. (2014) [23] described, in a systematic review, the different treatments used on the management of the individuals with osteoarthritis and the importance of the WBV exercise as a modality of exercise in this context. Moreover, Ferreira et al. (2018) [3] pointed out that there are several non-pharmacological and non-surgical interventions, including WBV and acupuncture, for the management of KOA individuals. Suen et al. (2016) [13] showed the effect of the AT with and without magnetotherapy in KOA individuals and observed improvements in their functionality. Costa-Cavalcanti and colleagues [36] evaluated the effect of auriculotherapy on biochemical parameters during 5 weeks, once a week. The study showed a significant reduction in direct bilirubin, aspartate aminotransferase, and triglycerides.

Considering the ATF, significant changes (improvements) were observed with the intervention of WBV alone or with the AT (acute and cumulative effect) (Figure 6A,C). In the acute effect of the last session, alterations were not found in the ATF, likely because the individuals had already reached the maximum of answers for the stimulus in the previous sessions (Figure 6B). These findings are in agreement with Sá-Caputo and colleagues [25,33], who showed improvements in the flexibility after the WBV exercise protocol in individuals with metabolic syndrome. Neto and colleagues [5] reported a significant decrease of the level of pain in the KOA individuals treated with WBV alone and with a combined intervention (WBV and AT). In addition, Moreira-Marconi and colleagues [37] reported that WBV alone, and with a combined intervention with AT, improved handgrip strength in cumulative response. These responses [25,33] could favor the findings presented in Figure 6A,B related to the ATF. The absence of the effect of the interventions in the acute effect of the last session could be associated with the adaptation phenomenon to the stimulus [38] and/or genetic characteristics of the individuals to the response to the stimulus [39]. These facts could justify the uncertainty of the findings of Figure 6B with elevated standard deviations of the measurements.

Using the protocol of the current study, for five weeks, once (AT) or twice (AT and WBV or only WBV) per week, no difference was observed in any group using the IKDC score (Table 2). Park and colleagues [40] described improvements related with the use of the Korean Western Ontario McMaster score (KWOMAC) in KOA individuals, who performed the WBV exercise for 8 weeks, three times per week. These differences could be associated with the protocol used in this study (five weeks/once or twice a week) whereas in the other study [40] the intervention was 8 weeks/three times a week. Moreover, there is a difference in the qualitative tools used in the evaluation, IKDC and KWOMAC.

Concerning the functionality (considering the SPPB) of the KOA individuals, in the cumulative effects, an improvement was found in the interventions with WBV alone and WBV and AT (Table 3). This finding is in agreement with Trans et al. (2009) [41] in a protocol of the WBV exercise twice a week for 8 weeks, with a progressive increase intensity (frequencies among 25–30 Hz), which performed unloaded static WBV exercise. They concluded that WBV is a safe method for rehabilitation of women with KOA, increasing the muscle strength. Wang et al., 2016 [42] showed improvements in the symptoms, physical function, and spatiotemporal parameters with a protocol of the WBV exercise using a frequency of 35 Hz and peak-to-peak displacement of 4–6 mm for 5 days a week for 12 weeks in a vertical VP. Avelar et al. (2011) [31] described that the addition of the WBV exercise to squat training failed to result in a significant improvement in functional performance and self-reported status of knee osteoarthritis in the elderly. However, the intragroup results suggest that the WBV exercise may present a feasible and effective way of improving the functionality and self-perception of the disease status in older adults with KOA. They used frequencies from 25–40 Hz and peak-to-peak displacement of 4 mm for 12 weeks in a vertical VP. The improvement observed on the functional performance could be associated with muscle stimulation due to exposition to mechanical vibration in the WBV exercise. Paineiras-Domingos et al. (2018) [32] found significant responses in the improvement of the functionality of Metabolic Sindrome individuals that performed WBV exercises, analyzing the overall score of the SPPB and the three tests (balance test, the gait speed, and the five chair stand test).

In the current study, an alternative unusual position of the patient was used, compared to other studies [5,41,42], involving the WBV exercise in KOA individuals. The selection of this position (seated in a chair, barefoot, on the basis of the SAVP) is related to reduction of the load and the impact in the knee joint so as to make these individuals able to practice the exercise [5].

An important finding of this investigation was the absence of any undesirable effect due to the interventions, either isolated (WBV or AT) or combined (WBV and AT). These considerations reinforce the relevance of these procedures, as useful, safe, and feasible. Moreover, they are inexpensive.

This study had some limitations and can be improved, such as the total time (5 weeks) of the protocol of the WBV exercise. The elevated ratio between females and males in the groups was also a limitation. Only three anthropometric characteristics of the participants were evaluated. Another limitation was that no specific tool was used to evaluate the level of physical activity of the KOA individuals. Concerning the AT interventions, only three points of the ear were stimulated. In addition, temporal analysis of the responses in the various sessions was not done. These considerations could help to guide future experimental designs.

7. Conclusions

In spite of the limitations, it is still possible to conclude that the WBV alone or associated with AT might induce physiological responses that favour the improvement of the flexibility with the evaluation of anterior flexibility of the trunk and the functionality (SPPB—overall score) of the KOA individuals. Moreover, the absence of effects of these interventions in the acute response of the last session could be associated with possible adaptation and/or genetic characteristics of the studied population. In consequence, a temporal analysis of these responses in the various sessions and the addition of other quantitative parameters in the protocols are ongoing.

Supplementary Materials: The following are available online at http://www.mdpi.com/2076-3417/9/23/5194/s1.

Author Contributions: Conceptualization, M.B.-F. and D.d.C.d.S.-C.; data curation, C.R.K., E.M.-M., Y.T.-S., M.C.M.-F., A.G.d.M., M.J.d.S.P., and L.L.P.-D.; formal analysis, M.J.d.S.P.; funding acquisition, M.B.-F.; investigation, C.R.K., E.M.-M., Y.T.-S., M.C.M.-F., A.G.d.M., and L.L.P.-D.; methodology, C.R.K., E.M.-M., Y.T.-S., M.C.M.-F., A.G.d.M., and L.L.P.-D.; project administration, D.d.C.d.S.-C.; writing—original draft, C.R.K. and M.B.-F.; writing—review and editing, S.C., J.A.B., R.T., and D.d.C.d.S.-C.

Funding: This research was funded by Coordenação de Aperfeiçoamento de Pessoal de Nível Superior-Brazil (CAPES)-Finance Code 001, the CNPq (Conselho Nacional de Desenvolvimento Científico e Tecnológico), and FAPERJ (Fundação Carlos Chagas Filho de Amparo à Pesquisa do Estado do Rio de Janeiro).

Conflicts of Interest: The authors declare no conflict of interest.

References

1. Hochberg, M.C.; Altman, R.D.; April, K.T.; Benkhalti, M.; Guyatt, G.; McGowan, J.; Towheed, T.; Welch, V.; Wells, G.; Tugwell, P. American College of Rheumatology 2012 recommendations for the use of nonpharmacologic and pharmacologic therapies in osteoarthritis of the hand, hip, and knee. *Arthritis Care Res.* **2012**, *64*, 465–474. [CrossRef] [PubMed]
2. Dulay, G.S.; Cooper, C.; Dennison, E.M. Knee pain, knee injury, knee osteoarthritis & work. *Best Pract. Res. Clin. Rheumatol.* **2015**, *29*, 454–461. [PubMed]
3. Ferreira, R.; Duarte, J.; Gonçalves, R. Non-pharmacological and non-surgical interventions to manage patients with knee osteoarthritis: An umbrella review. *Acta Reumatol. Port.* **2018**, *43*, 182–200. [PubMed]
4. Zhang, W.; Moskowitz, R.W.; Nuki, G.; Abramson, S.; Altman, R.D.; Arden, N.; Bierma-Zeinstra, S.; Brandt, K.D.; Croft, P.; Doherty, M.; et al. OARSI recommendations for the management of hip and knee osteoarthritis, Part II: OARSI evidence-based, expert consensus guidelines. *Osteoarthr. Cartil.* **2008**, *16*, 137–162. [CrossRef]
5. Neto, S.B.S.; Marconi, E.M.; Kutter, C.R.; Frederico, E.H.F.F.; de Paiva, P.d.C.; Meyer, P.F.; Chang, S.; Sá-Caputo, D.; Bernardo-Filho, M. Beneficial effects of whole body mechanical vibration alone or combined with auriculotherapy in the pain and in flexion of knee of individuals with knee osteoarthritis. *Acupunct. Electrother. Res.* **2017**, *42*, 185–201. [CrossRef]

6. ACR. *ACR OA Guidelines Non-Pharmacological—Knee and Hip*; American College of Rheumatology: Chicago, IL, USA, 2009; pp. 4–30.
7. Allen, K.D.; Bierma-Zeinstra, S.M.A.; Foster, N.E.; Golightly, Y.M.; Hawker, G. OARSI Clinical Trials Recommendations: Design and conduct ofimplementation trials of interventions for osteoarthritis. *Osteoarthr. Cartil.* **2015**, *23*, 826–838. [CrossRef]
8. Asher, G.N.; Jonas, D.E.; Coeytaux, R.R.; Reilly, A.C.; Loh, Y.L.; Motsinger-Reif, A.A.; Winham, S.J. Auriculotherapy for Pain Management: A Systematic Review and Meta-Analysis of Randomized Controlled Trials. *J. Altern. Complement. Med.* **2010**, *16*, 1097–1108. [CrossRef]
9. Hou, P.W.; Hsu, H.C.; Lin, Y.W.; Tang, N.Y.; Cheng, C.Y.; Hsieh, C.L. The history, mechanism, and clinical application of auricular therapy in traditional Chinese medicine. *Evid.-Based Complement. Altern. Med.* **2015**, *2015*, 495684. [CrossRef]
10. Huang, L. *Auricular Medicine: A Complete Manual of Auricular Diagnosis and Treatment*; Auricular International Research and Training: Orlando, FL, USA, 2005.
11. Rabischong, P.; Terral, C. Scientific Basis of Auriculotherapy: State of the Art. *Med. Acupunct.* **2014**, *26*, 84–96. [CrossRef]
12. Suen, L.K.P.; Wong, T.K.S.; Leung, A.W.N. Is there a place for auricular therapy in the realm of nursing? *Complement. Ther. Nurs. Midwifery* **2001**, *7*, 132–139. [CrossRef]
13. Suen, L.K.P.; Yeh, C.H.; Yeung, S.K.W. Erratum to Using auriculotherapy for osteoarthritic knee among elders: A double-blinded randomised feasibility study. *BMC Complement. Altern. Med.* **2016**, *16*, 257. [CrossRef] [PubMed]
14. Round, R.; Litscher, G.; Bahr, F. Auricular acupuncture with laser. *Evid.-Based Complement. Altern. Med.* **2013**, *2013*, 984763. [CrossRef] [PubMed]
15. Rubach, A. *Principles of Ear Acupuncture-Microsystem of the Auricle: Microsystem of the Auricle*, 2nd ed.; Thieme Publishers: Stuttgart, Germany, 2001.
16. Gori, L.; Firenzuoli, F. Ear acupuncture in European traditional medicine. *Evid.-Based Complement. Altern. Med.* **2007**, *4*, 13–16. [CrossRef] [PubMed]
17. Suen, L.K.P.; Wong, T.K.S.; Leung, A.W.N. Auricular therapy using magnetic pearls on sleep: A standardized protocol for the elderly with insomnia. *Clin. Acupunct. Orient. Med.* **2002**, *3*, 39–50. [CrossRef]
18. Alimi, D.; Rubino, C.; Pichard-Léandri, E.; Fermand-Brulé, S.; Dubreuil-Lemaire, M.L.; Hill, C. Analgesic effect of auricular acupuncture for cancer pain: A randomized, blinded, controlled trial. *J. Clin. Oncol.* **2003**, *21*, 4120–4126. [CrossRef]
19. Usichenko, T.I.; Dinse, M.; Hermsen, M.; Witstruck, T.; Pavlovic, D.; Lehmann, C. Auricular acupuncture for pain relief after total hip arthroplasty—A randomized controlled study. *Pain* **2005**, *114*, 320–327. [CrossRef]
20. Sitjà-Rabert, M.; Rigau, D.; Fort Vanmeerghaeghe, A.; Romero-Rodríguez, D.; Bonastre Subirana, M.; Bonfill, X. Efficacy of whole body vibration exercise in older people: A systematic review. *Disabil. Rehabil.* **2012**, *34*, 883–893. [CrossRef]
21. Rittweger, J. Vibration as an exercise modality: How it may work, and what its potential might be. *Eur. J. Appl. Physiol.* **2010**, *108*, 877–904. [CrossRef]
22. Rauch, F.; Sievanen, H.; Boonen, S.; Cardinale, M.; Degens, H.; Felsenberg, D.; Roth, J.; Schoenau, E.; Verschueren, S.; Rittweger, J. Reporting whole-body vibration intervention studies: Recommendations of the International Society of Musculoskeletal and Neuronal Interactions. *J. Musculoskelet. Neuronal Interact.* **2010**, *10*, 193–198.
23. da Costa, P.R.; da Cunha S&a-Caputo, D.U.; Arn&obio, A.; Pacheco, R.; Kutter, C.; Costa, R.; Giehl, P.M.; Paiva, D.N.; Marin, P.J.; Salmon, J.R.; et al. Whole-body vibration and benefits for people with osteoarthritis: A systematic review. *Int. J. Med. Med. Sci.* **2014**, *6*, 201–210.
24. Di Giminiani, R.; Manno, R.; Scrimaglio, R.; Sementilli, G.; Tihanyi, J. Effects of individualized whole-body vibration on muscle flexibility and mechanical power. *J. Sports Med.* **2010**, *50*, 139–151.
25. Sá-Caputo, D.D.; Ronikeili-Costa, P.; Carvalho-Lima, R.P.; Bernardo, L.C.; Bravo-Monteiro, M.O.; Costa, R.; Moraes-Silva, J.D.; Paiva, D.N.; Machado, C.B.; Mantilla-Giehl, P.; et al. Whole Body Vibration Exercises and the Improvement of the Flexibility in Patient with Metabolic Syndrome. *J. Rehabil. Res. Dev.* **2014**, *2014*, 628518. [CrossRef] [PubMed]
26. Yang, J.; Seo, D. The effects of whole body vibration on static balance, spinal curvature, pain, and disability of patients with low back pain. *J. Phys. Ther. Sci.* **2015**, *27*, 805–808. [CrossRef]

27. Wang, P.; Yang, X.; Yang, Y.; Yang, L.; Zhou, Y.; Liu, C.; Reinhardt, J.D.; He, C. Effects of whole body vibration on pain, stiffness and physical functions in patients with knee osteoarthritis: A systematic review and meta-analysis. *Clin. Rehabil.* **2015**, *29*, 939–951. [CrossRef] [PubMed]
28. Collins, N.J.; Misra, D. Measures of Knee Function. *Arthritis Care Res.* **2011**, *63*, 208–228. [CrossRef] [PubMed]
29. Marshall, C.J.; Dobson, F.; Hinman, R.S.; Sayer, T.; Anderson, C.; Newcomb, N.; Bennell, K.L. Reliability of the osteoarthritis research society international recommended performance-based tests of physical function in people with hip and knee osteoarthritis. *Osteoarthr. Cartil.* **2015**, *23*, A343–A344. [CrossRef]
30. Bennell, K.; Dobson, F.; Hinman, R. Measures of physical performance assessments: Self-Paced Walk Test (SPWT), Stair Climb Test (SCT), Six-Minute Walk Test (6MWT), Chair Stand Test (CST), Timed Up & Go (TUG), Sock Test, Lift and Carry Test (LCT), and Car Task. *Arthritis Care Res.* **2011**, *63*, S350–S370.
31. Avelar, N.C.P.; Simão, A.P.; Tossige-Gomes, R.; Neves, C.D.C.; Rocha-Vieira, E.; Coimbra, C.C.; Lacerda, A.C.R. The Effect of Adding Whole-Body Vibration to Squat Training on the Functional Performance and Self-Report of Disease Status in Elderly Patients with Knee Osteoarthritis: A Randomized, Controlled Clinical Study. *J. Altern. Complement. Med.* **2011**, *17*, 1149–1155. [CrossRef]
32. Paineiras-Domingos, L.L.; da Cunha Sá-Caputo, D.; Reis, A.S.; Francisca Santos, A.; Sousa-Gonçalves, C.R.; Dos Anjos, E.M.; dos Santos Pereira, M.J.; Sartorio, A.; Bernardo-Filho, M. Assessment through the short physical performance battery of the functionality in individuals with metabolic syndrome exposed to whole-body vibration exercises. *Dose-Response* **2018**, *16*. [CrossRef]
33. Sá-Caputo, D.C.; Paineiras-Domingos, L.L.; Oliveira, R.; Neves, M.F.T.; Brandão, A.; Marin, P.J.; Sañudo, B.; Furness, T.; Taiar, R.; Bernardo-Filho, M. Acute Effects of Whole-Body Vibration on the Pain Level, Flexibility, and Cardiovascular Responses in Individuals With Metabolic Syndrome. *Dose-Response* **2018**, *16*. [CrossRef]
34. Nogier, R. *Auriculotherapy*; Ed.Thieme: Stuttgart, Germany; New York, NY, USA, 2009; Volume 14.
35. Metsavaht, L.; Leporace, G.; Riberto, M.; de Mello Sposito, M.M.; Batista, L.A. Translation and Cross-Cultural Adaptation of the Brazilian Version of the International Knee Documentation Committee Subjective Knee Form. *Am. J. Sports Med.* **2010**, *38*, 1894–1899. [CrossRef] [PubMed]
36. Costa-Cavalcanti, R.G.; de Sá-Caputo, D.D.; Moreira-Marconi, E.; Küter, C.R.; Brandão-Sobrinho-Neto, S.; Paineiras-Domingos, L.L.; Moura-Fernandes, M.C.; da Costa, J.M.; de Moraes Carmo, J.M.; Nogueira-Neto, J.F.; et al. Effect of Auriculotherapy on the Plasma Concentration of Biomarkers in Individuals with Knee Osteoarthritis. *J. Acupunct. Meridian Stud.* **2018**, *11*, 145–152. [CrossRef] [PubMed]
37. Moreira-Marconi, E.; Dionello, C.F.; Morel, D.S.; Sá-Caputo, D.C.; Sousa-gonçalves, C.R.; José, M.; Bernardo-Filho, M. Whole body vibration and auriculotherapy improve handgrip strength in individuals with knee osteoarthritis. *J. Tradit. Chin. Med.* **2019**, *39*, 707–715.
38. Coffey, V.G.; Hawley, J.A. Concurrent exercise training: Do opposites distract? *J. Physiol.* **2017**, *595*, 2883–2896. [CrossRef]
39. Mann, T.N.; Lamberts, R.P.; Lambert, M.I. High Responders and Low Responders: Factors Associated with Individual Variation in Response to Standardized Training. *Sport. Med.* **2014**, *44*, 1113–1124. [CrossRef]
40. Park, Y.G.; Kwon, S.; Park, J.-W.; Cha, D.Y.; Nam, K.Y.; Sim, K.B.; Chang, J.; Lee, H.J. Therapeutic Effect of Whole Body Vibration on Chronic Knee Osteoarthritis. *Ann. Rehabil. Med.* **2013**, *37*, 505–515. [CrossRef]
41. Trans, T.; Aaboe, J.; Henriksen, M.; Christensen, R.; Bliddal, H.; Lund, H. Effect of whole body vibration exercise on muscle strength and proprioception in females with knee osteoarthritis. *Knee* **2009**, *16*, 256–261. [CrossRef]
42. Wang, P.; Yang, L.; Li, H.; Lei, Z.; Yang, X.; Liu, C.; Jiang, H.; Zhang, L.; Zhou, Z.; Reinhardt, J.D.D.; et al. Effects of whole-body vibration training with quadriceps strengthening exercise on functioning and gait parameters in patients with medial compartment knee osteoarthritis: A randomised controlled preliminary study. *Physiotherapy* **2016**, *102*, 86–92. [CrossRef]

applied sciences

MDPI

Article

Effects of Whole-Body Vibration Exercises on Parameters Related to the Sleep Quality in Metabolic Syndrome Individuals: A Clinical Trial Study

Claudia Figueiredo Azeredo [1,2], Patrícia de Castro de Paiva [1,2], Leandro Azeredo [3], Aline Reis da Silva [1,2], Arlete Francisca-Santos [2,4], Laisa Liane Paineiras-Domingos [2,4,5], Adriana Lírio Pereira da Silva [2], Camila Leite Bernardes-Oliveira [1,5], Juliana Pessanha-Freitas [2], Márcia Cristina Moura-Fernandes [2,6], Rubens Guimarães Mendonça [2], José Alexandre Bachur [7], Ygor Teixeira-Silva [2,5], Eloá Moreira-Marconi [2,6], Eliane de Oliveira Guedes-Aguiar [2,4,5], Bruno Bessa Monteiro de Oliveira [2,4], Mário Fritsch Neves [5], Luiz Felipe Ferreira-Souza [2], Vinicius Layter Xavier [8], Daniel Lago Borges [9], Ana Cristina Lacerda [10], Vanessa Amaral Mendonça [10], Anelise Sonza [11], Redha Taiar [12,*], Alessandro Sartorio [13], Mario Bernardo-Filho [2] and Danúbia da Cunha de Sá-Caputo [2,4,5]

1. Programa de Pós-graduação em Saúde, Medicina Laboratorial e Tecnologia Forense, Universidade do Estado do Rio de Janeiro, Rio de Janeiro, RJ 20950-003, Brazil; figueiredo.claudiaf@gmail.com (C.F.A.); patriciadecastrodepaiva@gmail.com (P.d.C.d.P.); fisioalinereis@gmail.com (A.R.d.S.); camilabernardes@hotmail.com (C.L.B.-O.)
2. Laboratório de Vibrações Mecânicas e Práticas Integrativas, Departamento de Biofísica e Biometria, Instituto de Biologia Roberto Alcântara Gomes e Policlínica Piquet Carneiro, Universidade do Estado do Rio de Janeiro, Rio de Janeiro, RJ 20950-003, Brazil; fisioarlete@gmail.com (A.F.-S.); laisanit@gmail.com (L.L.P.-D.); adrianaliriolavimpi@gmail.com (A.L.P.d.S.); ju.freitas.fisio@gmail.com (J.P.-F.); marciafernandesfisio@gmail.com (M.C.M.-F.); rubens_fisio@hotmail.com (R.G.M.); silvarogy@hotmail.com (Y.T.-S.); eloamarconi@gmail.com (E.M.-M.); ellianeguedes@gmail.com (E.d.O.G.-A.); bessa.oliveira@gmail.com (B.B.M.d.O.); lumadaragu@gmail.com (L.F.F.-S.); bernardofilhom@gmail.com (M.B.-F.); dradanubia@gmail.com (D.d.C.d.S.-C.)
3. Departamento de Fisioterapia, Hospital Central da Polícia Militar, Rio de Janeiro, RJ 20211-270, Brazil; azeredo.lm@gmail.com
4. Faculdade Bezerra de Araújo, Rio de Janeiro, RJ 23052-180, Brazil
5. Programa de Pós-Graduação em Ciências Médicas, Faculdade de Ciências Médicas, Universidade do Estado do Rio de Janeiro, Rio de Janeiro, RJ 20551-030, Brazil; mariofneves@gmail.com
6. Programa de Pós-Graduação em Fisiopatologia Clínica e Experimental, Instituto de Biologia Roberto Alcântara Gomes, Universidade do Estado do Rio de Janeiro, Rio de Janeiro, RJ 20550-900, Brazil
7. Cursos de Medicina e Fisioterapia da Universidade de Franca, Franca, SP 14401-426, Brazil; jabachur@hotmail.com
8. Departamento de Estatística, Instituto de Matemática e Estatística, Universidade do Estado do Rio de Janeiro, Rio de Janeiro, RJ 20550-900, Brazil; viniciuslx@ime.uerj.br
9. Hospital Universitário, Universidade Federal do Maranhão, São Luís do Maranhão, MA 65080-805, Brazil; dlagofisio83@hotmail.com
10. Faculdade de Ciências Biológicas e da Saúde, Universidade Federal dos Vales do Jequitinhonha e Mucuri, Diamantina, MG 39100-000, Brazil; lacerdaacr@gmail.com (A.C.L.); vaafisio@hotmail.com (V.A.M.)
11. Programa de Pós-graduação em Fisioterapia, Universidade do Estado de Santa Catarina, Florianópolis, SC 88080-350, Brazil; anelise.sonza@gmail.com
12. GRESPI, Université de Reims, 51100 Reims, France
13. Istituto Auxologico Italiano, IRCCS, Experimental Laboratory for Auxo-Endocrinological Research, Division of Metabolic Diseases & Auxology, 28824 Milan and Verbania, Italy; sartorio@auxologico.it

* Correspondence: redha.taiar@univ-reims.fr

Received: 24 October 2019; Accepted: 22 November 2019; Published: 29 November 2019

check for updates

Abstract: Metabolic syndrome (MetS) is an undesirable clinical condition with physiological, biochemical, clinical, and metabolic factors that contribute to increased cardiovascular risks (CR). A poor sleep quality might be found in obese and MetS individuals. Whole-body vibration (WBV) exercise has been used on the management of MetS individuals. This clinical trial investigated the effect of WBV exercise on parameters related to the sleep quality in MetS individuals. After randomization, nine individuals (seven women and two men) were exposed to a fixed frequency (FF) and ten individuals (eight women and two men) were exposed to a variable frequency (VF). Both groups performed the protocol twice a week, for 6 weeks. All of the evaluations were performed before the first and after the last sessions. Anthropometric and cardiovascular parameters were measured before and after the 6-week intervention. Pittsburgh Sleep Quality Index (PSQI), Epworth Sleepiness Scale (ESS), and Berlin Questionnaire were also used to evaluate the quality of the sleep. A significant ($p \leq 0.05$) reduction of the waist circumference in the VFG and an increase of the heart rate were found in the FFG and VFG group. The score of the PSQI of the both groups decreased significantly ($p = 0.01$). The score of the ESS decreased ($p = 0.04$) only in the VF group. The scores of the Berlin Questionnaire were not altered in both groups. In conclusion, WBV intervention was capable in interfering with physiological mechanisms with effects on the WC and HR, leading to the improvement of the quality of sleep in MetS individuals. WBV exercise might be an important clinical intervention to the management of some factors associated with poor quality of sleep (FFG and VFG) and in the daytime sleepiness in MetS individuals with variable frequencies (5–16 Hz) (VFG).

Keywords: metabolic syndrome; sleep quality; whole-body vibration exercise; Pittsburgh Sleep Quality Index; Epworth Sleepiness Scale; Berlin Questionnaire

---

1. Introduction

Metabolic syndrome (MetS) is an undesirable clinical condition with physiological, biochemical, clinical, and metabolic factors, including alterations on the level of the lipids on the plasma, arterial hypertension, central adiposity, insulin resistance (IR), and hyperglycemia. These conditions strongly contribute to increased cardiovascular risks (CR) [1], and they are interconnected by pathophysiological basis in low-grade chronic inflammation, increase of the risk of type 2 diabetes mellitus (T2DM), and all-cause mortality [2]. MetS is also related to the increased deposit of central adiposity and ectopic fat infiltration (muscle and liver) associated with overeating and/or sedentary lifestyles. These conditions are related to various deleterious consequences in late life [3,4].

Lian et al. [5], in a systematic review and meta-analysis, reported that the overall sleep quality as well as sleep complaints have significant positive associations with MetS. Massar et al. [6] and Tsai et al. [7] pointed out that this condition of the sleep is linked with cardiovascular disease. Massar et al. [6] reported in individuals with poor habitual sleep efficiency during the week before stress induction (Trier Social Stress Test) responded with higher stress-related elevations of blood pressure and cortisol levels as compared to subjects with high sleep efficiency. This relationship between poor sleep efficiency and elevated blood pressure persisted during the post-stress recovery period. In addition, poor sleep health is also associated with MetS and mental illness [8,9]. Iftikhar et al. [10], in a meta-analysis reported that some investigations demonstrated the association between sleep quality and MetS. It was verified that the poor sleep quality characterized by sleep fragmentation was related to impaired glucose metabolism, independent of sleep duration [11], however, the studies of the meta-analysis analyzed had not reached a consensus. Mesas et al. [12] reported that the difficulty falling asleep is associated with MetS and, in particular, with high blood pressure, and this association would be independent of sleep duration and would be not due to lifestyles related to poor sleep.

Ying et al. [5] reported in a systematic review and meta-analysis that the overall sleep quality, as well as sleep complaints have significant positive associations with MetS. Moreover, it is estimated [13,14] that 50% to 60% of obese and MetS individuals can present obstructive sleep apnea (OSA).

Some tools have been used to verify parameters (such as sleep duration, sleep inertia, sleep latency, snoring, disorientation or confusion during sleep, and daytime sleepiness) related to the quality of the sleep, such as the Pittsburgh Sleep Quality Index (PSQI) [15], Epworth Sleepiness Scale (ESS) [16] and the Berlin Questionnaire [17]. Moreover, the neck circumference (NC) [16] and waist circumference (WC) [18,19] have also been used.

There is moderate evidence supporting the use of physical exercise (PE) programs to counter MetS, although the optimal dose and type of PE is not well established. It is the main challenge for health care professionals to persuade individuals to have adherence to perform PE to the prevention and management of MetS [20]. In this case, it is relevant to consider the whole-body vibration (WBV) exercise, a modality of PE, an option. WBV exercise has been used as a clinical intervention in individuals with different clinical disorders, including MetS [21–26] and it can be considered a feasible, safe, and low-cost technique [27].

WBV exercise is generated in an individual who is on the base of a vibrating platform. It produces a mechanical vibration (MV) that is transmitted to the body inducing muscle contractions, with physiological responses like those produced by other types of PE, such as aerobic conditioning and strength training [28–30]. Biomechanical parameters, such as the frequency (f) and the peak-to-peak displacement (PPD) of the MV must be considered in the WBV exercise protocols [26,31,32].

As it is suggested that the sleep quality plays an important role in development of MetS [5] and putting together all of the previous considerations, to our knowledge, this is a first work aiming to study the effect of WBV exercise on the sleep quality of MetS individuals using the PSQI, ESS, and Berlin Questionnaire. It is hypothesized that the WBV exercise would be able in improving the sleep quality of MetS individuals.

2. Materials and Methods

2.1. Individuals

In this cross-sectional and randomized study, 31 subjects were recruited and 19 performed the protocol (58.79 ± 12.55 years old, 1.62 ± 0.09 m height, 86.27 ± 15.03 kg body mass). The recruitment of the participants occurred from January 2018 to January 2019, in the *Hospital Universitário Pedro Ernesto* (HUPE), in the *Universidade do Estado do Rio de Janeiro* (UERJ). The WBV protocol was performed in the *Laboratório de Vibrações Mecânicas e Práticas Integrativas* (LAVIMPI).

This project was approved by the Research Ethics Committee of HUPE-UERJ with the number CAAE 54981315.6.0000.5259, the registry in the Brazilian Registry of Clinical Trials (ReBEC) with the number RBR 2bghmh and UTN: U1111-1181-1177. The participants of both groups signed a consent form.

Consolidated Standards of Reporting Trials (CONSORT) was used to report all of the different steps of the interventions utilized in this work [33].

In the randomization, a blinded envelope was used for the cards with the name of the groups: fixed frequency group (FFG) (control group) or variable frequency group (VFG) (WBVE group).

The inclusion criteria were individuals of both sexes, over 18 years of age, with MetS according to the International Diabetes Federation.

The exclusion criteria were: (i) individuals with high blood pressure (BP) levels (systolic blood pressure (SBP) ≥ 180 mmHg and diastolic blood pressure (DBP) ≥ 110 mmHg); (ii) cardiovascular disease (CVD) clinically evident in the last 6 months, manifested by a myocardial infarction or a stroke; (iii) a neurological, muscular, or rheumatologic disease that difficulte the position of the individual on the vibrating platform; (iv) disabling clinical disease according to the evaluation; (v) speech therapy or respiratory physiotherapy in the last 3 months; (vi) body mass index (BMI) > 40 kg/m^2; (vii) orthodontic therapy with intraoral device; (viii) individuals who refuse to sign the consent form.

The participants declared that they were sedentary. They were instructed to continue their daily activities, dietary habits, and medications during the period of the investigation. The drugs used by the participants of both groups were diuretics, beta blockers, calcium channel blockers, angiotensin converting enzyme inhibitors, and angiotensin receptor antagonists.

2.2. Interventions

2.2.1. Fixed Frequency Group (Control Group)

After the randomization, nine individuals (seven women and two men) allocated in the FFG performed a protocol twice a week, for 6 weeks. In this protocol, the individuals were positioned in a squat position, barefoot and with 130° knee flexion (Figure 1A). The biomechanical parameters were 2.5, 5 and 7.5 mm (Figure 1B–D) of PPD and 5 Hz for 10 s of vibration and 110 s of non-vibration in each bout. From 1 to 4 weeks, were performed 3 bouts in each session, totaling 18 min of total time. From 5 to 8 weeks, were performed 4 bouts in each session, totaling 24 min of total time. From 9 to 12 weeks, were performed 5 bouts in each session, totaling 30 min of total time. The participants performed dynamic and static squats in the sessions.

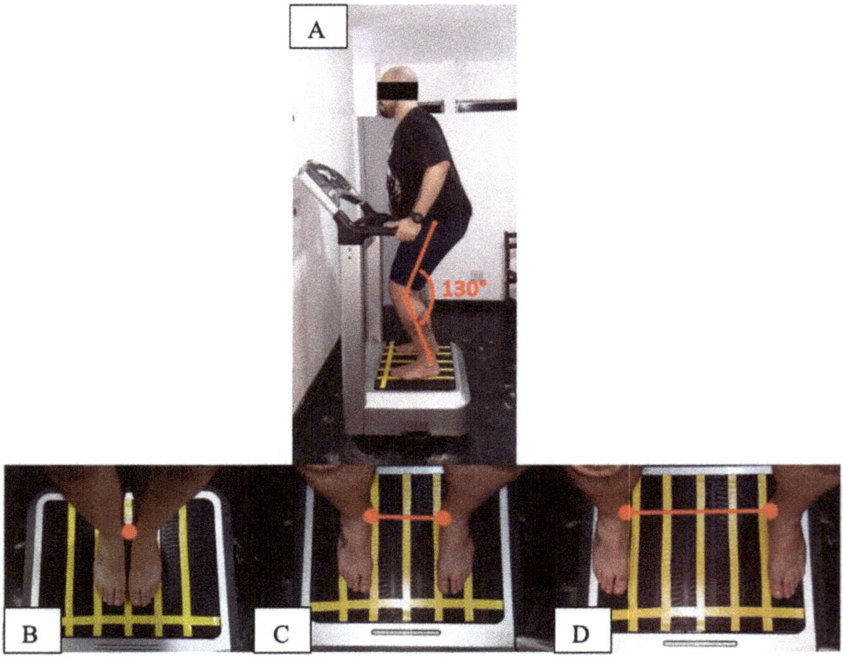

Figure 1. Biomechanical parameters during the intervention (fixed frequency (FF) and variable frequency (VF) groups) using the side alternating vibratory platform; (**A**) individual in a stand position, barefoot and with 130° knee flexion; (**B**) individual in 2.5 mm of peak-to-peak displacement (PPD) on the base of the vibrating platform with medial malleoli together; (**C**) individual in 5 mm of PPD on the base of the vibrating platform 21 cm apart between the medial malleoli; (**D**) individual in 7.5 mm of PPD on the base of the vibrating platform 38 cm apart between the medial malleoli.

2.2.2. Variable Frequency Group (WBVE Group)

After the randomization, ten individuals (eight women and two men) allocated in the VFG performed the protocol twice a week, for 6 weeks. In this protocol, the individuals were positioned in a squat position, barefoot and with 130° knee flexion (Figure 1) The biomechanical parameters of the MV were 2.5, 5 and 7.5 mm (Figure 1B–D) of the PPD; the frequency was 5 Hz in the first session and it increased by one Hz in each session, and it was 16 Hz in the last session. During the intervention, were performed 60 s of MV and 60 s of non-vibration in each bout. From 1 to 4 weeks were performed 3 bouts in each session, for a total time of 18 min. From 5 to 8 weeks, were performed 4 bouts in each session, for a total time of 24 min. From 9 to 12 weeks were performed 5 bouts in each session, for a total time of 30 min. Dynamic and static squats were performed in interspersed sessions.

2.3. Anthropometric Measurements and Physiological Parameters

All of the measurements were performed in the same location, at room temperature. Height and body mass were measured on a digital balance (MIC 200 PPA, Micheletti, São Paulo, Brazil). The BMI was calculated by dividing body mass (in kg) by square of height (in m) [34]. WC was obtained with a flexible measuring tape connecting the midpoints between the last costal arch and iliac crest, at the end of the soft exhalation and in the orthostatic position [34]. NC was obtained using a flexible metric tape just below the prominence of the larynx [35].

An automated device (OMRON, model HEM-7113, Dalian, China) was utilized to record the SBP and DBP (mm Hg), and the HR (beats/min) was measured on the left arm of seated patient after a 10-min rest. Three measurements were performed with 1 min of rest after each measurement. The mean of these 3 records of SBP and DBP and HR was used in the analyses of the WBV exercise group and the control group [36]. These evaluations were performed before and after the 6-week intervention in both groups.

The data of these parameters were expressed in mean and standard derivation and the Wilcoxon rank test was performed.

2.4. Questionnaires

The PSQI [15] was used to assess sleep quality and disturbances over a 1-month period. This questionnaire is composed of 19 self-rated questions and 5 questions that should be answered by bedmates or roommates. The 19 questions are separated into 7 categories and there is a score from 0 to 3. The components of this questionnaire are subjective sleep quality, sleep latency, sleep duration, habitual sleep efficiency, sleep disturbances, use of medication, and daytime dysfunction. The maximum PSQI score is 21 points. Higher scores indicate poor quality of sleep [37].

The ESS [16] is composed of 8 self-reported questions and it is used to assess the self-reported level of daytime sleepiness [38]. These items have a four-point scale (0 indicating "would never nod off" and 3 indicating a "strong chance of nodding off"). The questionnaire questions are related to daily activities. The questionnaire score ranges between 0 and 24. Higher total scores are related to more sleepiness [39].

The Berlin Questionnaire [17] has been proposed as a tool in screening OSA and validated in primary care. The subjects were classified into high risk (if there were 2 or more categories where the score was positive), and low risk (if there was only 1 or no categories where the score was positive). The percentage os individuals with high risk to OSA was calculated in both groups [40,41].

The PSQI, ESS, and Berlin Questionnaire were only measured before and after the 6-week intervention in both groups.

2.5. Statistical Analyses

For the sample size calculation, we used a standard deviation of 3.1, a margin of error of 2, the significance level of 5%, and a sample size of 9 individuals in each group, considering the PSQI [42].

The Wilcoxon signed-rank test with continuity correction was used in the analyses of the DBP, SBP, HR, WC, NC, the PSQI, and ESS. Descriptive statistics, median, and interquartile range (IQR) are also reported. Wilcoxon signed-rank test is a nonparametric test that was used to determine if paired samples, before and after intervention, have the same distribution. Fisher's exact test was applied to the qualitative variable, Berlin, to asses if the proportion of people in each class (high risk and low risk) have been affected by the experiment. In the analyses of the PSQI and ESS, the values are expressed in scores. The statistical analyses of the Berlin Questionnaire considered only the percentage of individuals with a high risk to OSA.

All statistical analysis was performed using R software, version 3.5.0 [43] and the R Librarie Table 1 [44]. Results are considered statistically significant if the p-value is under 0.05 ($p \leq 0.05$).

3. Results

The flow diagram with the enrolment of the study is shown in Figure 2. Thirty-one individuals were recruited, two individuals were excluded because they did not have risk of sleep apnea, three declined to participate, and seven individuals declined due to other reasons (economical problem related to the transportation).

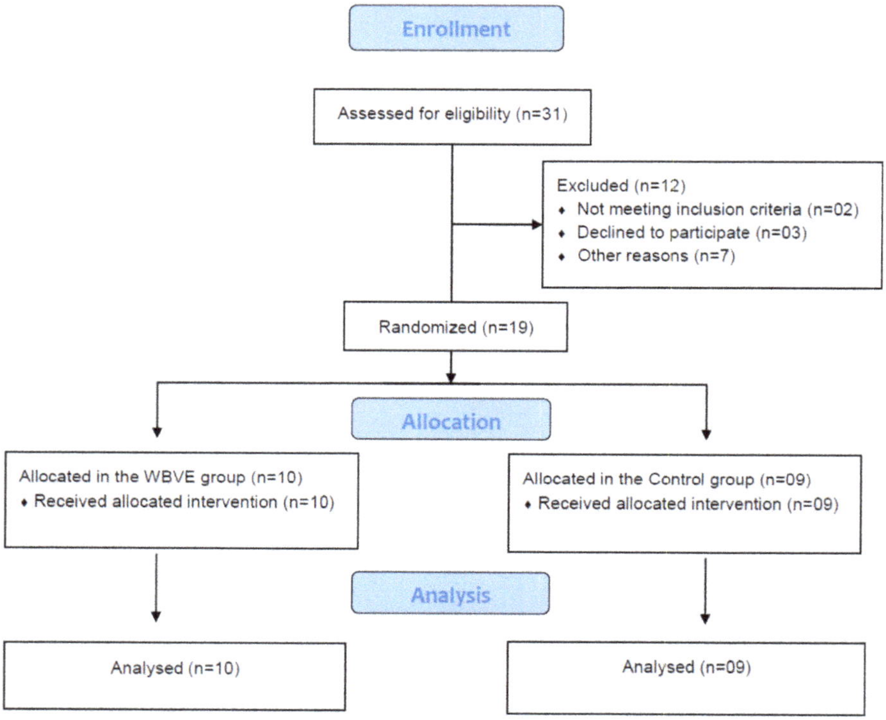

Figure 2. Flow diagram of the intervention.

Table 1 shows some anthropometric parameters of the individuals that were exposed to WBV exercise (FV group) and to only 5 Hz (FF group). An analysis is shown of both groups before and just after the treatments. It was verified that the BMI, NC, SBP, and DBP ($p > 0.05$) were not significantly altered in both groups. However, a significant increase in the HR ($p < 0.05$) was found in both groups exposed to mechanical vibration, only 5 Hz (FFG) and variable frequency 5–16Hz (VFG). Furthermore, a reduction was found in the group that was submitted to the variable frequency 5–16 Hz (VFG) in the WC.

Table 1. Anthropometrical parameters of the individuals exposed to variable frequency group (intervention 5–16 Hz) and to only 5 Hz (fixed frequency group).

	FFG before (Median [IQR])	FFG after (Median [IQR])	*p*	VFG before (Median [IQR])	VFG after (Median [IQR])	*p*
BMI (kg/m^2)	29.50 [29.0, 34.20]	29.60 [28.90, 34.10]	0.54	35.40 [32.40, 36.60]	34.70 [32.88, 36.08]	0.23
WC (cm)	105.00 [101.20, 106.30]	101.00 [100.00, 105.00]	0.23	108.90 [103.75, 116.10]	107.00 [102.17, 112.85]	0.02
NC (cm)	38.00 [36.40, 38.50]	36.60 [35.60, 39.50]	0.29	38.00 [37.58, 38.88]	38.30 [37.17, 39.00]	0.25
SBP (mm Hg)	129.00 [128.00, 130.00]	127.00 [121.00, 128.00]	0.12	130.00 [122.50, 132.50]	130.00 [120.00, 132.00]	0.76
DBP (mm Hg)	78.00 [75.00, 83.00]	75.00 [68.00, 87.00]	1.00	74.00 [72.25, 75.75]	70.00 [60.75, 74.75]	0.18
HR (bpm)	57.00 [56.00, 62.00]	62.00 [60.00, 66.00]	0.05	63.50 [59.75, 66.00]	72.00 [61.25, 83.00]	0.05

BMI—body mass index, WC—waist circumference, NC—neck circumference, SBP—systolic blood pressure, DBP—diastolic blood pressure, HR—heart rate, FF—fixed frequency group, VF—variable frequency group.

Figure 3 shows the values of the scores of the PSQI of the individuals of the group exposed only tc mechanical vibration of 5 Hz (FFG-control). A significant decrease was found (p = 0.01) of the score (after and before) of the individuals of the FF group (exposure to 5 Hz).

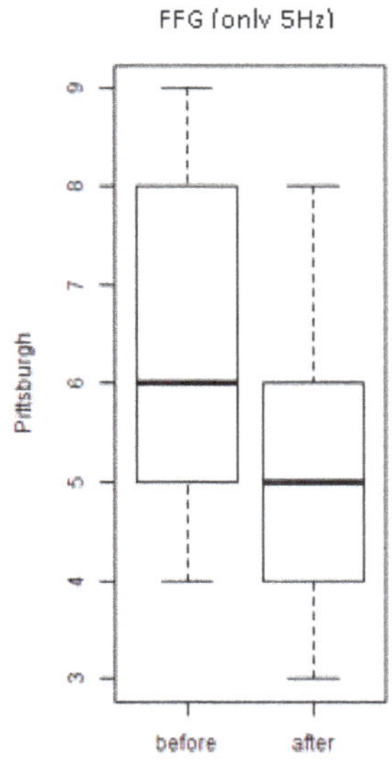

Figure 3. Scores of the Pittsburgh Sleep Quality Index (PSQI) of the individuals of the group exposed only to a mechanical vibration of 5 Hz (FFG) (before and after).

Figure 4 shows the values of the scores of the PSQI of the individuals of the groups exposed to the treatment with WBV exercises (VFG). It is observed that an important and significant ($p = 0.008$) decrease of score in the group after the exposure to various frequencies (intervention) was seen. This finding indicates an improvement of quality of the sleep.

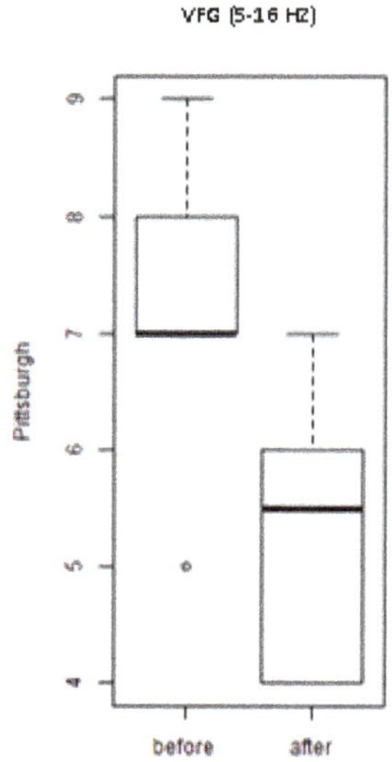

Figure 4. Scores of the PSQI of the individuals exposed to the whole-body vibration (WBV) intervention with variable frequency (VFG).

Figure 5 shows the values of the scores of the ESS of the individuals of the group exposed only to mechanical vibration of 5 Hz (FFG). It is observed that there was no significant decrease ($p = 1.0$) of the score (after and before) of the individuals. This finding indicates no reduction of daytime sleepiness.

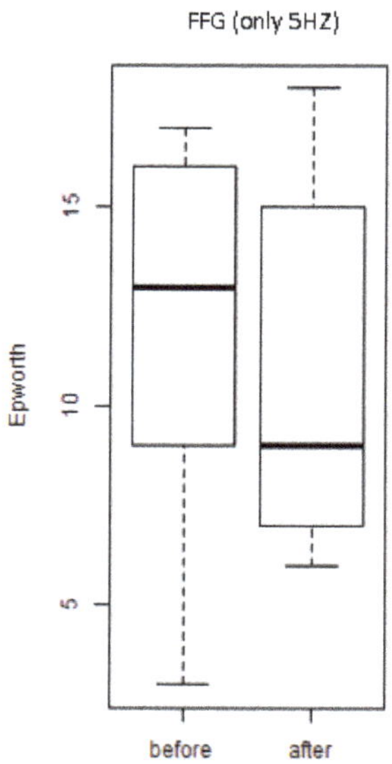

Figure 5. Scores of the Epworth Sleepiness Scale (ESS) of the individuals of the SG individuals of the FFG (before and after).

Figure 6 shows the values (before and after the interventions) of the scores of the ESS scale of the individuals of the groups exposed to the treatment with WBV exercises (VFG). It is observed that there was a significant ($p = 0.04$) decrease of the score in the group after the exposure to various frequencies (intervention). This finding indicates a reduction of daytime sleepiness.

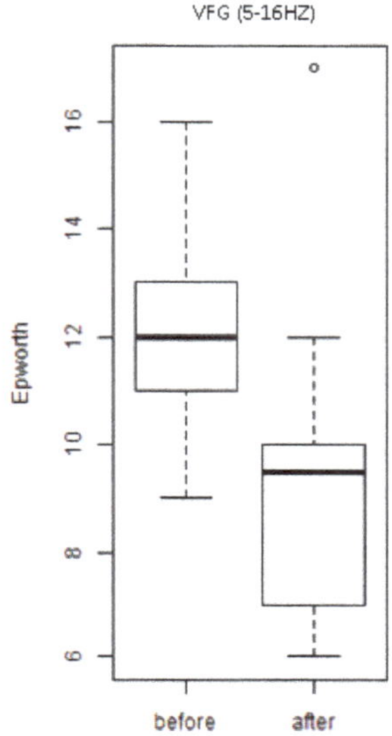

Figure 6. Scores of the ESS of the individuals of the group exposed to the intervention with WBV exercises (VFG).

Considering the Berlin Questionnaire, no alteration in ter percentage of individuals with risk to OSA were significantly found in both groups, the FF $p = 0.29$ and FVG $p = 1.00$.

4. Discussion

The underlying etiology of MetS is multifactorial, however, they are not well known, but sedentary lifestyles and unhealthy dietary habits favor the appearance of MetS [45]. Considering the high prevalence and strong and undesirable complications, early identifying and controlling the modified risk factors are key prevention methods to counter the development of MetS and its progression to CVD.

It is relevant to point out that, to our knowledge, this is the first work that was performed to evaluate the effect of WBV exercise in sleep quality of MetS individuals. As it was hypothesized that the WBV exercise would be able to improve the sleep quality of MetS individuals, the quality of sleep was improved (PSQI) and a reduction of daytime sleepiness (ESS) in the MetS was found due to a protocol with variable frequencies (5–16 Hz) of WBV. A protocol with fixed frequency (5 Hz) also improved the quality of sleep (PSQI).

It was shown (Table 1) that the 6-week intervention with WBV exercise with variable frequencies (from 5 up to 16 Hz) (VF group) did not alter significantly the BMI, NC, SBP, and DBP. The same finding was verified in the individuals with fixed frequency (5 Hz). These findings agree with other studies. Feairheller et al. [46], in a study of six months, have shown that aerobic exercise training is enough to elicit improvements in vascular structure and function in African Americans, even without alterations on BP measurements (SBP and DBP) in African Americans. Moreover, Sá-Caputo et al. [31] have reported that, in individuals with MetS exposed to acute WBV exercise, the SBP and DBP were not altered. Figueroa et al. [21] have reported that the BMI of overweight/obese women was not modified after 6 weeks of WBV exercises.

Considering the HR (Table 1), an increase was verified in FFG (only 5 Hz) and VFG (variable 5–16 Hz), and this is in agreement with Sunita et al. [47] that have observed an increase of the HR in response to exercise. Moreover, Kang et al. [48] have described an association between resting HR and cardiorespiratory fitness with an elevated resting HR.

The findings presented in Table 1 indicated no alterations on SBP and DBP due to WBVE (a possible stressor), however, Massar et al. [6] reported that individuals with poor habitual sleep efficiency during the week before stress induction (Trier Social Stress Test) responded with higher stress-related elevations of blood pressure and cortisol levels as compared to subjects with high sleep efficiency.

In the current study, no alterations on the NC were found. However, a decrease on the WC was found (Table 1). Sîrbu et al. [49] have compared some anthropometric parameters in trained and sedentary healthy young students. It was found that in moderately aerobic trained students, the BMI and BP were not significantly different between the two groups. However, a significantly lower NC was found in the trained students. Considering the WC, in agreement with our results, King et al. [50] have verified that a 12-week supervised aerobic exercise intervention has reduced significantly the WC of overweight/obese men and women. But, Sîrbu et al. [49], have not found alteration in the WC due to exercise.

WBV exercise is an intervention with several improvements in populations with various diseases [23–25,31]. In this current study, it was verified that WBV exercise has improved the quality of the sleep in MetS when we used the PSQI (Figures 3 and 4) and ESS (Figure 6). This finding is partially in agreement with Souza et al. [51] who have evaluated, in a population of 16 individuals with OSA, the effectiveness of inspiratory muscle training (IMT) on sleep and functional capacity to exercise. It was a significant improvement in sleep quality was seen with PSQI values, but no significant changes were seen in daytime sleepiness (ESS) after the intervention. Furthermore, Reid et al. [52] have verified that moderate aerobic PE improved the sleep quality of sedentary adults considering the PSQI and the ESS. Brandão et al. [53] have reported that home exercise improves the quality of sleep and daytime sleepiness of elderlies considering, respectively, the PSQI and the ESS.

The findings of the current work have verified that the sleep quality was not altered when the Berlin Questionnaire was used. This result agrees with Brandão et al. [53] who have reported that home exercise does not alter the risk to OSA when the same questionnaire was used.

Kline et al. [54], evaluated the utility of PE for improving daytime functioning in adults with OSA. Sleepiness and functional impairment due to sleepiness also were improved following exercise versus control, though these changes were not statistically significant. It is suggested that PE may be helpful for improving aspects of daytime functioning of adults with OSA. Moreover, Yilmaz Gokmen et al. [55] have investigated the effects of Tai Chi and Qigong training on severity of OSA for 12-weeks. In the intervention group, there was a statistically significant decrease in the ESS. However, Itoh et al. [56], in a cross-sectional study among non-obese male workers in Japan, found no significant association between physical activity and the risk of sleep-disordered breathing.

Considering the MetS, not all forms of exercise are equally effective and safe; although aerobic exercise [57] or resistance training have been associated with decreased cardiovascular disease risk factors, obesity, or MetS severity [58,59]. Due to the pain or even the low physical fitness, most individuals are unable or unwilling to perform these exercises [60]. Given the limitations for some types of exercises that many individuals report, different forms of exercise are being suggested for MetS individuals [61]. WBV exercise has improved several parameters in MetS individuals [26,31,32] and in the current study a decrease of the WC was found (Table 1).

This study has some limitations, no control of daily activity, daily working, lifestyle habits, smoking, PE, and daily energy intake. The follow-up data were not available after the intervention. The external validity of this intervention considering its generalizability to other settings (like the everyday-living condition) was not explored and only a small sample was used. The big ratio of female to male would be also considered as a limitation of the study. Moreover, a sham group and a group of individuals without MetS were not used. Furthermore, the analysis of the questionnaires was not done in the weeks in which there were changes of the frequency and duration of the intervention in the VFG. Some points of the questionnaires are influenced by factors that are not taken into account in the current study, as categories of individuals (good-sleepers and bad-sleepers), sleep duration, sleep inertia, and sleep latency. In consequence, further investigations are required. However, putting together all the considerations, the strength was to verify that a WBV exercise might be suggested as an option to the management of MetS individuals with some important outcomes, and it would be expected that a WBV exercise might lead to responses that can reflect in improvements of the sleep quality.

5. Conclusions

In conclusion, WBV intervention was capable of interfering with physiological mechanisms with effects on the WC and HR, leading to the improvement of the quality of sleep in MetS individuals. Although the calculated sample size determined a study with a small number of individuals, WBV exercise might be an important clinical intervention to the management of some factors associated with poor quality of sleep (FFG and VFG) and in the daytime sleepiness in MetS individuals with variable frequencies (5–16Hz) (VFG). However, further investigations are necessary to try to understand the mechanisms that underlines this positive effect of the WBV exercise in the studied population, as well as with individuals with sleep disturbances without MetS.

Author Contributions: Conceptualization, R.T., A.S. (Alessandro Sartorio), M.B.-F., and D.d.C.d.C.D.S.-C.; data curation, C.F.A., J.A.B., V.L.X., and D.L.B.; formal analysis, J.A.B., V.L.X., D.L.B., A.C.L., V.A.M., A.S. (Anelise Sonza), and R.T.; funding acquisition, M.B.-F.; Investigation, C.F.A., P.d.C.d.P., L.A., A.R.d.S., A.F.-S., L.L.P.-D., A.L.P.d.S., C.L.B.-O., J.P.-F., M.C.M.-F., R.G.M., Y.T.-S., E.M.-M., E.d.O.G.-A., B.B.M.d.O., M.F.N., and L.F.F.-S.; methodology, C.F.A., P.d.C.d.P., L.A., A.R.D.S., A.F.-S., L.L.P.-D., A.L.P.d.S., C.L.B.-O., J.P.-F., M.C.M.-F., R.G.M., J.A.B., Y.T.-S., E.M.-M., E.d.O.G.-A., B.B.M.d.O., M.F.N., L.F.F.-S., and A.S.(Alessandro Sartorio); project administration, M.B.-F. and D.d.C.d.C.D.S.-C.; resources, C.F.A., P.d.C.d.P., L.A., A.R.d.S., R.G.M., and B.B.M.d.O.; software, V.L.X.; supervision, M.B.-F. and D.d.C.d.C.D.S.-C.; validation, V.L.X., M.B.-F., and D.d.C.d.C.D.S.-C.; visualization, A.S.(Anelise Sonza) and R.T.; writing—original draft, C.F.A.; Writing—review and editing, A.C.L., V.A.M., A.S. (Anelise Sonza), A.S. (Alessandro Sartorio), M.B.-F., and D.d.C.d.C.D.S.-C.

Funding: This study was supported in part by the *Coordenação de Aperfeiçoamento de Pessoal de Nível Superior* - Brazil (CAPES) - Finance Code 001, the *Conselho Nacional de Desenvolvimento Científico e Tecnológico (CNPq) and the Fundação de Amparo à Pesquisa do Estado do Rio de Janeiro* (FAPERJ).

Conflicts of Interest: There is no conflict of interest.

References

1. Kaur, J. Retracted: A Comprehensive Review on Metabolic Syndrome. *Cardiol. Res. Pract.* **2019**, *2019*, 4301528.
2. Seaman, D.R.; Palombo, A.D. An Overview of the Identification and Management of the Metabolic Syndrome in Chiropractic Practice. *J. Chiropr. Med.* **2014**, *13*, 210–219. [CrossRef]
3. Dominguez, L.J.; Barbagallo, M. The Biology of the Metabolic Syndrome and Aging:2006. *Curr. Opin. Clin. Nutr. & Metab. Care* **2016**, *9*, 5–11.
4. Alberti, K.G.M.M.; Zimmet, P.; Shaw, J. Metabolic syndrome-a new world-wide definition. A Consensus Statement from the International Diabetes Federation. *Diabet. Med.* **2006**, *23*, 469–480. [CrossRef]
5. Lian, Y.; Yuan, Q.; Wang, G.; Tang, F. Association between sleep quality and metabolic syndrome: A systematic review and meta-analysis. *Psychiatry Res.* **2019**, *274*, 66–74. [CrossRef]
6. Massar, S.A.A.; Liu, J.C.J.; Mohammad, N.B.; Chee, M.W.L. Poor habitual sleep efficiency is associated with increased cardiovascular and cortisol stress reactivity in men. *Psychoneuroendocrinology* **2017**, *81*, 151–156. [CrossRef]
7. Tsai, H.-J.; Kuo, T.B.J.; Lin, Y.-C.; Yang, C.C.H. The association between prolonged sleep onset latency and heart rate dynamics among young sleep-onset insomniacs and good sleepers. *Psychiatry Res.* **2015**, *230*, 892–898. [CrossRef]
8. Salo, P.; Sivertsen, B.; Oksanen, T.; Sjösten, N.; Pentti, J.; Virtanen, M.; Kivimäki, M.; Vahtera, J. Insomnia symptoms as a predictor of incident treatment for depression: Prospective cohort study of 40,791 men and women. *Sleep Med.* **2012**, *13*, 278–284. [CrossRef]
9. Wang, S.; Wu, Y.; Ungvari, G.S.; Ng, C.H.; Forester, B.P.; Gatchel, J.R.; Chiu, H.F.K.; Kou, C.; Fu, Y.; Qi, Y.; et al. Sleep duration and its association with demographics, lifestyle factors, poor mental health and chronic diseases in older Chinese adults. *Psychiatry Res.* **2017**, *257*, 212–218. [CrossRef]

10. Iftikhar, I.H.; Donley, M.A.; Mindel, J.; Pleister, A.; Soriano, S.; Magalang, U.J. Sleep Duration and Metabolic Syndrome. An Updated Dose–Risk Metaanalysis. *Ann. Am. Thorac. Soc.* **2015**, *12*, 1364–1372. [CrossRef]
11. Stamatakis, K.A.; Punjabi, N.M. Effects of Sleep Fragmentation on Glucose Metabolism in Normal Subjects. *Chest* **2010**, *137*, 95–101. [CrossRef]
12. Mesas, A.E.; Guallar-Castillón, P.; López-García, E.; León-Muñoz, L.M.; Graciani, A.; Banegas, J.R.; Rodríguez-Artalejo, F. Sleep quality and the metabolic syndrome: The role of sleep duration and lifestyle. *Diabetes. Metab. Res. Rev.* **2014**, *30*, 222–231. [CrossRef]
13. Resta, O.; Foschino-Barbaro, M.P.; Legari, G.; Talamo, S.; Bonfitto, P.; Palumbo, A.; Minenna, A.; Giorgino, R.; De Pergola, G. Sleep-related breathing disorders, loud snoring and excessive daytime sleepiness in obese subjects. *Int. J. Obes.* **2001**, *25*, 669–675. [CrossRef]
14. Drager, L.F.; Lopes, H.F.; Maki-Nunes, C.; Trombetta, I.C.; Toschi-Dias, E.; Alves, M.J.N.N.; Fraga, R.F.; Jun, J.C.; Negrão, C.E.; Krieger, E.M.; et al. The Impact of Obstructive Sleep Apnea on Metabolic and Inflammatory Markers in Consecutive Patients with Metabolic Syndrome. *PLoS ONE* **2010**, *5*, e12065. [CrossRef]
15. Kline, C.E.; Crowley, E.P.; Ewing, G.B.; Burch, J.B.; Blair, S.N.; Durstine, J.L.; Davis, J.M.; Youngstedt, S.D. The Effect of Exercise Training on Obstructive Sleep Apnea and Sleep Quality: A Randomized Controlled Trial. *Sleep* **2011**, *34*, 1631–1640. [CrossRef]
16. Soler, X.; Liao, S.-Y.; Marin, J.M.; Lorenzi-Filho, G.; Jen, R.; DeYoung, P.; Owens, R.L.; Ries, A.L.; Malhotra, A. Age, gender, neck circumference, and Epworth sleepiness scale do not predict obstructive sleep apnea (OSA) in moderate to severe chronic obstructive pulmonary disease (COPD): The challenge to predict OSA in advanced COPD. *PLoS ONE* **2017**, *12*, e0177289. [CrossRef]
17. Qu, X.-X.; Esangbedo, I.C.; Zhang, X.-J.; Liu, S.-J.; Li, L.-X.; Gao, S.; Li, M. Obstructive Sleep Apnea Syndrome is Associated with Metabolic Syndrome among Adolescents and Youth in Beijing. *Chin. Med. J.* **2015**, *128*, 2278–2283. [CrossRef]
18. Hoffstein, V.; Mateika, S. Differences in abdominal and neck circumferences in patients with and without obstructive sleep apnoea. *Eur. Respir. J.* **1992**, *5*, 377–381.
19. Horner, R.L.; Shea, S.A.; McIvor, J.; Guz, A. Pharyngeal size and shape during wakefulness and sleep in patients with obstructive sleep apnoea. *QJM* **1989**, *72*, 719–735.
20. Paley, C.A.; Johnson, M.I. Abdominal obesity and metabolic syndrome: Exercise as medicine? *BMC Sports Sci. Med. Rehabil.* **2018**, *10*, 7. [CrossRef]
21. Figueroa, A.; Kalfon, R.; Madzima, T.A.; Wong, A. Whole-body vibration exercise training reduces arterial stiffness in postmenopausal women with prehypertension and hypertension. *Menopause* **2014**, *21*, 131–136. [CrossRef]
22. Wong, A.; Alvarez-Alvarado, S.; Kinsey, A.W.; Figueroa, A. Whole-Body Vibration Exercise Therapy Improves Cardiac Autonomic Function and Blood Pressure in Obese Pre- and Stage 1 Hypertensive Postmenopausal Women. *J. Altern. Complement. Med.* **2016**, *22*, 970–976. [CrossRef]
23. Anwer, S.; Alghadir, A.; Zafar, H.; Al-Eisa, E. Effect of whole body vibration training on quadriceps muscle strength in individuals with knee osteoarthritis: A systematic review and meta-analysis. *Physiotherapy* **2016**, *102*, 145–151. [CrossRef]
24. Alev, A.; Mihriban, A.; Bilge, E.; Ayça, E.; Merve, K.; Şeyma, C.; Uğur, E.; Adnan, B.; Zeynel, K.; Mahmut, G.S. Effects of whole body vibration therapy in pain, function and depression of the patients with fibromyalgia. *Complement. Ther. Clin. Pract.* **2017**, *28*, 200–203. [CrossRef]
25. Zhou, J.; Pang, L.; Chen, N.; Wang, Z.; Wang, C.; Hai, Y.; Lyu, M.; Lai, H.; Lin, F. Whole-body vibration training-better care for COPD patients: A systematic review and meta-analysis. *Int. J. COPD* **2018**, *13*, 3243–3254. [CrossRef]
26. Sá-Caputo, D.; Paineiras-Domingos, L.L.; Francisca-Santos, A.; dos Anjos, E.M.; Reis, A.S.; Neves, M.F.T.; Oigman, W.; Oliveira, R.; Brandão, A.; Machado, C.B.; et al. Whole-body vibration improves the functional parameters of individuals with metabolic syndrome: An exploratory study. *BMC Endocr. Disord.* **2019**, *19*, 6. [CrossRef]
27. Braz, D.S.; de Andrade, A.D.; Teixeira, A.S.; Cavalcanti, C.A.; Morais, A.B.; Marinho, P.E.M. Whole-body vibration improves functional capacity and quality of life in patients with severe chronic obstructive pulmonary disease (COPD): A pilot study. *Int. J. COPD* **2015**, *10*, 125–132.

28. Rønnestad, B.R. Comparing the Performance-Enhancing Effects of Squats on a Vibration Platform With Conventional Squats in Recreationally Resistance-Trained Men. *J. Strength Cond. Res.* **2004**, *18*, 839–845.
29. Merkert, J.; Butz, S.; Nieczaj, R.; Steinhagen-Thiessen, E.; Eckardt, R. Combined whole body vibration and balance training using Vibrosphere®. *Z. für Gerontol. und Geriatr.* **2011**, *44*, 256–261. [CrossRef]
30. Cardim, A.B.; Marinho, P.E.; Nascimento, J.F.; Fuzari, H.K.; de Andrade, A.D. Does Whole-Body Vibration Improve the Functional Exercise Capacity of Subjects With COPD? A Meta-Analysis. *Respir. Care* **2016**, *61*, 1552–1559. [CrossRef]
31. Sá-Caputo, D.C.; Paineiras-Domingos, L.L.; Oliveira, R.; Neves, M.F.T.; Brandão, A.; Marin, P.J.; Sañudo, B.; Furness, T.; Taiar, R.; Bernardo-Filho, M. Acute Effects of Whole-Body Vibration on the Pain Level, Flexibility, and Cardiovascular Responses in Individuals With Metabolic Syndrome. *Dose-Response* **2018**, *16*, 155932581880213. [CrossRef]
32. Paineiras-Domingos, L.L.; da Cunha Sá-Caputo, D.; Reis, A.S.; Francisca Santos, A.; Sousa-Gonçalves, C.R.; dos Anjos, E.M.; dos Santos Pereira, M.J.; Sartorio, A.; Bernardo-Filho, M. Assessment Through the Short Physical Performance Battery of the Functionality in Individuals With Metabolic Syndrome Exposed to Whole-Body Vibration Exercises. *Dose-Response* **2018**, *16*, 155932581879453. [CrossRef]
33. Moher, D.; Hopewell, S.; Schulz, K.F.; Montori, V.; Gøtzsche, P.C.; Devereaux, P.J.; Elbourne, D.; Egger, M.; Altman, D.G. CONSORT 2010 explanation and elaboration: Updated guidelines for reporting parallel group randomised trials. *Int. J. Surg.* **2012**, *10*, 28–55. [CrossRef]
34. Casadei, K.; Kiel., J. *Anthropometric Measurement*; StatPearls Publishing LLC: Treasure Island, FL, USA, 2019. Available online: https://www.ncbi.nlm.nih.gov/books/NBK537315/ (accessed on 24 October 2019).
35. Gasa, M.; López-Padrós, C.; Monasterio, C.; Salord, N.; Mayos, M.; Vilarrasa, N.; Fernandez-Aranda, F.; Montserrat, J.M.; Dorca, J.; López-Padrós, C.; et al. Anthropometrical phenotypes are important when explaining obstructive sleep apnea in female bariatric cohorts. *J. Sleep Res.* **2019**, *28*, e12830. [CrossRef]
36. De Cardiologia, S.B. III Diretrizes Brasileiras Sobre Dislipidemias e Diretriz de Prevenção da Aterosclerose do Departamento de Aterosclerose da Sociedade Brasileira de Cardiologia. *Arq. Bras. Cardiol.* **2001**, *77*, 48.
37. Kim, T.Y.; You, S.E.; Ko, Y.S. Association between Sasang constitutional types with obesity factors and sleep quality. *Integr. Med. Res.* **2018**, *7*, 341–350. [CrossRef]
38. Schwab, R.J.; Pasirstein, M.; Pierson, R.; Mackley, A.; Hachadoorian, R.; Arens, R.; Maislin, G.; Pack, A.I. Identification of Upper Airway Anatomic Risk Factors for Obstructive Sleep Apnea with Volumetric Magnetic Resonance Imaging. *Am. J. Respir. Crit. Care Med.* **2003**, *168*, 522–530. [CrossRef]
39. Duarte, R.L.; Mello, F.C.; Magalhães-da-Silveira, F.J.; Oliveira-e-Sá, T.S.; Rabahi, M.F.; Gozal, D. Comparative performance of screening instruments for obstructive sleep apnea in morbidly obese patients referred to a sleep laboratory: A prospective cross-sectional study. *Sleep Breath.* **2019**, *23*, 1123–1132. [CrossRef]
40. Netzer, N.C.; Stoohs, R.A.; Netzer, C.M.; Clark, K.; Strohl, K.P. Using the Berlin Questionnaire to Identify Patients at Risk for the Sleep Apnea Syndrome. *Ann. Intern. Med.* **1999**, *131*, 485. [CrossRef]
41. Ng, S.S.; Tam, W.; Chan, T.O.; To, K.W.; Ngai, J.; Chan, K.K.P.; Yip, W.H.; Lo, R.L.; Yiu, K.; Ko, F.W.; et al. Use of Berlin questionnaire in comparison to polysomnography and home sleep study in patients with obstructive sleep apnea. *Respir. Res.* **2019**, *20*, 40. [CrossRef]
42. Cizza, G.; de Jonge, L.; Piaggi, P.; Mattingly, M.; Zhao, X.; Lucassen, E.; Rother, K.I.; Sumner, A.E.; Csako, G. Neck Circumference Is a Predictor of Metabolic Syndrome and Obstructive Sleep Apnea in Short-Sleeping Obese Men and Women. *Metab. Syndr. Relat. Disord.* **2014**, *12*, 231–241. [CrossRef] [PubMed]
43. Team, R. R Development Core Team. *R A Lang. Environ. Stat. Comput.* **2013**, *3*, 201.
44. Yoshida, K.; Bohn, J.; Yoshida, M.K. *Package 'Tableone'*; R Foundation for Statistical Computing: Vienna, Austria, 2016.
45. Rao, D.P.; Orpana, H.; Krewski, D. Physical activity and non-movement behaviours: Their independent and combined associations with metabolic syndrome. *Int. J. Behav. Nutr. Phys. Act.* **2016**, *13*, 26. [CrossRef] [PubMed]
46. Feairheller, D.L.; Diaz, K.M.; Kashem, M.A.; Thakkar, S.R.; Veerabhadrappa, P.; Sturgeon, K.M.; Ling, C.; Williamson, S.T.; Kretzschmar, J.; Lee, H.; et al. Effects of Moderate Aerobic Exercise Training on Vascular Health and Blood Pressure in African Americans. *J. Clin. Hypertens.* **2014**, *16*, 504–510. [CrossRef]
47. Mahajan, A.; Mishra, T.; Singh, N.; Jain, A. Heart rate and blood pressure response to exercise and recovery in subclinical hypothyroid patients. *Int. J. Appl. Basic Med. Res.* **2013**, *3*, 106. [CrossRef]

48. Kang, S.J.; Ha, G.C.; Ko, K.J. Association between resting heart rate, metabolic syndrome and cardiorespiratory fitness in Korean male adults. *J. Exerc. Sci. Fit.* **2017**, *15*, 27–31. [CrossRef]
49. Sîrbu, E.; Buzaș, R.; Mihăescu, R.; Suceava, I.; Lighezan, D. Influence of exercise training and eating behavior on arterial stiffness in young healthy students. *Wien. Klin. Wochenschr.* **2015**, *127*, 555–560. [CrossRef]
50. King, N.A.; Hopkins, M.; Caudwell, P.; Stubbs, R.J.; Blundell, J.E. Beneficial effects of exercise: Shifting the focus from body weight to other markers of health. *Br. J. Sports Med.* **2009**, *43*, 924–927. [CrossRef]
51. Souza, A.K.F.; de Andrade, A.D.; de Medeiros, A.I.C.; de Aguiar, M.I.R.; de Souza Rocha, T.D.; Pedrosa, R.P.; de Lima, A.M.J. Effectiveness of inspiratory muscle training on sleep and functional capacity to exercise in obstructive sleep apnea: A randomized controlled trial. *Sleep Breath.* **2018**, *22*, 631–639. [CrossRef]
52. Reid, K.J.; Baron, K.G.; Lu, B.; Naylor, E.; Wolfe, L.; Zee, P.C. Aerobic exercise improves self-reported sleep and quality of life in older adults with insomnia. *Sleep Med.* **2010**, *11*, 934–940. [CrossRef]
53. Brandão, G.S.; Gomes, G.S.B.F.; Brandão, G.S.; Callou Sampaio, A.A.; Donner, C.F.; Oliveira, L.V.F.; Camelier, A.A. Home exercise improves the quality of sleep and daytime sleepiness of elderlies: A randomized controlled trial. *Multidiscip. Respir. Med.* **2018**, *13*, 2. [CrossRef] [PubMed]
54. Kline, C.E.; Ewing, G.B.; Burch, J.B.; Blair, S.N.; Durstine, J.L.; Davis, J.M.; Youngstedt, S.D. Exercise Training Improves Selected Aspects of Daytime Functioning in Adults with Obstructive Sleep Apnea. *J. Clin. Sleep Med.* **2012**, *8*, 357–365. [CrossRef] [PubMed]
55. Yilmaz Gokmen, G.; Akkoyunlu, M.E.; Kilic, L.; Algun, C. The Effect of T'ai Chi and Qigong Training on Patients with Obstructive Sleep Apnea: A Randomized Controlled Study. *J. Altern. Complement. Med.* **2019**, *25*, 317–325. [CrossRef] [PubMed]
56. Itoh, H.; Yokoyama, K.; Matsukawa, T.; Kitamura, F. Association between physical activity and sleep-disordered breathing in male Japanese workers: A cross-sectional study. *BMC Res. Notes* **2017**, *10*, 37. [CrossRef]
57. Dixit, S.; Maiya, A.; Shastry, B. Effect of aerobic exercise on quality of life in population with diabetic peripheral neuropathy in type 2 diabetes: A single blind, randomized controlled trial. *Qual. Life Res.* **2014**, *23*, 1629–1640. [CrossRef]
58. Farias, D.L.; Tibana, R.A.; Teixeira, T.G.; Vieira, D.C.L.; Tarja, V.; Nascimento, D.D.C.; Silva, A.D.O.; Funghetto, S.S.; Coura, M.A.D.S.; Valduga, R.; et al. Elderly women with metabolic syndrome present higher cardiovascular risk and lower relative muscle strength. *Einstein* **2013**, *11*, 174–179. [CrossRef]
59. World Health Organization. *GLOBAL STATUS REPORT on Noncommunicable Diseases 2014—"Attaining the Nine Global Noncommunicable Diseases Targets; A Shared Responsibility"*; WHO Libr. Cat. Data: Geneva, Switzerland, 2014; p. 298.
60. Callaghan, B.C.; Xia, R.; Reynolds, E.; Banerjee, M.; Rothberg, A.E.; Burant, C.F.; Villegas-Umana, E.; Pop-Busui, R.; Feldman, E.L. Association Between Metabolic Syndrome Components and Polyneuropathy in an Obese Population. *JAMA Neurol.* **2016**, *73*, 1468. [CrossRef]
61. Sossa, C.; Delisle, H.; Agueh, V.; Sodjinou, R.; Ntandou, G.; Makoutodé, M. Lifestyle and dietary factors associated with the evolution of cardiometabolic risk over four years in West-African adults: The Benin study. *J. Obes.* **2013**, *2013*, 298024. [CrossRef]

applied sciences

MDPI

Article

Evaluation of Whole-Body Vibration Exercise on Neuromuscular Activation Through Electromyographic Pattern of Vastus Lateralis Muscle and on Range of Motion of Knees in Metabolic Syndrome: A Quasi-Randomized Cross-Over Controlled Trial

Cintia Renata de Sousa-Gonçalves [1,2], Laisa Liane Paineiras-Domingos [1,2,3], Ygor Teixeira-Silva [1,2], Thais Amadeu [1], Adriana Pereira da Silva Lírio[2], Arlete Francisca-Santos [2], Luiz Felipe Ferreira De Souza [2], Mario José dos Santos Pereira [2], Maria Eduarda Souza Melo-Oliveira [2,4], Alexandre Gonçalves de Meirelles [2,4], Glória Maria Guimarães-Lourenço [2,4], Aline Reis-Silva [2,4], Eloá Moreira-Marconi [2,5], Marcia Cristina Moura-Fernandes [2,5], Vinicius Layter Xavier [6], Alessandra da Rocha Pinheiro Mulder [7], Ana Cristina Rodrigues Lacerda [8], Vanessa Amaral Mendonça [8], José Alexandre Bachur [9], Redha Taiar [10,*], Alessandro Sartorio [11], Danúbia da Cunha de Sá-Caputo [1,2,3] and Mario Bernardo-Filho [2]

1 Programa de Pós-Graduação em Ciências Médicas, Faculdade de Ciências Médicas, Universidade do Estado do Rio de Janeiro, Rio de Janeiro 20.551-030, Brazil; ci.renata@gmail.com (C.R.d.S.-G.); laisanit@gmail.com (L.L.P.-D.); silvarogy@hotmail.com (Y.T.-S.); tpamadeu@gmail.com (T.A.); dradanubia@gmail.com (D.d.C.d.S.-C.)

2 Laboratório de Vibrações Mecânicas e Práticas Integrativas, Departamento de Biofísica e Biometria, Instituto de Biologia Roberto Alcântara Gomes e Policlínica Américo Piquet Carneiro, Universidade do Estado do Rio de Janeiro, Rio de Janeiro 20950-003, Brazil; adrianaliriolavimpi@gmail.com (A.P.d.S.L.); fisioarlete@hotmail.com (A.F.-S.); lumadaragu@gmail.com (L.F.F.D.S.); mariojsp@gmail.com (M.J.d.S.P.); mariaeduardaoliveira@hotmail.com (M.E.S.M.-O.); meirelles.ale@gmail.com (A.G.d.M.); dra.glorialourenco@gmail.com (G.M.G.-L.); fisioalinereis@hotmail.com (A.R.-S.); eloamarconi@gmail.com (E.M.-M.); marciafernandesfisio@hotmail.com (M.C.M.-F.); bernardofilhom@gmail.com (M.B.-F.)

3 Faculdade Bezerra de Araújo, Rio de Janeiro 23052-180, Brazil

4 Programa de Pós-graduação em Saúde, Medicina Laboratorial e Tecnologia Forense, Universidade do Estado do Rio de Janeiro, Rio de Janeiro 20950-003, Brazil

5 Programa de Pós-Graduação em Fisiopatologia Clínica e Experimental, Instituto de Biologia Roberto Alcântara Gomes, Universidade do Estado do Rio de Janeiro, Rio de Janeiro 20.551-030, Brazil

6 Departamento de Estatística, Instituto de Matemática e Estatística, Universidade do Estado do Rio de Janeiro, Rio de Janeiro 20550-900, Brazil; viniciuslx@ime.uerj.br

7 Departamento de Nutrição, Instituto de Nutrição e Policlínica Américo Piquet Carneiro, Universidade do Estado do Rio de Janeiro, Rio de Janeiro 20550-900, Brazil; alessandra.mulder@gmail.com

8 Faculdade de Ciências Biológicas e da Saúde, Universidade Federal dos Vales do Jequitinhonha e Mucuri, Diamantina 39.100.000, Brazil; lacerdaacr@gmail.com (A.C.R.L.); vaafisio@hotmail.com (V.A.M.)

9 Centro Universitário de Franca, Franca 14401-426, SP, Brazil; jabachur@hotmail.com

10 GRESPI, Université de Reims, 51100 Reims, France

11 Istituto Auxologico Italiano, IRCCS, Experimental laboratory for Auxo-endocrinological Research, Division of Metabolic Diseases & Auxology, 28824 Verbania, Italy; sartorio@auxologico.it

* Correspondence: redha.taiar@univ-reims.fr; Tel.: +33-677-944628

Received: 16 October 2019; Accepted: 15 November 2019; Published: 20 November 2019

check for updates

Abstract: Metabolic syndrome (MetS) is related to overweight and obesity, and contributes to clinical limitations. Exercise is used for the management of MetS individuals, who are often not motivated to perform this practice. Whole body vibration exercise (WBVE) produces several biological effects, besides being safe, effective, and feasible for MetS individuals. This pseudo-randomized and cross-over controlled trial study aimed to analyze the effects of WBVE on MetS individuals' neuromuscular activation using the surface electromyography (sEMG) pattern (root mean square (RMS)) of the vastus lateralis (VL) muscle and on the range of motion (ROM) of the knees. Participants (n = 39) were allocated to two groups: the treatment group (TG), which was exposed to WBVE, and the control group (CG). WBVE interventions were performed twice a week, for a period of 5 weeks. ROM and sEMG were analyzed at baseline, after the first session, and before and after the last session. sEMG (%RMS) significantly increased in the acute effect of the last session of WBVE (108.00 ± 5.07, $p < 0.008$, right leg; 106.20 ± 3.53, $p < 0.02$, left leg) compared to the CG. ROM did not significantly change in TG or CG. In conclusion, 5 weeks of WBVE exerted neuromuscular effects capable of increasing VL muscle RMS in individuals with MetS, this effect being potentially useful in the physical rehabilitation of these individuals.

Keywords: metabolic syndrome; whole body vibration exercise; range of motion of the knees; surface electromyographic pattern; neuromuscular activation; biomechanics

---

1. Introduction

In the modern world, unhealthy lifestyles, including physical inactivity and bad dietary habits, contribute to the development and spread of diseases associated with metabolic commitment, such as metabolic syndrome (MetS). MetS is defined as a clustering of metabolic abnormalities, such as central obesity (increased waist circumference), dyslipidemia, hyperglycemia, and hypertension, according to the guidelines of the International Diabetes Federation (IDF) [1,2].

Yang et al. pointed out that overweight and obesity are emerging major health issues, which are closely related MetS [3]. Body mass excess can promote adipose tissue stored in the abdominal cavity and intramuscular adipose tissue (IMAT) [4,5]. It is known that these conditions directly affect the biomechanics of movements [6–8], leading to decreased mobility, strength, and ability to perform common activities of daily living [6,9,10]. In obese individuals, lower extremity overuse injury is caused by multiple mechanisms, including increased load bearing with ambulation, altered gait biomechanics due to abnormal body mass distribution, a systemic pro inflammatory state [7,11–13]. The affected joint kinetics may increase the risk of musculoskeletal injury, with the development of osteoarthritis, especially in the knees [7,14], and the impairment of the range of motion (ROM) [7,15]. Reduced ROM can imply the limitation of the knee flexion [15], which is required to perform activities of daily living [16]. ROM is a parameter of physical evaluation, since it can allow the identification of joint and muscle limitations. Moreover, it is used for the evaluation of individuals during the rehabilitation process [17].

In obese individuals, the lower limb muscles (flexors and extensors) are frequently committed [9,10]. In consequence, decrease of muscle strength and power, and premature fatigue [6] are observed, leading to impaired motor performance [18]. As these individuals are not motivated to practice regular exercise, methods to increase adherence including exercises that minimize the impact on the joints are necessary [19]. Among the different kinds of exercises, whole body vibration exercise (WBVE) may be an option for the management of obese individuals [20,21].

WBVE involves exposing individuals to mechanical vibrations that are produced in a vibrating platform (VP). These vibrations are transmitted to the body of the individual that is in contact with the base of the VP. Some parameters must be adjusted (frequency, peak-to-peak displacement, and peak acceleration) considering the clinical condition of the individual [22]. Similarly, if the person is sitting on an auxiliary conventional chair [23] or wheelchair [24] with the feet on the base of VP, the individual will also experience WBVE.

Authors evaluated the effects of WBVE on MetS individuals, and important responses were observed, such as improvement of quality of life and flexibility [25,26]. Moreover, WBVE is safe and effective to treat muscle complications in populations with different disorders [27–31] as well as healthy individuals [32,33].

Surface electromyography (sEMG) is an adequate tool to evaluate neuromuscular effects of WBVE [27,34–36]. An increased sEMG activity after WBVE indicates that more motor units are recruited [34].

It is well described that WBVE can improve joint mobility [29,37,38]. Moreover, Wang et al. found that WBVE in combination with quadriceps resistance exercise compared with quadriceps resistance exercise alone showed significantly greater improvement in knee flexion and extension in individuals with knee osteoarthritis [39].

The aim of this study was to analyze the effects of WBVE on neuromuscular activation through the electromyographic pattern of the vastus lateralis muscle and on range of motion of the knees in MetS individuals. The study hypothesis was that WBVE would increase neuromuscular activation and range of motion of the knees of MetS individuals.

2. Materials and Methods

This was a quasi-randomized cross-over controlled trial, where 39 individuals with MetS were selected to evaluate the effect of WBVE on neuromuscular activation and range of motion of the knees. Individuals were allocated, as they arrived, to the control (CG) or treatment group (TG). The study protocol was approved by the Research Ethics Committee of the Hospital Universitário Pedro Ernesto (HUPE), Universidade do Estado do Rio de Janeiro (UERJ), with the number CAAE 54981315.6.0000.5259, and was registered in the Registro Brasileiro de Ensaios clínicos (ReBEC) (RBR 2bghmh).

Recruitment took place from April 2014 to January 2016 following screening of MetS individuals by the medical staff at HUPE-UERJ, Brazil. The WBVE protocol was performed in the Laboratório de Vibrações Mecânicas e Práticas Integrativas (LAVIMPI), UERJ. Participants signed a consent form before any procedures, and the principles embodied in the Declaration of Helsinki were followed.

2.1. Sample Size

The sample size was determined based on a study by Sá-Caputo et al. [26], involving individuals with MetS exposed to WBVE, using an online calculator of the Laboratório de Epidemiologia e Estatística (LEE), Faculdade de Medicina, USP, São Paulo, Brazil [40]. For a statistical power of 95%, a sample size of 13 individuals was calculated [26,40]

2.2. Inclusion and Exclusion Criteria

For this study, the inclusion criteria were age over 40 years (male and female gender) and a diagnosis of MetS. All participants were selected by a clinical physician, who diagnosed MetS according to the International Diabetes Federation criteria [1].

The exclusion criteria comprised very high blood pressure (≥180/110 mmHg), cardiovascular disease (coronary artery disease or stroke), neurological, musculoskeletal, or rheumatologic disease that do not permit to be on the VP, refusal to sign the consent form for participation in the study, and fear of being on the VP due to its movements.

2.3. Interventions

Eligible participants (n = 39) were allocated to one of two groups: TG and CG. In the TG, the individuals performing the protocol with the VP turned on, while in the CG, the VP was turned off. After a 2-month washout (i.e., the period of time between the two interventions), the participants were crossed-over to the other group [41]. All procedures were performed by health professionals previously trained by experienced trainers to perform this protocol. Intra-rater reliability was not performed. The VP (Novaplate Fitness Evolution, DAF Produtos Hospitalares Ltda., São Paulo, Brazil) was a side alternating platform, in which the right site is displaced down while the left side is displaced up, and vice versa [26,42].

In the first session of the TG, all the individuals were seated on a chair placed in front of the platform with a 130° knee flexion. Their feet, shoeless, were placed on the base of the platform, alternately in three positions (peak to peak displacements of 2.5, 5.0, and 7.5 mm) and with a frequency of 5 Hz. The working time in each position was 1 minute followed by 1 minute of rest. This sequence was performed three times. This procedure is shown in Figure 1.

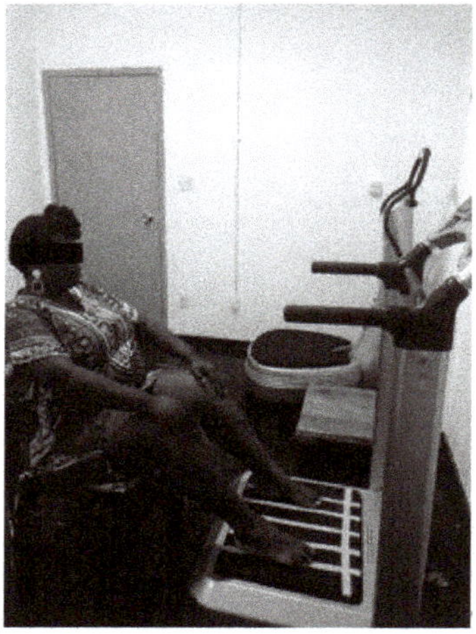

Figure 1. Individual performing whole-body vibration exercise seated on a chair placed in front of the vibrating platform.

From the second to the last session (10th session), the individuals were subjected to exactly the same protocol of the first session; however, they were standing on the base of the VP in a squat position (130° knee flexion), and the frequency was progressively increased by 1 unit per session up to 14 Hz in the last session. The WBVE session was performed twice per week during 5 weeks. This procedure is shown in Figure 2.

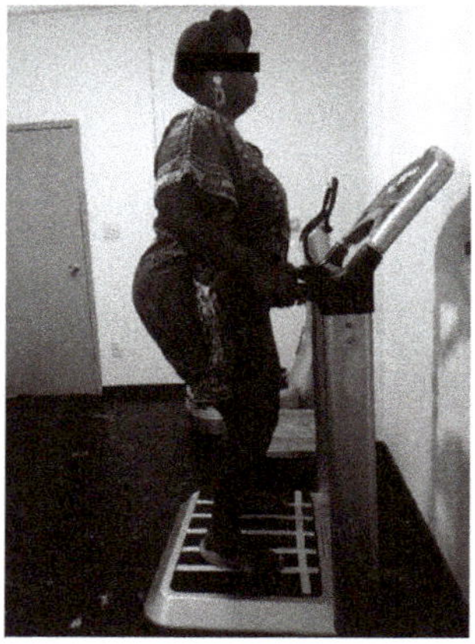

Figure 2. Individual performing WBVE in a standing position on the vibrating platform.

2.4. Outcome Measures

2.4.1. Anthropometric Evaluation

Anthropometric data (body mass, height) of the participants were assessed by a single operator using a balance with a stadiometer (Micheletti MIC 200PPA, São Paulo, Brazil). The waist circumference (WC) was measured with a tape in the horizontal plane, midway between the inferior margin of the ribs and the superior border of the iliac crest, in agreement with the IDF recommendations [1]. The body mass index (BMI) was calculated dividing body mass in kilograms by squared height in meters [43].

2.4.2. Evaluation of Knee Range of the Motion

A standard digital goniometer (EMG 830RF, EMG*System*®, São Paulo, Brazil) was used to measure the active ROM [44–46] of the knees during flexion from extension. The goniometer was properly fixed on the skin with tape on the lateral side of the knee joint, alternately right and left, to measure the flexion angle. Knee angles related to ROM were recorded on a computer. These procedures are shown in Figure 3.

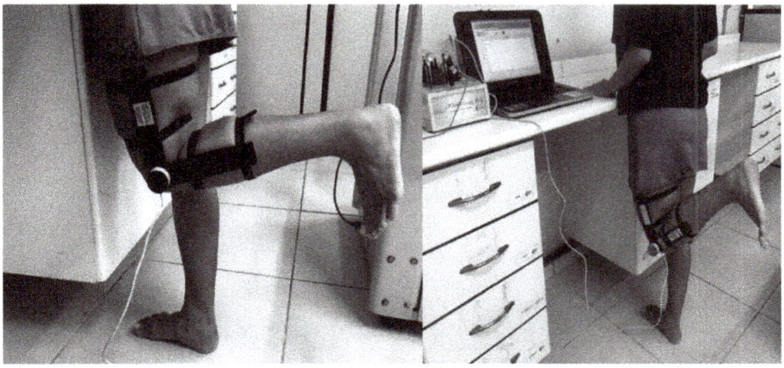

Figure 3. Position of the goniometer on the lower limb for measurement of knee range of motion (ROM).

ROM was determined before and after the first session (acute effect of the first session). The same procedure was performed in the last session (acute effect of the last session). In addition, comparisons were made before the first session and before the last session (cumulative effect) [47]. The percentage of the alteration of ROM (%AROM) in each condition was calculated dividing the ROM after the session by the ROM before the intervention (acute effect of the first and of the last session) multiplied by 100. The %ROM to the cumulative effect in each condition was calculated by dividing the value of ROM before the last session by the ROM before the first session multiplied by 100.

2.5. *Surface Electromyography (sEMG) Instrumentation and Measurement*

The neuromuscular activity of the vastus lateralis (VL) muscle was evaluated following the recommendations of the Surface ElectroMyoGraphy for the Non-Invasive Assessment of Muscles (SENIAM) regarding skin preparation and fixation of the electrodes in the specified positions related to the VL muscle [48], and the reference electrode was positioned on the spinous process of C7. The sEMG signal was collected in microvolts (μV) in a computer and was analyzed using the root mean square (RMS) (EMG832WF, EMG*System*®, São Paulo, Brazil). In the sEMG assessment, the individuals were asked to sit on a chair with the back straight, feet approximately shoulder-width apart and placed on the floor, and the arms crossed over the chest. Then, they were instructed to perform five repetitions of sitting and standing [49].

The RMS amplitude signal was recorded before and after the first session (acute effect of the first session), and before and after the last session (acute effect of the last session). The %RMS in each condition was calculated dividing the RMS after the session by the RMS before the session (acute effect of the first and of the last session) multiplied by 100. The %RMS to the cumulative effect was calculated in each condition by dividing the value of the RMS before the last session by the RMS before the first session multiplied by 100.

2.6. Statistical Analysis

Statistical analyses were performed using GraphPad Prism 6. The normality of the data was evaluated with the Shapiro-Wilk test. For the comparison of paired nonparametric values, the Wilcoxon Signed-Rank test was used. For the comparison of the two groups (TG and CG), the Mann Whitney test was used. The comparison of the cumulative effect of ROM between the groups was performed using the difference (Δ) between the value before the first session and that after the last session. The difference was considered significant when $p \leq 0.05$. The descriptive statistics used were mean and standard error, since the focus of the work was the effect of the intervention.

3. Results

The anthropometric data of the participants are reported in Table 1. There were no significant differences between the two groups ($p > 0.05$). According to IDF [2], participants with increase in WC values have more probability to develop MetS.

Table 1. Anthropometrics data of the individuals of the study.

Anthropometrics	CG (n = 17) Mean ± SE	TG (n = 22) Mean ± SE	P
Height (cm)	1.62 ± 0.01	1.63 ± 0.01	0.58
Body mass (kg)	88.8 ± 4.08	83.1 ± 3.65	0.46
BMI (kg/m^2)	33.5 ± 1.56	31.3 ± 1.12	0.25
Age (years)	58.1 ± 2.07	60.7 ± 1.91	0.40
WC (cm)	108 ± 3.58	103 ± 2.57	0.11

CG: control Group; TG: treatment group; SE: standard error; BMI: body mass index; WC: waist circumference.

The acute effects of WBVE on the ROM of the right and left knees in the first session are shown in Table 2. No significant changes in ROM were observed in either of the groups (CG and TG) or in the comparison between them. Although no significant differences ($p > 0.05$) were detected, the %AROM of the right knee of the participants in the CG decreased slightly (85.65%), while that of the left knee increased slightly (107.21%). The %AROM of the right and left knees of the participants in the TG decreased slightly (98.09% and 98.10%, respectively).

Table 2. Acute effects on range of the motion of right and left knees in the first session.

Knee	CG			TG			CG × TG
	Before Mean ± SE (Degrees)	After Mean ± SE (Degrees)	P	Before Mean ± SE (Degrees)	After Mean ± SE (Degrees)	P	P
Right knee	94.13 ± 6.89	80.63 ± 9.50	0.29	99.47 ± 7.12	97.58 ± 6.86	0.62	0.59
Left knee	94.11 ± 6.91	100.90 ± 2.74	0.25	99.37 ± 7.16	97.49 ± 6.88	0.62	0.90

CG: control group; TG: treatment group; SE: standard error.

Table 3 shows the acute effects of WBVE on the ROM of the right and left knees in the last session. No significant changes were observed in either of groups (CG and TG) or in the comparison between CG and TG. Although no significant differences ($p > 0.05$) were detected, the %AROM of the right knee and of the left knee of the participants in the CG increased slightly (102.17% and 103.96%, respectively).

The %AROM of the right knee of the participants of the TG decreased slightly (98.67%), and that of left knee increased (102.04%).

Table 3. Acute effects on range of the motion of right and left knees in the last session.

Knee	CG			TG			CG × TG
	Before Mean ± SE (Degree)	After Mean ± SE (Degree)	P	Before Mean ± SE (Degree)	After Mean ± SE (Degree)	P	P
Right knee	101.20 ± 3.30	103.40 ± 3.49	0.49	105.50 ± 3.40	104.10 ± 3.55	0.15	0.10
Left knee	100.90 ± 2.53	104.90 ± 3.76	0.12	102.90 ± 3.22	105.00 ± 2.72	0.10	0.43

CG: control group; TG: treatment group; SE: standard error.

The cumulative effects of WBVE on the ROM of both knees (i.e., value of ROM before the first session vs. the value before the last session) is reported in Table 4. Although no significant differences were detected in the TG or CG, an improvement of 7.5 and 13.51 degrees in ROM was observed in the TG on the left and right knee, respectively. Moreover, the %AROM of the right knee of the participants in the CG decreased slightly (98.53%), and that of the left knee increased (105.79%). The %AROM of the right and left knees of the participants in the TG increased importantly (114.29% and 107.44%, respectively).

Table 4. Cumulative effects on range of the motion of right and left knees before the first session vs. before the last session.

Knee	CG			TG			CG × TG
	Before Mean ± SE (Degree)	After Mean ± SE (Degree)	P	Before Mean ± SE (Degree)	After Mean ± SE (Degree)	P	P
Right knee	102.70 ± 3.01	101.20 ± 3.30	0.62	94.49 ± 4.43	108.00 ± 3.39	0.09	0.07
Left knee	95.37 ± 4.28	100.90 ± 2.53	0.64	100.80 ± 2.64	108.30 ± 2.59	0.07	0.27

CG: control group; TG: treatment group; SE: standard error.

Using sEMG, Table 5 shows the response of the vastus lateralis muscle of both legs to the acute effect of WBVE in the first session, in %RMS. No significant effects were found (left or right legs) in TG or CG.

Table 5. Acute effects on electromyography (%RMS) of vastus lateralis muscles (right and left) in the first session.

Muscle	CG	TG	P
	Mean ± SE %RMS	Mean ± SE %RMS	
VL right	93.59 ± 5.50	102.10 ± 3.58	0.28
VL left	98.65 ± 3.21	94.91 ± 3.66	0.99

CG: control group; TG: treatment group; SE: standard error; %RMS: percentage of the root mean square; VL: vastus lateralis muscle.

The acute effects of WBVE on sEMG of the vastus lateralis muscles in the last session are shown in Table 6. Significant increases ($p < 0.05$) were observed on the %RMS of vastus lateralis muscles on both legs of the TG relative to the CG.

Table 6. Acute effects on electromyography (%RMS) of vastus lateralis muscles (right and left) in the last session.

Muscle	CG	TG	P
	Mean ± SE % RMS	Mean ± SE % RMS	
VL right	95.22 ± 3.18	108.00 ± 5.07	0.008 *
VL left	96.57 ± 3.69	106.20 ± 3.53	0.02 *

CG: control group; TG: treatment group; SE: standard error; %RMS: percentage of the root mean square; VL: vastus lateralis muscle; $p \leq 0.05$ *.

The cumulative effects on sEMG (i.e., %RMS of the vastus lateralis muscle of right and left legs measured before the first session and before the last session) are shown in Table 7. The vastus lateralis muscle activity was not altered in either leg for either of the groups.

Table 7. Cumulative effects on electromyography (%RMS) of vastus lateralis muscle (right and left).

Muscle	CG	TG	P
	Mean ± SE % RMS	Mean ± SE % RMS	
VL right	99.38 ± 5.76	105.70 ± 5.40	0.63
VL left	97.10 ± 4.03	110.60 ± 7.87	0.26

CG: control group; TG: treatment group; SE: standard error; %RMS: percentage of the root mean square; VL: vastus lateralis muscle.

4. Discussion

All the individuals recruited in the current study presented an increased waist circumference (Table 1); according to the IDF, this is the first criteria followed by a set of two or more metabolic abnormalities to characterize MetS [1].

There is an association between the negative effect of body adiposity and muscle function, with more severe impairment in metabolic disease with IMAT accumulation impacting physical performance [5]. Insulin resistance and type 2 diabetes mellitus are highly prevalent in individuals with MetS [1], and are associated with an infiltration of adipose tissue in skeletal muscles [50]. These features lead to a reduction in the power output and strength per unit of muscle mass. Hilton et al. [51] showed that marked IMAT was inversely correlated with muscle power, strength, and physical performance scores. Lafortuna et al. also reported a negative effect of metabolic abnormalities related to impaired glucose homeostasis on motor performance [52]. IMAT is now recognized as an important predictor of muscle metabolism and function, and also appears to be a modifiable muscle risk factor. Exercise and physical activity appear to be effective countermeasures against increases in IMAT [4]. In addition, exercise is also reported to be an important intervention to improve the physical condition of MetS individuals [26]. Moreover, a simple and useful kind of exercise generated in individuals exposed to mechanical vibration, WBVE, could prove useful for these individuals.

4.1. *Effect of Whole Body Vibration Exercise (WBVE) on Range of Motion (ROM)*

In our study, there was no improvement in knee ROM in individuals with MetS exposed to WBVE. No increase in ROM was observed in a single session or even in a 5-weeks treatment. Our results are in agreement with the findings of Neto et al. (2017), who did not find an increase in ROM in knee flexion in individuals with knee osteoarthritis (KOA) [47]. Wang et al., comparing quadriceps resistance exercise alone and associated with WBVE, observed a significantly greater improvement in active knee flexion and extension at 2 and 4 weeks after quadriceps resistance exercise combined with WBVE in individuals with KOA [53]. Authors have reported that higher body fat is also associated with poor physical performance and subsequent disability than is muscle mass reduction [4,54]. This could justify the results of the current study. In addition, obesity can modify the movements and force response of these individuals. Del Porto et al. reported that some adaptations of shifts in the body's total ROM are difficult to adjust without compromising normal patterns of movement in obese individuals [55]. Putting together these considerations, it is possible to justify no improvements in the ROM of the knees in the population evaluated in our work. By contrast, Yang et al., in a longitudinal study of 8 weeks with WBVE in individuals with multiple sclerosis, observed a significant increase in the active ROM assessed at the ankle joint bilaterally, in plantar flexion ($p = 0.002$) and dorsiflexion ($p = 0.003$). Yang et al. concluded that WBVE increased the flexibility of the ankle joint, reduced the fear of falling, and strengthened the bones [38]. Although Krause et al. observed no changes in the ankle joint excursion immediately after a 1-min bout of WBVE, they found an increase in the knee joint active angular excursion during flexion and extension ($p < 0.01$) in individuals with cerebral palsy [29]. In another study performed in children with cerebral palsy, a significant increase in ROM was detected immediately after treatment with WBVE [44]. These conflicting results about the effect of WBVE on the ROM of joints could be associated with the variability of the protocols and the populations evaluated.

4.2. *Effect of WBVE on sEMG*

In the present study, significant changes in the electromyographic signal of the VL muscle on both legs were observed in the acute effect of the last session of the 5-week WBVE intervention (Table 6). This is in agreement with other authors that have shown improvement in muscle activity due to WBVE intervention in different muscles in individuals with various diseases [27], such as acute and chronic stroke [30,56], spinal cord injury [24], Friedreich's ataxia [57], and cerebral palsy [29]. Liao et al. detected changes in muscle activity even in individuals with paresis [30]. Furthermore, increased muscular activity was also observed in healthy individuals [34,58]. Borges et al. found that WBVE increased the sEMG amplitude of the VL muscle during an isometric semi-squat exercise in active woman [34]. A systematic review has analyzed the effects of WBVE in lower limbs' neuromuscular activity, suggesting that there was a greater recruitment of motor units after WBVE [27].

In the current study, no effects were observed in %RMS in the acute effect of the first session or in the cumulative effect (Tables 5 and 7). Considering the acute effect, Annino et al. (2017) also reported no alteration in the sEMG of the VL muscle when healthy individuals were exposed to 10 minutes of WBV at 35 Hz [59]. Borges et al. (2016) evaluated the immediate effects of WBV in healthy individuals with frequencies of 30 and 50 Hz, and no significant differences in the value of RMS were found [60]. Considering the cumulative effect (Table 7), the present study found no significant changes in the electromyographic signal of the VL muscle in either leg. Rubio-Arias et al. (2018) also observed no statistically significant differences in body composition or muscle architecture variables, nor changes in muscle activity during the take-off phase of the vastus lateralis pre- versus post-training in a 6-week WBVE treatment with a frequency between 30 and 45 Hz in healthy individuals [61].

Some limitations of this study should be pointed out, such as the relatively small number of participants; however, the statistical power was 95 %. Moreover, although we evaluated both lower limbs, we did not register which lower limb was the dominant one. Nevertheless, a strength of this work is the improvement in neuromuscular activity; this finding may have positive clinical implications by improving daily activities in individuals with MetS.

In conclusion, WBVE can be a modality of exercise to increase the neuromuscular activity of the VL muscle using a 5-week protocol. An increase in ROM of the knees in individuals with MetS was not observed with the same protocol. WBVE appears to be an adequate strategy to improve neuromuscular activity in individuals with MetS, overweight, and obesity, being a potential opportunity for the management of physical impairment in these individuals. Nevertheless, further additional studies with larger samples and more prolonged periods of WBVE exposure are needed to confirm our preliminary findings.

Author Contributions: Conceptualization, M.B.-F.; Data curation, C.R.d.S.-G., D.d.C.d.S.-C., L.L.P.-D., E.M.-M., M.E.S.M.-O., Y.T.-S., A.P.d.S.L., J.A.B., M.J.d.S.P., V.L.X., A.G.d.M., L.F.F.D.S., A.R.-S., A.F.-S., G.M.G.-L., M.C.M.-F., A.d.R.P.M., T.A., A.C.R.L., V.A.M., A.S. and M.B.-F.; Formal analysis, M.J.d.S.P., V.L.X., and R.T.; Funding acquisition, M.B.-F.; Investigation, C.R.d.S.-G., D.d.C.d.S.-C., L.L.P.-D., E.M.-M., M.E.S.M.-O., Y.T.-S., A.P.d.S.L., A.G.d.M., L.F.F.D.S., A.R.-S., A.F.-S., G.M.G.-L., M.C.M.-F. and A.d.R.P.M.; Methodology, C.R.d.S.-G., D.d.C.d.S.-C., L.L.P.-D., E.M.-M., M.E.S.M.-O., Y.T.-S., A.P.d.S.L., A.G.d.M., L.F.F.D.S., A.R.-S., A.F.-S., G.M.G.-L., M.C.M.-F. and A.d.R.P.M.; Supervision, D.d.C.d.S.-C.; Writing—original draft, C.R.d.S.-G. and Y.T.-S.; Writing—review & editing, D.d.C.d.S.-C., J.A.B., R.T., T.A., A.C.R.L., V.A.M., A.S. and M.B.-F.

Funding: This research was funded by Coordenação de Aperfeiçoamento de Pessoal de Nível Superior (CAPES)—Finance Code 001, and the Conselho Nacional de Desenvolvimento Científico e Tecnológico (CNPq).

Acknowledgments: The authors are grateful for the support of the Fundação Carlos Chagas Filho de Amparo à Pesquisa do Estado do Rio de Janeiro (FAPERJ), the Coordenação de Aperfeiçoamento de Pessoal de Nível Superior (CAPES)—Finance Code 001, and the Conselho Nacional de Desenvolvimento Científico e Tecnológico (CNPq).

Conflicts of Interest: The authors declare that there is no conflict of interests regarding the publication of this paper.

Abbreviations

BMI	body mass index
CG	control group
HUPE	Hospital Universitário Pedro Ernesto
IDF	International Diabetes Federation
IMAT	intramuscular adipose tissue
KOA	knee osteoarthritis
LAVIMPI	Laboratório de Vibrações Mecânicas e Práticas Integrativas
LEE	Laboratório de Epidemiologia e Estatística
MetS	Metabolic syndrome
ReBEC	Registro Brasileiro de Ensaios clínicos
RMS	root mean square
ROM	range of motion
sEMG	surface electromyography
SENIAM	Surface ElectroMyoGraphy for the Non-Invasive Assessment of Muscles
TG	treatment group
UERJ	Universidade do Estado do Rio de Janeiro
VL	vastus lateralis
VP	vibrating platform
WBVE	Whole body vibration exercise
WC	waist circumference
μV	microvolts

References

1. Alberti, K.G.M.M.; Eckel, R.H.; Grundy, S.M.; Zimmet, P.Z.; Cleeman, J.I.; Donato, K.A.; Fruchart, J.C.; James, W.P.T.; Loria, C.M.; Smith, S.C. Harmonizing the metabolic syndrome: A joint interim statement of the international diabetes federation task force on epidemiology and prevention; National heart, lung, and blood institute; American heart association; World heart federation; International Atherosclerosis Society; and International Association for the Study of Obesity. *Circulation* **2009**, *120*, 1640–1645. [PubMed]
2. Worachartcheewan, A.; Schaduangrat, N.; Prachayasittikul, V.; Nantasenamat, C. Data mining for the identification of metabolic syndrome status. *EXCLI J.* **2018**, *17*, 72–88. [PubMed]
3. Yang, C.S.; Wang, H.; Sheridan, Z.P. Studies on prevention of obesity, metabolic syndrome, diabetes, cardiovascular diseases and cancer by tea. *J. Food Drug Anal.* **2018**, *26*, 1–13. [CrossRef] [PubMed]
4. Addison, O.; Marcus, R.L.; Lastayo, P.C.; Ryan, A.S. Intermuscular fat: A review of the consequences and causes. *Int. J. Endocrinol.* **2014**, *2014*, 309570. [CrossRef]
5. Tuttle, L.J.; Sinacore, D.R.; Mueller, M.J. Intermuscular adipose tissue is muscle specific and associated with poor functional performance. *J. Aging Res.* **2012**, *2012*, 172957. [CrossRef]
6. Bollinger, L.M. Potential contributions of skeletal muscle contractile dysfunction to altered biomechanics in obesity. *Gait Posture* **2017**, *56*, 100–107. [CrossRef]

7. Capodaglio, P.; Castelnuovo, G.; Brunani, A.; Vismara, L.; Villa, V.; Maria Capodaglio, E. Functional limitations and occupational issues in obesity: A review. *Int. J. Occup. Saf. Ergon.* **2010**, *16*, 507–523. [CrossRef]
8. Ghroubi, S.; Kossemtini, W.; Mahersi, S.; Elleuch, W.; Chaabene, M.; Elleuch, M.H. Contribution of isokinetic muscle strengthening in the rehabilitation of obese subjects. *Ann. Phys. Rehabil. Med.* **2016**, *59*, 87–93. [CrossRef]
9. Fabris, S.M.; Faintuch, J.; Brienze, S.L.A.; Brito, G.B.; Sitta, I.S.; Mendes, E.L.P.; Fonseca, I.C.B.; Cecconello, I. Are knee and foot orthopedic problems more disabling in the superobese? *Obes. Surg.* **2013**, *23*, 201–204. [CrossRef]
10. Gadducci, A.; de Cleva, R.; Santarém, G.; Silva, P.; Greve, J.; Santo, M. Muscle strength and body composition in severe obesity. *Clinics* **2017**, *72*, 272–275. [CrossRef]
11. Abella, V.; Scotece, M.; Conde, J.; López, V.; Lazzaro, V.; Pino, J.; Gómez-Reino, J.J.; Gualillo, O. Adipokines, metabolic syndrome and rheumatic diseases. *J. Immunol. Res.* **2014**, *25*, 343746. [CrossRef] [PubMed]
12. Nye, N.S.; Kafer, D.S.; Olsen, C.; Carnahan, D.H.; Crawford, P.F. Abdominal Circumference Versus Body Mass Index as Predictors of Lower Extremity Overuse Injury Risk. *J. Phys. Act. Health* **2018**, *15*, 127–134. [CrossRef] [PubMed]
13. Wearing, S.C.; Hennig, E.M.; Byrne, N.M.; Steele, J.R.; Hills, A.P. Musculoskeletal disorders associated with obesity: A biomechanical perspective. *Obes. Rev.* **2006**, *7*, 239–250. [CrossRef] [PubMed]
14. Sowers, M.R.; Karvonen-Gutierrez, C.A. The evolving role of obesity in knee osteoarthritis. *Curr. Opin. Rheumatol.* **2010**, *22*, 533–537. [CrossRef]
15. Holla, J.F.M.; Steultjens, M.P.M.; van der Leeden, M.; Roorda, L.D.; Bierma-Zeinstra, S.M.A.; den Broeder, A.A.; Dekker, J. Determinants of range of joint motion in patients with early symptomatic osteoarthritis of the hip and/or knee: An exploratory study in the CHECK cohort. *Osteoarthr. Cart.* **2011**, *19*, 411–419. [CrossRef]
16. Rowe, P.J.; Myles, C.M.; Walker, C.; Nutton, R. Knee joint kinematics in gait and other functional activities measured using flexible electrogoniometry: How much knee motion is sufficient for normal daily life? *Gait Posture* **2000**, *12*, 143–155. [CrossRef]
17. Clarkson, H. *Musculoskeletal Assessment: Joint Range of Motion and Manual Muscle Strength*, 2th ed.; Lippincott Wiliams and Wilkin: Philadelphia, PA, USA, 2002.
18. Pajoutan, M.; Ghesmaty Sangachin, M.; Cavuoto, L.A. Central and peripheral fatigue development in the shoulder muscle with obesity during an isometric endurance task. *BMC Musculoskelet Disord* **2017**, *18*, 314. [CrossRef]
19. Zdziarski, L.A.; Wasser, J.G.; Vincent, H.K. Chronic pain management in the obese patient: A focused review of key challenges and potential exercise solutions. *J. Pain Res.* **2015**, *8*, 63–77.
20. Alvarez-Alvarado, S.; Jaime, S.J.; Ormsbee, M.J.; Campbell, J.C.; Post, J.; Pacilio, J.; Figueroa, A. Benefits of whole-body vibration training on arterial function and muscle strength in young overweight/obese women. *Hypertens. Res.* **2017**, *40*, 487–492. [CrossRef]
21. Patel, V.S.; Ete Chan, M.; Rubin, J.; Rubin, C.T. Marrow Adiposity and Hematopoiesis in Aging and Obesity: Exercise as an Intervention. *Curr. Osteoporos. Rep.* **2018**, *16*, 105–115. [CrossRef]
22. Rauch, F.; Sievanen, H.; Boonen, S.; Cardinale, M.; Degens, H.; Felsenberg, D.; Roth, J.; Schoenau, E.; Verschueren, S.; Rittweger, J. Reporting whole-body vibration intervention studies: Recommendations of the International Society of Musculoskeletal and Neuronal Interactions. *J. Musculoskelet. Neuronal Interact.* **2010**, *10*, 193–198. [PubMed]
23. Pinto, N.S.; Monteiro, M.B.; Arthur, A.P.; Paiva, D.N.; Meyer, P.F.; Santos-Filho, S.D.; Marín, P.J.; Bernardo-Filho, M. Effectiveness of a protocol involving acute whole-body vibration exercises in an adult and health individual with delayed-onset muscle soreness observed after running: A case repor. *J. Med Med. Sci.* **2011**, *2*, 612–617.
24. Menéndez, H.; Ferrero, C.; Martín-Hernández, J.; Figueroa, A.; Marín, P.J.; Herrero, A.J. Acute effects of simultaneous electromyostimulation and vibration on leg blood flow in spinal cord injury. *Spinal Cord* **2016**, *54*, 383–389. [CrossRef] [PubMed]
25. Carvalho-Lima, R.P.; Sá-Caputo, D.C.; Moreira-Marconi, E.; Dionello, C.; Paineiras-Domingos, L.L.; Sousa-Gonçalves, C.R.; Morel, D.S.; Frederico, E.H.; Neves, M.F.; Oliveira, R.; et al. Quality of life of patients with metabolic syndrome is improved after Whole Body Vibration Exercises. *Afr. J. Tradit. Complement. Altern. Med.* **2017**, *14*, 59–65. [CrossRef]

26. Sá-Caputo, D.d.C.; Ronikeili-Costa, P.; Carvalho-Lima, R.P.; Bernardo, L.C.; Bravo-Monteiro, M.O.; Costa, R.; Moraes-Silva, J.d.; Paiva, D.N.; Machado, C.B.; Mantilla-Giehl, P.; et al. Whole Body Vibration Exercises and the Improvement of the Flexibility in Patient with Metabolic Syndrome. *Rehabil. Res. Prac.* **2014**, *2014*, 1–10. [CrossRef]
27. Dionello, C.F.; De Souza, P.L.; Sá-Caputo, D.; Morel, D.S.; Moreira-Marconi, E.; Paineiras-Domingos, L.L.; Frederico, E.H.F.F.; Guedes-Aguiar, E.; Paiva, P.D.C.; Taiar, R.; et al. Do whole body vibration exercises affect lower limbs neuromuscular activity in populations with a medical condition? *Restor. Neurol. Neurosc.* **2017**, *35*, 667–681. [CrossRef]
28. Herrero, A.J.; Martín, J.; Martín, T.; García-López, D.; Garatachea, N.; Jiménez, B.; Marín, P.J. Whole-body vibration alters blood flow velocity and neuromuscular activity in Friedreich's ataxia. *Clin. Physiol. Funct. Imaging* **2011**, *31*, 139–144. [CrossRef]
29. Krause, A.; Schoenau, E.; Gollhofer, A.; Duran, I.; Ferrari-Malik, A.; Freyler, K.; Ritzmann, R. Alleviation of Motor Impairments in Patients with Cerebral Palsy: Acute Effects of Whole-body Vibration on Stretch Reflex Response, Voluntary Muscle Activation and Mobility. *Front. Neurol.* **2017**, *8*, 1–11. [CrossRef]
30. Liao, L.-R.; Ng, G.Y.F.; Jones, A.Y.M.; Huang, M.-Z.; Pang, M.Y.C. Whole-Body Vibration Intensities in Chronic Stroke: A Randomized Controlled Trial. *Med. Sci. Sports Exerc.* **2016**, *48*, 1227–1238. [CrossRef]
31. Van Ruymbeke, B.; Boone, J.; Coorevits, P.; Vanderstraeten, G.; Bourgois, J. Whole-body vibration in breast cancer survivors: A pilot study exploring its effects on muscle activity and subjectively perceived exertion. *Int. J. Rehabil. Res.* **2014**, *37*, 371–374. [CrossRef]
32. Perchthaler, D.; Horstmann, T.; Grau, S. Variations in neuromuscular activity of thigh muscles during whole-body vibration in consideration of different biomechanical variables. *J. Sci. Med. Sport* **2013**, *12*, 439–446.
33. Fattorini, L.; Tirabasso, A.; Lunghi, A.; Di Giovanni, R.; Sacco, F.; Marchetti, E. Muscular forearm activation in hand-grip tasks with superimposition of mechanical vibrations. *J. Electromyogr. Kinesiol.* **2016**, *26*, 143–148. [CrossRef] [PubMed]
34. Borges, D.T.; Macedo, L.B.; Lins, C.A.A.; Sousa, C.O.; Brasileiro, J.S. Effects of Whole Body Vibration on the Neuromuscular Amplitude of Vastus Lateralis Muscle. *J. Sci. Med. Sport* **2017**, *16*, 414–420.
35. Mosier, E.M.; Herda, T.J.; Trevino, M.A.; Miller, J.D. The influence of prolonged vibration on motor unit behavior. *Muscle Nerve* **2017**, *55*, 500–507. [CrossRef]
36. Simsek, D. Different fatigue-resistant leg muscles and EMG response during whole-body vibration. *J. Electromyogr. Kinesiol.* **2017**, *37*, 147–154. [CrossRef]
37. Cheng, H.-Y.K.; Ju, Y.-Y.; Chen, C.-L.; Chuang, L.-L.; Cheng, C.-H. Effects of whole body vibration on spasticity and lower extremity function in children with cerebral palsy. *Hum. Mov. Sci.* **2015**, *39*, 65–72. [CrossRef]
38. Yang, F.; Finlayson, M.; Bethoux, F.; Su, X.; Dillon, L.; Maldonado, H.M. Effects of controlled whole-body vibration training in improving fall risk factors among individuals with multiple sclerosis: A pilot study. *Disabil. Rehabil.* **2018**, *40*, 553–560. [CrossRef]
39. Wang, P.; Yang, L.; Li, H.; Lei, Z.; Yang, X.; Liu, C.; Jiang, H.; Zhang, L.; Zhou, Z.; Reinhardt, J.D.; et al. Effects of whole-body vibration training with quadriceps strengthening exercise on functioning and gait parameters in patients with medial compartment knee osteoarthritis: A randomised controlled preliminary study. *Physiotherapy (UK)* **2016**, *102*, 86–92. [CrossRef]
40. LEE—Laboratório de Epidemiologia e Estatística. Available online: http://www.lee.dante.br/pesquisa.html (accessed on 10 October 2018).
41. O'Keefe, K.; Orr, R.; Huang, P.; Selvadurai, H.; Cooper, P.; Munns, C.F.; Singh, M.A.F. The effect of whole body vibration exposure on muscle function in children with cystic fibrosis: A pilot efficacy trial. *J. Clin. Med. Res.* **2013**, *5*, 205–216. [CrossRef]
42. Cardinale, M.; Wakeling, J. Whole body vibration exercise: Are vibrations good for you? *J. Sports Med.* **2005**, *39*, 585–589. [CrossRef]
43. WHO. Physical status: The use and interpretation of anthropometry. Report of a WHO Expert Committee. *WHO Tech. Rep. Ser.* **1995**, *854*, 1–452.
44. Dos Santos, R.A.; Derhon, V.; Brandalize, M.; Brandalize, D.; Rossi, L.P. Evaluation of knee range of motion: Correlation between measurements using a universal goniometer and a smartphone goniometric application. *J. Bodyw. Mov. Ther.* **2017**, *21*, 699–703. [CrossRef] [PubMed]

45. Milanese, S.; Gordon, S.; Buettner, P.; Flavell, C.; Ruston, S.; Coe, D.; O'Sullivan, W.; McCormack, S. Reliability and concurrent validity of knee angle measurement: Smart phone app versus universal goniometer used by experienced and novice clinicians. *Man. Ther.* **2014**, *19*, 569–574. [CrossRef] [PubMed]
46. Marks, R. Knee Osteoarthritis and Exercise Adherence: A Review. *Curr. Aging Sci.* **2012**, *5*, 72–83. [CrossRef] [PubMed]
47. Neto, S.B.S.; Marconi, E.M.; Kutter, C.R.; Frederico, E.H.F.F.F.; De Paiva, P.d.C.; Meyer, P.F.; Chang, S.; Sá-Caputo, D.; Bernardo-Filho, M. Beneficial effects of whole body mechanical vibration alone or combined with auriculotherapy in the pain and in flexion of knee of individuals with knee osteoarthritis. *Acupunct. Electro Ther. Res.* **2017**, *42*, 185–201. [CrossRef]
48. Hermens, H.J.; Freriks, B.; Disselhorst-Klug, C.; Rau, G. Development of recommendations for SEMG sensors and sensor placement procedures. *J. Electromyogr. Kinesiol.* **2000**, *10*, 361–374. [CrossRef]
49. Lee, T.-H.; Choi, J.-D.; Lee, N.-G. Activation timing patterns of the abdominal and leg muscles during the sit-to-stand movement in individuals with chronic hemiparetic stroke. *J. Phys. Ther. Sci.* **2015**, *27*, 3593–3595. [CrossRef]
50. Tomlinson, D.J.; Erskine, R.M.; Winwood, K.; Morse, C.I.; Onambélé, G.L. The impact of obesity on skeletal muscle architecture in untrained young vs. old women. *J. Anat.* **2014**, *225*, 675–684. [CrossRef]
51. Hilton, T.N.; Tuttle, L.J.; Bohnert, K.L.; Mueller, M.J.; Sinacore, D.R. Excessive Adipose Tissue Infiltration in Skeletal Muscle in Individuals With Obesity, Diabetes Mellitus, and Peripheral Neuropathy: Association With Performance and Function. *Phys. Ther.* **2008**, *88*, 1336–1344. [CrossRef]
52. Lafortuna, C.L.; Prinelli, F.; Adorni, F.; Agosti, F.; De Col, A.; Sartorio, A. Effect of mechanical and metabolic factors on motor function and fatigue in obese men and women: A cross-sectional study. *J. Endocrinol. Investig.* **2013**, *36*, 1062–1068.
53. Wang, P.; Yang, L.; Liu, C.; Wei, X.; Yang, X.; Zhou, Y.; Jiang, H.; Lei, Z.; Reinhardt, J.D.; He, C. Effects of Whole Body Vibration Exercise associated with Quadriceps Resistance Exercise on functioning and quality of life in patients with knee osteoarthritis: A randomized controlled trial. *Clin. Rehabil.* **2016**, *30*, 1074–1087. [CrossRef] [PubMed]
54. Rolland, Y.; Lauwers-Cances, V.; Cristini, C.; van Kan, G.A.; Janssen, I.; Morley, J.E.; Vellas, B. Difficulties with physical function associated with obesity, sarcopenia, and sarcopenic-obesity in community-dwelling elderly women: The EPIDOS (EPIDemiologie de l'OSteoporose) Study 1–3. *Am. J. Clin. Nutr.* **2009**, *89*, 1895–1900. [CrossRef] [PubMed]
55. Porto, H.C.D.; Pechak, C.M.; Smith, D.R.; Reed-Jones, R.J. Biomechanical Effects of Obesity on Balance. *Int. J. Exerc. Sci.* **2012**, *5*, 301–320.
56. Tihanyi, T.K.; Horváth, M.; Fazekas, G.; Hortobágyi, T.; Tihanyi, J. One session of whole body vibration increases voluntary muscle strength transiently in patients with stroke. *Clin. Rehabil.* **2007**, *21*, 782–793. [CrossRef]
57. Herrero, A.J.; Menéndez, H.; Gil, L.; Martín, J.; Martín, T.; García-López, D.; Gil-Agudo, Á.; Marín, P.J. Effects of whole-body vibration on blood flow and neuromuscular activity in spinal cord injury. *Spinal Cord* **2011**, *49*, 554–559. [CrossRef]
58. Avelar, N.C.P.; Ribeiro, V.G.C.; Mezêncio, B.; Fonseca, S.F.; Tossige-Gomes, R.; da Costa, S.J.; Szmuchrowski, L.; Gripp, F.; Coimbra, C.C.; Lacerda, A.C.R. Influence of the knee flexion on muscle activation and transmissibility during whole body vibration. *J. Electromyogr. Kinesiol.* **2013**, *23*, 844–850. [CrossRef]
59. Annino, G.; Iellamo, F.; Palazzo, F.; Fusco, A.; Lombardo, M.; Campoli, F.; Padua, E. Acute changes in neuromuscular activity in vertical jump and flexibility after exposure to whole body vibration. *Medicine (US)* **2017**, *96*, e7629. [CrossRef]
60. Borges, D.T.; Macedo, L.B.; Lins, C.A.A.; Brasileiro, J.S. Immediate effects of whole-body vibration on neuromuscular performance of quadriceps and oscillation of the center of pressure: A randomized controlled trial. *Man. Ther.* **2016**, *25*, 62–68. [CrossRef]
61. Rubio-Arias, J.Á.; Ramos-Campo, D.J.; Esteban, P.; Martínez, F.; Jiménez, J.F. Effect of 6-weeks WBVT on the behaviour of the lower limb muscle fibres during vertical jumping. *J. Sports Sci.* **2018**, *36*, 398–406. [CrossRef]

applied sciences

MDPI

Article

Test-Retest Reliability of Kinematic Parameters of Timed Up and Go in People with Type 2 Diabetes

Francisco J. Dominguez-Muñoz [1], Miguel A. Hernández-Mocholi [1], Luis J. Manso [2], Daniel Collado-Mateo [3,*], Santos Villafaina [1,*], Jose C. Adsuar [4] and Narcis Gusi [1,5]

1 Physical Activity and Quality of Life Research Group (AFYCAV), Faculty of Sport Science, University of Extremadura, 10003 Cáceres, Spain; fjdominguez@unex.es (F.J.D.-M.); mhmocholi@unex.es (M.A.H.-M.); ngusi@unex.es (N.G.)

2 School of Engineering and Applied Sciences, Aston University, Birmingham B4 7ET, UK; l.manso@aston.ac.uk

3 Centre for Sport Studies, Rey Juan Carlos University, 28943 Fuenlabrada, Spain

4 Health Economy Motricity and Education (HEME), Faculty of Sport Science, University of Extremadura, 10003 Cáceres, Spain; jadssal@unex.es

5 CIBER de Fragilidad y Envejecimiento Saludable, 28029 Madrid, Spain

* Correspondence: danicolladom@gmail.com (D.C.-M.); svillafaina@unex.es (S.V.); Tel : +34-9-2725-7460

Received: 13 September 2019; Accepted: 1 November 2019; Published: 5 November 2019

Abstract: Diabetes mellitus is a chronic disease defined as a state of hyperglycaemia in fasting or postprandial states. Patients with type 2 diabetes mellitus (T2DM) often show reduced physical function, including low levels of strength, balance or mobility. In this regard, the timed up and go (TUG) is a widely used physical fitness test in people with T2DM. However, there is a lack of studies evaluating the properties TUG in this population. The present study aimed to evaluate the test-retest reliability of kinetic and kinematic parameters obtained from TUG in the diabetic population with different levels of diabetic neuropathy. A total of 56 patients with T2DM participated in the study. They were divided into three groups according to the vibration threshold: (a) severe neuropathy, (b) moderate neuropathy and (c) normal perception. The TUG was performed using two force platforms to assess kinematic measurements. The results show that both kinetic and kinematic variables had good to excellent reliability. The reliability of TUG was excellent for the whole sample and the groups with non-severe neuropathy. However, it was just good for the group with severe neuropathy.

Keywords: forefoot; Gait; Heel; TUG; Type 2 diabetes mellitus

1. Introduction

Diabetes mellitus (DM) is a chronic disease defined as a state of hyperglycaemia in fasting or postprandial states [1]. DM is one of the largest global public health problems and affects approximately 415 million people in the world among adults aged 20–79 years-old. It is estimated that, in the year 2040, there will be 642 million persons (confidence interval 521–829 million) with DM in the world [2]. The International Diabetes Federation also estimates that, globally, 46.5% of people suffering from diabetes were still undiagnosed in 2015, which may markedly increase the reported prevalence. According with the American Diabetes Association (ADA), the total costs for diabetes care in the United States were approximately 245 billion dollars due to medical costs, lost productivity and disability [3]. Much of the burden of this disease comes from vascular complications, which include cardiovascular disease, retinopathy and nephropathy. Another complication is diabetic peripheral neuropathy, affecting more than 50% of long-term diabetic cases [4].

Diabetic neuropathy is characterised by progressive degeneration that primarily affects small-diameter cutaneous nociceptive fibres [5]. It may also affect motor fibres, which can cause muscular weakness. In this regard, persons with DM have a reduction of 17% and 14% in the strength of the flexor and extensor muscles of the knee, respectively [6]. Somatosensory feedback is a relevant factor to maintain balance and there is strong evidence showing that diabetic neuropathy affects this source of information, leading to alterations in postural and gait performance [7]. In this regard, previous studies have demonstrated that the main sources of deterioration in the balance in persons with type 2 DM (T2DM) are deficits in the proprioception of the foot and the ankle [8], or a loss of sensitivity in the feet [9].

The neurologic exam of the lower limb is the most important aspect in the clinical diagnosis of diabetic neuropathy [10]. The loss of foot vibration perception is associated with an increased risk of foot ulceration in people with diabetes [11]. In this regard, Abbott et al. [12] showed that each one-unit increment in the foot vibration threshold increases the risk of foot ulceration by more than 5% in a single year period. Therefore, the foot vibration threshold is a very relevant variable in the diabetic population.

Previous studies have demonstrated a relationship between the foot vibration threshold and the risk of falling [13], gait speed [14] and mobility disability [15]. Although balance and mobility tests are of great interest in DM studies due to the association with the risk of falling and the ability to perform activities of daily living, a recent systematic review showed that there is a lack of studies evaluating the properties of these tests when they are conducted in the diabetic population [16].

Therefore, the main objective of this study was to evaluate the test-retest reliability of the kinetic and kinematic parameters obtained from one of the most widely used tests for assessing balance and mobility, the timed up and go (TUG), in people suffering from T2DM. The second objective was to calculate the reliability of TUG according to the severity of peripheral neuropathy (assessed through an evaluation of the foot vibration threshold).

2. Materials and Methods

2.1. Participants

A total of 56 patients with DM participated in the study. Of these, 40 were men and 16 were women. The following inclusion criteria were: (a) diagnosed with T2DM, (b) affected by at least one risk factor of diabetic neuropathy: (1) being overweight, (2) a former smoker, (3) diagnosed with diabetic nephropathy and (4) diagnosed with diabetic retinopathy, (c) levels of glycated haemoglobin higher than 5.7%, and (d) have read and signed the written informed consent. In addition to these inclusion criteria, some exclusion criteria were defined: (a) pregnancy, (b) the use of psychotropic or chemotherapeutic medications, (c) affected by other diseases that may influence balance and gait, such as Parkinson's disease, and (d) patients with a high risk of non-diabetic neuropathy (HIV or uraemia). The protocol of the present study was approved by the Committee of Bioethics of the University and was developed in accordance with the updated Helsinki Declaration and the national legislation on bioethics, biomedical research and personal data confidentiality.

2.2. Procedure

After reading and signing the written informed consent, participants were measured and weighed. They were also asked about their age and years since diagnosis. Then, the vibration threshold was evaluated and, finally, the TUG was conducted after a light warm-up.

The foot vibration threshold was assessed using a Biothensiometer Vibratron II (Physitemp Instruments, Inc. Clifton; New Jersey; USA). This device drives vibration to modules A and B placed under the feet of the participant. Each module has a vibrating pole on the top, which vibrates at a frequency of 120 Hz. Therefore, the vibration threshold is determined by modifying the amplitude. These vibration units are related to the amplitude of the movement in microns according to the formula:

$A = x^2/2$ (where x is the vibration units (vu) and A is the amplitude in microns (μm)). The present study used the protocol the 'force two alternative choices procedure', which is one of the two methods suggested by the manufacturer. Participants were asked to place their first toe on the vibrating pole. The procedure started when the participant felt the vibration in the left or right toe. After that, the amplitude was reduced progressively until the participant was not able to tell which pole was vibrating. When the participant failed to detect vibration, the amplitude of the vibration was increased. The vibration threshold was then calculated using the last five rights and wrongs, but omitting the extreme low and high values. The average of the remaining eight values was then computed to calculate the vibration threshold [17].

TUG was performed three times, with a 5-min rest in between. The first repetition was for familiarisation, the second was the test measure and the third was the retest measure. All participants performed a light warm-up which included walking and joint mobility for 5 min. In the TUG, two force platforms (Kistler, NY, USA) were placed between the chair and the mark where participants had to turn around. Therefore, participants stepped on the platforms before and after reaching the mark placed at 3 m. The time required to complete the full test was assessed manually with a stopwatch by an expert rater.

Variables obtained from the force platforms included the duration of (a) the double support phase (both feet on the platforms), (b) left support (only the left foot on the platforms) and (c) right support (only the right foot on the platforms), as well as the left and right peak forces from the forefoot and the heel.

The results are presented for the whole sample (n = 56), and also according to the degree of neuropathy based on vibration perception, i.e., severe neuropathy (n = 22), moderate neuropathy (n = 22) and normal perception (n = 12). To classify patients into one of the three groups, the cut-off points suggested by normative values of the manual of the measuring device were considered.

2.3. Statistical Analysis

Descriptive statistics included mean and standard deviation (SD) of age, weight, glycated haemoglobin, years since T2DM diagnosis, body mass index and vibration threshold were calculated for the whole sample and divided into women or men. Parametric and non-parametric tests were conducted based on the results of Shapiro-Wilk and Kolmogorov-Smirnov tests.

Differences between test and retest were evaluated using the paired samples t-test or Wilcoxon signed rank test when appropriate. The time spent to complete the TUG, the duration of phases and the forces registered by the force platforms were included in those analyses to compare the test and retest results.

Reliability analyses were conducted in accordance with the recommendations of Weir [18]. An intraclass correlation coefficient (ICC) of 3.1 (two-way mixed, single measures) with a 95% CI for test and retest [19] was selected. Both absolute and relative reliability were computed. The standard error of measurement (SEM) was calculated as $SEM = SD\ \sqrt{1-ICC}$ where SD is the mean SD of the three repetitions, while the smallest real difference (SRD) was $SRD = 1.96 \times SEM \times \sqrt{2}$. These measures were converted into percentages (%SEM and %SRD, respectively) to enable comparisons with other investigations.

3. Results

3.1 Participant Characteristics

Table 1 summarises the participant characteristics. A total of 40 men and 16 women aged 64.52 (8.41) and 67.43 (8.86), respectively, participated in the present study. The mean weight of the whole sample was 80.08 (17.59) kg and the body mass index (BMI) was 28.81 (4.40) kg/m^2. The mean vibration threshold was 4.27 (1.90) for women and 5.75 (2.50) for men.

Table 1. Participant characteristics.

	All Participants (n = 56)	Men (n = 40)	Women (n = 16)
Age (years)	65.35 (8.56)	64.52 (8.41)	67.43 (8.86)
Weight (kg)	80.08 (17.59)	85.26 (18.03)	67.11 (6.14)
Glycated haemoglobin (%)	6.66 (0.91)	6.67 (0.90)	6.63 (0.94)
Years since diagnosis	9.10 (7.43)	9.07 (6.52)	9.18 (9.59)
BMI (kg/m^2)	28.81 (4.40)	29.24 (4.84)	27.73 (2.85
Vibration threshold (vu)	5.33 (2.42)	5.75 (2.50)	4.27 (1.90)

BMI: body mass index; vu: vibration units.

3.2. Kinematic Variables of the Timed Up and Go

Table 2 summarises the mean duration of the walking phases in the TUG. The results from the paired samples t-test or Wilcoxon signed rank test showed that there were significant differences between the test and retest values in some of the variables, including the time required to complete the TUG and the duration of the left support before reaching the mark for the whole sample. These differences were also observed in the group with severe neuropathy.

Table 2. Differences between test and retest on the duration of the walking phases in the entire sample and in participants with (a) severe neuropathy, (b) moderate neuropathy and (c) normal perception.

	Test Measurement	Test	Retest	p-Value [a]
GENERAL (n = 56)	TUG (S)	8.10 (1.37)	7.80 (1.28)	<0.001
	Double support phase duration (before reaching the mark)	0.09 (0.02)	0.09 (0.02)	0.341 [a]
	Double support phase duration (after reaching the mark)	0.10 (0.02)	0.10 (0.02)	0.483 [a]
	Right support duration (before reaching the mark)	0.62 (0.09)	0.60 (0.09)	0.047 [a]
	Left support duration (before reaching the mark)	0.62 (0.09)	0.60 (0.08)	0.005 [a]
	Right support duration (after reaching the mark)	0.62 (0.08)	0.61 (0.08)	0.261
	Left support duration (after reaching the mark)	0.63 (0.09)	0.63 (0.10)	0.399 [a]
SEVERE (n = 22)	TUG (S)	8.36 (1.40)	7.94 (1.10)	0.007
	Double support phase duration (before reaching the mark)	0.09 (0.02)	0.09 (0.02)	0.463 [a]
	Double support phase duration (after reaching the mark)	0.10 (0.02)	0.10 (0.02)	0.637 [a]
	Right support duration (before reaching the mark)	0.62 (0.09)	0.60 (0.08)	0.117 [a]
	Left support duration (before reaching the mark)	0.63 (0.08)	0.60 (0.06)	0.020 [a]
	Right support duration (after reaching the mark)	0.62 (0.06)	0.62 (0.06)	0.610
	Left support duration (after reaching the mark)	0.62 (0.06)	0.63 (0.08)	0.986 [a]
MODERATE (n = 22)	TUG (S)	7.92 (1.47)	7.69 (1.54)	0.016
	Double support phase duration (before reaching the mark)	0.09 (0.03)	0.09 (0.02)	0.930 [a]
	Double support phase duration (after reaching the mark)	0.10 (0.02)	0.10 (0.03)	0.974 [a]
	Right support duration (before reaching the mark)	0.61 (0.10)	0.60 (0.11)	0.443 [a]
	Left support duration (before reaching the mark)	0.60 (0.11)	0.60 (0.11)	0.750 [a]
	Right support duration (after reaching the mark)	0.61 (0.10)	0.61 (0.10)	0.697
	Left support duration (after reaching the mark)	0.64 (0.12)	0.64 (0.12)	0.650 [a]
NORMAL (n = 12)	TUG (S)	7.94 (1.16)	7.74 (1.15)	0.095
	Double support phase duration (before reaching the mark)	0.09 (0.01)	0.08 (0.01)	0.260 [a]
	Double support phase duration (after reaching the mark)	0.10 (0.01)	0.09 (0.02)	0.164 [a]
	Right support duration (before reaching the mark)	0.63 (0.09)	0.61 (0.07)	0.153 [a]
	Left support duration (before reaching the mark)	0.62 (0.06)	0.59 (0.06)	0.025 [a]
	Right support duration (after reaching the mark)	0.61 (0.05)	0.60 (0.06)	0.235
	Left support duration (after reaching the mark)	0.63 (0.08)	0.62 (0.09)	0.312 [a]

[a] p-values obtained from the Wilcoxon signed rank test.

Reliability parameters for the total time required to complete the TUG and duration of phases can be observed in Table 3. The ICC was good (0.70 to 0.90) or excellent (>0.90) for almost every variable and group. The best reliability was obtained for the time required to complete the TUG, which was excellent in the two groups with non-severe neuropathy, while it was just good in the group with severe neuropathy.

Table 3. Test-retest analyses of the duration of the walking phases in the entire sample and in participants with (a) severe neuropathy, (b) moderate neuropathy and (c) normal perception.

	Test Measurement	ICC (95% CI)	SEM (s)	SEM (%)	SRD (s)	SRD (%)
GENERAL (n = 56)	TUG (S)	0.927 (0.878–0.956)	0.36	4.52	0.99	12.55
	Double support phase duration (going)	0.760 (0.623–0.852)	0.01	14.07	0.03	39.02
	Double support phase duration (return)	0.801 (0.683–0.878)	0.01	10.73	0.03	29.74
	Right support duration (before reaching the mark)	0.865 (0.781–0.919)	0.03	5.74	0.09	15.93
	Left support duration (before reaching the mark)	0.845 (0.749–0.906)	0.03	5.80	0.09	16.08
	Right support duration (after reaching the mark)	0.876 (0.798–0.925)	0.02	4.71	0.08	13.05
	Left support duration (after reaching the mark)	0.860 (0.772–0.915)	0.03	5.77	0.10	16.01
SEVERE (n= 22)	TUG (S)	0.870 (0.714–0.944)	0.45	5.54	1.25	15.37
	Double support phase duration (going)	0.855 (0.684–0.937)	0.01	10.75	0.02	29.81
	Double support phase duration (return)	0.853 (0.678–0.936)	0.01	7.60	0.02	21.07
	Right support duration (before reaching the mark)	0.782 (0.544–0.903)	0.04	6.58	0.11	18.25
	Left support duration (before reaching the mark)	0.730 (0.454–0.879)	0.04	6.63	0.11	18.38
	Right support duration (after reaching the mark)	0.853 (0.678–0.936)	0.02	4.07	0.07	11.30
	Left support duration (after reaching the mark)	0.727 (0.448–0.877)	0.03	6.01	0.10	16.66
MODERATE (n = 22)	TUG (S)	0.963 (914–0.985)	0.29	3.727	0.80	10.31
	Double support phase duration (going)	0.723 (0.441–0.875)	0.01	18.06	0.04	50.07
	Double support phase duration (return)	0.861 (0.695–0.940)	0.01	10.83	0.03	30.03
	Right support duration (before reaching the mark)	0.914 (0.805–0.964)	0.03	5.37	0.09	14.90
	Left support duration (before reaching the mark)	0.923 (0.824–0.967)	0.03	5.11	0.08	14.17
	Right support duration (after reaching the mark)	0.903 (0.781–0.959)	0.03	5.43	0.09	15.06
	Left support duration (after reaching the mark)	0.912 (0.800–0.962)	0.03	5.71	0.10	15.83
NORMAL (n = 12)	TUG (S)	0.938 (0.800–0.982)	0.28	3.67	0.79	10.17
	Double support phase duration (going)	0.565 (0.018–0.852)	0.01	11.93	0.02	33.08
	Double support phase duration (return)	0.511 (0.059–0.829)	0.01	15.04	0.04	41.69
	Right support duration (before reaching the mark)	0.880 (0.636–0.964)	0.03	4.82	0.08	13.38
	Left support duration (before reaching the mark)	0.881 (0.639–0.964)	0.02	3.70	0.06	10.25
	Right support duration (after reaching the mark)	0.774 (0.387–0.929)	0.02	4.64	0.07	12.88
	Left support duration (after reaching the mark)	0.843 (0.543–0.952)	0.03	5.68	0.09	15.76

3.3. Peak Forces from the Heel and the Forefoot in the Timed Up and Go

Regarding the peak forces from the heel and the forefoot, differences between test and retest were only observed for the right heel and left forefoot forces in the group without neuropathy. For the rest of the variables, the retest was not significantly different than the test (see Table 4).

Table 4. Differences between the test and retest on the peak forces from the heel and the forefoot forces in the entire sample and in participants with (a) severe neuropathy, (b) moderate neuropathy and (c) normal perception.

	Test Measurement in Newtons	Test	Retest	p-Value
GENERAL (n = 56)	Right forefoot forces (before reaching the mark)	846.91 (185.68)	841.62 (192.78)	0.525 [a]
	Left forefoot forces (before reaching the mark)	827.33 (195.94)	828.06 (187.55)	0.562 [a]
	Right forefoot forces (after reaching the mark)	823.69 (181.65)	819.28 (174.34)	0.707 [a]
	Left forefoot forces (after reaching the mark)	773.50 (179.97)	767.44 (179.11)	0.458 [a]
	Right heel forces (before reaching the mark)	881.64 (201.23)	902.94 (209.41)	0.058 [a]
	Left heel forces (before reaching the mark)	934.64 (217.76)	944.97 (217.63)	0.098 [a]
	Right heel forces (after reaching the mark)	938.52 (226.32)	940.87 (211.95)	0.579 [a]
	Left heel forces (after reaching the mark)	937.35 (210.51)	943.70 (214.06)	0.458 [a]
SEVERE (n= 22)	Right forefoot forces (before reaching the mark)	868.77 (198.23)	867.59 (215.78)	0.833 [a]
	Left forefoot forces (before reaching the mark)	879.37 (208.31)	857.73 (204.77)	0.291 [a]
	Right forefoot forces (after reaching the mark)	855.99 (204.95)	849.56 (204.22)	0.615 [a]
	Left forefoot forces (after reaching the mark)	821.39 (199.56)	820.55 (194.15)	0.783 [a]
	Right heel forces (before reaching the mark)	937.91 (209.96)	958.33 (237.64)	0.485 [a]
	Left heel forces (before reaching the mark)	994.51 (229.16)	1004.19 (232.83)	0.338 [a]
	Right heel forces (after reaching the mark)	991.13 (263.60)	973.35 (237.58)	0.548 [a]
	Left heel forces (after reaching the mark)	983.31 (226.01)	979.73 (219.78)	0.783 [a]

Table 4. *Cont.*

	Test Measurement in Newtons	Test	Retest	*p*-Value
MODERATE (n = 22)	Right forefoot forces (before reaching the mark)	846.14 (212.30)	827.06 (207.90)	0.178 [a]
	Left forefoot forces (before reaching the mark)	809.70 (215.55)	819.77 (205.52)	0.638 [a]
	Right forefoot forces (after reaching the mark)	820.03 (187.24)	809.08 (174.89)	0.527 [a]
	Left forefoot forces (after reaching the mark)	747.40 (193.17)	741.54 (185.78)	0.200 [a]
	Right heel forces (before reaching the mark)	860.34 (204.46)	877.55 (202.41)	0.408 [a]
	Left heel forces (before reaching the mark)	919.46 (221.58)	934.12 (221.63)	0.168 [a]
	Right heel forces (after reaching the mark)	935.64 (208.33)	938.26 (211.80)	0.961 [a]
	Left heel forces (after reaching the mark)	929.66 (206.01)	947.18 (225.00)	0.200 [a]
NORMAL (n = 12)	Right forefoot forces (before reaching the mark)	808.24 (93.51)	820.69 (110.33)	0.136 [a]
	Left forefoot forces (before reaching the mark)	764.25 (100.23)	788.90 (107.78)	0.006 [a]
	Right forefoot forces (after reaching the mark)	771.20 (112.84)	782.47 (102.16)	0.388 [a]
	Left forefoot forces (after reaching the mark)	733.56 (83.56)	717.58 (113.36)	0.754 [a]
	Right heel forces (before reaching the mark)	817.53 (163.63)	847.94 (150.04)	0.099 [a]
	Left heel forces (before reaching the mark)	852.69 (167.82)	856.27 (153.98)	0.695 [a]
	Right heel forces (after reaching the mark)	847.34 (161.18)	886.09 (160.69)	0.034 [a]
	Left heel forces (after reaching the mark)	867.18 (182.55)	871.26 (178.58)	0.754 [a]

[a] *p*-values obtained from the Wilcoxon signed rank test.

Table 5 summarises the reliability analyses of the kinetic variables for the whole sample and according to the neuropathy classification. Reliability was excellent (ICC > 0.90) in almost every variable and group, except for the left forefoot forces after reaching the mark in the group with normal vibration perception.

Table 5. Test-retest analyses on the peak forces from the heel and the forefoot forces in the entire sample and in participants with (a) severe neuropathy, (b) moderate neuropathy and (c) normal perception.

	Test Measurement in Newtons	ICC (95% CI)	SEM (N)	SEM (%)	SRD (N)	SRD (%)
GENERAL (n = 56)	Right forefoot forces (before reaching the mark)	0.937 (0.894–0.962)	47.49	5.62	131.65	15.59
	Left forefoot forces (before reaching the mark)	0.944 (0.906–0.967)	45.37	5.48	125.77	15.19
	Right forefoot forces (after reaching the mark)	0.955 (0.925–0.974)	37.75	4.59	104.66	12.74
	Left forefoot forces (after reaching the mark)	0.942 (0.903–0.966)	43.23	5.61	119.85	15.55
	Right heel forces (before reaching the mark)	0.946 (0.910–0.968)	47.71	5.34	132.25	14.82
	Left heel forces (before reaching the mark)	0.941 (0.901–0.965)	52.87	5.62	146.57	15.59
	Right heel forces (after reaching the mark)	0.935 (0.892–0.961)	55.86	5.94	154.86	16.48
	Left heel forces (after reaching the mark)	0.955 (0.925–0.974)	45.03	4.78	124.82	13.27
SEVERE (n= 22)	Right forefoot forces (before reaching the mark)	0.952 (0.888–0.980)	45.35	5.22	125.71	14.48
	Left forefoot forces (before reaching the mark)	0.937 (0.854–0.973)	51.84	5.96	143.69	16.54
	Right forefoot forces (after reaching the mark)	0.968 (0.925–0.987)	36.59	4.29	101.44	11.89
	Left forefoot forces (after reaching the mark)	0.973 (0.936–0.989)	32.34	3.94	89.66	10.92
	Right heel forces (before reaching the mark)	0.947 (0.878–0.978)	51.52	5.43	142.81	15.06
	Left heel forces (before reaching the mark)	0.959 (0.903–0.983)	46.77	4.68	129.64	12.97
	Right heel forces (after reaching the mark)	0.933 (0.846–0.972)	64.86	6.60	179.79	18.30
	Left heel forces (after reaching the mark)	0.959 (0.904–0.983)	45.13	4.59	125.10	12.74
MODERATE- (n = 22)	Right forefoot forces (before reaching the mark)	0.925 (0.828–0.968)	57.53	6.87	159.49	19.06
	Left forefoot forces (before reaching the mark)	0.953 (0.890–0.980)	45.64	5.60	126.51	15.52
	Right forefoot forces (after reaching the mark)	0.944 (0.871–0.976)	42.84	5.26	118.77	14.58
	Left forefoot forces (after reaching the mark)	0.939 (0.858–0.974)	46.79	6.28	129.71	17.42
	Right heel forces (before reaching the mark)	0.947 (0.878–0.978)	46.83	5.38	129.81	14.93
	Left heel forces (before reaching the mark)	0.913 (0.803–0.963)	65.36	7.05	181.18	19.54
	Right heel forces (after reaching the mark)	0.945 (0.873–0.977)	49.26	5.25	136.55	14.57
	Left heel forces (after reaching the mark)	0.960 (0.906–0.983)	43.10	4.59	119.47	12.73
NORMAL (n = 12)	Right forefoot forces (before reaching the mark)	0.931 (0.778–0.980)	26.77	3.28	74.20	9.11
	Left forefoot forces (before reaching the mark)	0.976 (0.920–0.993)	16.11	2.07	44.66	5.75
	Right forefoot forces (after reaching the mark)	0.926 (0.765–0.978)	29.24	3.76	81.06	10.43
	Left forefoot forces (after reaching the mark)	0.697 (0.234–0.902)	54.20	7.47	150.24	20.70
	Right heel forces (before reaching the mark)	0.915 (0.732–0.975)	45.72	5.49	126.74	15.22
	Left heel forces (before reaching the mark)	0.934 (0.789–0.981)	41.33	4.83	114.57	13.40
	Right heel forces (after reaching the mark)	0.931 (0.778–0.980)	42.27	4.87	117.17	13.51
	Left heel forces (after reaching the mark)	0.928 (0.770–0.979)	48.45	5.57	134.30	15.45

4. Discussion

The present study aimed to evaluate the test-retest reliability of kinetic and kinematic parameters obtained from TUG in the diabetic population. The results were analysed according to the severity of neuropathy. The main finding was that almost each variable achieved good (ICC between 0.70 and 0.90) or excellent (>0.90) reliability considering the classification by Munro et al. [20].

The reliability of TUG seemed to be conditioned by the severity of neuropathy since the ICC was higher than 0.90 (excellent) for participants with moderate neuropathy and patients with normal foot vibration perception, but just good (0.70 to 0.90) for patients with severe neuropathy. Therefore, although the TUG is reliable in T2DM patients, changes in the duration of the phases (single and double support) must be taken with caution since the %SEM may be relatively high in some cases (over 10%). Furthermore, we can observe how the SRD of the required time to complete the TUG was higher in patients with severe neuropathy. These results must be considered by clinicians and researchers in order to interpret their results when using TUG in the T2DM population. This is relevant since a previous study showed that patients with diabetic neuropathy have worse health-related quality of life and lower functional status than patients without diabetic neuropathy [21]. Neuropathy leads to balance impairments [22], gait and mobility alterations [23,24] and an increased risk of falls [25]. In addition, patients with T2DM seem to be more susceptible to falls and consequently to bone fractures [26–28]. Due to the relevance of diabetic neuropathy in patients with T2DM, future studies aimed at improving physical conditioning variables should consider that the smallest clinically relevant improvement may be higher in T2DM patients with this complication.

To our knowledge, this is the first study to assess the reliability of TUG according to the severity of neuropathy. Only one previous study has evaluated the reliability of TUG in the diabetic population [29]. That study was conducted on a sample of 18 older adults with T2DM, and no classification of the participants was performed. Furthermore, that study only evaluated the time needed to complete the TUG and not any other measure such as the duration of walking phases or kinetic parameters. Comparing results from the present study with those obtained by Alfonso-Rosa, Del Pozo-Cruz, Del Pozo-Cruz, Sanudo and Rogers [29], both studies report excellent reliability. However, the ICC and %SEM from that study were 0.98 and 3.5%, respectively, whereas the present study reports an ICC of 0.927 and a %SEM of 4.52. Therefore, the reliability is slightly lower in this study, which may be linked to the greater heterogeneity of the sample in the current study. In this regard, it must be noted that the study by Alfonso-Rosa, Del Pozo-Cruz, Del Pozo-Cruz, Sanudo and Rogers [29] was conducted on 18 older adults with T2DM, while the current one was conducted with 56 adults aged 65.35 (8.56) years.

The current study reported not only the SEM but also the SRD, which is extremely relevant for clinicians and researchers since it indicates whether the differences obtained as a consequence of an intervention program could be considered clinically important [30]. A previous study in T2DM [29] established the SRD for the TUG at 9.8%, whereas in the present study the SRD was 12.55%. This difference could be also explained by the heterogeneity of the sample in the current study. In this regard, the results from the previous study were limited to older adults. Thus, those studies conducted with samples comprised of adults (but not limited older adults) had no information about the reliability and the minimal clinically important difference to interpret their results. Therefore, the current study was needed since it is the first to provide reliability information in a sample comprised of adults (not limited to older adults).

Test-retest analyses have been also reported in kinematic parameters, taking into account the heel and the forefoot forces during the TUG. The results revealed that the reliability of the heel and forefoot forces could be considered as excellent, so future studies can confidently use this as a measurement to evaluate gait patterns in T2DM patients. People with diabetic neuropathy often have balance problems while performing common activities such as walking or ascending/descending stairs [31]. Furthermore, the postural mechanisms at the ankle joints are impaired in diabetic neuropathy patients during quiet standing [32]. Thus, postural instability and gait imbalance in diabetic neuropathy may contribute to a high risk of fall incidence, especially in the geriatric population [7,31]. Therefore, future studies should be focused on the implementation of interventions aimed to modify gait parameters to reduce the risk of falling in T2DM patients. The current study provides useful information in order to interpret changes achieved after a specific program, stating the minimal clinically relevant change for kinetic and kinematic variables.

The present study has two main limitations. First, although the sample size ($n = 56$) was sufficient to conduct this test-retest reliability analysis, the group with no alteration in the foot vibration threshold was comprised of only 12 people. The second limitation may be related to the inclusion criteria and the difficulty in determining that peripheral neuropathy is caused by diabetes. Even though the inclusion and exclusion criteria were very restrictive (excluding people with other diseases and people who were taking drugs that may potentially affect balance and gait), there could be other non-diagnosed diseases, environmental factors or healthy/unhealthy habits that may increase or reduce neuropathy. In spite of these two limitations, this study succeeded at reporting reliability parameters according to the severity of neuropathy in patients with T2DM.

5. Conclusions

The reliability of TUG was excellent for the whole sample and the groups with non-severe neuropathy. However, it was just good for the group with severe neuropathy. Regarding the kinetic and kinematic parameters, the reliability was good or excellent for almost every variable and group. The present study reports the minimal change that may be considered real (SRD) and the SEM, which should be considered by future studies aimed to assess the effects of interventions on different variables related to the TUG.

Author Contributions: F.J.D.-M., J.C.A., M.A.H.-M. conceived the study. F.J.D.-M., S.V. and D.C.-M. collected the data. F.J.D.-M., N.G., L.J.M., D.C.-M. analysed the data. S.V., N.G. and D.C.-M. designed the figures and tables. F.J.D.-M., J.C.A. and D.C.-M. wrote the manuscript. S.V., M.A.H.-M., L.J.M., and N.G. provided critical revisions on successive drafts. All authors approved the manuscript in its final form.

Acknowledgments: The author SV is supported by a grant from the regional Department of Economy and Infrastructure of the Government of Extremadura and European Social Fund (PD16008). We are grateful to the Primary Care Centre "Manuel Encinas" in Cáceres for helping recruit the participants for this study. The funders had no role in study design, data collection and analysis, decision to publish, or preparation of the manuscript.

Conflicts of Interest: The authors declare no competing interests.

Abbreviations

ADA	American Diabetes Association
BMI	Body mass index
DM	Diabetes mellitus
ICC	Intraclass correlation coefficient
T2DM	Type 2 diabetes mellitus
TUG	Timed up and go
SEM	Standard error of measurement
SPSS	Statistical Package for Social Sciences
SRD	Smallest real difference

References

1. Alam, U.; Asghar, O.; Azmi, S.; Malik, R.A. General aspects of diabetes mellitus. *Handb. Clin. Neurol.* **2014**, *126*, 211–222. [CrossRef] [PubMed]
2. Ogurtsova, K.; da Rocha Fernandes, J.D.; Huang, Y.; Linnenkamp, U.; Guariguata, L.; Cho, N.H.; Cavan, D.; Shaw, J.E.; Makaroff, L.E. IDF Diabetes Atlas: Global estimates for the prevalence of diabetes for 2015 and 2040. *Diabetes Res. Clin. Pract.* **2017**, *128*, 40–50. [CrossRef] [PubMed]
3. ADA. American Diabetes Association: Economic costs of diabetes in the US in 2012. *Diabetes Care* **2013**, *36*, 1033–1046.
4. Boulton, A.J.; Malik, R.A.; Arezzo, J.C.; Sosenko, J.M. Diabetic somatic neuropathies. *Diabetes Care* **2004**, *27*, 1458–1486. [CrossRef]

5. Thomas, P.K. Classification, differential diagnosis, and staging of diabetic peripheral neuropathy. *Diabetes* **1997**, *46* (Suppl. 2), S54–S57. [CrossRef]
6. Andersen, H.; Nielsen, S.; Mogensen, C.E.; Jakobsen, J. Muscle strength in type 2 diabetes. *Diabetes* **2004**, *53*, 1543–1548. [CrossRef] [PubMed]
7. Mustapa, A.; Justine, M.; Mohd Mustafah, N.; Jamil, N.; Manaf, H. Postural control and gait performance in the diabetic peripheral neuropathy: A systematic review. *BioMed Res. Int.* **2016**, *2016*, 9305025. [CrossRef] [PubMed]
8. Van Deursen, R.; Simoneau, G.G. Foot and ankle sensory neuropathy, proprioception, and postural stability. *J. Orthop. Sports Phys. Ther.* **1999**, *29*, 718–726. [CrossRef]
9. Ducic, I.; Short, K.W.; Dellon, A.L. Relationship between loss of pedal sensibility, balance, and falls in patients with peripheral neuropathy. *Ann. Plast. Surg.* **2004**, *52*, 535–540. [CrossRef]
10. Boulton, A.J.M. Chapter 8—Diabetic neuropathy and foot complications. In *Handbook of Clinical Neurology*; Zochodne, D.W., Malik, R.A., Eds.; Elsevier: Amsterdam, The Netherlands, 2014; Volume 126, pp. 97–107.
11. Young, M.J.; Breddy, J.L.; Veves, A.; Boulton, A.J. The prediction of diabetic neuropathic foot ulceration using vibration perception thresholds. A prospective study. *Diabetes Care* **1994**, *17*, 557–560. [CrossRef]
12. Abbott, C.A.; Vileikyte, L.; Williamson, S.; Carrington, A.L.; Boulton, A.J. Multicenter Study of the Incidence of and Predictive Risk Factors for Diabetic Neuropathic Foot Ulceration. *Diabetes Care* **1998**, *21*, 1071–1075. [CrossRef]
13. Macgilchrist, C.; Paul, L.; Ellis, B.M.; Howe, T.E.; Kennon, B.; Godwin, J. Lower-limb risk factors for falls in people with diabetes mellitus. *Diabetic Med.* **2010**, *27*, 162–168. [CrossRef] [PubMed]
14. Allet, L.; Armand, S.; de Bie, R.A.; Golay, A.; Pataky, Z.; Aminian, K.; de Bruin, E.D. Clinical factors associated with gait alterations in diabetic patients. *Diabetic Med.* **2009**, *26*, 1003–1009. [CrossRef] [PubMed]
15. Ward, R.E.; Boudreau, R.M.; Caserotti, P.; Harris, T.B.; Zivkovic, S.; Goodpaster, B.H.; Satterfield, S.; Kritchevsky, S.B.; Schwartz, A.V.; Vinik, A.I.; et al. Sensory and motor peripheral nerve function and incident mobility disability. *J. Am. Geriatr. Soc.* **2014**, *62*, 2273–2279. [CrossRef] [PubMed]
16. Dixon, C.J.; Knight, T.; Binns, E.; Ihaka, B.; O'Brien, D. Clinical measures of balance in people with type two diabetes: A systematic literature review. *Gait Posture* **2017**, *58*, 325–332. [CrossRef] [PubMed]
17. Deng, H.; He, F.; Zhang, S.; Calleman, C.J.; Costa, L.G. Quantitative measurements of vibration threshold in healthy adults and acrylamide workers. *Int. Arch. Occup. Environ. Health* **1993**, *65*, 53–56. [CrossRef] [PubMed]
18. Weir, J.P. Quantifying test-retest reliability using the intraclass correlation coefficient and the SEM. *J. Strength Cond. Rres.* **2005**, *19*, 231–240. [CrossRef]
19. Shrout, P.E.; Fleiss, J.L. Intraclass correlations: Uses in assessing rater reliability. *Psychol. Bull.* **1979**, *86*, 420–428. [CrossRef]
20. Munro, B.; Visintainer, M.; Page, E. *Statistical Methods for Health Care Research*; Lippincott Williams & Wilkins: Philadelphia, PA, USA, 1986.
21. Riandini, T.; Wee, H.L.; Khoo, E.Y.; Tai, B.C.; Wang, W.; Koh, G.C.; Tai, E.S.; Tavintharan, S.; Chandran, K.; Hwang, S.W. Functional status mediates the association between peripheral neuropathy and health-related quality of life in individuals with diabetes. *Acta Diabetol.* **2018**, *55*, 155–164. [CrossRef]
22. Boucher, P.; Teasdale, N.; Courtemanche, R.; Bard, C.; Fleury, M. Postural stability in diabetic polyneuropathy. *Diabetes Care* **1995**, *18*, 638–645. [CrossRef]
23. Martinelli, A.R.; Mantovani, A.M.; Nozabieli, A.J.L.; Ferreira, D.M.A.; Barela, J.A.; de Camargo, M.R.; Fregonesi, C.E.P.T. Muscle strength and ankle mobility for the gait parameters in diabetic neuropathies. *Foot* **2013**, *23*, 17–21. [CrossRef]
24. Manor, B.; Li, L. Characteristics of functional gait among people with and without peripheral neuropathy. *Gait Posture* **2009**, *30*, 253–256. [CrossRef] [PubMed]
25. Richardson, J.K.; Hurvitz, E.A. Peripheral neuropathy: A true risk factor for falls. *J. Gerontol. Ser. A Biol. Sci. Med. Sci.* **1995**, *50*, M211–M215. [CrossRef] [PubMed]
26. Wongsurawat, N.; Armbrecht, H.J. Insulin modulates the stimulation of renal 1, 25-dihydroxyvitamin D3 production by parathyroid hormone. *Acta Endocrinol.* **1985**, *109*, 243–248. [CrossRef] [PubMed]
27. Pietschmann, P.; Schernthaner, G.; Woloszczuk, W. Serum osteocalcin levels in diabetes mellitus: Analysis of the type of diabetes and microvascular complications. *Diabetologia* **1988**, *31*, 892–895. [CrossRef] [PubMed]

28. Schwartz, A.V.; Sellmeyer, D.E.; Ensrud, K.E.; Cauley, J.A.; Tabor, H.K.; Schreiner, P.J.; Jamal, S.A.; Black, D.M.; Cummings, S.R.; Study of Osteoporotic Fractures Research Group. Older women with diabetes have an increased risk of fracture: A prospective study. *J. Clin. Endocrinol. Metab.* **2001**, *86*, 32–38. [CrossRef] [PubMed]
29. Alfonso-Rosa, R.M.; Del Pozo-Cruz, B.; Del Pozo-Cruz, J.; Sanudo, B.; Rogers, M.E. Test-retest reliability and minimal detectable change scores for fitness assessment in older adults with type 2 diabetes. *Rehabil. Nurs.* **2014**, *39*, 260–268. [CrossRef]
30. Beckerman, H.; Roebroeck, M.E.; Lankhorst, G.J.; Becher, J.G.; Bezemer, P.D.; Verbeek, A.L. Smallest real difference, a link between reproducibility and responsiveness. *Qual. Life Res.* **2001**, *10*, 571–578. [CrossRef]
31. Brown, S.J.; Handsaker, J.C.; Bowling, F.L.; Boulton, A.J.; Reeves, N.D. Diabetic peripheral neuropathy compromises balance during daily activities. *Diabetes Care* **2015**, *38*, 1116–1122. [CrossRef]
32. Lafond, D.; Corriveau, H.; Prince, F. Postural control mechanisms during quiet standing in patients with diabetic sensory neuropathy. *Diabetes Care* **2004**, *27*, 173–178. [CrossRef]

applied sciences

MDPI

Article

No Evidence That Frontal Optical Flow Affects Perceived Locomotor Speed and Locomotor Biomechanics When Running on a Treadmill

Martina Caramenti [1,2,3,*], Claudio L. Lafortuna [4], Elena Mugellini [3], Omar Abou Khaled [3], Jean-Pierre Bresciani [1,5,†] and Amandine Dubois [1,6,†]

1 Department of Neuroscience and Movement Science, University of Fribourg, 1700 Fribourg, Switzerland; jean-pierre.bresciani@unifr.ch (J.-P.B.); amandine.dubois@univ-lorraine.fr (A.D.)
2 Istituto di Bioimmagini e Fisiologia Molecolare, CNR, 20090 Segrate (Milano), Italy
3 HumanTech Institute, University of Applied Sciences and Arts Western Switzerland, 1705 Fribourg, Switzerland; elena.mugellini@hes-so.ch (E.M.); omar.aboukhaled@hes-so.ch (O.A.K.)
4 Istituto di Fisiologia Clinica, CNR, 20162 Milano, Italy; claudio.lafortuna@cnr.it
5 Université Grenoble-Alpes, LPNC, F-38000 Grenoble, France
6 Université de Lorraine, 2LPN-CEMA Group (Cognition-EMotion-Action), EA 7489, F-57070 Metz, France
* Correspondence: martina.caramenti@ibfm.cnr.it
† These authors contributed equally to this work.

Received: 9 October 2019; Accepted: 24 October 2019; Published: 29 October 2019

Abstract: We investigated how the presentation and the manipulation of an optical flow while running on a treadmill affect perceived locomotor speed (Experiment 1) and gait parameters (Experiment 2). In Experiment 1, 12 healthy participants were instructed to run at an imposed speed and to focus on their sensorimotor sensations to be able to reproduce this running speed later. After a pause, they had to retrieve the reference locomotor speed by manipulating the treadmill speed while being presented with different optical flow conditions, namely no optical flow or a matching/slower/faster optical flow. In Experiment 2, 20 healthy participants ran at a previously self-selected constant speed while being presented with different optical flow conditions (see Experiment 1). The results did not show any effect of the presence and manipulation of the optical flow either on perceived locomotor speed or on the biomechanics of treadmill running. Specifically, the ability to retrieve the reference locomotor speed was similar for all optical flow conditions. Manipulating the speed of the optical flow did not affect the spatiotemporal gait parameters and also failed to affect the treadmill running accommodation process. Nevertheless, the virtual reality conditions affected the heart rate of the participants but without affecting perceived effort.

Keywords: virtual reality; visual speed perception; treadmill running; self-motion perception; optical flow; locomotion; biomechanics

1. Introduction

Human locomotion is controlled through a complex sensorial integration of different types of information, in particular about the relative motion between the body and the surrounding environment [1]. Visual, vestibular, proprioceptive, and auditory information is integrated by the central nervous system, allowing the generation of the perception of motion [2]. Though walking is influenced by different factors [3], the main contribution derives from visual, vestibular and proprioceptive signals. Using static cues, object-motion cues and optical flow cues, the visual system allows us to discriminate self-motion from movement of objects of the environment [4]. Vestibular cues provide motion data about changes of velocity or direction-based data about accelerations [5]. Lastly, the proprioceptive system uses joint- and muscle-related information to provide feedback about the status of the whole musculoskeletal system, which greatly contributes to the perception and control of motion during active movements [3]. Speed perception, goal-directed movements, navigation, and collision avoidance all rely on the congruency and integration of the signals provided by these different sensory systems.

Visual information plays an important role in motion perception. In particular, optical flow information is used to estimate direction [6], egocentric speed [7], and time-to-collision [8], and it heavily contributes to heading perception [9–11]. The absolute travelled distance can also be calculated by integrating this information [12,13]. However, when visual cues are in conflict with other sensory information about self-motion, visual information not always constitutes the "dominant" source of information [4,14]. In fact, when walking at constant speed, people tend to weight the information deriving from proprioceptive cues higher than visual cues [15–17]. Nevertheless, visual input seems to have a great impact on self-motion perception while walking, and studies have shown how the manipulation of the optical flow affects visuomotor recalibration [11,15,17–19]. In particular, it was found that if optical flow is manipulated to create sensory conflicts when walking, the locomotor speed decreases as the optical flow speed increases [15,20], suggesting a response to the relative weightings assigned to different sensory cues, in this case proprioceptive and visual cues [15]. Some studies have also shown that the manipulation of optical flow can significantly influence locomotor patterns [15,17,21,22].

All above-mentioned studies were performed with walking participants and, to our knowledge, the influence of optical flow on self-motion perception and on locomotor patterns has never been investigated with running individuals. Yet, locomotor patterns are modulated differently in walking and running [23–25], with automated spinal programs allowing path integration in running, while walking is more dependent on visual control [24]. For this reason, one might wonder whether the results obtained with walking individuals can be generalized to running individuals.

Here we investigated how the presence and manipulation of a virtual optical flow influences the control of locomotion while running on a treadmill. In Experiment 1, we tested how the presence and manipulation of a virtual optical flow influences self-motion speed perception. Specifically, participants were instructed to retrieve a previously presented reference running speed while being exposed to different optical flow conditions. In Experiment 2, we investigated how the presence and manipulation of an artificial optical flow influences gait parameters while treadmill running at a self-selected constant speed with different optical flow conditions.

2. Experiment 1

2.1. Introduction

In this experiment, we tested if the perception of self-motion speed is influenced by the presence and the manipulation of an optical flow. For walking, studies have shown that movement influences the perception of the optical flow speed in virtual environments [26], and that the rate of optical flow has a modulating effect on walking speed, with an inverse linear relationship between optical flow and walking velocity [15]. Optical flow has also been shown to affect stride length, cadence, and velocity [17]. All these results suggest an internal calibration between locomotion and visual perception. Here we investigated if participants were able to retrieve a previously experienced running speed in different optical flow conditions while running on a treadmill.

2.2. Methods

2.2.1. Participants

12 healthy participants (4 female, 8 male) with a mean age of 25.0 (±1.2 SD) participated in this study. They were unaware of the purpose of the study, were moderately trained to trained, had normal

or corrected-to-normal vision, and none had a history of cardiovascular disease. All participants gave their informed and written consent prior to the inclusion in the study. This study was performed in accordance with the ethical standards specified by the 1964 Declaration of Helsinki and approved by the Ethics Committee of the University of Fribourg.

2.2.2. Experimental Setup

Participants ran on a HP Cosmos Mercury treadmill with a running surface of 150 × 50 cm. The treadmill was positioned in front of a 4.30 × 2.70 m screen at a distance of 1.7 m, leading to an effective field of view of 90°. To simulate optic flow, a virtual environment was projected onto the screen using a Barco F50 WUXGA projector with a 1920 × 1200 pixels resolution. The VR scene for which judgements were made and that was subsequently manipulated was created using Unity and depicted a neutral open-air hallway presented at constant-velocity motion (Figure 1). Rich optical flow information was provided by a granular texture on the floor and by the random pattern on the walls, without giving any landmarks or usable spatial information. The room was darkened during the experiment, with the display screen was the only source of light. While running, participants wore a soundproof headset (Hearing Protection type Pamir—Swiss Army).

Figure 1. Experimental setup.

2.2.3. Procedure

Prior to the running task, the participants filled out a custom-made questionnaire about their sports activity.

Before starting the actual test, participants familiarized themselves with treadmill running and experimental trials were initiated when the participant felt comfortable. Every single trial would start with the participant running at constant speed. Participants were instructed to gaze at the fixation cross, that was positioned straight ahead and was visible for the whole duration of the test.

For the running task, participants were asked to run at 10 km/h for 2 min in front of the black screen focusing on the sensations perceived while running. The participants wore a soundproof headset to prevent them from using the sound of the treadmill as an information about the running speed. The treadmill display was covered so that participants did not have any feedback on their actual running speed. After 2 min of running at 10 km/h, the participants were asked to pause for 1 min. After this short pause, they started running at 7 km/h and had to manipulate the treadmill speed until they found a running speed that they perceived as matching the reference speed of 10 km/h.

This task was proposed in four different blocks in which the participants were asked to regulate the speed of the treadmill to match the previous speed while being presented with:

- No optical flow (noOF)
- An optical flow matching the treadmill speed (matchOF)
- A faster optical flow (+5 km/h) (fastOF)
- A slower optical flow (−5 km/h) (slowOF).

The order of presentation of the conditions was randomly selected. Note that the fixation cross was positioned so that it corresponded to the focus of expansion of the visual scene in the conditions with optical flow.

The treadmill speed perceived by participants as matching the reference speed and the time necessary to define this speed were recorded at the end of each block.

In between blocks, participants were asked to walk for about 2 min.

2.3. Results

For each condition, data distribution was assessed using the Shapiro–Wilk test (please refer to Supplementary Materials S1-T1 for the data). We then compared the selected running speed to the reference running speed using in each case either a *t*-test when data was normally distributed or a Wilcoxon signed rank test. For all four tests, the alpha level was corrected for multiple comparisons using Bonferroni correction (i.e., 0.05/4 = 0.0125). For all conditions, the selected running speed was not significantly different from the reference running speed after correction.

We then tested whether the selected running speed was different based on the optical flow condition. Since data was not normally distributed, a Friedman rank sum test for repeated measures was used. No differences were found for the selected running speed in the four conditions $\left[\chi^2(3) = 3.1,\ p = 0.3765\right]$ (see Figure 2). Because we did not find any significant difference between the means (i.e., we could not reject H0), we also computed the Bayes factor. The Bayes factor was 0.37, a value which can be interpreted as an 'anecdotal' evidence in favour of H0 [27].

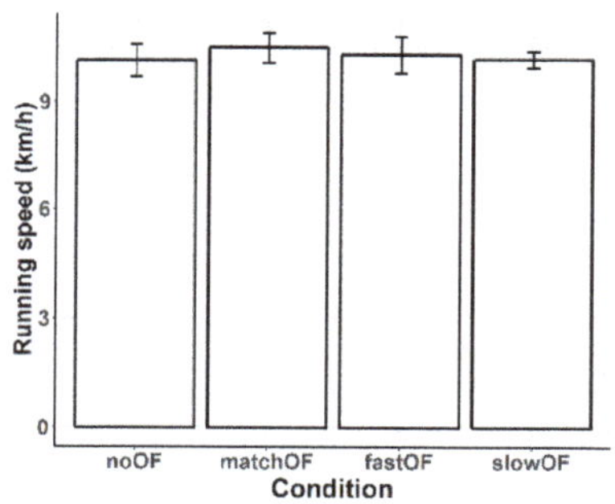

Figure 2. Running speed selected as matching the reference running speed in different optical flow conditions. The bars represent the means and the error bars the 95% confidence interval of those means.

Lastly, we tested whether the duration of the selection phase was different based on the optical flow condition. Data was not normally distributed, so a Friedman rank sum test for repeated measures was used. Here again, we did not find any significant difference between optical flow conditions regarding the duration of the running phase $\left[\chi^2(3) = 1.084,\ p = 0.7809\right]$ (see Figure 3). As for the previous analysis, we computed the Bayes factor, which was 0.11, thereby constituting a moderate to strong evidence in favour of H0 [27].

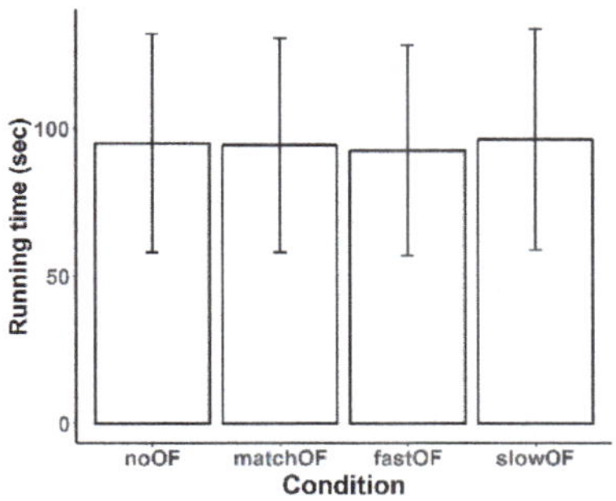

Figure 3. Task duration. Time needed for the participants to retrieve the running speed they perceived as matching the previously experienced reference running speed. The bars represent the means and the error bars the 95% confidence interval of those means.

2.4 *Discussion*

Participants running on a treadmill were asked to match their running speed to a previously experienced reference running speed while being presented with different optical flow conditions. For all optical flow conditions, participants selected a running speed that did not differ significantly from the reference speed, and there was no difference in the selected speed between the four optical flow conditions. Also, the duration of the selection task did not differ significantly between the four optical flow conditions.

Previous studies with walking participants showed an influence of the optical flow rate on the control of locomotor speed [15,17,28], suggesting an internal calibration between locomotion and visual perception. Here, the adaptation of running speed to the presented optical flow speed was not possible, as the treadmill was not self-driven, but should have influenced the ability to retrieve the reference running speed due to the disruption of this internal calibration. Our results show that this is not the case for running individuals. Moreover, for walking individuals, previous studies suggested the presence of a predetermined expectation of the visual effect that should be associated with a certain walking speed [26]. This should also have led to expecting an influence on the ability to retrieve an imposed speed in the different optical flow conditions. It seems that visual incongruences do not have the same influence on running speed control as they do on walking speed control, possibly due to the higher automatization of running compared to walking as well as to the fact that it requires less cortical control [25,29].

Note that our results might be consistent with a statistically optimal integration model [30], which predicts that sensory information from multiple sources is weighted according to the estimation of the reliability of each sensory source. In our task, because running at the reference speed was performed without optical flow, participants might have given more weight to proprioceptive information than to visual information also in the second part of the running task.

3. Experiment 2

3.1. *Introduction*

In this experiment, we tested if treadmill running is influenced by the presence and the manipulation of an optical flow. Studies show that the manipulation of the optical flow has a destabilizing effect on postural stability [31–34], and a significant influence on locomotor patterns when walking [15,17,21,22]. The results regarding modifications of gait parameters and their variability as an effect of a virtual optical flow are contradicting, suggesting a great influence of the specific setup, the walking mode (i.e., fixed speed vs. self-paced treadmill) and the time to adapt to the new environment. Specifically, some studies found a lower stride length, with an increased step width [35,36], an increased walking speed variability, and an increased step width variability [15,36,37]. On the other hand, other authors found a stabilizing effect on locomotion patterns [38]. Here we investigated if the presence and manipulation of an optical flow could influence spatiotemporal gait parameters during treadmill running.

3.2. Methods

3.2.1. Participants

20 healthy participants (13 female, 7 male) with a mean age of 25.4 (±2.7 SD) participated in this study. They were unaware of the purpose of the study, were moderately trained to trained, had normal or corrected-to-normal vision and none had a history of cardiovascular disease. All participants gave their informed and written consent prior to the inclusion in the study. This study was performed in accordance with the ethical standards specified by the 1964 Declaration of Helsinki and approved by the Ethics Committee of the University of Fribourg.

3.2.2. Experimental Setup

The experimental setup was the same as in Experiment 1, except for the position of the treadmill, which was at a distance of 2.5 m from the screen, leading to an effective field of view of 70°.

While running, participants' heart rate (HR) was constantly monitored (Polar Team2 System, Polar Electro Oy—Kempele, Finland). Moreover, the NaturalPoint Optitrack system (NaturalPoint, Inc. DBA OptiTrack—Corvallis, OR, USA) with 16 OptiTrack cameras (Prime 17W model) was used to monitor the position of the markers that were used for the estimation of the spatiotemporal gait parameters.

3.2.3. Procedure

As in Experiment 1, prior to the running task, the participants filled out a custom-made questionnaire about their sports activity. For this experiment, also the short version of the International Physical Activity Questionnaire (IPAQ—French version) was used.

At the beginning of the experiment and prior to the testing phase, the participants spent 4 min familiarizing themselves with treadmill running. During this phase, they were instructed to "select an exercise intensity that you prefer and can be sustained for 20 min" [39,40]. The selected running speed was then used for the subsequent testing phase. To ensure that the participant's speed selection was solely based on perception, the speed display on the treadmill was covered so that it was only visible to the investigator. Self-selected running speed was used to reduce inter-individual differences, because fitness level and running experience have an influence on the parameters measured for this study.

The first part of the experiment allowed us to assess the visual speed perceived as matching the chosen running speed for each participant. Participants were presented with the visual scene while running at their preferred running speed. They were instructed to estimate if the optical flow was slower or faster than the actual running speed. Four consecutive tests were proposed in a random order using a one up–one down staircase method [41,42]. The optical flow started two times at a higher speed than the actual treadmill speed (+4 km/h) and two times at a lower speed (−4 km/h). The order of starting speed was randomized. The speed of the visual scene was adjusted according to the one up–one down staircase method, with an increase/decrease of the visual speed of 0.5 km/h until the first inversion of the participant's response, followed by steps of 0.3 km/h. This allowed us to determine the perceptual threshold that indicated the visual speed perceived as matching the actual running speed, i.e., the point of subjective equality (PSE). Prior to the task, participants were familiarized with the experimental setting and task using a training program. The single trials started with the participant running at the previously chosen speed. The visual scene was presented for 2 s before the participants were challenged with a black screen presenting the question "up or down", thus asking them to decide if they wanted to increase (i.e., up) or decrease (i.e., down) the speed of the visual scene. For each trial, participants gave their response while continuing their treadmill run by pressing a switch put on top of a light custom-made plastic cylinder (115 × 30 mm, 15 g) they were holding in each hand. The responses (i.e., left for up and right for down) were sent to the computer via Bluetooth and directly integrated into the staircase test. Once the participant pressed the chosen button, the graphics returned to the visual scene of the following trial (see Figure 4). Each staircase ended when 15 inversions of the responses were reached and the participants were free to take a pause before starting the subsequent test.

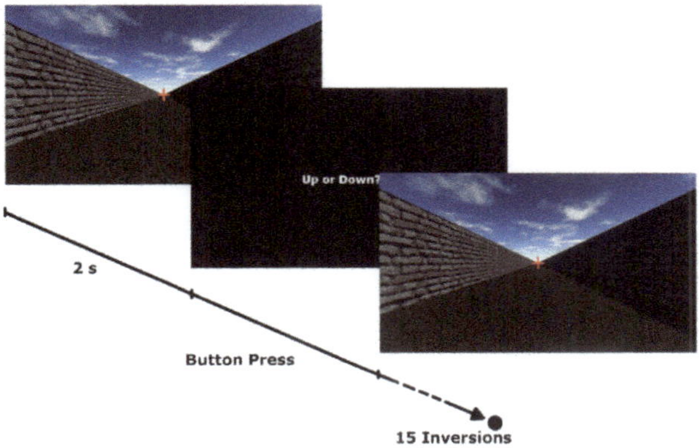

Figure 4. Experimental design.

The second part of the experiment consisted of 4 different blocks in which the participants were asked to run at constant speed for 3 min and 30 s while being presented with different optical flow conditions. The reference value for the optical flow speed was personalized for each participant using the PSE found with the staircase test. The four conditions were:

- no optical flow (noOF)
- matching optical flow with a visual speed corresponding to the PSE (matchOF)
- faster optical flow with a visual speed that was 40% higher than the PSE (fastOF)
- slower optical flow with a visual speed that was 40% lower than the PSE (slowOF).

The order of conditions was randomly assigned. At the end of each block, the participants took a 2-min pause before starting the next block.

For all the parts involving the presentation of a visual speed, the participants were instructed to gaze at the fixation cross that was visible in all conditions.

At the end of the running phase, the participants estimated their perceived exertion using the Borg RPE Scale (6 to 20 scale), with values ranging from 'no exertion at all' to 'maximal exertion'.

3.2.4. Spatiotemporal Parameter Analysis

For the spatiotemporal gait parameter estimation, the information derived from the OptiTrack system was used. The participants were equipped with two reflective markers, one positioned on each heel. The data analysis was performed on steady gait by removing the first and the last 15 s of each block.

The different gait parameters were estimated using the heel marker trajectory and were averaged for each block, as described in Dubois and Bresciani [43]. Table 1 shows which spatiotemporal gait parameters were estimated and how they were calculated.

Table 1. Gait parameters and their method of estimation.

Spatiotemporal Parameters	Estimation
Step duration (s)	The duration between the local minima of the left and right heel
Step length (m)	Treadmill speed (m/s) × Step duration (s)
Step frequency (steps/s)	1/step duration
Step width (cm)	The distance on the x-axis between the left and right heel

For analysis, these values were averaged both for the whole duration of the block as also for each minute of each block.

3.3. Results

We first compared the perceived visual speed with the actual running speed using a paired *t*-test, as both running and perceived visual speed were normally distributed (please refer to Supplementary Materials S1-T2 for the data.) This allowed us to determine how visual speed was estimated compared to the actual running speed, i.e., treadmill speed. Participants set visual speed significantly higher than the actual treadmill speed ($t(19) = -4.8743$, $p < 0.001$). Figure 5 shows the difference between running speed and perceived visual speed.

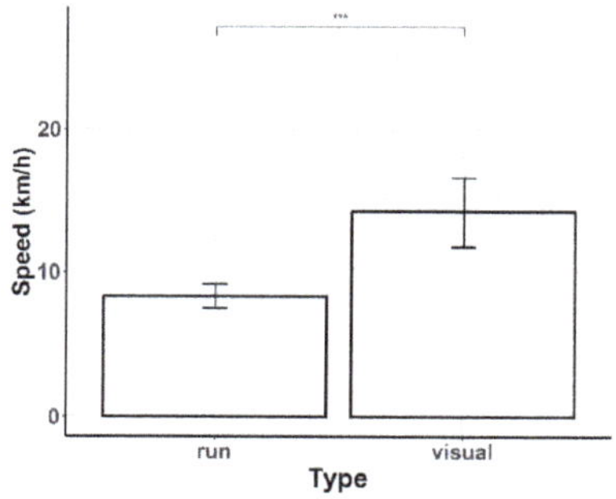

Figure 5. Visual speed underestimation. Speed of the visual scene that was perceived as matching running/treadmill speed. The bars represent the means and the error bars the 95% confidence interval of those means.

We then tested whether the presentation and manipulation of the optical flow affected spatiotemporal parameters in treadmill running using a 4 × 3 (Condition[noOF, matchOF, fastOF, slowOF]*minute[1,2,3]) mixed analysis of variance (ANOVA). The results showed no main effect of the condition for all the spatiotemporal parameters, i.e., step length, step duration, step frequency, and step width. Even if there was no influence of the condition, the results showed a significant effect of the minute of the acquisition. There was no interaction between the main factors. Table 2 reports all the results.

Table 2. Results of the 4 x3 mixed ANOVA. Significant difference: * = $p < 0.05$, ** = $p < 0.01$, *** = $p < 0.001$.

Variable	Condition	Minute	Interaction
Step length	$F(3,57) = 1.695$, $p = 0.178$	$F(2,38) = 10.000$, $p < 0.001$ ***	$F(6,114) = 1.423$, $p = 0.212$
Step duration	$F(3,57) = 1.747$, $p = 0.168$	$F(2,38) = 10.085$, $p < 0.001$ ***	$F(6,114) = 1.357$, $p = 0.238$
Step frequency	$F(3,57) = 1.977$, $p = 0.128$	$F(2,38) = 9.492$, $p < 0.001$ ***	$F(6,114) = 1.534$, $p = 0.173$
Step width	$F(3,57) = 1.011$, $p = 0.395$	$F(2,38) = 11.514$, $p < 0.001$ ***	$F(6,114) = 0.255$, $p = 0.956$

A Tukey's post-hoc test indicated that the difference was significant when comparing minute 1 to minute 3 for step duration ($p < 0.05$), step frequency ($p < 0.05$) and step width ($p < 0.01$), while there was only a tendency for step length ($p = 0.061$).

For the HR, since a plateau was observed in all participants, the mean value was calculated as average of 30 s once the plateau was reached. A repeated-measures ANOVA showed a significant difference for mean HR [F(3,57) = 5.7465, $p < 0.01$], while the starting HR did not differ between conditions $\left[\chi^2(3) = 0.6528, p = 0.8842\right]$. A Tukey's post-hoc test indicated a significant difference between the noOF and the VR conditions (noOF-matchOF: $p < 0.01$, noOF-fastOF: $p < 0.05$, noOF-slowOF: $p < 0.001$) as shown in Figure 6.

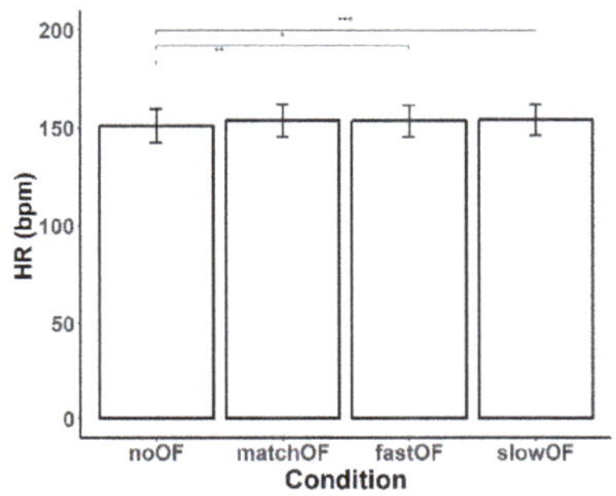

Figure 6. Mean heart rate. The bars represent the means and the error bars the 95% confidence interval of those means.

Last, a repeated-measures ANOVA performed on perceived effort values (as measured with the Borg RPE scale) showed no difference between the different OF conditions $\left[\chi^2(3) = 2.4296,\ p = 0.4882\right]$.

3.4 Discussion

First, participants running on a treadmill while being presented with a moving virtual scene were asked to match the visual speed of the VR scene to their actual running speed. Participants set visual speed significantly higher than the actual running speed, i.e., treadmill speed. In other words, the visual scene had to move significantly faster than the treadmill speed for the two speeds to be perceived as equivalent, indicating and underestimation of visual speed relative to treadmill speed. Then, spatiotemporal gait parameters and HR were measured with participants running on a treadmill at their preferred running speed while being presented with different optical flow conditions. At the end of each condition, perceived effort was recorded. We did not observe any significant difference between optical flow conditions regarding the spatiotemporal gait parameters. However, we observed an effect of the time (i.e., minute) of acquisition. In addition, the HR was significantly different in the noOF condition than in the VR conditions, while there was no difference between conditions regarding perceived effort.

Our results confirm the general tendency to underestimate visual speed relative to locomotor speed in virtual environments, both when walking [18,44–46] and running [47–49]. Compared to previous work with running participants, our results show that visual speed underestimation relative to running speed also occurs with participants running at their preferred running speed. Specifically, previous work was always based on imposed running speeds [47–49], which were not necessarily in the range of preferred speeds of all participants. Here, every participant selected his/her running speed, reducing a possible influence of fatigue. This likely also reduced the attentional demand and fear of falling off of the treadmill [50] thanks to an increased felt safety compared to an imposed running speed.

Previous studies on walking showed that the manipulation of the optical flow can significantly affect locomotor patterns [15,17,21,22]. In particular, optical flow can modulate parameters known to be reflective of gait instability, such as step width, step length, stride cycle, and stride velocity variability [51,52]. Our results show that this influence on locomotor patterns cannot be extended to individuals running on a treadmill. In fact, none of the spatiotemporal gait parameters (i.e., step length, step duration, step frequency, and step width) was significantly affected by the optical flow condition. This suggests that when running at a self-selected speed, the presence and manipulation of the optical flow has no stabilizing or destabilizing effect. The difference between our results and those previously reported with walking participants could be linked to the fact that running seems to be characterized by a higher automatization and a lower cortical control compared to walking [25,29]. The observed effect of the minute of acquisition could be linked to treadmill running accommodation, which can be different from individual to individual [53]. Moreover, our results show that the presence and manipulation of the optical flow does not influence this accommodation process, since no difference was found between the four optical flow conditions.

Even if there was no significant effect of the optical flow conditions on spatiotemporal running parameters, VR seems to influence HR. Specifically, we found significant differences between the noOF and the VR conditions. Because the order of presentation of the conditions was randomized, exercise intensity was the same for each subject in all conditions, and starting HR did not differ between blocks. Therefore, the observed difference should not be linked to the running task per se. The higher HR values in the VR conditions may therefore be linked to an emotional activation, HR being a possible indicator of excitement [54]. This would also explain the absence of difference in perceived effort as measured with the Borg RPE Scale, even with the higher HR values in the VR conditions. In fact, even if there was no significant difference, there was a slight tendency toward lower RPE values for the VR conditions compared to the noOF condition. This could confirm the results of previous studies that suggest that VR exercise equipment could act as a distraction from the exercise intensity [55,56].

4. General Discussion

The control of locomotion by the central nervous system relies on the integration of sensory information provided by different systems such as the visual, vestibular, proprioceptive, and auditory systems. When the information provided by one of these sensory inputs is considered as less reliable or is in conflict with other cues, the central nervous system tends to weight sensory information taking into account its estimated reliability. Our results suggest that when running in a treadmill-mediated virtual environment, participants tend to provide a higher weight to proprioceptive information than to visual information. Specifically, the presence and manipulation of a simulated optical flow did not affect the ability of participants to retrieve a 'proprioceptive' reference speed previously experienced. In addition, optical flow did not affect the spatiotemporal gait parameters known to be indicators of gait stability.

The results of Experiment 2 confirm that the speed of an artificial optical flow is underestimated relative to treadmill speed when running in treadmill-mediated virtual environments. In particular, our results show that this relative underestimation applies not only to situations in which the running speed is imposed, as shown in previous studies [47–49], but also to situations in which participants are allowed to run at their preferred running speed. Specifically, in our study, the participants were free to select their running speed for the tests. This likely increased their feeling of safety compared to tests performed at an imposed running speed. It also likely lowered the influence of fatigue and the fear of falling off of the treadmill [50]. This means that the visual underestimation previously reported with participants running in virtual environments still applies at running speeds for which past experience from practice could have been used for recalibration [57,58]. This could be linked to treadmill running in itself, since it has been shown to influence speed perception. In fact, treadmill running has been shown to be perceived as faster compared to the same overground running speed [59–62]. It has also been shown to influence the capacity to discriminate actual running speed due to differences in kinetics and kinematics compared to overground running [53,63]. Another factor that could have influenced the results is the specific experimental setup. In fact, studies suggest that visual speed is perceived more accurately as peripheral flow increases [44,49,64–66]. Moreover, a 3D stereoscopic environment would have been more immersive than the simple 2D on the projection screen and studies show that 3D environments induce a stronger sense of presence during spatial navigation tasks [67]. Nevertheless, here we used a 2D setup since more immersive 3D virtual reality situations inducing a strong sense of presence have been shown to have a greater capability of inducing postural instability compared to 2D VR situations [68] presenting also a greater possibility that participants might have experienced VR induced symptoms and effects [69].

Though our study confirms the tendency to misperceive the speed of a simulated optical flow when running in treadmill-mediated virtual environments, it shows that this misperception might not have an influence on the control of treadmill running itself. This result suggests that the findings of previous studies that reported an influence of the optical flow rate on the control of locomotor speed might not apply when running on a treadmill. This difference between previous results and ours could result from a higher automatization of running compared to walking, with the former requiring less cortical control [25,29]. This would in turn lead to a reduced influence of visual incongruences on the control of running speed. This would be in line with the results of Jahn et al. [24,70], who found larger gait deviations with vestibular and visual stimulation during walking compared to running. Those results can probably be attributed to a suppression of sensory signals during unhindered running in order to avoid potential disturbances of the optimized spinal program controlling running motor pattern [71,72]. In fact, studies have shown that locomotion without disturbances or obstacles is more independent of cortical control, the latter being mainly required when exact foot placement is needed, or when the system must react to external perturbations, notably by changing muscle activity to modify limb trajectories to step over obstacles [73]. This would also explain the lower impact of visual incongruences on balance and locomotor patterns while treadmill running, as shown in Experiment 2. Our results contradict previous studies on walking, which showed an influence of an artificial optical flow on postural stability [31–34] and on locomotor patterns. In particular, Pailhous et al. [17] found that when walking overground, stride length decreases in presence of an artificial optical flow. Prokop et al. [15] found that when walking on a treadmill at constant speed, stride-cycle variability increases and the optical flow significantly modulates walking velocity. Hollman et al. [36] found that when walking at pre-determined speeds on a treadmill in VR, step width increases and step length decreases, whereas the variability of stride velocity and step width increases. All these parameters are known to characterize gait instability [51,52], suggesting that the presence of VR during treadmill locomotion may lead to a more instable gait. In fact, a shorter step length and an increased step width are associated with a more conservative gait pattern, often adopted in case of instability or fear of falling. Other studies show that this influence on locomotor patterns extends also to the walk–run and run–walk transition, leading to changes in the transition speed based on the speed of the optical flow [38,74] Our results show that this influence on locomotor patterns cannot be extended to running individuals, and that for running at a self-selected speed, the presence and manipulation of the optical flow has no stabilizing or destabilizing effect. This could also be due to the fact that when visual cues are in conflict with other sensory information about self-motion, which is the case when manipulating optical flow speed without having the possibility to vary locomotor speed, visual information ceases to be dominant compared to other types of information [4,14]. In fact, studies show that when participants are asked to walk at a constant speed, they tend to weight information from proprioceptive cues higher than cues derived from the visual system [15–17].

Our results also indicate that the simulated optical flow does not influence the process of accommodation to treadmill running, because no differences were found between the four optical flow conditions. This accommodation process has been shown to vary between individuals [53]. Healthy adults devoid of experience with treadmill running can familiarize themselves with the task after 6 min in a single session, and angular kinematics as well as spatiotemporal parameters are highly reliable after 2 min [75]. White et al. [76] found that there was no significant difference in vertical force after 30 s of treadmill running. The differences we found between minute 1 and 3 could be linked to the new accommodation required after each pause in between the blocks. This would be in line with the fact that previous studies have shown that minimal amounts of treadmill training are necessary before full accommodation to treadmill locomotion [77].

Surprisingly, our results suggest that the presentation of an artificial optical flow can influence HR. Specifically, the results of Experiment 2 show a significant difference between the HR monitored in the noOF condition and the HR monitored in the VR conditions. Because the HR is an indicator of excitement [54], the observed results could be linked to an emotional activation, which would also explain the absence of difference between the four optical flow conditions regarding perceived effort. This excitement could be due to the novelty of the virtual scene and to some sense of presence deriving from the optical flow compared to the traditional treadmill-running settings, which are often considered as boring [78]. This would be in line with previous studies showing that VR can act as a distraction from exercise intensity, favoring a dissociation of the attentional focus [55,56], and improving some of the mood benefits of physical exercise [79]. If confirmed, this effect could be exploited to increase engagement and adherence to physical activity, notably in the context of health and disease prevention.

Supplementary Materials: The following are available online at http://www.mdpi.com/2076-3417/9/21/4589/s1, Archive S1: Data tables.

Author Contributions: Conceptualization, M.C., J.-P.B. and A.D.; Data curation, M.C. and A.D.; Formal analysis, M.C., J.-P.B. and A.D.; Investigation, M.C. and A.D.; Methodology, M.C. and A.D.; Resources, C.L., E.M., O.A.K. and J.-P.B.; Software, J.-P.B. and A.D.; Supervision, C.L., E.M., O.A.K., J.-P.B. and A.D.; Validation, M.C. and J.-P.B.; Visualization, M.C., J.-P.B. and A.D.; Writing—original draft, M.C.; Writing—review and editing, M.C., C.L., E.M., O.A.K., J.-P.B. and A.D.

Funding: This research received no external funding.

Acknowledgments: The authors thank Florian Blunschi and Lisa Perissinotto (Department of Neuroscience and Movement Science, University of Fribourg, Fribourg, Switzerland) for recruiting the participants and for providing support in data collection, Marco Scodeggio (Istituto di Bioimmagini e Fisiologia Molecolare, Consiglio Nazionale delle Ricerche, Segrate, Milano, Italy) for technical assistance in constructing the response device, and Thibaut Le Naour (Department of Neuroscience and Movement Science, University of Fribourg, Fribourg, Switzerland) for his help in the creation of the representation of the experimental set-up (Figure 1).

Conflicts of Interest: The authors declare no conflict of interest.

References

1. Sun, H.-J.; Campos, J.L.; Chan, G.S.W. Multisensory integration in the estimation of relative path length. *Exp. Brain Res.* **2004**, *154*, 246–254. [CrossRef]
2. Mergner, T.; Rosemeier, T. Interaction of vestibular, somatosensory and visual signals for postural control and motion perception under terrestrial and microgravity conditions—A conceptual model. *Brain Res. Rev.* **1998**, *28*, 118–135. [CrossRef]
3. Dietz, V. Proprioception and locomotor disorders. *Nat. Rev. Neurosci.* **2002**, *3*, 781. [CrossRef] [PubMed]
4. Sun, H.J.; Lee, A.J.; Campos, J.L.; Chan, G.S.W.; Zhang, D.H. Multisensory integration in speed estimation during self-motion. *Cyberpsychol. Behav.* **2003**, *6*, 509–518. [CrossRef] [PubMed]
5. Angelaki, D.E.; Cullen, K.E. Vestibular system: The many facets of a multimodal sense. *Annu. Rev. Neurosci.* **2008**, *31*, 125–150. [CrossRef]
6. Warren, W.H.; Hannon, D.J. Direction of self-motion is perceived from optical flow. *Nature* **1988**, *336*, 162–163. [CrossRef]
7. Larish, J.F.; Flach, J.M. Sources of optical information useful for perception of speed of rectilinear self-motion. *J. Exp. Psychol. Hum. Percept. Perform.* **1990**, *16*, 295. [CrossRef]
8. Lee, D.N. A theory of visual control of braking based on information about time-to-collision. *Perception* **1976**, *5*, 437–459. [CrossRef]
9. Bruggeman, H.; Zosh, W.; Warren, W.H. Optic flow drives human visuo-locomotor adaptation. *Curr. Biol.* **2007**, *17*, 2035–2040. [CrossRef]

10. Rushton, S.K.; Harris, J.M.; Lloyd, M.R.; Wann, J.P. Guidance of locomotion on foot uses perceived target location rather than optic flow. *Curr. Biol.* **1998**, *8*, 1191–1194. [CrossRef]
11. Warren, W.H.; Kay, B.A.; Zosh, W.D.; Duchon, A.P.; Sahuc, S. Optic flow is used to control human walking. *Nat. Neurosci.* **2001**, *4*, 213. [CrossRef] [PubMed]
12. Bremmer, F.; Lappe, M. The use of optical velocities for distance discrimination and reproduction during visually simulated self motion. *Exp. Brain Res.* **1999**, *127*, 33–42. [CrossRef] [PubMed]
13. Riecke, B.E.; Veen, H.A.H.C.V.; Bülthoff, H.H. Visual homing is possible without landmarks: A path integration study in virtual reality. *Presence Teleoper. Virtual Environ.* **2002**, *11*, 443–473. [CrossRef]
14. Harris, L.R.; Jenkin, M.; Zikovitz, D.C. Visual and non-visual cues in the perception of linear self motion. *Exp. Brain Res.* **2000**, *135*, 12–21. [CrossRef]
15. Prokop, T.; Schubert, M.; Berger, W. Visual influence on human locomotion modulation to changes in optic flow. *Exp. Brain Res.* **1997**, *114*, 63–70. [CrossRef]
16. Varraine, E.; Bonnard, M.; Pailhous, J. Interaction between different sensory cues in the control of human gait. *Exp. Brain Res.* **2002**, *142*, 374–384. [CrossRef]
17. Pailhous, J.; Ferrandez, A.-M.; Flückiger, M.; Baumberger, B. Unintentional modulations of human gait by optical flow. *Behav. Brain Res.* **1990**, *38*, 275–281. [CrossRef]
18. Durgin, F.H.; Fox, L.F.; Schaffer, E.; Whitaker, R. The perception of linear self-motion. In Proceedings of the Electronic Imaging 2005, San Jose, CA, USA, 18 March 2005; pp. 503–514.
19. Mohler, B.J.; Thompson, W.B.; Creem-Regehr, S.; Pick, H.L.; Warren, W.; Rieser, J.J.; Willemsen, P. Visual motion influences locomotion in a treadmill virtual environment. In Proceedings of the 1st Symposium on Applied Perception in Graphics and Visualization, Los Angeles, CA, USA, 7–8 August 2004; pp. 19–22.
20. Konczak, J. Effects of optic flow on the kinematics of human gait: A comparison of young and older adults. *J. Mot. Behav.* **1994**, *26*, 225–236. [CrossRef]
21. Baumberger, B.; Flückiger, M.; Roland, M. Walking in an environment of moving ground texture. *Jpn. Psychol. Res.* **2000**, *42*, 238–250. [CrossRef]
22. Zijlstra, W.; Rutgers, A.W.F.; Hof, A.L.; Van Weerden, T.W. Voluntary and involuntary adaptation of walking to temporal and spatial constraints. *Gait Posture* **1995**, *3*, 13–18. [CrossRef]
23. Brandt, T. Vestibulopathic gait: you're better off running than walking. *Curr. Opin. Neurol.* **2000**, *13*, 3–5. [CrossRef] [PubMed]
24. Jahn, K.; Strupp, M.; Schneider, E.; Dieterich, M.; Brandt, T. Visually induced gait deviations during different locomotion speeds. *Exp. Brain Res.* **2001**, *141*, 370–374. [CrossRef] [PubMed]
25. Jahn, K.; Deutschländer, A.; Stephan, T.; Strupp, M.; Wiesmann, M.; Brandt, T. Brain activation patterns during imagined stance and locomotion in functional magnetic resonance imaging. *Neuroimage* **2004**, *22*, 1722–1731. [CrossRef] [PubMed]
26. Durgin, F.H.; Gigone, K.; Scott, R. Perception of visual speed while moving. *J. Exp. Psychol. Hum. Percept. Perform.* **2005**, *31*, 339. [CrossRef] [PubMed]
27. Van Doorn, J.; van den Bergh, D.; Bohm, U.; Dablander, F.; Derks, K.; Draws, T.; Evans, N.J.; Gronau, Q.F.; Hinne, M.; Kucharský, Š. The JASP Guidelines for Conducting and Reporting a Bayesian Analysis. *PsyArXiv* **2019**. [CrossRef]
28. Powell, W.; Hand, S.; Stevens, B.; Simmonds, M.J. Optic flow with a stereoscopic display: Sustained influence on speed of locomotion. *Annu. Rev. Cyber Ther. Telemed.* **2006**, *4*, 65–70.
29. Brandt, T.; Strupp, M.; Benson, J. You are better off running than walking with acute vestibulopathy. *Lancet* **1999**, *354*, 746. [CrossRef]
30. Ernst, M.O.; Banks, M.S. Humans integrate visual and haptic information in a statistically optimal fashion. *Nature* **2002**, *415*, 429. [CrossRef]
31. Keshner, E.A.; Kenyon, R.V. The influence of an immersive virtual environment on the segmental organization of postural stabilizing responses. *J. Vestib. Res.* **2000**, *10*, 207–219.
32. Wright, W.G. Using virtual reality to induce cross-axis adaptation of postural control: Implications for rehabilitation. In Proceedings of the 2013 International Conference on Virtual Rehabilitation (ICVR), Philadelphia, PA, USA, 26–29 August 2013; pp. 289–294.
33. Wright, W.G. Using virtual reality to augment perception, enhance sensorimotor adaptation, and change our minds. *Front. Syst. Neurosci.* **2014**, *8*. [CrossRef]

34. Slobounov, S.; Sebastianelli, W.; Newell, K.M. Incorporating virtual reality graphics with brain imaging for assessment of sport-related concussions. In Proceedings of the 2011 Annual International Conference of the IEEE Engineering in Medicine and Biology Society, Boston, MA, USA, 30 August–3 September 2011; pp. 1383–1386.
35. Hollman, J.H.; Brey, R.H.; Bang, T.J.; Kaufman, K.R. Does walking in a virtual environment induce unstable gait? An examination of vertical ground reaction forces. *Gait Posture* **2007**, *26*, 289–294. [CrossRef]
36. Hollman, J.H.; Brey, R.H.; Robb, R.A.; Bang, T.J.; Kaufman, K.R. Spatiotemporal gait deviations in a virtual reality environment. *Gait Posture* **2006**, *23*, 441–444. [CrossRef]
37. Kastavelis, D.; Mukherjee, M.; Decker, L.M.; Stergiou, N. The effect of virtual reality on gait variability. *Nonlinear Dyn. Psychol. Life Sci.* **2010**, *14*, 239–256.
38. Guerin, P.; Bardy, B.G. Optical modulation of locomotion and energy expenditure at preferred transition speed. *Exp. Brain Res.* **2008**, *189*, 393–402. [CrossRef]
39. Parfitt, G.; Rose, E.A.; Markland, D. The effect of prescribed and preferred intensity exercise on psychological affect and the influence of baseline measures of affect. *J. Health Psychol.* **2000**, *5*, 231–240. [CrossRef]
40. Parfitt, G.; Rose, E.A.; Burgess, W.M. The psychological and physiological responses of sedentary individuals to prescribed and preferred intensity exercise. *Br. J. Health Psychol.* **2006**, *11*, 39–53. [CrossRef] [PubMed]
41. Kingdom, F.A.A.; Prins, N. *Psychophysics: A Practical Introduction*; Elsevier Science: Amsterdam, The Netherlands, 2010.
42. Leek, M.R. Adaptive procedures in psychophysical research. *Percept. Psychophys.* **2001**, *63*, 1279–1292. [CrossRef] [PubMed]
43. Dubois, A.; Bresciani, J.-P. Validation of an ambient system for the measurement of gait parameters. *J. Biomech.* **2018**, *69*, 175–180. [CrossRef] [PubMed]
44. Banton, T.; Stefanucci, J.; Durgin, F.; Fass, A.; Proffitt, D.R. The perception of walking speed in a virtual environment. *Presence* **2005**, *14*, 394–406. [CrossRef]
45. Powell, W.; Stevens, B.; Hand, S.; Simmonds, M. Blurring the boundaries: The perception of visual gain in treadmill-mediated virtual environments. In Proceedings of the 3rd IEEE VR 2011 Workshop on Perceptual Illusions in Virtual Environments, Singapore, 19 March 2011.
46. Kassler, L.; Feasel, J.; Lewek, M.D.; Brooks, F.P., Jr.; Whitton, M.C. Matching actual treadmill walking speed and visually perceived walking speed in a projection virtual environment. In Proceedings of the 7th Symposium on Applied Perception in Graphics and Visualization, Los Angeles, CA, USA, 23–24 July 2010; p. 161.
47. Caramenti, M.; Lafortuna, C.L.; Mugellini, E.; Abou Khaled, O.; Bresciani, J.-P.; Dubois, A. Matching optical flow to motor speed in virtual reality while running on a treadmill. *PLoS ONE* **2018**, *13*, e0195781. [CrossRef]
48. Caramenti, M.; Lafortuna, C.L.; Mugellini, E.; Abou Khaled, O.; Bresciani, J.-P.; Dubois, A. Regular physical activity modulates perceived visual speed when running in treadmill-mediated virtual environments. *PLoS ONE* **2019**, *14*, e0219017. [CrossRef] [PubMed]
49. Caramenti, M.; Pretto, P.; Lafortuna, C.; Bresciani, J.-P.; Dubois, A. Influence of the size of the field of view on visual perception while running in a treadmill-mediated virtual environment. *Front. Psychol.* **2019**, *10*, 2344. [CrossRef]
50. Abernethy, B.; Hanna, A.; Plooy, A. The attentional demands of preferred and non-preferred gait patterns. *Gait Posture* **2002**, *15*, 256–265. [CrossRef]
51. Menz, H.B.; Lord, S.R.; Fitzpatrick, R.C. Age-related differences in walking stability. *Age Ageing* **2003**, *32*, 137–142. [CrossRef]
52. Krebs, D.E.; Goldvasser, D.; Lockert, J.D.; Portney, L.G.; Gill-Body, K.M. Is base of support greater in unsteady gait? *Phys. Ther.* **2002**, *82*, 138–147. [CrossRef]
53. Nigg, B.M.; De Boer, R.W.; Fisher, V. A kinematic comparison of overground and treadmill running. *Med. Sci. Sports Exerc.* **1995**, *27*, 98–105. [CrossRef]
54. Obrist, P.A. *Cardiovascular Psychophysiology: A Perspective*; Springer Science & Business Media: Berlin, Germany, 2012. [CrossRef]
55. Annesi, J.J.; Mazas, J. Effects of virtual reality-enhanced exercise equipment on adherence and exercise-induced feeling states. *Percept. Mot. Skills* **1997**, *85*, 835–844. [CrossRef]
56. Mestre, D.; Dagonneau, V.; Mercier, C.-S. Does virtual reality enhance exercise performance, enjoyment, and dissociation? An exploratory study on a stationary bike apparatus. *Presence* **2011**, *20*, 1–14. [CrossRef]

57. Bingham, G.; Romack, J.L. The rate of adaptation to displacement prisms remains constant despite acquisition of rapid calibration. *J. Exp. Psychol. Hum. Percept. Perform.* **1999**, *25*, 1331. [CrossRef]
58. Redding, G.M.; Wallace, B. Prism adaptation during target pointing from visible and nonvisible starting locations. *J. Mot. Behav.* **1997**, *29*, 119–130. [CrossRef]
59. Kong, P.W.; Candelaria, N.G.; Tomaka, J. Perception of self-selected running speed is influenced by the treadmill but not footwear. *Sports Biomech.* **2009**, *8*, 52–59. [CrossRef] [PubMed]
60. Kong, P.W.; Koh, T.M.C.; Tan, W.C.R.; Wang, Y.S. Unmatched perception of speed when running overground and on a treadmill. *Gait Posture* **2012**, *36*, 46–48. [CrossRef] [PubMed]
61. Marsh, A.P.; Katula, J.A.; Pacchia, C.F.; Johnson, L.C.; Koury, K.L.; Rejeski, W.J. Effect of treadmill and overground walking on function and attitudes in older adults. *Med. Sci.Sports Exerc.* **2006**, *38*, 1157. [CrossRef] [PubMed]
62. White, S.C.; Yack, H.J.; Tucker, C.A.; Lin, H.-Y. Comparison of vertical ground reaction forces during overground and treadmill walking. *Med. Sci. Sports Exerc.* **1998**, *30*, 1537–1542. [CrossRef] [PubMed]
63. Riley, P.O.; Dicharry, J.; Franz, J.A.S.O.N.; Croce, U.D.; Wilder, R.P.; Kerrigan, D.C. A kinematics and kinetic comparison of overground and treadmill running. *Med. Sci. Sports Exerc.* **2008**, *40*, 1093. [CrossRef] [PubMed]
64. Thurrell, A.E.I.; Pelah, A.; Distler, H.K. The influence of non-visual signals of walking on the perceived speed of optic flow. *Perception* **1998**, *27*, 147–148.
65. Thurrell, A.E.I.; Pelah, A. Reduction of perceived visual speed during walking: Effect dependent upon stimulus similarity to the visual consequences of locomotion. *J. Vis.* **2002**, *2*, 628. [CrossRef]
66. Nilsson, N.C.; Serafin, S.; Nordahl, R. Establishing the range of perceptually natural visual walking speeds for virtual walking-in-place locomotion. *IEEE Trans. Vis. Comput. Graph.* **2014**, *20*, 569–578. [CrossRef]
67. Kober, S.E.; Kurzmann, J.; Neuper, C. Cortical correlate of spatial presence in 2D and 3D interactive virtual reality: An EEG study. *Int. J. Psychophysiol.* **2012**, *83*, 365–374. [CrossRef]
68. Slobounov, S.M.; Ray, W.; Johnson, B.; Slobounov, E.; Newell, K.M. Modulation of cortical activity in 2D versus 3D virtual reality environments: An EEG study. *Int. J. Psychophysiol.* **2015**, *95*, 254–260. [CrossRef]
69. Sharples, S.; Cobb, S.; Moody, A.; Wilson, J.R. Virtual reality induced symptoms and effects (VRISE): Comparison of head mounted display (HMD), desktop and projection display systems. *Displays* **2008**, *29*, 58–69. [CrossRef]
70. Jahn, K.; Strupp, M.; Schneider, E.; Dieterich, M.; Brandt, T. Differential effects of vestibular stimulation on walking and running. *Neuroreport* **2000**, *11*, 1745–1748. [CrossRef] [PubMed]
71. Armstrong, D.M. The supraspinal control of mammalian locomotion. *J. Physiol.* **1988**, *405*, 1–37. [CrossRef] [PubMed]
72. Dietz, V. Spinal cord pattern generators for locomotion. *Clin. Neurophysiol.* **2003**, *114*, 1379–1389. [CrossRef]
73. Drew, T.; Prentice, S.; Schepens, B. Cortical and brainstem control of locomotion. In *Progress in Brain Research*; Elsevier: Amsterdam, The Netherlands, 2004; Volume 143, pp. 251–261.
74. Mohler, B.J.; Thompson, W.B.; Creem-Regehr, S.H.; Pick, H.L., Jr.; Warren, W.H., Jr. Visual flow influences gait transition speed and preferred walking speed. *Exp. Brain Res.* **2007**, *181*, 221–228. [CrossRef]
75. Lavcanska, V.; Taylor, N.F.; Schache, A.G. Familiarization to treadmill running in young unimpaired adults. *Hum. Mov. Sci.* **2005**, *24*, 544–557. [CrossRef]
76. White, S.C.; Gilchrist, L.A.; Christina, K.A. Within-day accommodation effects on vertical reaction forces for treadmill running. *J. Appl. Biomech.* **2002**, *18*, 74–82. [CrossRef]
77. Schieb, D.A. Kinematic accommodation of novice treadmill runners. *Res. Q. Exerc. Sport* **1986**, *57*, 1–7. [CrossRef]
78. Thompson Coon, J.; Boddy, K.; Stein, K.; Whear, R.; Barton, J.; Depledge, M.H. Does participating in physical activity in outdoor natural environments have a greater effect on physical and mental wellbeing than physical activity indoors? A systematic review. *Environ. Sci. Technol.* **2011**, *45*, 1761–1772. [CrossRef]
79. Plante, T.G.; Aldridge, A.; Bogden, R.; Hanelin, C. Might virtual reality promote the mood benefits of exercise? *Comput. Hum. Behav.* **2003**, *19*, 495–509. [CrossRef]

applied sciences

MDPI

Article

Patellofemoral Joint Loads during Running Immediately Changed by Shoes with Different Minimalist Indices: A Cross-sectional Study

Chenhao Yang [1], Songlin Xiao [1], Yang Yang [1], Xini Zhang [1], Junqing Wang [1] and Weijie Fu [1,2,*]

1 School of Kinesiology, Shanghai University of Sport, Shanghai 200438, China; chyang9610@163.com (C.Y.); xiao_songlin@126.com (S.X.); 18049922807@163.com (Y.Y.); zhangxini1129@163.com (X.Z.); wangjunqingray@163.com (J.W.)

2 Key Laboratory of Exercise and Health Sciences of Ministry of Education, Shanghai University of Sport, Shanghai 200438, China

* Correspondence: fuweijie@sus.edu.cn or fuweijie315@163.com; Tel.: +86-21-65507368

Received: 14 September 2019; Accepted: 2 October 2019; Published: 6 October 2019

Abstract: Purpose: Given the high incidence of patellofemoral pain syndrome (PFPS) in runners, this study aimed to investigate the immediate effect of shoes with different minimalist indices (MI) on the mechanical loads of the patellofemoral joint. Methods: Fifteen healthy male rearfoot strike runners were recruited to complete overground running trials at 3.33 m/s (±5%) in two running shoe conditions (MI = 26% versus MI = 86%). The amount of ten Vicon infrared cameras (100 Hz) and two Kistler force plates (1000 Hz) were used to collect kinematic and ground reaction force (GRF) data simultaneously. Quadriceps strength, patellofemoral contact force, patellofemoral contact area, and patellofemoral contact stress were calculated. Results: No significant differences were observed in the impact force and the second peak of the vertical GRF between the two shoe conditions. Compared to wearing low-MI shoes, wearing high-MI shoes showed that the maximum flexion angle of the knee, the contact area of patellofemoral joint and the peak knee extension moment reduced significantly ($p < 0.01$), and the peak patellofemoral contact force and stress decreased significantly ($p < 0.05$). Conclusion: These findings suggest that wearing high-MI shoes significantly decreases the patellofemoral contact force and patellofemoral joint stress by reducing the moment of knee extension, thus effectively reducing the load of the patellofemoral joint during the stance phase of running and potentially lowering the risk of PFPS.

Keywords: minimalist index; patellofemoral contact force; patellofemoral contact stress; footwear; patellofemoral joint pain syndrome

1. Introduction

Hundreds of millions of citizens participate in sports, among which running is widely known. Injury remains high with the popularity of running, with 19.4% to 79.3% sustaining a running-related injury annually [1]; patellofemoral pain syndrome (PFPS) is one of the most common running injuries, exhibiting the highest incidence of 17% in the specific pathologies of running-related injuries [2]. Increased patellofemoral joint stress (PFJS) was determined as an important pathogenic factor [3]. The knee joint features a complex structure; the soft tissue and muscles maintain the stability of the tibia [4]. Thus, any change in mechanics may influence the force distribution around the patellofemoral joint. Studies have shown that patients with patellofemoral joint pain exhibit different biomechanical characteristics during running [5]. Therefore, changes in knee joint torque, patellofemoral contact force and stress during this activity may affect the risk of PFPS. Correspondingly, the mechanism of overwork injury is due to the cumulative effect of repeated high-load work of the muscles and bones [6];

therefore, reducing the knee load to a certain extent may considerably decrease the cumulative effect of long-term running on the knee joint.

With the development of technology, new materials and designs have been applied to sports shoes, especially midsole structures, to reduce the load on the lower extremities. The unique mechanical structures of different sports shoes can change the force characteristics of the human running gait [7]. For healthy runners, wearing different sneakers can affect the biomechanical characteristics of the lower extremities, such as stride frequency, ground contact angle, vertical load rate and joint force [8]. In recent years, barefoot running has become increasingly popular, and many runners have been opting for minimalist shoes, which provide basic protection but limited cushioning. The standardised definition of minimalist shoes, the Minimalist Index (MI), was developed to allow running shoes to be distinguished based on their degree of minimalism, and may help to decrease injuries related to footwear transition [9]. Numerous studies have verified [7,10,11] that minimalist shoes can change the landing posture of the lower limbs, mainly focusing on the adaptation of the ankle joints, that is changing the angle between the foot and the ground, the tendency of forefoot pattern strike and the adaptability of the calf triceps and the Achilles tendon. Relatively few studies have focused on knee load and this condition limits the understanding of the relationship between sneakers and knee joints, especially the patellofemoral joint.

This study aimed to determine the biomechanical characteristics of the patellofemoral joint during running. To achieve this the immediate effects of wearing running shoes with different minimalist indices on patellofemoral contact force (PFCF) and patellofemoral contact stress (PFJS) were explored. Based on the previous observations, it was hypothesized that PFCF and PFJS would decrease when wearing running shoes with high Minimalist Index (MI), e.g., minimalist shoes.

2. Materials and Methods

2.1. Participants

A total of fifteen recreational and healthy male runners (age: 31.4 ± 6.6 years; height: 174.7 ± 6.3 cm; body mass: 73.2 ± 9.8 kg; weekly running volume: 30.6 ± 9.5 km) were recruited. Inclusion criteria were as follows: 1) The weekly running volume in the past month was over 20 km; 2) rearfoot strikers and to be accustomed to wearing cushioned shoes; 3) no experience in barefoot running nor wearing minimalist shoes or special sneakers (e.g., five-finger shoes and racing spikes); 4) suffered no musculoskeletal injuries within three months prior to the tests. A two-tailed t-test was executed via the G*Power 3.1 software (Univ. Kiel, Kiel, Germany) to determine whether a sample size of 15 was sufficient to minimize the probability of type II error for all the variables ($P = 80\%$ at $\alpha = 0.05$).

This study, with detailed guidelines for participants' safety and experimental protocols, was approved by the Institutional Review Board of the Shanghai University of Sport (No. 2017007). The study was conducted in accordance with the declaration of Helsinki. Specifically, all procedures and potential hazards were clarified to the participants in nontechnical terms, and informed consent was signed prior to the tests. All participants had full knowledge of test procedures and requirements.

2.2. Shoe Conditions

A rating scale for the running shoes was developed in accordance with the work of Esculier et al. [9] and the MI was defined as the minimalism of different running shoes. Specifically, the MI of running shoes was calculated according to guidelines by weight, heel-to-toe drop, stack height, motion control/stability technologies and flexibility. The motion control/stability technologies included multi-density midsole, rigid heel counter, elevated medial insole under the foot arch, and tensioned medial upper. Flexibility included longitudinal and torsional flexibility. A high MI indicated high minimalism of running shoes.

In this experiment, the size of the experimental shoes ranged from EUR 41 to 43 based on the foot size of the participants. The used minimalist shoes were the INOV-8 Bare-XF 210 V2 series

(Figure 1). The outsole of the shoe was made of INOV Corp.'s patented viscous rubber, which measured approximately 3 mm in thickness, and the shoes contained no cushioning midsole. The upper was made of a mesh composite material, which was light, breathable, fitted the foot surface and was good for toe movement during running. The heel-to-toe drop of the shoe was 0 mm, the weight (EUR size 42) was 227 g, and the MI = 86%. The used cushioned shoes were the Nike Air Zoom Pegasus 34. EVA foam was used in the midsole of this brand, and a Zoom Air cushion was set in the heel and the forefoot. The heel-to-toe drop was 7 mm, the weight (EUR size 42) was 285 g, and the MI = 26%. The runners were required to wear uniform Nike Dri-FIT socks to eliminate interference. Table 1 shows the specific MI of the two experimental shoes.

Figure 1. Shoes used in the experiment. Minimalist shoe: INOV-8 Bare-XF 210 V2 (**left**). Cushioned shoe: Nike Air Zoom Pegasus 34 (**right**).

Table 1. Detailed comparison of the Minimalist Index (MI) of the two shoes.

Shoe	Weight (g)	Stack Height (mm)	Heel-to-Toe Drop (mm)	Stability and Motion Control Technologies	Flexibility	MI (%)
INOV-8 Bare-XF 210 V2	227	3	0	None	longitudinal 360° torsional 360°	86
Subscore	2	5	5	5	4.5	
Nike Air Zoom Pegasus 34	285	30	7	Four devices	longitudinal 45°–90° torsional 0°–45°	26
Subscore	1	1	2	1	1.5	

Notes: Four stability and motion control technologies of the Nike Air Zoom Pegasus 34: multi-density midsole, rigid heel counter, supportive tensioned medial upper and medial flare. MI = sum of subscore × 4 × 100%.

2.3. Data Collection

First, the experimental process was explained to the participants and they signed the relevant questionnaire and informed consent. The same uniform sportswear was replaced, and forty infrared retroreflective markers (diameter: 14.0 mm) were attached bilaterally to both of the lower extremities to define hip, knee, and ankle joints according to the plug-in gait marker set [12]. Before the experiment, the participants performed a 5-min warm-up on a treadmill at an optional running speed only with cushioned shoes, followed by a 1-minute 3.33 m/s experimental speed adaptation. Meanwhile, it was the first time that they had worn minimalist shoes, which was considered as "immediately". After completing the preparation, the participants began the formal experimental tests.

Two 90 × 60 × 10 cm Kistler 3D force platforms (9287B, Kistler Corporation, Switzerland) were used to collect ground reaction force (GRF) data at a sampling rate of 1000 Hz. A 10-camera infrared 3D motion capture system (Vicon T40, Oxford Metrics, UK) was used to collect the trajectory markers (14.0 mm in diameter) of the bone landmarks of the lower extremities to define the knee joint at a sampling rate of 100 Hz (Figure 2). The participants randomly wore one of the shoes. A 5–10 minute break was provided between two shoe tests to eliminate the potential interference. Three running trials were collected per condition at 3.33 m/s (±5%). This target speed was common for recreational runners and all participants completed running trials without any discomfort. Running speed was monitored by the use of a Witty-Manual grating timing system (Micro gate, Italy) positioned 4 m apart along a 20 m runway. Meanwhile, the participants were required to run back and forth until three successful steps (trials) in which the foot of the dominant lower extremity (defined as preferred kicking leg [13]) was completely on the force plate were collected for each shoe condition. Each trial that met the following criteria was deemed successful: 1) The speed was 3.33 m/s (±5%) within the capture volume; 2) no relative sliding occurred between the shoes and the force plate.

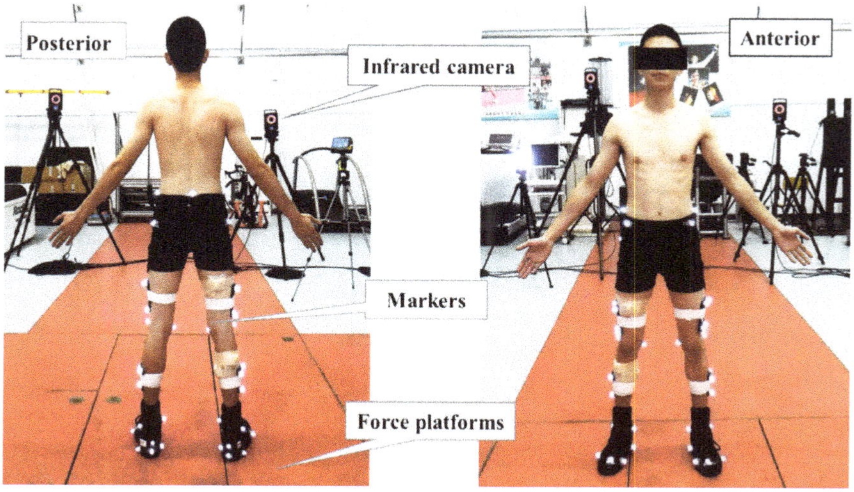

Figure 2. Set of reflective markers used in the study and the experimental setup.

2.4. Data Processing

Knee load during running was generally assessed by the peak torque of knee extension, patellofemoral contact force (PFCF) and patellofemoral contact stress (PFJS) [14,15]. First, marker trajectories were filtered with a cut-off frequency of 7 Hz via the gait analysis software Visual 3D (v5, C-Motion, Inc., Germantown, MD, USA). We established a 3D model in Visual 3D to calculate the GRF, knee flexion angle, torque of knee extension, angle of the knee joint when the runners touch the ground, angle of the ankle joint and the angle between the foot and the ground (foot inclination angle). The relevant parameters of the stance phase were standardised in time (0–100%), and a curve was plotted using the averages to identify the differences between the biomechanical parameters of the knee and various MI running shoes in the entire stand phase. In addition, PFCF, PFJS and related parameters were calculated by referring to previous studies. The details of the cited model were as follows.

The contact area (mm^2) between the patella and femur was a function of the sagittal knee angle and expressed as follows [16]:

$$S_{PFCA} = 0.0781 \times {\theta_i}^2 + 0.06763 \times \theta_i + 151.75 \quad (1)$$

where S_{PFCA} represents the contact area between the patella and the femur, and θ_i (°) was the sagittal angle of knee extension and flexion (Figure 3).

The effective arm of the quadriceps force (L_A, m) was a function of the sagittal knee angle and expressed as follows [17]:

$$L_A = \begin{cases} 0.036\theta_i + 3.0 \ (0 \leq \theta_i < 30°) \\ -0.043\theta_i + 5.4 \ (30 \leq \theta_i < 60°) \\ -0.027\theta_i + 4.3 \ (60 \leq \theta_i < 90°) \\ 2.0 \ (90 \leq \theta_i < 120°) \end{cases} . \quad (2)$$

Quadriceps force (N) was calculated as follows [14]:

$$F_Q\,(\theta_i) = M_{EXT}(\theta_i)/L_A\,(\theta_i) \tag{3}$$

where F_Q is the quadriceps force, and M_{EXT} is the extension moment (N·m).

The patellofemoral contact force (F_{PF}, N) was calculated as follows [17]:

$$F_{PF} = 2F_Q\sin(\beta/2) \tag{4}$$

where $\beta = 30.46 + 0.53(\theta_i)$, F_{PF} (N) is PFCF, and β (°) is the angle of the quadriceps line and patellar ligament (Figure 3).

The patellofemoral contact stress (MPa) was calculated as follows:

$$P_{PFS} = F_{PF}\,/S_{PFCA}\,(\theta_i) \tag{4}$$

where P_{PFS} is the PFJS.

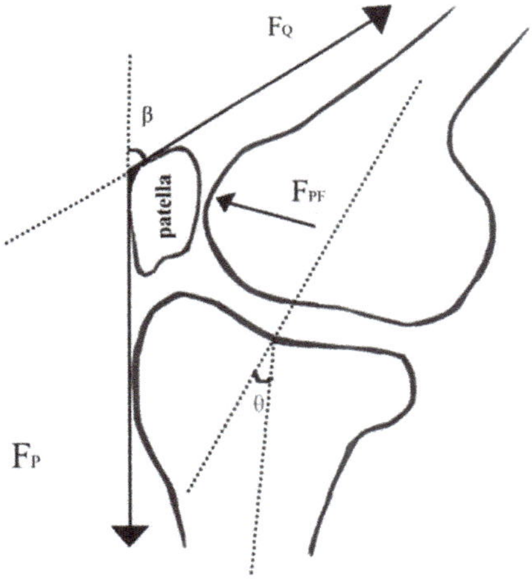

Figure 3. Free-body diagram of the patellofemoral joint. Notes: F_Q refers to the quadriceps muscle strength, F_{PF} denotes the patellofemoral contact force, F_P represents the patellar ligament tension line, β stands for the angle between the quadriceps muscle line and the patellar ligament tension line, and θ is the knee flexion angle.

2.5. Statistics

All the dependent variables were normally distributed, as indicated by the Shapiro–Wilk test. The paired *t*-test was used to determine the differences in all variables (GRF and joint mechanics) between the MI 86% and MI 26% of shoes (25.0, SPSS Inc., Chicago, IL, USA). The significance level was set as $\alpha = 0.05$.

3. Results

3.1. GRF

The vertical GRF (vGRF) of all the runners using both shoes exhibited two peaks. For the Minimal Index (MI) 86% and MI 26% shoes, no significant differences were observed between the impact force (first peak, FP) and the peak vGRF (second peak, SP) during the stance phase ($p > 0.05$) (Table 2).

3.2. Kinematics

The foot inclination angle at foot strike was positive regardless of shoe conditions, indicating that all the runners ran with a rearfoot strike. No statistical differences were identified between the knee flexion angle and the ankle plantar flexion angle at the moment of foot strike ($p > 0.05$). However, with the MI 86% shoes, the foot inclination angle significantly reduced by 35.6% ($p < 0.05$). Moreover, the peak knee flexion angle and the peak contact area of the patellofemoral joint were significantly lower than those with the MI 26% shoes, in which the peak knee flexion angle decreased by 6.5% and the peak contact area of the patellofemoral joint reduced by 5.4% ($p < 0.05$) (Table 2). The curve of the entire stance phase with the minimalist shoes indicated a considerably lower knee flexion angle than that with the cushioned shoes in the middle and late phases (Figure 4A).

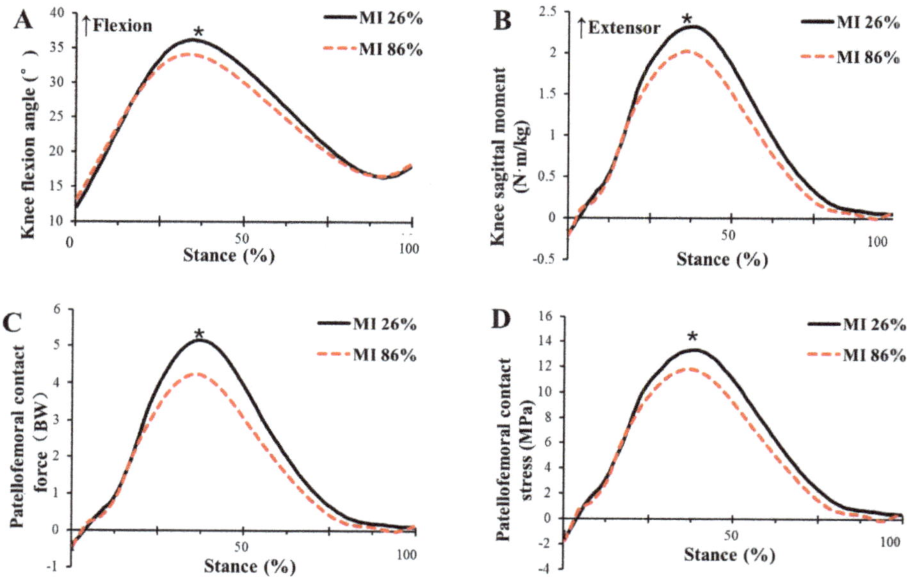

Figure 4. Comparison of the mean values of knee kinematics and kinetics from all participants between shoe conditions (MI 26%: cushioned shoes; MI 86%: minimalist shoes). (**A**) knee flexion angle during stance phase; (**B**) knee sagittal moment during stance phase; (**C**) patellofemoral contact force during stance phase; (**D**) patellofemoral contact stress during stance phase. *$P < 0.05$.

3.3. Kinetics

At the peak knee extension moment, the peaks of PFCF and PFJS were significantly lower with the MI 86% shoes than with the MI 26% shoes, in which, at the peak knee extension moment, the peaks of PFCF and PFJS reduced by 12.3%, 17.0% and 10.4%, respectively ($p < 0.05$) (Table 2). Consistent with the knee flexion angle curve, the curves of the abovementioned knee loading parameters were lower under the MI 86% shoe condition during the middle- and late-stance phases than those under the MI 26% shoe condition (Figure 4B–D).

Table 2. Sagittal plane kinematics and kinetics during running with two different shoe conditions ($\bar{x} \pm$ SD.

Phase	Parameters	MI 26%	MI 86%	p
Touch-down	Knee flexion angle (°)	12.64 ± 4.56	13.56 ± 5.60	0.637
	Ankle plantar flexion angle (°)	0.13 ± 7.92	0.13 ± 4.29	0.835
	Foot inclination angle (°)	12.54 ± 2.18	8.07 ± 4.64	<0.001
Stance phase	FP (BW)	1.74 ± 0.23	1.78 ± 0.16	0.701
	SP (BW)	2.59 ± 0.31	2.61 ± 0.25	0.559
	Peak knee flexion angle (°)	36.29 ± 3.41	33.93 ± 1.74	0.008
	Peak patellofemoral joint contact area (mm^2)	280.10 ± 21.96	265.08 ± 10.20	0.008
	Peak knee extension moment (N × m/kg)	2.36 ± 0.41	2.07 ± 0.47	<0.001
	Peak patellofemoral contact force (BW)	5.23 ± 1.12	4.34 ± 0.97	<0.001
	Peak patellofemoral contact stress (MPa)	13.48 ± 2.64	12.08 ± 3.27	0.015

Notes: Foot inclination angle is the angle between the foot and the ground. The FP (the first peak of the vGRF) and SP (the second peak of the vGRF), knee extension moment and patellofemoral contact force are standardised by body weight (BW). MI 86%: minimalist shoes. MI 26%: cushioned shoes.

4. Discussion

This study aimed to determine the immediate effects of wearing shoes with different Minimal Index (MI) on the patellofemoral joints and to explore whether directly changing shoes can reduce knee load and the risk of patellofemoral pain syndrome (PFPS). The conventional belief is that the cushioned structure of running shoes can reduce the impact force during running and protect the lower limbs. In this study, however, the MI 26% shoes caused no change in the impact force during the stance phase and the ankle/knee flexion angle at touchdown compared with those with the MI 86% shoes. Moreover, during the transition and propulsive phases of the stance phase (20–100%), wearing the MI 86% shoes can effectively reduce the patellofemoral contact force (PFCF) and patellofemoral contact stress (PFJS) by changing the parameters of knee loads, such as the knee flexion angle and knee extension moment. The mechanism of PFPS is that the patellofemoral joints are subjected to excessively high loads [14,18,19]. Therefore, the results of this study suggest that running in high-MI running shoes might be beneficial for reducing the risk of PFPS.

This finding is consistent with those in previous studies [20,21]. Sinclair and Esculier [7,8] reported that the body actively changed the lower limb strategy while wearing minimalist shoes, i.e., increased step rates, lowered foot inclination angle at foot strike and reduced knee maximum flexion angle and joint range of motion (ROM). Moreover, according to the equation for the knee joint angle and the patellofemoral joint contact area, a decreased knee ROM decreased the patellofemoral joint contact area during the stance phase of runners wearing minimalist shoes; this condition is adverse to the reduction of PFJS. By contrast, runners with high-MI shoes showed a considerably smaller knee extension moment during the propulsive phase, which substantially reduced PFCF and, eventually, the PFJS decreased by 10.4%. Furthermore, we explored the characteristics of PFCF and PFJS during the entire stance phase with minimalist shoes. Findings showed that in the 0–20% stance phase (braking phase), the curves were coincident on different shoe conditions (Figure 4), and the differences between the two shoes mostly occurred from 20% to 100% of the stance phase. First, the initial 20% of the stance phase is the braking phase. At this phase, the lower limbs were subjected to the impact of the ground, and the vGRF reached the FP (impact force). The results of the current study showed no significant difference in the impact or propulsive force (SP of vGRF) between the low- and high-MI shoes. The braking phase accounts for 1/4 of the entire stance phase, and this study suggests that wearing high-MI shoes is beneficial for decreasing knee load in the propulsive phase (20–100%). According to recent research supporting barefoot running, running shoes not only reduce the impact force and loading rate and filter high-frequency impact signals [22] but also provide different neural inputs to the lower limbs, resulting in different responses and changes in the lower limb strategy [23]. As shown in this study, the lower limbs actively changed the knee flexion angle and extension moment when the runner was wearing minimalist shoes during the propulsive phase. This phase accounts for a large proportion and may be important for reducing the patellofemoral joint loads.

Using the rating scale for minimalist shoes, this study compared minimalist shoes (MI = 86%) with cushioned shoes (MI = 26%) to further explore the relationship between knee joint load and the MI. Esculier [8] tested shoes with different MI with regards to running mechanics and observed that a higher MI shoe indicates a low foot inclination angle at foot strike and a low peak PFCF. Sinclair [7] compared minimalist, cushioned and maximalist shoes in terms of patellofemoral kinetics and observed that the knee ROM decreased when the runner was wearing minimalist shoes. Interestingly, the patellofemoral joint loads were greater with the low-MI shoes, i.e., maximalist shoes, than with the others. The above findings are similar to the results of this study. Shoes with low MI caused no reduction in the patellofemoral joint loading despite the thick midsole. By contrast, the neural inputs caused by the cushioning midsole resulted in a reduced limb movement strategy with increased knee flexion angle and knee extension moment, which may increase the risk of knee joint load and PFPS.

In previous studies, subjects were recruited regardless of their foot strike patterns and researchers failed to rule out runners who had experience with wearing minimalist shoes [13]. More than 70% of long-distance runners adopt the heel strike pattern [24,25]. Horvais noted that heel strikers use the midfoot or forefoot strike pattern during the adaptation process when wearing minimalist shoes for the first time [26]. In the current study, the vGRF graph showed two peaks, and the foot inclination angle at the point of foot strike was positive. Therefore, the runners still used the heel strike pattern after wearing the minimalist shoes [27], but the foot inclination angle at the point of foot strike still reduced by 35.6%.

Furthermore, adopting the heel strike pattern while wearing minimalist shoes on hard ground would cause heel pain. The runners who are used to rearfoot running have difficulties in switching to forefoot strike immediately when wearing minimalist shoes for a short distance. Correspondingly, runners will actively attempt to use the midfoot or forefoot strike pattern to avoid pain (various reasons lead to changes in foot strike patterns when transitioning to minimalist shoes) during longer running distances [28]. Numerous studies have shown that the reduced PFCF and PFJS in forefoot strike running can also reduce knee loads and PFPS [16,21,29]. Therefore, the analysis of the long-term effects of switching to high-MI running shoes on patellofemoral joint loads still needs to be confirmed.

In general, an adaptive process occurs when we first wear high-MI shoes. Wearing high-MI shoes, e.g., minimalist shoes, is similar to barefoot running, which has certain requirements (increased strength of the plantar muscles) [30]. Lieberman [31] stated that the function of the foot in human evolution is to adapt to barefoot walking or running, and humans began wearing cushioned shoes only in the last hundred years. Thus, the biomechanical characteristics of the foot may not be fully adapted to current cushioned shoes. The plantar muscles of runners who are accustomed to wearing cushioned shoes may lack strength to wear minimalist shoes for a long period, consequently resulting in fatigue and foot injury. Meanwhile, minimalist shoes change the runner's strike patterns to the forefoot, which requires increased calf triceps strength [32]. Therefore, in switching to high-MI shoes, the runner should gradually increase the duration of running in minimalist shoes based on previous running time and distances, thereby gradually improving the related muscle strength and the lower limb/foot adaptability.

Several limitations should be considered in this investigation that could be addressed in follow-up studies. First, we only considered the immediate shoe influence on patellofemoral joint loads. Therefore, the assessment of knee muscle forces or activation was warranted to provide further evidence of neuro-musculoskeletal reactions. Second, a long-term effect of wearing different MI shoes and the gender effects should ideally be taken into account. Finally, one should be cautioned that the differences in leg length were not considered in this study.

5. Conclusions

Patellofemoral joint loads during running were immediately changed by wearing shoes with different Minimalist Index (MI). In particular, while maintaining similar running velocity, wearing high-MI shoes significantly decreased patellofemoral contact force (PFCF) and patellofemoral contact stress (PFJS)—compared with those with low-MI shoes—by reducing the knee extension moment. This condition effectively reduced the loads on the patellofemoral joint during the mid-stance phase of running, thereby possibly reducing the risk of patellofemoral pain syndrome (PFPS).

Author Contributions: C.Y., S.X., contributed equally. Conceptualization, W.F.; methodology, C.Y., S.X.; formal analysis, C.Y., Y.Y., X.Z., J.W. and S.X.; investigation, C.Y., S.X., Y.Y., X.Z., J.W. and W.F.; resources, W.F.; data curation, X.Z.; writing—original draft preparation, C.Y., S.X.; writing—review and editing, W.F.; project administration, W.F.; funding acquisition, W.F.

Funding: This work was supported by the National Natural Science Foundation of China (11772201, 11572202); Talent Development Fund of Shanghai Municipal (2018107); the National Key Technology Research and Development Program of the Ministry of Science and Technology of China (2019YFF0302100), and the "Dawn" Program of Shanghai Education Commission, China.

Conflicts of Interest: The authors declare no conflict of interest.

References

1. Van Gent, R.N.; Siem, D.; van Middelkoop, M.; van Os, A.G.; Bierma-Zeinstra, S.M.; Koes, B.W. Incidence and determinants of lower extremity running injuries in long distance runners: A systematic review. *Br. J. Sports Med.* **2007**, *41*, 469–480. [CrossRef] [PubMed]
2. Ceyssens, L.; Vanelderen, R.; Barton, C.; Malliaras, P.; Dingenen, B. Biomechanical Risk Factors Associated with Running-Related Injuries: A Systematic Review. *Sports Med.* **2019**, *49*, 1095–1115. [CrossRef] [PubMed]
3. Dutton, R.A.; Khadavi, M.J.; Fredericson, M. Patellofemoral Pain. *Phys. Med. Rehabil. Clin. N. Am.* **2016**, *27*, 31–52. [CrossRef] [PubMed]
4. Jordan, M.J.; Aagaard, P.; Herzog, W. A comparison of lower limb stiffness and mechanical muscle function in ACL-reconstructed, elite, and adolescent alpine ski racers/ski cross athletes. *J. Sport Health Sci.* **2018**, *7*, 416–424. [CrossRef]
5. Neal, B.S.; Barton, C.J.; Gallie, R.; O'Halloran, P.; Morrissey, D. Runners with patellofemoral pain have altered biomechanics which targeted interventions can modify: A systematic review and meta-analysis. *Gait Posture* **2016**, *45*, 69–82. [CrossRef] [PubMed]
6. Del Coso, J.; Herrero, H.; Salinero, J.J. Injuries in Spanish female soccer players. *J. Sport Health Sci.* **2018**, *7*, 183–190. [CrossRef]
7. Sinclair, J.; Richards, J.; Selfe, J.; Fau-Goodwin, J.; Shore, H. The Influence of Minimalist and Maximalist Footwear on Patellofemoral Kinetics During Running. *J. Appl. Biomech.* **2016**, *32*, 359–364. [CrossRef]
8. Esculier, J.F.; Dubois, B.; Bouyer, L.J.; McFadyen, B.J.; Roy, J.S. Footwear characteristics are related to running mechanics in runners with patellofemoral pain. *Gait Posture* **2017**, *54*, 144–147. [CrossRef]
9. Esculier, J.F.; Dubois, B.; Dionne, C.E.; Leblond, J.; Roy, J.S. A consensus definition and rating scale for minimalist shoes. *J. Foot Ankl. Res.* **2015**, *8*, 42. [CrossRef]
10. Squadrone, R.; Rodano, R.; Hamill, J.; Preatoni, E. Acute effect of different minimalist shoes on foot strike pattern and kinematics in rearfoot strikers during running. *J. Sports Sci.* **2015**, *33*, 1196–1204. [CrossRef]
11. Rice, H.M.; Jamison, S.T.; Davis, I.S. Footwear Matters: Influence of Footwear and Foot Strike on Load Rates during Running. *Med. Sci. Sports Exerc.* **2016**, *48*, 2462–2468. [CrossRef] [PubMed]
12. Xia, R.; Zhang, X.; Wang, X.; Sun, X.; Fu, W. Effects of Two Fatigue Protocols on Impact Forces and Lower Extremity Kinematics during Drop Landings: Implications for Noncontact Anterior Cruciate Ligament Injury. *J. Healthc. Eng.* **2017**, *2017*, 5690519. [CrossRef] [PubMed]
13. Sun, X.; Yang, Y.; Wang, L.; Zhang, X.; Fu, W. Do Strike Patterns or Shoe Conditions have a Predominant Influence on Foot Loading? *J. Hum. Kinet.* **2018**, *64*, 13–23. [CrossRef] [PubMed]
14. Heino Brechter, J.; Powers, C.M. Patellofemoral stress during walking in persons with and without patellofemoral pain. *Med. Sci. Sports Exerc.* **2002**, *34*, 1582–1593. [CrossRef] [PubMed]

15. Dos Santos, A.F.; Nakagawa, T.H.; Serrao, F.V.; Ferber, R. Patellofemoral joint stress measured across three different running techniques. *Gait Posture* **2019**, *68*, 37–43. [CrossRef] [PubMed]
16. Vannatta, C.N.; Kernozek, T.W. Patellofemoral joint stress during running with alterations in foot strike pattern. *Med. Sci. Sports Exerc.* **2015**, *47*, 1001–1008. [CrossRef] [PubMed]
17. Bressel, E. The influence of ergometer pedaling direction on peak patellofemoral joint forces. *Clin. Biomech.* **2001**, *16*, 431–437. [CrossRef]
18. LaBella, C. Patellofemoral pain syndrome: Evaluation and treatment. *Prim. Care* **2004**, *31*, 977–1003. [CrossRef]
19. Besier, T.F.; Gold, G.E.; Beaupre, G.S.; Delp, S.L. A modeling framework to estimate patellofemoral joint cartilage stress in vivo. *Med. Sci. Sports Exerc.* **2005**, *37*, 1924–1930. [CrossRef]
20. Sinclair, J. Effects of barefoot and barefoot inspired footwear on knee and ankle loading during running. *Clin. Biomech.* **2014**, *29*, 395–399. [CrossRef]
21. Kulmala, J.P.; Avela, J.; Pasanen, K.; Parkkari, J. Forefoot strikers exhibit lower running-induced knee loading than rearfoot strikers. *Med. Sci. Sports Exerc.* **2013**, *45*, 2306–2313. [CrossRef] [PubMed]
22. Nigg, B.M.; Baltich, J.; Hoerzer, S.; Enders, H. Running shoes and running injuries: Mythbusting and a proposal for two new paradigms: 'preferred movement path' and 'comfort filter'. *Br. J. Sports Med.* **2015**, *49*, 1290–1294. [CrossRef]
23. Holowka, N.B.; Wynands, B.; Drechsel, T.J.; Yegian, A.K.; Tobolsky, V.A.; Okutoyi, P.; Mang'eni Ojiambo, R.; Haile, D.W.; Sigei, T.K.; Zippenfennig, C.; et al. Foot callus thickness does not trade off protection for tactile sensitivity during walking. *Nature* **2019**, *571*, 261–264. [CrossRef] [PubMed]
24. Hasegawa, H.; Yamauchi, T.; Kraemer, W.J. Foot strike patterns of runners at the 15-km point during an elite-level half marathon. *J. Strength Cond. Res.* **2007**, *21*, 888–893. [CrossRef] [PubMed]
25. Kasmer, M.E.; Liu, X.C.; Roberts, K.G.; Valadao, J.M. Foot-strike pattern and performance in a marathon. *Int. J. Sports Physiol. Perform.* **2013**, *8*, 286–292. [CrossRef]
26. Bonacci, J.; Hall, M.; Saunders, N.; Vicenzino, B. Gait retraining versus foot orthoses for patellofemoral pain: A pilot randomised clinical trial. *J. Sci. Med. Sport* **2018**, *21*, 457–461. [CrossRef]
27. Lieberman, D.E.; Venkadesan, M.; Werbel, W.A.; Daoud, A.I.; D'Andrea, S.; Davis, I.S.; Mang'eni, R.O.; Pitsiladis, Y. Foot strike patterns and collision forces in habitually barefoot versus shod runners. *Nature* **2010**, *463*, 531–535. [CrossRef]
28. Perl, D.P.; Daoud, A.I.; Lieberman, D.E. Effects of footwear and strike type on running economy. *Med. Sci. Sports Exerc.* **2012**, *44*, 1335–1343. [CrossRef]
29. Roper, J.L.; Harding, E.M.; Doerfler, D.; Dexter, J.G.; Kravitz, L.; Dufek, J.S.; Mermier, C.M. The effects of gait retraining in runners with patellofemoral pain: A randomized trial. *Clin. Biomech.* **2016**, *35*, 14–22. [CrossRef]
30. Ervilha, U.F.; Mochizuki, L.; Figueira, A., Jr.; Hamill, J. Are muscle activation patterns altered during shod and barefoot running with a forefoot footfall pattern? *J. Sports Sci.* **2017**, *35*, 1697–1703. [CrossRef]
31. Holowka, N.B.; Wallace, I.J.; Lieberman, D.E. Foot strength and stiffness are related to footwear use in a comparison of minimally- vs. conventionally-shod populations. *Sci. Rep.* **2018**, *8*, 3679. [CrossRef] [PubMed]
32. Takabayashi, T.; Edama, M.; Nakamura, M.; Nakamura, E.; Inai, T.; Kubo, M. Gender differences associated with rearfoot, midfoot, and forefoot kinematics during running. *Eur. J. Sport Sci.* **2017**, *17*, 1289–1296. [CrossRef] [PubMed]

applied sciences

MDPI

Article

Simple Assessment of Height and Length of Flight in Complex Gymnastic Skills: Validity and Reliability of a Two-Dimensional Video Analysis Method

Christoph Schärer [1,2,*], Luca von Siebenthal [1], Ishbel Lomax [1], Micah Gross [1], Wolfgang Taube [2] and Klaus Hübner [1]

1 Department of Elite Sport, Swiss Federal Institute of Sport (SFISM), 2532 Magglingen, Switzerland; lvsiebenthal@gmail.com (L.v.S.); ishbel.lomax@gmail.com (I.L.); micah.gross@baspo.admin.ch (M.G.); huebner@baspo.admin.ch (K.H.)

2 Department of Neuroscience and Movement Science, University of Fribourg, 1700 Fribourg, Switzerland; wolfgang.taube@unifr.ch

* Correspondence: christoph.schaerer@baspo.admin.ch; Tel.: +41-58-467-65-04

Received: 28 August 2019; Accepted: 18 September 2019; Published: 23 September 2019

Abstract: In artistic gymnastics, the possibility of using 2D video analysis to measure the peak height (h_{peak}) and length of flight (L) during routine training in order to monitor the execution and development of difficult elements is intriguing. However, the validity and reliability of such measurements remain unclear. Therefore, in this study, the h_{peak} and L of 38 vaults, performed by top-level gymnasts, were assessed by 2D and 3D analysis in order to evaluate criterion validity and both intrarater and interrater reliability of the 2D method. Validity calculations showed higher accuracy for h_{peak} (±95% LoA: ±3.6% of average peak height) than for L (±95% LoA: ±7.6% of average length). Minor random errors, but no systematic errors, were observed in the examination of intrarater reliability (h_{peak}: $CV\% = 0.44\%$, $p = 0.81$; L: $CV\% = 0.87\%$, $p = 0.14$) and interrater reliability (h_{peak} $CV\% = 0.51\%$, $p = 0.55$; L: $CV\% = 0.72\%$, $p = 0.44$). In conclusion, the validity and reliability of the 2D method are deemed sufficient (particularly for h_{peak}, but with some limitations for L) to justify its use in routine training of the vault. Due to its simplicity and low cost, this method could be an attractive monitoring tool for gymnastics coaches.

Keywords: two-dimensional video analysis; validity; reliability; quantitative biomechanical parameters; artistic gymnastics

1. Introduction

Video analysis is common in elite sports and covers different areas of application. In individual sports, especially acrobatic sports, video analysis is used to compare and improve complex skills [1]. For quantifying biomechanical aspects of such skills, such as score-relevant kinematic variables of vaults in artistic gymnastics, 3D motion capture (3D video analysis) is used. [2,3]. By means of video analyses, Brehmer and Naundorf [4] created reference values for kinematic parameters, indicating the safe execution of vaults. To make use of such reference values, measuring certain kinematic parameters in routine training becomes crucial, so that gymnasts' performances can be compared with the requirements of certain vaults. In regard to monitoring training progress, using video analysis during routine training is intriguing [2,3].

The gold standard for complex kinematic analyses is 3D motion capture, with which movements can be analyzed in sagittal, frontal, and transversal planes simultaneously [5]. However, precise 3D motion capture is associated with considerable financial, spatial, and temporal issues, and is therefore uncommon in routine training settings [6]. Two-dimensional video analysis (2D analysis), if sufficiently accurate, could offer a more practical alternative for measuring kinematics during training, especially since advancements in smartphone and tablet technology are continually making high-definition video capture and analysis more convenient. Aside from convenience while capturing motion, 2D video is faster, cheaper, and requires less prior knowledge compared to 3D analysis [1,7]. Nonetheless, although multiple studies showed moderate validity compared to 3D motion capture and high intrarater reliability of 2D analysis when examining joint angles during simple movements, such as single-leg squats and running [1,5–7], the validity and reliability of 2D video analysis for complex skills in acrobatic sports are unclear and must be scientifically assessed before this method can be recommended for gymnastics. The validity and reliability of 2D analysis were deemed sufficient for measuring the flight time and height of vertical jumps, when compared to optical measurement systems (Opto Jump, Microgate, I) [8] or to a force plate [9]. Further, 2D analysis was deemed valid (ICC = 0.8) and reliable (ICC > 0.85) for measuring joint angles during the more complex skill of baseball pitching, as long as the camera was placed properly [1], so the possibility of it being useful in gymnastics seems worthy of investigation.

In summary, the studies mentioned above offer justification for using simple, time-saving, and cost-effective 2D analysis for simple lower-extremity movements based on good validity and reliability compared to gold-standard methods. However, in contrast to movements in previous studies, artistic gymnastics involve highly complex movements that are performed with high accelerations and velocities. Therefore, the aim of this study was to investigate criterion validity (comparison of 2D to 3D analysis) and both intrarater (test-retest) and interrater (equivalence of two assessors) reliability of 2D video analysis for the peak height (h_{peak}) and length of flight (L) of vaults in artistic gymnastics.

2. Materials and Methods

Twenty-two junior and international elite gymnasts (female: n = 5; male: n = 17) volunteered to participate in the study. All subjects gave written informed consent before participating in the study. All study procedures were approved by the ethics committee Bern (17.01.2017; Project-ID: 2016-01970) and conducted in accordance with the current version of the Declaration of Helsinki, the ICH-GCP, ISO EN 14155, and all national legal and regulatory requirements.

In total, 38 vaults were used for assessment. These included vaults from the three most important vault categories (handspring, Tsukahara, and Yurchenko) and comprised a variety of different sagittal plane rotations (tuck/double tuck, pike/double pike, and layout), which were combined with up to three turns around the longitudinal axis. These vaults were simultaneously recorded with conventional 2D video and a 3D motion capture system (Vicon Motion System, Denver, CO, USA) and analyzed to determine h_{peak} and L in the second flight phase.

For 2D analysis, vaults were filmed using an iPad (iPad Pro 9.7", Apple Corporation, Cupertino, CA, USA) at 100 frames/s. The lens of the iPad was placed at a height of 1.55 m and at a distance of 10 m from the vaulting table, so that the take-off, first and second flight phases, and landing were all visible. The height and width of the image were calibrated within the analytical software (Dartfish SA, Fribourg, CH) by using a rod of known length (2.78 m) that was held vertically and horizontally along the landing zone (Figure 1). With the Dartfish software, h_{peak} was measured as the vertical distance between the landing mat and the gymnast's center of gravity at its highest point during the second flight phase of the vault. For this, the gymnast's center of gravity was estimated visually in the video frame at which the maximal height was deemed to have occurred. L was determined as the horizontal distance between the ankle at the foot's first contact with the mat upon landing and the end of the vaulting table (Figure 2). The software then automatically calculated h_{peak} and L by using the reference height and length.

For the 3D analysis, all vaults were captured by 14 Vicon Vantage Cameras (Vicon Motion System, Denver, CO, USA) that were arranged in two planes: eight cameras were placed at a height of 5.50 m and six were placed at a height of 1.70 m (Figure 3) above ground level. Forty-three reflective markers were placed on the gymnasts' bodies, according to the Vicon Plug-in Gait model [10] (Figure 4). For capturing and, afterward, determining h_{peak} and L, Vicon Nexus, version 2.6, Vicon Motion System, Denver, CO, USA) software was used.

The 2D videos were analyzed by a first-time assessor and an experienced assessor. The first-time assessor was only briefly introduced to the relevant functions of the software (reference measurements) and the important aspects of the measurement of h_{peak} and L (e.g., determination of the center of gravity at the maximum height of flight) before evaluating the trials.

The parameters determined by the first-time assessor ($2D_1$) were compared to those obtained by 3D analysis to assess criterion validity of 2D analysis ($2D_1$ vs. 3D). Further, the parameters were determined with Dartfish (Dartfish SA, Fribourg, CH) by the same assessor at two different points in time (first measurement: $2D_1$; second measurement: $2D_2$) and by an experienced assessor ($2D_e$) to verify intrarater ($2D_1$ vs. $2D_2$) and interrater reliability ($2D_1$ vs. $2D_e$).

Figure 1. 2D calibration: calibration of measuring range by definition of reference height and width with the calibration rod (2.78 m) in the 2D-video software Dartfish (Dartfish SA, Fribourg, CH).

Figure 2. 2D measurements: determination of peak height (h_{peak}) and length of flight (L) for a "Yeo" Vault (CoG: position of center of gravity at the highest point of the second flight phase; A: position of the ankle at first foot contact).

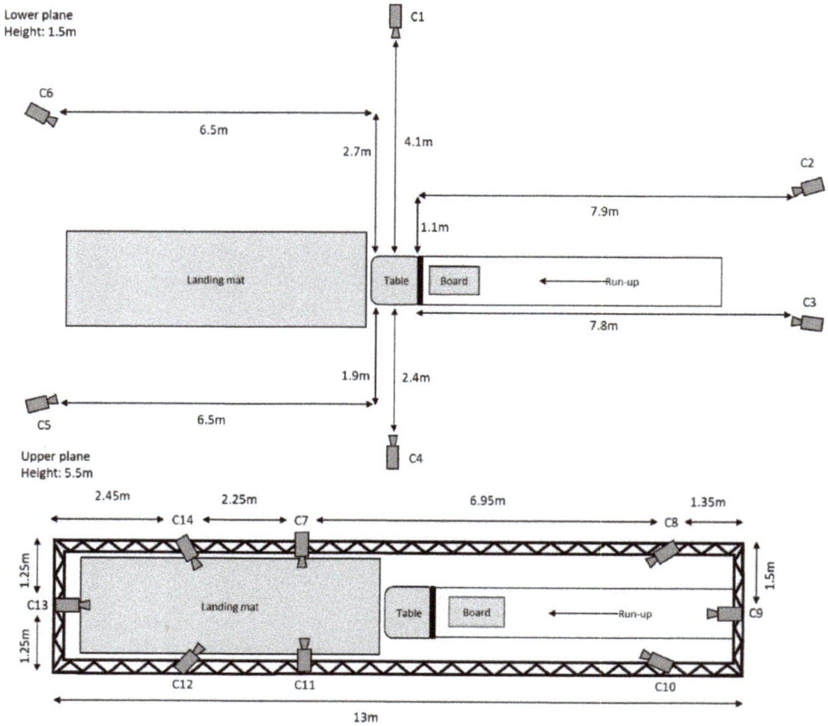

Figure 3. Placement of 3D cameras: positioning of the 14 Vicon Vantage Cameras (C1–C14) in an upper and lower plane for three-dimensional recording of vaults.

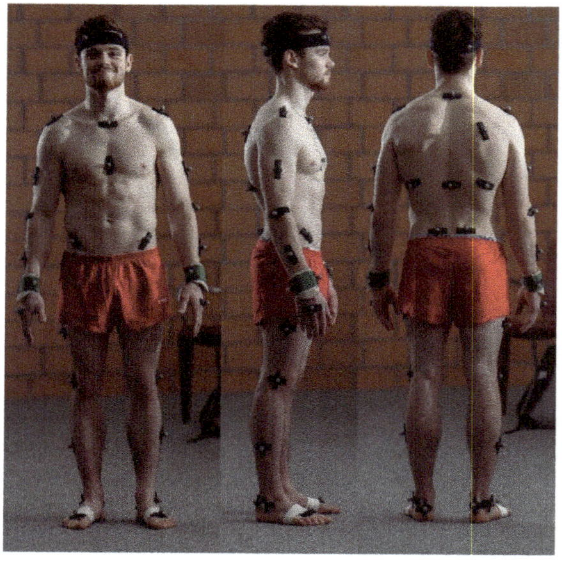

Figure 4. Plug-in Gait model: positioning of 43 markers for 3D analysis, attached to the athlete according to Vicon Plug-in Gait model [10].

Mean, standard deviation, Pearson's correlation coefficient (r), random error ($CV\%$), typical error (TE), and systematic error (t-test: p) of h_{peak} and L measurements were calculated to determine criterion validity (3D vs. $2D_1$) and both intrarater ($2D_1$ vs. $2D_2$) and interrater reliability ($2D_1$ vs. $2D_e$), according to Hopkins et al. [11]. The comparison between the 3D and $2D_1$ measurements (criterion validity) is displayed in a Bland–Altman diagram [12]. In order to determine the influence of lateral displacement at landing on the accuracy of the 2D measurements, the relationship between the difference in L (2D–3D measurement) and displacement along the x axis (to the left or right) was calculated. The level of statistical significance was set to $p < 0.05$. Data analysis was conducted using Microsoft Excel spreadsheets (Microsoft Excel 2016, Microsoft Corporation, Redmond, WA, USA).

3. Results

Two-dimensional analysis ($2D_1$) showed small differences in mean compared to 3D analysis (Table 1). The ±95% limits of agreement corresponded to measurement errors of ±3.6% for h_{peak} and ±7.6% for L (Figure 5). Regarding the validity and reliability of 2D analysis, high correlation coefficients and minor random errors for the h_{peak} and L of vaults were found (Table 2). In contrast, there was a tendency for a systematic error for h_{peak} with 2D compared to 3D analysis. Further, we found a significant correlation between the lateral displacement (x axis) at landing and the difference in L between 2D and 3D analysis ($r = 0.58$; $p < 0.01$). The ex post facto power analysis revealed a power of ~1 for all investigated correlations. The data of all measurements in this study can be found in the supplementary material (Table S1).

Table 1. Mean (± standard deviation) of peak height (h_{peak}) and length of flight (L) of all recorded vaults (n = 38) for 3D (Vicon Motion System, Denver, USA) and 2D analyses (Dartfish SA, Fribourg, CH; $2D_1$: first-time assessor; $2D_2$: second measurement of first-time assessor; $2D_e$: expert assessor).

	h_{peak} ± SD (m)	L ± SD (m)
3D	2.73 ± 0.24	2.29 ± 0.49
$2D_1$	2.75 ± 0.24	2.28 ± 0.51
$2D_2$	2.75 ± 0.24	2.27 ± 0.51
$2D_e$	2.75 ± 0.24	2.27 ± 0.52

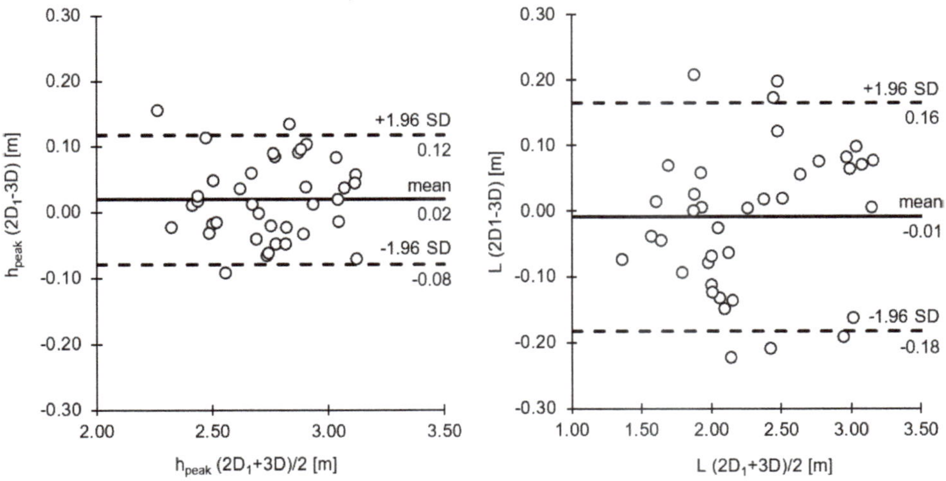

Figure 5. Validity: Bland–Altman diagrams comparing 2D and 3D analyses of peak height (h_{peak}) and length of flight (L) of vaults (n = 38). $2D_1$: two-dimensional video analysis with Dartfish SA of a first-time assessor); 3D: three-dimensional motion capture with Vicon Motion System).

Table 2. Random error (CV%), typical error (TE), correlation coefficient, and systematic error (p) of peak height (h_{peak}) and length of flight (L) of vauls, when comparing first 2D video analysis ($2D_1$) evaluations to 3D motion capture (validity) and to second 2D analysis (intrarater reliability), as well as second 2D video analysis ($2D_2$) and expert 2D video analysis ($2D_e$) (interrater reliability).

	h_{peak}				L			
	CV%	TE (cm)	r	p	CV%	TE (cm)	r	p
Validity	2.24	6.16	0.97	0.06	4.64	10.59	0.98	0.43
Intrarater reliability	0.44	1.21	0.99	0.81	0.87	1.98	0.99	0.14
Interrater reliability	0.51	1.40	0.99	0.55	0.72	1.63	0.99	0.44

4. Discussion

This paper is the first to evaluate the validity and reliability of 2D video analysis of the h_{peak} and L of vaults in artistic gymnastics. Compared to 3D analysis (validity), 2D analysis showed smaller ±95% limits of agreement for h_{peak} (±3.6%) than for L (±7.6%). Further, we found smaller random errors for h_{peak} (CV% = 2.24%) than for L (CV% = 4.64%). In contrast, there was a tendency toward a systematic error of h_{peak} ($p = 0.06$), but not for L ($p = 0.43$). Regarding reliability, the h_{peak} and L of vaults can be repeatedly (intrarater reliability: h_{peak}: CV% = 0.44%, $r = 0.99$; L: CV% = 0.87%, $r > 0.99$) and independently (interrater reliability: h_{peak}: CV% = 0.51%, $r = 0.99$; L: CV% = 0.72%, $r > 0.99$) determined by 2D analysis.

The results support 2D video analysis as a valid measurement tool—particularly for determining the h_{peak}, but slightly less so for the L, of vaults in gymnastics. Our findings are in line with those of Balsalobre-Fernandez, Tejero-Gonzalez, del Campo-Vecino, and Bavaresco [9], who have shown that 2D analysis is a valid technique for accurately measuring the flight height of vertical jumps. Thus, under the described measurement conditions, an increase of at least 6.16 cm in h_{peak} or of at least 10.59 cm in L, measured by 2D video analysis, can be considered a true performance increase.

Differences in the validity between the h_{peak} and L of vaults, as shown in the Bland–Altman diagram, may have several reasons. The most obvious is that 2D analysis measures the height and length of flight in the sagittal plane only. Accordingly, lateral displacement at the landing cannot be detected by 2D analysis and presents a possible confounding factor. This was shown by the significant relationship between lateral displacement upon landing and the difference between 2D and 3D determinations of L. This result supports research by Oyama, Sosa, Campbell, and Correa [1], who compared 2D to 3D analysis of joint angles during complex movements. They found that 2D analysis is a valid measure only when cameras are placed perpendicular to the segment of interest. This limitation of 2D analysis was less important for the measurement of h_{peak}, because the lateral displacement is at its maximum at landing, long after the attainment of h_{peak}. Further, since L was determined as the distance between the vaulting table and the ankle, even straddled legs at landing may result in a lateral displacement and may lead to a measurement error of 2D analysis. In contrast, h_{peak} was measured vertically from the landing mat to the center of gravity. Therefore, the determination of h_{peak} did not depend on the position of one single extremity but on the highest point on the center of gravity's smooth trajectory.

Another reason for the less accurate measurement of L could be the quality of the 2D videos (resolution and frame rate). The still frame at landing was often slightly blurred due to the velocity of the recorded movement. Therefore, it was difficult in some cases to determine the position of the ankle precisely, and, for this reason, small measurement errors may have occurred. In contrast, the slightly blurred image did not affect the determination of the center of gravity for the measurement of h_{peak}, since the determination of the center of gravity depends on a global view of the body and not on a small single part of the body.

One more reason for the larger inaccuracy in the determination of L may be the slight bend in the calibration rod when it was held horizontally to scale the image width (Figure 1). This would have caused the reference length to be slightly too long, although this effect was too small to be apparent in our results (nonsignificant p-value).

In this study, we observed a tendency toward a systematic underestimation of h_{peak} with 2D analysis. Since the measurements of the experienced assessor were not statistically different from the values of the first-time assessor, it can be assumed that both assessors encountered the same basic problem that may have led to this tendency. In particular, it was difficult to determine the zero-point on the landing mat vertically underneath the highest point of the center of gravity during the second flight phase. We assume that this was the reason for the (almost significant) underestimation of h_{peak}. Nonetheless, considering the small measurement error when determining h_{peak}, 2D analysis may be considered a valid method. This knowledge is especially worthwhile since the height of flight dictates the potential to perform somersaults and twists during the second flight phase, and it is therefore an important performance-determining factor for vaults.

When comparing 2D analysis at two consecutive points in time, the variation coefficients of the h_{peak} and L of vaults were very low. For instance, the differences of 2D analysis were only 1.21 cm (h_{peak}) and 1.98cm (L), at a mean height of 2.75 m and a mean length of 2.28 m. These minor random errors are likely due to slightly different definitions of the zero-points (on the landing mat or at the end of the vaulting table), or of the gymnast's center of gravity (h_{peak}) or ankle (L), rather than a different determination of the frame at which h_{peak} or L were determined. Therefore, the almost perfect correlation of the first and second 2D analysis and the low variation coefficient values demonstrate a very high intrarater reliability. Thus, 2D analysis using Dartfish (Dartfish SA, Fribourg, CH) is a reliable and reproducible measure for the h_{peak} and L of vaults. This result is in line with the findings of Maykut, Taylor-Haas, Paterno, DiCesare, and Ford (2015), who, among other things, examined intrarater reliability of 2D analyses of joint angles during running. They measured kinematic variables during running on a treadmill with the same video analytical software we used (Dartfish SA, Fribourg, CH), and they reported excellent intrarater reliability [7]. Other studies have also shown 2D analyses to be reliable for measuring joint angles [6] and flight height [9].

Lastly, we compared the values of h_{peak} and L evaluated by an experienced assessor with those evaluated by a first-time assessor. Our results show that the h_{peak} and L of vaults are similar, as the coefficients of variance are low and the correlations are high. In this context, it should be mentioned that a brief introduction to the measurement software and determination of center of gravity is sufficient but indispensable. Therefore, 2D analysis is a straightforward measurement tool, where values of an inexperienced assessor, for example, those evaluated by a coach, are comparable to those of an expert. Therefore, our results show good interrater reliability with 2D analysis.

When evaluating data-collection procedures, it is important to evaluate practicality and not only validity and reliability. As there are few studies regarding the validity and reliability of 2D analysis, comparisons are difficult. Nevertheless, our results are similar to the results reached by Brehmer and Naundorf [4], as well as Schurr, Marshall, Resch, and Saliba [5], in regard to validity. As such, 2D analysis seems to be a reasonable, inexpensive, and portable alternative to 3D motion capture analysis. Furthermore, 2D analysis is time efficient, as video analysis of the h_{peak} and L of complex skills only takes about one minute for practiced evaluators. Additionally, the equipment needed for 2D analysis (video camera, tripod, analytical software, and calibration rod) is normally readily available in an artistic-gymnastics facility, making the described method an easy and cost-effective analytical tool. Lastly, even first-time assessors can precisely determine the important parameters of complex skills if they are briefly introduced to the measurement method. Thus, 2D analysis is a useful analytical tool for practical use in training and for scientific research, as the financial, spatial, and temporal costs are minimal.

5. Conclusions

This study evaluated the validity and reliability of 2D video analysis of the h_{peak} and L of vaults in artistic gymnastics. We conclude that 2D video analysis is a valid and reliable alternative to 3D motion capture, particularly for determining the h_{peak}, but slightly less so for the assessment of the L, of vaults. Thus, the ease of use and cost-effectiveness of 2D analysis, along with the results from this study, support the use of 2D analysis in routine training and scientific research.

Supplementary Materials: The following are available online: http://www.mdpi.com/2076-3417/9/19/3975/s1. Table S1. Supplementary material (raw data).

Author Contributions: Conceptualization, C.S. and L.v.S.; methodology, C.S. and L.v.S.; formal analysis, C.S. and L.v.S.; investigation, C.S., L.v.S., and M.G.; data curation, C.S., L.v.S., and M.G.; writing—original-draft preparation, C.S. and I.L.; writing—review and editing, K.H. and W.T.; supervision, K.H. and W.T.; project administration, C.S.

Funding: This research received no external funding.

Conflicts of Interest: The authors declare no conflicts of interest.

References

1. Oyama, S.; Sosa, A.; Campbell, R.; Correa, A. Reliability and validity of quantitative video analysis of baseball pitching motion. *J. Appl. Biomech.* **2017**, *33*, 64–68. [CrossRef] [PubMed]
2. Farana, R.; Uchytil, J.; Zahradnik, D.; Jandacka, D. The "akopian" vault performed by elite male gymnasts: Which biomechanical variables are related to a judge's score? *Acta. Gymnica.* **2015**, *45*, 33–40. [CrossRef]
3. Farana, R.; Vaverka, F. The effect of biomechanical variables on the assessment of vaulting in top-level artistic female gymnasts in world cup competitions. *Acta. Univ. Palacki. Olomuc. Gymn.* **2012**, *42*, 49–57. [CrossRef]
4. Brehmer, S.; Naundorf, F. Key parameters of the 2nd flight phase of the tsukahara with salto backward piked. In Proceedings of the 32nd International Conference of Biomechanics in Sports, Johnson City, TN, USA, 13 July 2014.
5. Schurr, S.A.; Marshall, A.N.; Resch, J.E.; Saliba, S.A. Two-dimensional video analysis is comparable to 3d motion capture in lower extremity movement assessment. *Int. J. Sports. Phys. Ther.* **2017**, *12*, 163–172. [PubMed]
6. Gribble, P.; Hertel, J.; Denegar, C.; Buckley, W. Reliability and validity of a 2-d video digitizing system during a static and a dynamic task. *J. Sport Rehabil.* **2005**, *14*, 137–149. [CrossRef]
7. Maykut, J.N.; Taylor-Haas, J.A.; Paterno, M.V.; DiCesare, C.A.; Ford, K.R. Concurrent validity and reliability of 2d kinematic analysis of frontal plane motion during running. *Int. J. Sports Phys. Ther.* **2015**, *10*, 136–146. [PubMed]
8. Balsalobre-Fernandez, C.; Tejero-Gonzalez, C.M.; del Campo-Vecino, J.; Bavaresco, N. The concurrent validity and reliability of a low-cost, high-speed camera-based method for measuring the flight time of vertical jumps. *J. Strength Cond. Res.* **2014**, *28*, 528–533. [CrossRef] [PubMed]
9. Balsalobre-Fernandez, C.; Glaister, M.; Lockey, R.A. The validity and reliability of an iphone app for measuring vertical jump performance. *J. Sports Sci.* **2015**, *33*, 1574–1579. [CrossRef] [PubMed]
10. Vicon. Vicon plug-in gait product guide—foundation notes revision. Vicon Motion Systems Ltd.: Oxford, UK, 2010.
11. Hopkins, W.G.; Schabort, E.J.; Hawley, J.A. Reliability of power in physical performance tests. *Sports Med.* **2001**, *31*, 211–234. [CrossRef] [PubMed]
12. Bland, J.M.; Altman, D.G. Statistical methods for assessing agreement between two methods of clinical measurement. *Lancet* **1986**, *1*, 307–310. [CrossRef]

applied sciences

MDPI

Article

Shoe Cushioning Effects on Foot Loading and Comfort Perception during Typical Basketball Maneuvers

Xini Zhang [1], Zhen Luo [1], Xi Wang [1], Yang Yang [1], Jiaxin Niu [1] and Weijie Fu [1,2,*]

1 School of Kinesiology, Shanghai University of Sport, Shanghai 200438, China; zhangxini1129@163.com (X.Z.); 15621573815@163.com (Z.L.); zwx252@163.com (X.W.); 18049922807@163.com (Y.Y.); jfhou_sus@163.com (J.N.)

2 Key Laboratory of Exercise and Health Sciences of Ministry of Education, Shanghai University of Sport, Shanghai 200438, China

* Correspondence: fuweijie@sus.edu.cn or fuweijie315@163.com; Tel.: +86-21-65507368

Received: 21 August 2019; Accepted: 12 September 2019; Published: 17 September 2019

check for updates

Abstract: Purpose: This study aimed to explore the relationship between foot loading and comfort perception in two basketball shoes during basketball-specific maneuvers. Methods: Twelve male collegiate basketball players were required to complete three basketball maneuvers (i.e., side-step cutting, 90° L-direction running, and lay-up jumping) in two basketball shoe conditions (shoe L and shoe N, with different midsole cushioning types). Two Kistler force plates and a Medilogic insole plantar pressure system were used to collect kinetic data (i.e., impact force, peak loading rate, and plantar pressure variables). Perception scales were used to evaluate comfort perception. Results: No significant difference was observed between the two shoes during maneuvers in terms of ground reaction force. However, the plantar pressure of shoe L in the midfoot and lateral foot regions was significantly greater than that of shoe N during side-step cutting and lay-up jumping. Shoe N was significantly superior to shoe L, especially in dynamic scale in terms of the perception of comfort. The plantar pressure and perception characteristics in the two shoes were significantly different but inconsistent with each other. Conclusion: The biomechanical characteristics of the shoes themselves and the perception evaluation of the athletes should be considered in comprehensive shoe-cushioning design and evaluation.

Keywords: basketball shoe; comfort perception; foot loading; plantar pressure; maneuver

1. Introduction

In basketball, the lower limbs of athletes are subjected to large impact forces during each landing [1]. Players complete 70 jumps/landings in a single game and attenuate impacts of up to nine times their body weight every time [2], which increases the risk of knee and ankle injuries [3]. Reducing impact forces (which includes both magnitude and loading rate characteristics) may help prevent foot injuries [4]. The shock absorption characteristics and comfort of basketball shoes, as core equipment of the sport, have important influences not only on the performance of the players but also on the prevention of lower extremity injuries [5]. The impact force and pressure distribution characteristics between the feet and the shoes must be understood to effectively optimize the technical movements, reduce foot injuries, and improve the design of specialized shoes [6]. However, the existing studies on foot loading have mainly focused on common gait characteristics, such as walking and running [7,8], and studies on specific sports maneuvers, such as cross-over running and lay-up jumping [9–11], are limited.

Meanwhile, comfort and stability have been identified as the principal factors of specialized sports shoes [12], in addition to meeting the functional requirements of specific sports and strengthening the foot protection function [13]. However, these factors are easily neglected. The comfort of basketball shoes is closely related to the performance of basketball athletes, as well as to ankle injury and its prevention [2]. The subjective perception scale has been proven to be an effective and credible method for assessing the aforementioned variables [14,15]. Hennig et al. [16] found that a close relationship exists between runners' subjective perception of shoes' cushioning performance and their impacts on plantar pressure during running. Meanwhile, foot comfort is also closely related to the impact load on the lower limbs at initial contact [17]. Thus, the functionality and comfort of basketball shoes for typical maneuvers in basketball must be explored through both the subjective perception and foot biomechanical tests [18].

Therefore, the current study aims to determine the effect of different basketball shoes on the ground reaction force (GRF) and plantar pressure characteristics in three typical basketball maneuvers and to further understand foot loading characteristics in basketball and their relationship with comfort through the subjective perception scale evaluation of basketball shoes. It was hypothesized that wearing different shoes would affect comfort perception, and correspondingly change GRF and plantar pressure characteristics during basketball maneuvers.

2. Materials and Methods

2.1. Participants

Twelve healthy male collegiate basketball players (age: 23.1 ± 2.0 years; height: 176.3 ± 4.5 cm; body mass: 70.5 ± 7.5 kg) with an average of 8.4 years of experience in basketball events were recruited for this experiment. An observational cross-sectional research design according to the Strengthening The Reporting of OBservational Studies in Epidemiology (STROBE) criteria [19,20]. The inclusion criteria were: (1) at least five years of experience in basketball events; (2) none has suffered musculoskeletal injuries of the lower extremity over the last six months; (3) none has engaged in strenuous training within 24 h. A two-tailed t-test was executed via the G*Power 3.1 software to determine whether a sample size of 12 was sufficient to minimize the probability of type II errors for all the variables (P = 80% at $\alpha = 0.05$). All the participants signed an informed consent form, and ethical approval was granted by the Institutional Review Board of Shanghai University of Sports prior to the study (2017007).

2.2. Shoes and Instrumentations

Two types of U.S. size nine basketball shoes were used in this study. One of the shoes was a new sample provided by a local sports science laboratory (hereinafter referred to as shoe L), and the other (Figure 1) was a commercially available type with a popular international brand (hereinafter referred to as shoe N). The following are some details of the property of Shoe N: (1) upper: Black synthetic leather and Phyposite technology—Breathable tongue inner sleeve with a traditional lacing system; (2) midsole: Phylon midsole design that minimizes weight while maximizing cushioning with animpact absorption system; (3) outsole: Non-traditional outsole to reduce weight and optimize traction. Overall, the two types of shoes were similar in the abovementioned materials, design, color, weight (≈530 g), and so on, and only differed in the impact absorption systems of the midsole.

Two 90 cm × 60 cm three-dimensional force plates (9287B, Kistler Corporation, Switzerland) were utilized to collect GRF data, with a sampling rate of 1200 Hz.

The insole measurement system (Medilogic Corporation, Germany) was used to capture the plantar pressure of different regions. This system has been validated [21]. Each insole was calibrated using the manufacturer's calibration device (T&T medilogic Medizintechnik GmbH, Schönefeld, Germany) prior to the study. The size of the pressure insole was 8/9 according to the participants' foot size. The pressure insole consisted of 225 pressure sensors (0.6 cm × 0.4 cm), with a pressure range of

0-64 N/cm^2 and a maximum sampling rate of 300 Hz (Figure 2a). The plantar regions were divided into five parts, namely, the forefoot, the midfoot, the heel, the lateral, and the medial (Figure 2b).

Figure 1. Basketball shoes (shoe N) from a famous international manufacturer.

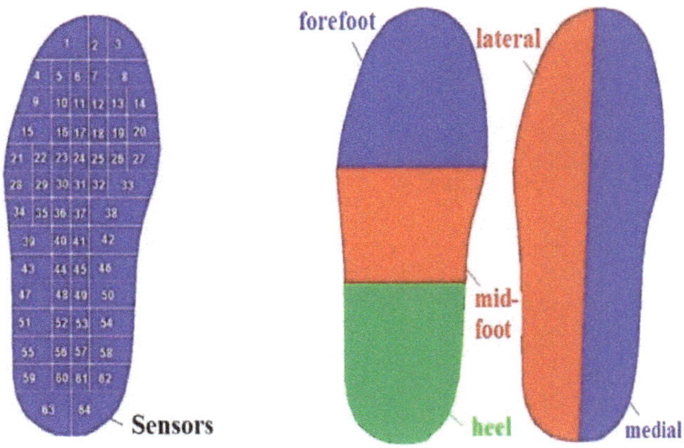

(**a**) (**b**)

Figure 2. (**a**) Composition of insole pressure sensors with a pressure range of 0–64 N/cm^2; (**b**) region division of plantar pressure, including the forefoot, the midfoot, the heel, the lateral, and the medial parts.

2.3. Experimental Protocol

Three typical maneuvers, which are the most frequently used in basketball games according to the video observation and the coach interview, were selected, namely, side-step cutting (SS), 90° L-direction running (90° LR), and lay-up toe-off (LU_{TO}) and touch-down (LU_{TD}). The sports surface used in this study is the most common wooden floor used in basketball games. The participants first set their pace according to their personal movement habits and then tried their best to complete the movements. They were required to achieve the three typical maneuvers (SS, 90° LR, and lay-up) in the two different shoes in random order. Five successful trials were obtained for each condition. The pressure insoles were placed flat in the shoes before the experiment, and data were transmitted to the computer in real-time via a wireless receiver. The participants were required to familiarize themselves with the maneuvers before the formal tests began. The familiarization period for each participant was 10–15 min. To avoid the influence of fatigue on the results, 1–2 min breaks were given between each trial [22].

The perception-comfort scale used in this experiment was adapted from the perception scale provided by a famous shoe research center in the US. The scale includes two parts, namely, general fit and dynamic scale (Figure 3). The indices of the general fit test include the toe-box height, the toe-box width, the ball girth, the waist/instep, the elasticity at the heel, and the shoe length. The index scores range from 1 to 9: 1 means too low (toe-box height), too narrow (toe-box width), too tight (ball girth, waist/instep, heel), or too short (length); 9 means too high, too wide, too loose, or too long; and 5 means just right (Figure 3a). A dynamic scale was used to evaluate the fit of the basketball shoes during the maneuvers, including overall liking, heel cushioning, heel responsiveness, heel stability, heel-to-toe transition, and forefoot cushioning, which were also scored 1–9 points (1 for extremely dislike, 5 for neutral, and 9 for extremely like). In addition to overall liking, the intensity aspects were also rated 1–9: 1 means soft (heel and forefoot cushioning), no response (heel responsiveness), very unstable (heel stability), no smoothness (heel-toe transition), and so on; and 9 means hard, very reactive, very stable, very smooth, and so on (Figure 3b). The experiment process included a 10-minute regular-intensity basketball maneuver practice (including SS, 90° LR, and lay-up). The tongue and logos of the basketball shoes were completely covered before testing to avoid the brand effect and the influence of other factors on the scoring results.

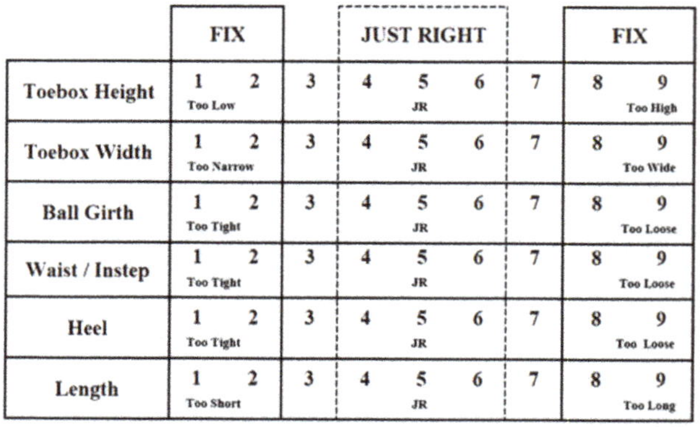

(**a**)

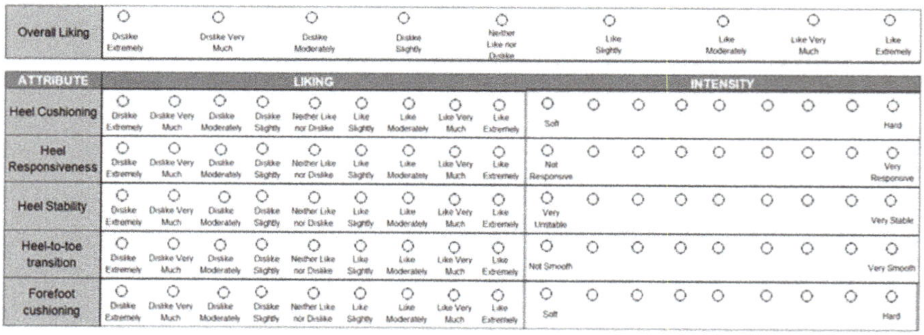

(**b**)

Figure 3. (**a**) General fit scale includes the toe-box height, the toe-box width, the ball girth, the waist/instep, the elasticity at the heel, and the shoe length. The index scores range from 1 to 9 points; (**b**) dynamic scale includes overall liking, heel cushioning, heel responsiveness, heel stability, heel-to-toe transition, and forefoot cushioning, which were scored 1–9 points.

2.4. Data Processing and Analysis

The main variables of the GRF included the following: (1) peak vertical GRF (Fz) and appearance time (t_F) and (2) peak loading rate (G_z) and appearance time (t_G). Fz and Gz were normalized by bodyweight (BW).

According to Bontrager's study on the settings of the insole area and the structure of the pressure insoles [20], the peak pressure (normalized by BW) and the peak pressure distribution (contact area) were measured for six plantar areas, namely, the entire sole, the forefoot, the midfoot, the rearfoot, the medial, and the lateral (Figure 2).

The shoe comfort characteristics included (1) general fit: toe-box height, toe-box width, ball girth, waist/instep, heel, and length; and (2) dynamic fit: overall liking, heel cushioning, heel responsiveness, heel stability, heel-to-toe transition, and forefoot cushioning. Liking and intensity were involved in each dynamic fit variable.

2.5. Statistics

All data were normally distributed based on the Shapiro–Wilk test. The paired sample t-test was used to determine the effects of different basketball shoe cushioning on the GRF and the plantar pressure characteristics. The comfort perception variables were determined by the Wilcoxon rank-sum test (SPSS 19.0, SPSS Inc., Chicago, IL, USA). The significance level was set at 0.05.

3. Results

3.1. Vertical GRF

The passive impact phase in SS and 90° LR occurred within 100 ms after ground contact, with the Fz approximately twice that of the BW (Figure 4a). The vertical GRF increased rapidly in LUTO and LUTD during contact, that is, the Fz during the push-off phase and after landing could be as large as four and eight-times the BW, respectively (Figure 4b). However, no significant differences in Fz, G_z, t_F, and t_G were observed between the two basketball shoes during SS, 90° LR, LU_{TO}, and LU_{TD} (Table 1).

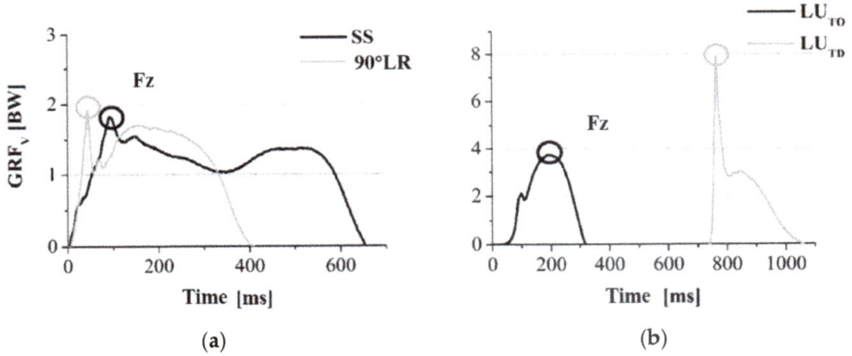

Figure 4. (**a**) Vertical ground reaction force-time curves during side-step cutting (SS) and 90° L-direction running (90° LR) maneuvers; (**b**) vertical ground reaction force-time curves during lay-up toe-off (LU_{TO}) and lay-up touch-down (LU_{TD}) maneuvers. Notes: GRF is ground reaction force, Fz is peak vertical GRF, and BW is bodyweight.

Table 1. Effect of different basketball shoes on ground reaction force during side-step cutting, 90° L-direction running, and lay-up toe-off and lay-up touch-down maneuvers.

Maneuvers	Shoes	Fz (BW)	t_F (ms)	G_z (BW/ms)	t_G (ms)
SS	L	1.92 ± 0.32	74.4 ± 30.2	0.26 ± 0.18	27.2 ± 21.8
	N	2.02 ± 0.43	63.3 ± 28.1	0.17 ± 0.06	19.7 ± 11.4
90° LR	L	2.47 ± 0.61	55.4 ± 25.6	0.35 ± 0.08	13.4 ± 12.6
	N	2.25 ± 0.71	54.8 ± 26.4	0.21 ± 0.07	13.2 ± 9.7
LU_{TO}	L	3.53 ± 0.43	130.7 ± 47.6		
	N	3.55 ± 0.67	124.8 ± 42.1		
LU_{TD}	L	6.33 ± 3.10	26.4 ± 9.7	1.13 ± 0.87	10.3 ± 4.2
	N	9.72 ± 3.42	28.6 ± 13.8	0.93 ± 0.32	11.9 ± 4.8

Notes: GRF is ground reaction force, Fz is peak vertical GRF, G_z is peak loading rate, tF is time to peak vertical GRF, t_G is time to peak loading rate, BW is bodyweight, SS is side-step cutting, 90° LR is 90° L-direction running, LU_{TO} is lay-up toe-off, and LU_{TD} is lay-up touch-down.

3.2. Maximum Plantar Pressure

Shoe L showed a lower maximum pressure in the entire sole ($p < 0.05$) and midfoot ($p < 0.01$) regions during SS (Table 2) compared with shoe N. Although shoe L showed a low plantar pressure in the entire sole and in the forefoot, the heel, the medial, and in the lateral regions, no significant differences were observed between the two shoes during 90° LR.

Table 2. Effect of different basketball shoes on maximum pressure of each plantar region during side-step cutting, 90° L-direction running, and lay-up toe-off and lay-up touch-down maneuvers.

Maneuvers	Shoe	Maximum Pressure of Each Plantar Region (N/kg/cm^2)					
		Entire	Forefoot	Midfoot	Heel	Lateral	Medial
SS	L	0.028 ± 0.008 *	0.037 ± 0.018	0.011 ± 0.012 **	0.024 ± 0.014	0.026 ± 0.013	0.033 ± 0.005
	N	0.032 ± 0.006	0.043 ± 0.022	0.023 ± 0.016	0.029 ± 0.014	0.031 ± 0.015	0.035 ± 0.008
90° -LR	L	0.028 ± 0.018	0.048 ± 0.031	0.020 ± 0.015	0.026 ± 0.018	0.026 ± 0.018	0.033 ± 0.023
	N	0.030 ± 0.019	0.051 ± 0.034	0.023 ± 0.021	0.025 ± 0.017	0.027 ± 0.019	0.035 ± 0.024
LU_{TO}	L	0.061 ± 0.008	0.098 ± 0.014	0.048 ± 0.009 **	0.080 ± 0.018	0.072 ± 0.009	0.054 ± 0.010
	N	0.063 ± 0.012	0.093 ± 0.018	0.061 ± 0.014	0.082 ± 0.020	0.075 ± 0.014	0.053 ± 0.012
LU_{TD}	L	0.064 ± 0.011	0.081 ± 0.019	0.050 ± 0.017 **	0.093 ± 0.038	0.063 ± 0.011 **	0.067 ± 0.019
	N	0.071 ± 0.012	0.082 ± 0.022	0.064 ± 0.017	0.099 ± 0.038	0.070 ± 0.011	0.076 ± 0.023

Notes: SS is side-step cutting, 90° LR is 90° L-direction running, LU_{TO} is lay-up toe-off, and LU_{TD} is lay-up touch-down. * $p < 0.05$; ** $p < 0.01$.

Notably, shoe L showed a lower maximum pressure in the midfoot and in the lateral regions during LU_{TO} ($p < 0.01$) and LU_{TD} ($p < 0.01$) compared with shoe N. Although no statistical differences were observed in plantar pressure on the rest of the regions, the maximum pressure was evidently lower in shoe L than in shoe N.

3.3. Foot Pressure Distribution

In general, the overall pressure distribution in the two shoes was similar but wider in shoe L than in shoe N, and the pressure value of each region was lower in shoe L than in shoe N. Specifically, less fore-lateral pressure distribution was noted in shoe L during SS; a wide heel pressure distribution was observed in shoe L during 90° LR, with pressure concentrated at the first metatarsal head, the first phalanges, and the fifth metatarsal head (Figure 5a); the foot pressure of each region in shoe L was smaller than that in shoe N during LU_{TO}; and a smaller pressure at the first phalanges and a wider heel pressure distribution were noted in shoe N compared with shoe L during LU_{TD} (Figure 5b).

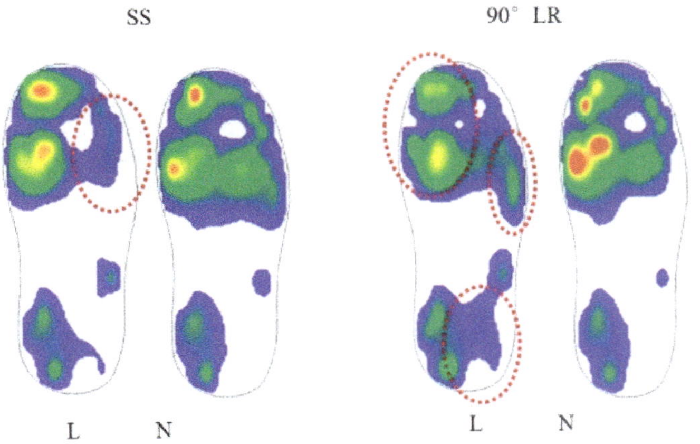

(a)

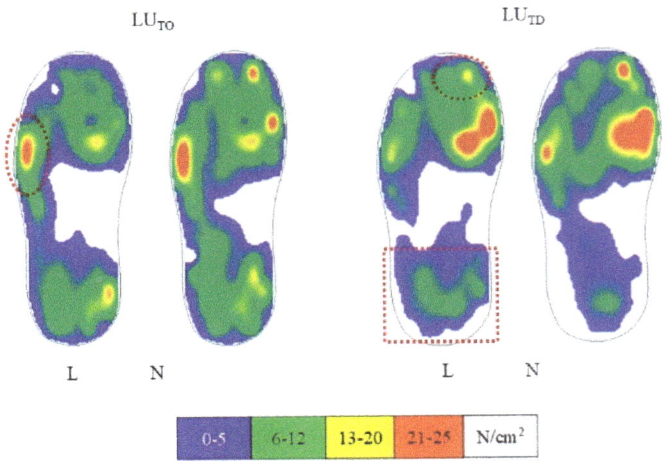

(b)

Figure 5. (**a**) Effect of different basketball shoes (L vs. N) on plantar pressure distribution during side-step cutting (SS) and 90° L-direction running (90° LR) maneuvers; (**b**) effect of different basketball shoes (L vs. N) on plantar pressure distribution during lay-up toe-off (LU$_{TO}$), and lay-up touch-down (LU$_{TD}$) maneuvers. Different colors represented different pressure values. The pressure values from small to large were blue, green, yellow and red.

3.4. Comfort Perception

No significant differences were observed in the general fit between shoe L and shoe N (Figure 6).

From the dynamic comfort scale perspective, the overall liking of shoe N was significantly higher than that of shoe L ($p < 0.01$) owing to forefoot and heel cushioning ($p < 0.01$). The heel responsiveness and stability of shoe N were greater than those of shoe L ($p < 0.05$), but no significant differences were observed in the heel-to-toe transition between the two shoes (Figure 7). The forefoot and heel cushioning of shoe L were significantly higher than those of shoe N ($p < 0.05$) (Figure 7) in terms of dynamic intensity. Meanwhile, no significant difference was observed in the heel-to-toe transition, heel response, and heel stability between the two types of shoes.

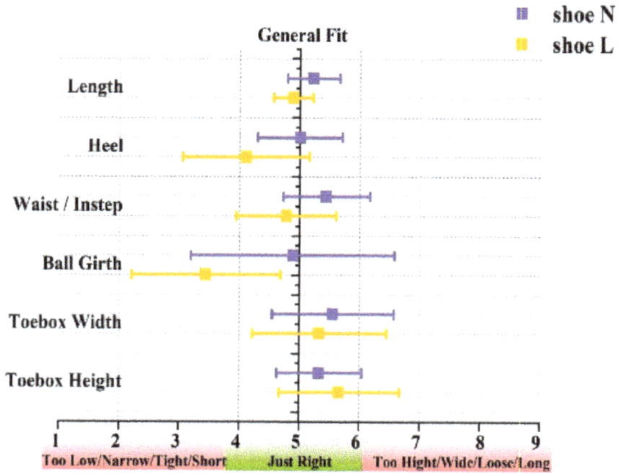

Figure 6. Effect of different basketball shoes (N vs. L) on general fit.

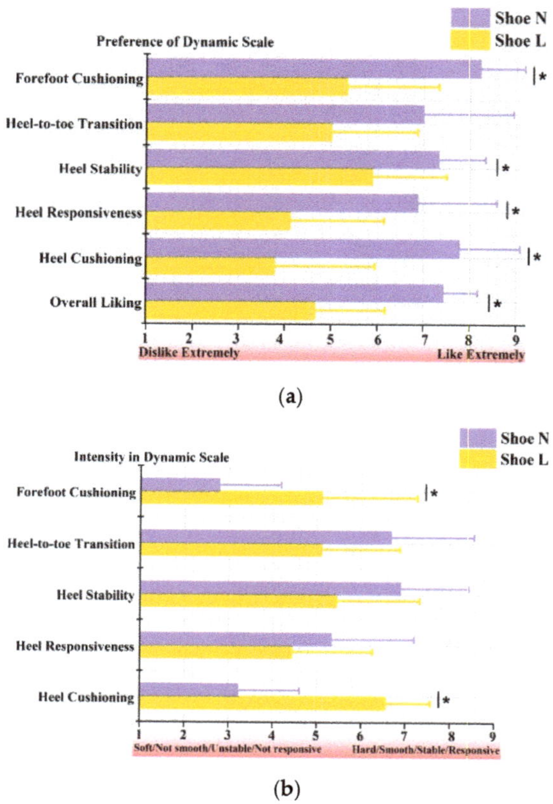

Figure 7. (**a**) Effect of different basketball shoes (N vs. L) on preference of dynamic scale; (**b**) effect of different basketball shoes (N vs. L) on intensity in dynamic scale.

4. Discussion

This study aimed to investigate the effect of two different shoe cushionings on foot loading (GRF and plantar pressure) and the perception of comfort during three basketball-specific maneuvers (i e., SS, 90° LR, and LU_{TO}/LU_{TD}). Consistent with our hypothesis, compared to shoe N, shoe L showed a lower maximum pressure in the entire and midfoot regions during SS and a lower maximum pressure in the midfoot and lateral regions during LU_{TO} and LU_{TD}. The overall pressure distribution was wider in shoe L than in shoe N. In terms of the perception of comfort, shoe N was significantly superior to shoe L, especially in the dynamic scale. However, contrary to our hypothesis, no significant differences in Fz, G_Z, t_F, and t_G were noted between the two basketball shoes during SS, 90° LR, LU_{TO}, and LU_{TD}.

4.1. Vertical GRF

In general, two peaks appeared in the vertical GRF on the lower limbs when running or jumping, namely, the impact force (first peak) and the active force (second peak). The impact force is closely related to sports injuries [2,23]. This study analyzed the characteristics of GRF during three kinds of basketball-specific maneuvers, namely, side-step cutting (SS), 90° varied-direction running (90° LR), and lay-up toe-off (LU_{TO}) and lay-up touch-down (LU_{TD}). Notably, the impact forces, which were approximately twice the BW, were identified as passive forces that appeared within 100 ms at the initial contact during SS and 90° LR, and the active forces were regarded as the principal components However, the results indicated that no significant differences in GRF existed between the two pairs of shoes during SS and 90° LR.

Among the numerous basketball maneuvers, the lay-up is one of the most representative and relatively complex movements that include the acceleration, take-off, and landing phases. The results of the present study showed that the GRF curves during LU_{TO} and LU_{TD} were very similar to the those when running with a heel strike pattern [24,25]. Although the force would not cause impact damage to the lower extremity at take-off, which was nearly four times the BW, the participants were subjected to up to eight times of the BW impact force during landing because of the high jumps performed in this study. The repeated impacts loaded on the lower limbs can easily lead to injuries, especially when high impacts cannot be loaded [26]. Therefore, the results indicated that the high impacts during LU_{TD} caused overuse injuries. However, we found no significant differences in Fz, G_Z, t_F, and t_G regardless of shoe conditions. This finding suggests that though the two midsole materials were different, the impacts of the active movements on the lower extremities made a small difference.

4.2. Plantar Pressure Characteristics

We found that shoe L exhibited a significantly lower maximum pressure on the entire sole and on the midfoot regions compared with shoe N. Although shoe L showed a lower plantar pressure on the entire sole and on the forefoot, the heel, the medial, and the lateral regions, no significant differences were noted between the two shoes during 90° LR. In addition, shoe L showed a lower maximum pressure on the midfoot and lateral regions during LU_{TO} and LU_{TD} compared with shoe N. Although no statistical differences were observed in the rest of the plantar pressure regions, the maximum pressure was lower in shoe L than in shoe N.

The overall plantar pressure distribution in the two shoes was similar but wider in shoe L than in shoe N, and the pressure value of each region was lower in shoe L than in shoe N. Specifically, the fore-lateral pressure distribution of shoe L during SS was low; the heel pressure distribution of shoe L during 90° LR was wide, with the pressure was concentrated at the first metatarsal head, the first phalanges, and the fifth metatarsal head; the foot pressure of each region of shoe L was smaller than that of shoe N during LU_{TO}; the pressure was smaller at the first phalanges; and the heel pressure distribution was wider for shoe N than those for shoe L during LU_{TD}. At present, plantar pressure distribution has been measured for efficiency in studies on the characteristics and biomechanical mechanism of plantar pressure in various sports. These studies have provided key technologies for sports shoes of different events, especially in terms of individualized design and manufacturing. The material of the sole and its structure for energy absorption and release are important factors in attenuating impacts and protecting the lower extremity from injury [27].

A significant difference was noted in the present study on the plantar pressure of the two shoes during the different maneuvers. However, no significant differences were observed in the GRF or loading rates, which suggests that the mechanical relationship between the foot and the ground became an indirect one owing to the intervention of shoe conditions. This change was mainly due to the special medium of the midsole. A special patented cushioning material was used in the midsole of shoe L. The materials and structures of the midsoles differed in the two shoes, thereby resulting in significant differences in plantar pressure during the three maneuvers despite the similar GRF. This finding suggests that the mechanical performance of shoes and feet and the relationship between feet and shoes caused by different midsole materials should be considered when exploring the function of sports shoes [28]. These factors should then be combined with plantar pressure to create a comprehensive design and to improve sports shoes effectively.

4.3. Comfort Perception

In addition to meeting the functional requirements of specific sporting events and strengthening foot protection, the most important element in specialized sport's shoes is comfort [12,29]. At present, subjective scoring is still the most frequently applied method for internationally applicable comfort testing; however, the scales and questionnaires used in each study differ. The perception test which was used in this experiment mainly includes two parts, namely, the general fit and the dynamic scale, which have been proven reliable.

For the general fit, the two shoes provided comfort with no significant differences. However, shoe N outperformed shoe L significantly on each index for liking in the dynamic scale, especially in heel stability, heel response, heel cushioning, forefoot cushioning, and overall impression. Moreover, shoe N had above five scores in all the indices. The other feedback from the participants regarding shoe L included poor appearance, dehumanized overall design, thick sponge behind the shoes, and lack of details. In terms of dynamic comfort intensity, shoe L was insufficiently stable at the heel, while shoe N was not only stable but also responsive. Although the cushioning of shoe N was softer at the forefoot, the forefoot and heel cushioning of shoe L were better than those of shoe N, based on the participants' feedback. This finding is probably due to the fact that shoe N was too soft to provide sufficient cushioning. The general fit in the perceptual test was preferred for assessing the appropriateness of the shoes, while the dynamic scale was preferred for evaluating the perceptual likings and performance of the two shoes from a functional aspect. The deficiency of shoe L was mainly due to its lack of ergonomic design for fit or may be related to the materials and structural design.

Notably, the results of plantar pressure were inconsistent with the perceptual comfort of shoe L and shoe N in the above-mentioned plantar pressure and perceptual comfort scale tests [30]. Although shoe L has a lower plantar pressure on the forefoot, it is not as comfortable as shoe N. The results of plantar pressure indicated the advantage of the forefoot cushioning performance of shoe L. However, the midsole structure of shoe L should be improved to avoid excessive force concentration under intense impact from the pressure distribution. Che et al. [31] found that the plantar pressure index affects comfort assessment. Similarly, the midsole material and structure could affect individual comfort assessment [32]. Therefore, we assumed that the plantar pressure distribution was related to perceptual comfort to some extent because the human being, as an active organism, self-evaluates the perceptual information of sneakers [33]. Differences were observed in the midsole cushioning performance obtained by the mechanical/biomechanical test or the perception of comfort, which could directly affect people's choice of sports shoes. Ignoring these differences would inevitably affect the evaluation of sports shoes in terms of function. However, relevant studies on the reasons for these differences are limited. Therefore, further studies are required.

4.4. Limitations

In the present study, it is noteworthy that marker trajectories and surface electromyographic data were not collected to simplify the design by focusing on the foot loading and mimicking basketball maneuvers by limiting the experimental devices that were attached to the participants. However, future studies including electromyography results with different sports shoes in basketball players should be carried out in order to determine muscle activity [19,20,34]. Besides, it is considered that different trajectories performed by players could influence the trial times, as well as the impact on shoes [35].

5. Conclusions

No differences were observed in the impact forces and the average maximum plantar pressure between the two shoes during the three basketball-specific maneuvers. However, compared to shoe N, the plantar pressure range of shoe L was wider and showed a lower pressure at the midfoot and the lateral foot. Moreover, the comfort perception results indicated that though the general fit of the two shoes was equal, shoe L provided less overall comfort, forefoot flexibility, heel cushioning, heel stability, and heel response. Interestingly, the plantar pressure results were inconsistent with the perceptual comfort of the two shoes, which suggests that the biomechanical characteristics of the shoes themselves and the perception evaluation of the athletes should be considered in comprehensive shoe cushioning design and evaluation.

Author Contributions: X.Z., Z.L. and X.W. contributed equally. Conceptualization, W.F.; methodology, X.Z., Z.L. and X.W.; formal analysis, X.Z., Z.L., Y.Y., J.N. and X.W.; investigation, X.Z., Z.L., X.W., Y.Y., J.N. and W.F.; resources, W.F.; data curation, X.Z.; writing—original draft preparation, X.Z., Z.L., X.W. and J.N.; writing—review and editing, W.F.; project administration, W.F.; funding acquisition, W.F.

Funding: This work was supported by the National Natural Science Foundation of China (11772201, 11572202), the Talent Development Fund of Shanghai Municipal, China (2018107), the National Key Technology Research and Development Program of the Ministry of Science and Technology of China and the "Dawn" Program of Shanghai Education Commission, China.

Conflicts of Interest: The authors declare no conflict of interest.

References

1. Randers, M.B.; Hagman, M.; Brix, J.; Christensen, J.F.; Pedersen, M.T.; Nielsen, J.J.; Krustrup, P. Effects of 3 months of full-court and half-court street basketball training on health profile in untrained men. *J. Sport Health Sci.* **2018**, *7*, 132–138. [CrossRef] [PubMed]
2. Nin, D.Z.; Lam, W.K.; Kong, P.W. Effect of body mass and midsole hardness on kinetic and perceptual variables during basketball landing manoeuvres. *J. Sport. Health Sci.* **2016**, *34*, 756–765. [CrossRef] [PubMed]
3. Meeuwisse, W.H.; Sellmer, R.; Hagel, B.E. Rates and risks of injury during intercollegiate basketball. *Am. J. Sports Med.* **2003**, *31*, 379–385. [CrossRef] [PubMed]
4. Irmischer, B.S.; Harris, C.; Pfeiffer, R.P.; DeBeliso, M.A.; Adams, K.J.; Shea, K.G. Effects of a knee ligament injury prevention exercise program on impact forces in women. *J. Strength Cond. Res.* **2004**, *18*, 703–707. [PubMed]
5. Chiu, H.T.; Shiang, T.Y. Effects of insoles and additional shock absorption foam on the cushioning properties of sport shoes. *J. Appl. Biomech.* **2007**, *23*, 119–127. [CrossRef] [PubMed]
6. Morlock, M.; Nigg, B.M. Theoretical considerations and practical results on the influence of the representation of the foot for the estimation of internal forces with models. *Clin. Biomech.* **1991**, *6*, 3–13. [CrossRef]
7. Blackmore, T.; Ball, N.; Scurr, J. The effect of socks on vertical and anteroposterior ground reaction forces in walking and running. *Foot* **2011**, *21*, 1–5. [CrossRef]
8. Nilsson, J.; Thorstensson, A. Ground reaction forces at different speeds of human walking and running. *Acta Physiol.* **2010**, *136*, 217–227. [CrossRef]
9. Lam, W.K.; Qu, Y.; Yang, F.; Cheung, R.T.H. Do rotational shear-cushioning shoes influence horizontal ground reaction forces and perceived comfort during basketball cutting maneuvers? *PeerJ* **2017**, *5*, 1–13. [CrossRef]
10. Wei, Q.; Wang, Z.; Woo, J.; Liebenberg, J.; Park, S.K.; Ryu, J.; Lam, W.K. Kinetics and perception of basketball landing in various heights and footwear cushioning. *PLoS ONE* **2018**, *13*, e0201758. [CrossRef]
11. Lam, W.K.; Ng, W.X.; Kong, P.W. Influence of shoe midsole hardness on plantar pressure distribution in four basketball-related movements. *Res. Sports Med.* **2017**, *25*, 37–47. [CrossRef] [PubMed]

12. Nigg, B.M.; Bahlsen, H.A.; Luethi, S.M.; Stokes, S. The influence of running velocity and midsole hardness on external impact forces in heel-toe running. *J. Biomech.* **1987**, *20*, 951–959. [CrossRef]
13. Brizuela, G.; Llana, S.; Ferrandis, R.; Garcia-Belenguer, A.C. The influence of basketball shoes with increased ankle support on shock attenuation and performance in running and jumping. *J. Sports Sci.* **1997**, *15*, 505–515. [CrossRef] [PubMed]
14. Jordan, C.; Payton, C.; Bartlett, R. Perceived comfort and pressure distribution in casual footwear. *Clin. Biomech.* **1997**, *12*, 215–220. [CrossRef]
15. Robbins, S.; Waked, E.; Krouglicof, N. Improving balance. *J. Am. Geriatr. Soc.* **1998**, *46*, 1363–1370. [CrossRef] [PubMed]
16. Hennig, E.M.; Valiant, G.A.; Qi, L. Biomechanical Variables and the Perception of Cushioning for Running in Various Types of Footwear. *J. Appl. Biomech.* **1996**, *12*, 143–150. [CrossRef]
17. Dinato, R.C.; Ribeiro, A.P.; Butugan, M.K.; Pereira, I.L.R.; Onodera, A.N.; Sacco, I.C.N. Biomechanical variables and perception of comfort in running shoes with different cushioning technologies. *J. Sci. Med. Sport* **2015**, *18*, 93–97. [CrossRef] [PubMed]
18. Lam, W.K.; Sterzing, T.; Cheung, J.T.M. Reliability of a basketball specific testing protocol for footwear fit and comfort perception. *Footwear Sci.* **2011**, *3*, 151–158. [CrossRef]
19. Roca-Dols, A.; Elena Losa-Iglesias, M.; Sanchez-Gomez, R.; Becerro-de-Bengoa-Vallejo, R.; Lopez-Lopez, D.; Palomo-Lopez, P.; Rodriguez-Sanz, D.; Calvo-Lobo, C. Electromyography activity of triceps surae and tibialis anterior muscles related to various sports shoes. *J. Mech. Behav. Biomed. Mater.* **2018**, *86*, 158–171. [CrossRef]
20. Roca-Dols, A.; Losa-Iglesias, M.E.; Sanchez-Gomez, R.; Becerro-de-Bengoa-Vallejo, R.; Lopez-Lopez, D.; Rodriguez-Sanz, D.; Martinez-Jimenez, E.M.; Calvo-Lobo, C. Effect of the cushioning running shoes in ground contact time of phases of gait. *J. Mech. Behav. Biomed. Mater.* **2018**, *88*, 196–200. [CrossRef]
21. Price, C.; Parker, D.; Nester, C. Validity and repeatability of three in-shoe pressure measurement systems. *Gait Posture* **2016**, *46*, 69–74. [CrossRef] [PubMed]
22. Nigg, B.M.; Anton, M. Energy aspects for elastic and viscous shoe soles and playing surfaces. *Med. Sci. Sports Exerc.* **1995**, *27*, 92–97. [CrossRef] [PubMed]
23. Patrek, M.F.; Kernozek, T.W.; Willson, J.D.; Wright, G.A.; Doberstein, S.T. Hip-abductor fatigue and single-leg landing mechanics in women athletes. *J. Athl. Train.* **2011**, *46*, 31–42. [CrossRef] [PubMed]
24. Thompson, M.A.; Lee, S.S.; Seegmiller, J.; Mcgowan, C.P. Kinematic and kinetic comparison of barefoot and shod running in mid/forefoot and rearfoot strike runners. *Gait Posture* **2015**, *41*, 957–959. [CrossRef] [PubMed]
25. Lieberman, D.E.; Venkadesan, M.; Werbel, W.A.; Daoud, A.I.; D'Andrea, S.; Davis, I.S.; Mang'eni, R.O.; Pitsiladis, Y. Foot strike patterns and collision forces in habitually barefoot versus shod runners. *Nature* **2010**, *463*, 531–535. [CrossRef] [PubMed]
26. Yu, L.; Fu, W.; Wei, S. Effects of basketball shoe on impact force and quadriceps vibrations during active and passive landings. *Footwear Sci.* **2011**, *3*, 96–98.
27. Zhang, S.; Clowers, K.; Kohstall, C.; Yu, Y.J. Effects of various midsole densities of basketball shoes on impact attenuation during landing activities. *J. Appl. Biomech.* **2005**, *21*, 3–17. [CrossRef]
28. Kong, P.W.; Lam, W.K.; Ng, W.X.; Aziz, L.; Leong, H.F. In-shoe plantar pressure profiles in amateur basketball players implications for footwear recommendations and orthosis use. *J. Am. Podiatr. Med. Assoc.* **2018**, *108*, 215–224. [CrossRef]
29. Nigg, B.M.; Nurse, M.A.; Stefanyshyn, D.J. Shoe inserts and orthotics for sport and physical activities. *Med. Sci. Sports Exerc.* **1999**, *31*, 421–428. [CrossRef]
30. Leong, H.F.; Lam, W.K.; Ng, W.X.; Kong, P.W. Center of Pressure and Perceived Stability in Basketball Shoes with Soft and Hard Midsoles. *J. Appl. Biomech.* **2018**, *34*, 284–290. [CrossRef]
31. Che, H.; Nigg, B.M.; de Koning, J. Relationship between plantar pressure distribution under the foot and insole comfort. *Clin. Biomech.* **1994**, *9*, 335–341. [CrossRef]
32. Park, S.K.; Lam, W.K.; Yoon, S.; Lee, K.K.; Ryu, J. Effects of forefoot bending stiffness of badminton shoes on agility, comfort perception and lower leg kinematics during typical badminton movements. *Sports Biomech.* **2017**, *16*, 374–386. [CrossRef]
33. Scanlan, A.T.; Fox, J.L.; Borges, N.R.; Tucker, P.S.; Dalbo, V.J. Temporal changes in physiological and performance responses across game-specific simulated basketball activity. *J. Sport Health Sci.* **2018**, *7*, 176–182. [CrossRef]

34. Condello, G.; Schultz, K.; Tessitore, A. Assessment of sprint and change-of-direction performance in college football players. *Int. J. Sports Physiol. Perform.* **2013**, *8*, 211–212. [CrossRef]
35. Roca-Dols, A.; Losa-Iglesias, M.E.; Sanchez-Gomez, R.; Lopez-Lopez, D.; Becerro-de-Bengoa-Vallejo, R.; Calvo-Lobo, C. Electromyography comparison of the effects of various footwear in the activity patterns of the peroneus longus and brevis muscles. *J. Mech. Behav. Biomed. Mater.* **2018**, *82*, 126–132. [CrossRef]

applied sciences

MDPI

Article

Integrative Neuromuscular Training in Young Athletes, Injury Prevention, and Performance Optimization: A Systematic Review

Borja Sañudo [1,*], Juan Sánchez-Hernández [1], Mario Bernardo-Filho [2], Ellie Abdi [3], Redha Taiar [4] and Javier Núñez [5]

1 Departamento de Educación Física y Deporte, Universidad de Sevilla, E-41013 Seville, Spain; juan_28256@hotmail.com
2 Departamento de Biofísica e Biometria, Universidade do Estado do Rio de Janeiro, 22050-032 Rio de Janeiro, Brazil; bernardofilhom@gmail.com
3 Center of Pedagogy, Montclair State University, Upper Montclair, NJ 07043, USA; ellieabdi@verizon.net
4 Département d'Education Physique et sportive (EPS), Université de Reims, 51687 Reims, France; redha.taiar@univ-reims.fr
5 Departamento de Deporte e Informática, Universidad Pablo de Olavide, 41704 Seville, Spain; fjnunsan@upo.es
* Correspondence: bsancor@us.es; Tel.: +34-652-387090

Received: 6 August 2019; Accepted: 10 September 2019; Published: 12 September 2019

Abstract: The aim of this systematic review was to evaluate the current evidence by assessing the effectiveness of integrative neuromuscular training programs in injury prevention and sports performance in young athletes. Different data sources were analyzed up to January 2018. Eligible studies contained information on population (young athletes), intervention (neuromuscular training), comparator (control group or another exercise intervention), outcomes (injury prevention or sport performance), and study design (randomized trials or prospective studies). The trials were restricted based on the language (English) and for publication date (after 1 January 2007). Fourteen randomized controlled trials were included: Seven included dynamic stability-related outcomes. Three assessed the coordination performing fundamental movements and sport-specific skills, while other five studies analyzed muscle strength and two assessed plyometric tests. Agility was evaluated in three studies and speed tests were also considered by four studies. Finally, fatigue resistance in three studies and injury risk in four were assessed. This review provides evidence that integrative neuromuscular training programs can enhance performance and injury prevention in young athletes, taken into account that adherence to the training program is adequate. Collectively, well-designed, randomized studies are necessary to collaborate with the present findings.

Keywords: neuromuscular training; strength; injury prevention; young athletes

1. Introduction

Integrative neuromuscular training (INT) is defined as a training program aimed to enhance physical fitness and prevent the aggregation of the neuromuscular deficits, along with the improvements of the motor competence, especially in youth with a lower level of motor skills [1]. This type of training in sync with the improved sports-related movement skills could also have an impact on injury prevention [2]. Further, recent literature has displayed that the absence of this training's type before or during the adolescence can lead to imbalances and incorrect movement patterns. This fact might be associated with a greater injury risk [3–5], especially in young female athletes [6] and in those sports involving a high number of landings or change of direction maneuvers, such as football or basketball [7]. Although young athletes present a lower injury ratio in comparison with adults, these injuries entail a longer recovery period. The injuries could even lead to a cessation of the practice due to fear of recurrence [8], with consequent effect on their careers as athletes [9,10].

Recently, Fort-Vanmeerhaeghe et al. [11] developed a classification of the INT components that included two blocks: (a) The first block focused on the development of fundamental movement skills, which includes coordination, strength, plyometric, agility, dynamic stability, speed, and fatigue resistance training, and (b) in the second block sport-specific movement skills were included [11]. It is suggested that all these contents must be exercised in order to improve performance and reduce injury risks. By coordination, within the fundamental movement skills, authors usually refer to locomotor, manipulation, and stability skills (variety of movements and multitasking, including unanticipated reactions). It is believed that these abilities allow control and optimize different sport-specific movements, thus decreasing the injury risks in youth athletes [11]. Additional aspects, such as strength, have been extensively analyzed in the literature. Resistance training might include core and lower limb positions and stability but also the application of upper-body and lower-body exercises and pushing/pulling strength exercises [11]. Strength deficits have been associated with less neuromuscular control and increasing the appearance of injuries in the young population [12].

Moreover, plyometric [13], considered as the development of stretch shortening cycle ability and agility training (development of skills at maximum speed integrating changes of direction actions) [11] also have been suggested in order to improve the performance and decrease injury risks in this population group. Literature also highlighted dynamic stability training (balance training that includes dynamic actions) as a means to manage neuromuscular control because of the improvements in the sensorimotor system, which enhances joint dynamic stability and may also reduce the risk of injury in youth athletes [14]. Furthermore, it seems that speed is the main determinant in the performance of many sports. The inability to produce speed in sprint, optimally during childhood, may decrease the possibility of achieving high competitive levels [15]. This term is closely related with agility and would include the development of skills at maximum speed [11]. On the other hand, fatigue resistance training could be defined as the development of skills under fatigue conditions [11]. It is known that neuromuscular fatigue is an important risk factor in numerous sports injuries [16].

Despite the importance of this type of training, especially in young athletes, the number of studies that include all INT contents in their intervention programs is scarce. The knowledge on its effect of injury prevention and performance has to be determined in this population. Therefore, this systematic review focuses on the analysis of the current evidence assessing the effectiveness of INT programs on injury prevention and sports performance in young athletes.

2. Methods

2.1. Eligibility Criteria

For this systematic review, the Preferred Reporting Items for Systematic Reviews, Meta-Analyses (PRISMA) statement, and checklist were used. The inclusion criteria were studies performed in the young population that assessed the efficacy of INT training compared to a control group, with other types of training or with no training. The population considered was both young and athletes. Children up to 11 years old in girls and 13 years old in boys in addition to adolescents between ages of 12 and 18 years old for girls and 14–18 years old for boys, as previously defined by Lloyd et al. [17] were examined. Exclusively, the studies including two or more components of INT (i.e., fundamental and specific movement skills, strength, plyometrics, speed, agility, coordination, dynamic stability, and fatigue resistance) in the training program, which were compared to a control group (i.e., randomized controlled trial or prospective study) were considered. Further, only studies that clearly detailed parameters of exercise containing a description of at least training intensity or volume were included. Finally, only studies published in English after 1 January 2007 were assessed.

2.2. Search Strategy

A literature search was performed in the following electronic databases: PubMed, Cochrane Central Register of Controlled Trials, Web of Science, CINAHL, MEDLINE, and SPORTDiscus. The last search was conducted in January 2018. The search strategies varied according to the different databases and used the following systematic search terms: Neuromuscular control or neuromuscular training or integrative neuromuscular training and strength training, or plyometric training or speed training or agility training or fundamental movement skills or specific movements skills or coordination training or dynamic stability or fatigue resistance and youth athletes or young or adolescents and injury prevention or sport performance. Language was limited to English and participants were all human.

2.3. Data Collection Process and Quality Analysis

Two authors (J.S. and B.S.) independently screened titles and abstracts to choose potential eligible studies. Full-text articles were obtained and independently evaluated for inclusion in the present review based on the inclusion criteria. Disagreements were resolved by consensus and, if necessary, the study's authors were contacted for clarification guidance. The flow chart can be observed in Figure 1. Each study that met the inclusion criteria was abstracted for information regarding: Number of trained and untrained athletes, age and sex, sport, main and secondary outcomes, main results, INT contents included, frequency, duration, and intensity of the training program.

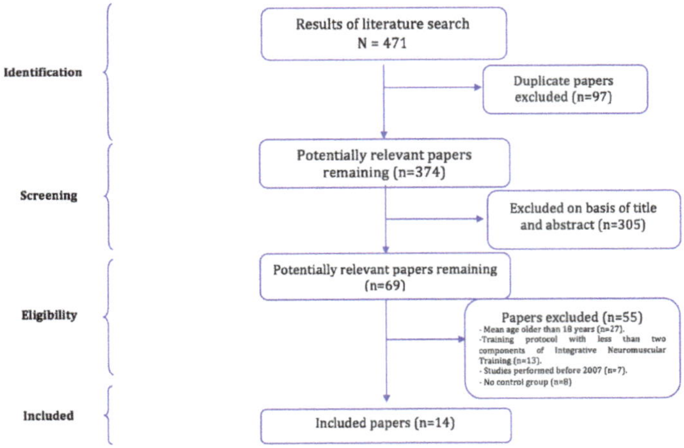

Figure 1. Flow diagram of the search process of integrative neuromuscular training effects in injury prevention and performance on youth athletes.

The methodological quality was individually assessed by using an adaptation of the Cochrane methods [18], which included the generation of randomization sequence, the allocation concealment, blinding of participants or assessors and description of withdrawals and dropouts (see Table 1).

Table 1. Quality assessment of studies.

Study (Author, Year)	Generation of Randomized Sequence	Allocation Concealment	Blinding of Participants or Assessors	Description of Withdrawals and Dropouts
Barendrecht [19]	No	No	No	Yes
Brown [20]	Yes	No	No	Yes
Chaouachi [21]	Yes	No	No	No
Emery [22]	Yes	Yes	No	Yes
Faigenbaum [23]	Yes	No	No	Yes
Fernandez [24]	No	No	No	Yes
Filipa [25]	No	No	No	Yes
Klugman [26]	No	No	No	Yes
Lindblom [27]	Yes	No	No	Yes
Mcleod [28]	No	No	Participants and assessors	Yes
Richmond [29]	Yes	Yes	No	Yes
Steffen [30]	Yes	No	No	Yes
Steffen [31]	Yes	No	No	Yes
Steffen [32]	Yes	No	No	Yes

3. Results

The literature search identified 471 potentially relevant studies (Figure 1). Following the screening of the titles, 305 studies were excluded in addition to 97 duplicates. Concerning the eligibility criteria, the remaining 69 studies were analyzed. Ultimately, 55 studies were excluded, 27 were due to the mean age which was higher than 18 years old, 13 included only one component of INT in their training protocols, seven were performed before 2007 and eight studies did not include a control group. This resulted in 14 included studies to be analyzed in the present review.

3.1. *Characteristics of the Included Studies*

Table 2 shows that most studies were executed exclusively on young female population [20,25–28, 30–32], four studies analyzed both male and female athletes [19,22,23,29], while two studies [21,24] included only male athletes.

Seven studies analyzed soccer players [22,25–27,30–32] and other studies were performed in handball [19], tennis [24] and basketball players [28]. Moreover, we found one study in players of different sports (i.e., field hockey, volleyball, soccer, and basketball) [20], and three remaining studies were developed in unspecified sports [21,23,29]. Regarding the age of the athletes and the peak of height velocity (PHV), which normally occurs around the age 12 in females and 14 in males [33], in twelve studies [19–22,24–28,30–32] mean age was post-PHV (>14 years old), while only in two studies [23,29] the mean age of the athletes was pre-PHV (<14 years old).

Table 2. Characteristics of the included studies.

Author	Participants	Sport	Age	Main and Secondary Outcomes	Main Results
Brown [20]	N = 43 female athletes. NMT, n = 10, CORE, n = 7, PLYO, n = 13, CG, n = 13.	Basketball, field hockey, soccer, and volleyball.	NMT: 14.1 ± 1.2 years, CORE: 15.0 ± 0.6 years, PLYO: 14.8 ± 0.6 years, CG: 14.7 ± 2.6 years.	**FUNDAMENTAL MOVEMENT SKILLS:** Sagittal and frontal planes hip and knee kinematics and kinetics.	Greater peak knee flexion following training, during both bilateral ($p = 0.027$) and unilateral landings ($p = 0.076$, d = 0.6) was reported in NMT. PLYO demonstrated reduced hip adduction ($p = 0.010$) and greater knee flexion ($p = 0.065$, d = 0.6) during bilateral landings following training. A significant time interaction was observed in CG showing a significant reduction in peak stance knee abduction moment ($p = 0.003$).
Chaouachi [21]	N = 42 male athletes. PLYO, n = 14. Balance and plyometric group (Bal-PLYO), n = 14. CG, n = 14.	Public school students with no after school activities or formalized strength and conditioning training programs.	PLYO: 13.7 ± 0.8 years, Bal-PLYO: 13.3 ± 0.8 years, CG: 13.5 ± 0.8 years.	**STREGNTH, SPEED, DYNAMIC STABILITY, AND AGILITY:** Lower-body strength test (one repetition maximum -1RM- leg press), power test (horizontal and vertical jumps, triple hop for distance, reactive strength, and leg stiffness), running speed (10-m and 30-m sprint), static and dynamic balance (standing Stork test and SEBT), and agility (shuttle run).	Bal-PLYO training was considered likely to be superior to PLYO training in leg stiffness (d = 0.7), 10-m sprint (d = 0.6), and shuttle run (d = 0.5). In 8 of the 11 dependent variables, the difference between Bal-PLYO vs. PLYO training was unclear. Bal-PLYO training was considered likely to be meaningfully better than CG in all dependent variables (d = 0.6–1.4). PLYO training was considered likely to be meaningfully better than CG in all dependent variables (d = 0.3–1.2) except leg stiffness and 10-m sprint.
Emery [22]	N = 744 male and females. NMT, n = 380. CG, n = 364	Soccer	NMT: U13-15 = 46.6%, U16-U18 = 53.4%, Female = 42.4%, Male = 57.6%, CG: U13-15 = 48.9%, U16-U18 = 51.1%, Female = 69%, Male = 31%.	**INJURY SURVEILLANCE:** Injury definition was soccer injury resulting in medical attention and/or removal from a session and/or time loss.	Significant differences were found in injury rates between NMT (2.08 injuries/1000 player-hours) and CG (3.35 injuries/1000 player-hours).
Faigenbaum [23]	N = 40 males and females. NMT, n = 10 boys and 11 girls. CG, n = 6 boys and 13 girls.	Children from two second-grade PE classes in a public school	All groups: Boys = 7.6 ± 0.3 years, Girls = 7.5 ± 0.3 years. Total = 7.6 ± 0.3 years.	**STRENGTH, DYNAMIC STABILITY, AGILITY, AND FATIGUE RESISTANCE:** Push-up, curl-up, standing long jump, single-legged hop, single-legged balance, sit-and-reach flexibility test, shuttle run, and 0.8-km run.	Intervention effects in the girls enhanced NMT-induced gains in performance relative to CG on the curl-up, long jump, single-legged hop, and 0.8-km run ($P < 0.05$) after controlling for baseline. Boys did not demonstrate similar adaptations from the NMT program ($P \geq 0.05$).
Fernández [24]	N = 16 male athletes. NMT, n = 8, CG, n = 8.	Tennis	All groups: 16.9 ± 0.5 years.	**STRENGTH, SPEED, AND FATIGUE RESISTANCE:** Sprint test (10 and 30 meters) CMJ, RSA, and maximal aerobic performance using a 30-15 intermittent fitness test (30-15 IFT).	After training, 10-m sprint, CMJ, and RSA were significantly improved in NMT ($P \leq 0.05$).

Table 2. *Cont.*

Author	Participants	Sport	Age	Main and Secondary Outcomes	Main Results
Filipa [25]	N = 20 female athletes. NMT, n = 13. CG, n = 7.	Soccer	NMT: 15.4 ± 1.5 years, CG: 14.7 ± 0.8 years.	**DYNAMIC STABILITY:** SEBT	Significant improvements in SEBT were observed in both limbs for NMT in the posterolateral direction and in the posteromedial direction of the left limb.
Klugman [26]	N = 49 female athletes. NMT, n = 15. CG, n = 34.	Soccer	ETG: 14.1 ± 0.4 years. CG: 14.7 ± 1.0 years.	**FUNDAMENTAL MOVEMENT SKILLS:** Tuck jump test.	Both groups significantly reduced their landing and jumping deficits. There was not a significant interaction of training status between groups.
Lindblom [27]	N = 41 female athletes. NMT, n = 23, CG, n = 18.	Soccer	NMT: 14.2 ± 0.7 years. CG: 14.2 ± 1.1 years.	**DYNAMIC STABILITY, STRENGTH, PLYOMETRICS, AGILITY, ADN SPEED:** SEBT, CMJ, triple-hop for distance test, a modified Illinois agility test, and 10-m and 20-m sprint tests.	Non-significant time effects were observed in NMT from baseline to follow-up. Changes were observed in CG compared to NMT for a sub-score of the SEBT and in the modified Illinois agility test ($p < 0.05$).
Mcleod [28]	N = 62 female athletes. NMT, n = 37, CG, n = 25.	Basketball	NMT: 15.6 ± 1.1 years, CG: 16.0 ± 1.3 years.	**DYNAMIC STABILITY:** BESS and SEBT.	A significant decrease in total BESS errors in NMT compared with CON (P = 0.003) was found. NMT subjects also reported significantly fewer BESS errors in the single-foam and tandem-foam conditions than CG and demonstrated improvements in the single-foam compared with their pre-test (P = 0.033). Between-group changes at posttest were observed in the lateral, anteromedial, medial, and posterior directions. They also improved in NMT compared with CG (P < 0.05) using the SEBT.
Richmond [29]	N = 725. NMT, n = 353 (female 57.2%, male 42.8%), CG, n = 372 (female 57.0%, male 43%).	Physical education students.	NMT: 13 years, CG: 13 years.	**INJURY SURVEILLANCE AND FATIGUE RESISTANCE:** Sports injuries (total, lower extremity injury, time loss, ankle, and knee sprain injury) secondary outcomes also included: waist circumference, and aerobic fitness (Leger 20-meter shuttle run test).	The NMT program was efficacious in reducing sport-related injury rates and improving measures of adiposity and fitness in the intervention group.
Steffen [30]	N = 2020 female athletes. NMT, n = 1073. CG, n = 947.	Soccer.	All groups: Ages between 13 and 17 (with 43% in NMT and 45% in CG with 15 years of age).	**INJURY SURVEILLANCE:** Overall injury rate (proportion of injured players, the incidence of ankle, knee, groin, hamstrings, and other injuries).	Neither a difference was observed in the overall injury rate between NMT (3.6 injuries/1000 h) and CG (3.7/1000 h) nor in the incidence for any type of injury. Only 14 out of 58 intervention teams completed more than 20 prevention training sessions, therefore, the study reported low adherence to the program.

Table 2. Cont.

Author	Participants	Sport	Age	Main and Secondary Outcomes	Main Results
Steffen [31]	N = 34 female athletes. NMT, n = 18. CG, n = 16.	Soccer.	All players were aged 16–18 years (17.1 ± 0.8).	**STRENGTH, SPEED, AND SPORT-SPECIFIC SKILLS:** Isokinetic and isometric strength protocols for the quadriceps and hamstrings, isometric hip adduction, and abduction strength, vertical jump tests, sprint running and soccer skill tests.	There was no difference between the NMT and CG for any of the outcome measured.
Steffen [32]	N = 226 female athletes. CTG, n = 78. RG, n = 68. CG, n = 80.	Soccer.	All players between 13 and 18 years old. NMT: U16 (57.7%)-U18 (42.3%), RG: U16 (26.5%)-U18 (73.5%), CG: U16 (22.5%)-U18 (77.5%).	**DYNAMIC STABILITY, PLYOMETRICS AND INJURY SURVEILLANCE:** SEBT, single-leg balance, triple hop, and jumping over a bar test. Injury risk.	Compared to CG, single-leg balance (OR = 2.8, 95% CI 1.1 to 4.6) and the anterior direction of the SEBT improved significantly in NMT (OR = 4.7, 95% CI 2.2 to 7.1), while two-leg jumping performance decreased (OR = −5.1, 95% CI −9.9 to −0.2). Significant improvements in five of six reach distances in the SEBT were found, favoring players who highly adhered to the program (FIFA 11+). Moreover, injury risk was lower for those players (IRR = 0.28, 95% CI 0.10 to 0.79).

NMT: Neuromuscular training group; PLYO: Plyometric group; CORE: Core group; CTG: Comprehensive group; RG: Regular group; CG: Control group; DJ: Drop jump; SEBT: Star excursion balance test; CMJ: Countermovement jump test; RSA: Repeated sprint ability; BESS: Balance error scoring system; IRR: Injury rate ratio.

3.2. Outcomes Measures

Dynamic stability related outcomes were assessed in seven studies [19,21,23,25,27,28,32]. As defined by Fort-Vanmeerhaeghe et al. [11], the term "dynamic stability" was considered as the training of lower limb dynamic stabilization (with three categories of progression: Static balance, dynamic balance, and dynamic stabilization) and core dynamic stability. Further studies assessed coordination performing fundamental movements [19,20,26] and sport-specific skills [31], while other studies analyzed muscle strength [21,23,24,27,31] and plyometric [27,32] tests. Agility [21,23,27] and speed [21,24,27,31] tests were also considered by different studies. Finally, fatigue resistance [23,24,29] and injury risk [22,29,30,32] were also examined.

Among the selected studies, five [19,21,25,28,32] found significant improvements in dynamic stability. Four of them reported significant increments in Star excursion balance test (SEBT) in the intervention group compared with control participants [21,25,28,32]. Furthermore, positive changes in single leg stability (one-leg hop test) were found [19]. Nevertheless, two studies [23,27] did not report changes between-groups in this outcome.

Significant changes compared to the control group were also observed in coordination performing fundamental movement skills (i.e., jumping and landing technique) in two studies [19,20]. Barendrecth et al. [19] showed that the in-season training group had the greatest benefits on knee kinematics and single leg stability. Although, Brown et al. [20] showed a greater peak knee flexion of landings in the neuromuscular training group in both bilateral (p = 0.027) and unilateral landings (p = 0.076). By contrast, non-significant between-group differences were observed by Klugman et al. [26]. With respect to the sport-specific skills, the only study that assessed this outcome [31] observed no differences between group performances of soccer players after a period of neuromuscular training.

Among the five studies that assessed muscle strength, three reported significant improvements [21,23,24]. Fernandez et al. [24] showed a 2% increment in countermovement jump (CMJ) in the experimental group. Moreover, Chaouachi et al. [21] found improvements in 1RM leg press test, CMJ and maximal hopping test in the neuromuscular training group compared with the control group. This agrees with Faigenbaum et al. [23] who reported significant changes in the training group (girls in physical education classes) in the curl up, long jump and single leg hop tests. Notwithstanding, two studies [27,31] did not find any improvement in strength after the training protocol. Neither of the two studies which analyzed plyometric performance reported positive changes between training and control groups [27,32].

Once the agility tests were analyzed, again contradictory results were found. Uniquely, the study performed by Chaouachi et al. [21] reported increments in shuttle run performance (d = 0.52). However, non-significant improvements were seen on agility tests in two additional studies [23,27]. This controversy persists in the speed tests results, while two studies found significant differences in 10-m sprint [21,24] and one in 30-m sprint [21], the remaining two studies did not show changes in this outcome [27,31].

With respect to the outcome of fatigue resistance, one study [23] conducted in girls, found positive changes. Moreover, a different study [24] observed significant changes in repeated sprint ability (RSA), with performance increments of ≈1.5%, but did not find changes in the 30-15 intermittent fitness test. Contrarily, Richmond et al. [29] showed that the experimental group, submitted to strength and dynamic stability training, had significant improvements in oxygen uptake (VO_{2max}) in the Leger 20-meter shuttle run test.

With respect to the injury risk, three studies [22,29,32] found significant improvements in the training group compared to the control group. In their study, Emery et al. [22] reported lower injury rates in the training group (2.08 injuries/1000 h) when compared to the control group (3.35 injuries/1000 h). Moreover, Richmond et al. [29] presented an injury rate of 7.1 per 100 students in the experimental group and 14.5 per 100 students in the control group. In a study, Steffen et al. [32] included an experimental group that performed a protocol combining strength, plyometrics, dynamic stability, and agility during four months with two, three sessions per week. It was reported that the injury risk is significantly lower in the high adherence group in lower extremities (IRR = 0.28, 95% CI 0.10 to 0.79) when compared with the control group. Uniquely, one of the analyzed studies [30] described no significant changes between groups in injury risk after a period of strength, plyometric, and dynamic stability training (training group = 3.6 injuries/1000 h, control group = 3.7/1000 h).

3.3. Characteristics of the Intervention Programs

The strength and dynamic stability showed in Table 3 were the contents of INT and were included in the analyzed studies [19–23,25–32]. Plyometric training was used in ten studies [19–22,24,26,28,30–32]. Three studies [19,20,27] included coordination performance fundamental movement skills training in their intervention programs, while four [22,24,28,32] used agility training and two-speed training [20,24].

The length of the exercise programs varied in the included studies from six weeks to eight months. Study duration was six weeks in two studies [20,28], eight weeks in four [21,23–25], and 10 weeks in other three studies [19,26,31]. The remaining studies had a duration of: 11 weeks [27], 12 weeks [29], 16 weeks [32], 20 weeks [22], and eight months [30].

The frequency of the sessions for most studies was two times per week [19,23–25,27,28]. The exceptions were four studies in which participants had three training sessions per week [20,21,26,31] and two studies that had a variable duration between two and three sessions per week [29,32]. Conjointly, in one study [30], the intervention group was trained 15 consecutive sessions and thereafter one session per week. Information regarding the number of training sessions per week was not provided in one study [22].

The duration of the sessions was between 15 and 20 min in most studies [19,22,23,27,29–32], throughout the time, in other three studies, the sessions lasted 60 min or more (20,25,28). Three studies [21,24,26] did not provide any information related to the duration of the sessions. In these studies, participants performed five to six exercises of strength, plyometric and dynamic stability training [21,26], or plyometric, agility and speed training after eight minutes of dynamic warm-up [24].

Table 3. Characteristics of the intervention programs.

Author	INT Contents	Duration	Frequency	Program
Barendrecht [19]	Strength, plyometrics, dynamic stability, and coordination.	10 wk	2x/wk	NMT: 20 min/session, 6 min warm-up, including agility exercises, coordination (4 min), balance (4 min), and strength-plyometric exercises (6 min). CG (RT group): Received their usual handball training.
Brown [20]	Strength, plyometrics, dynamic stability, speed, and coordination.	6 wk	3x/wk	NMT = 60 min standard neuromuscular training (core strength, balance, plyometrics, resistance, and speed training). PLYO = 20 min plyometric exercises. CORE = 20 min of core and balance training. CG = Daily activities. All training programs included an active warm-up (jogging, stretching, side-shuffle, and backward run) and self-selected stretching (≈10 min) immediately before and after each session.
Chaouachi [21]	Strength, plyometrics, and dynamic stability.	8 wk	3x/wk	Bal-PLYO and PLYO did five different exercises per session (strength, plyometrics and dynamic stability), 1/2 sets of 8–15 reps. CG: Regularly scheduled physical education class.
Emery [22]	Strength, agility, plyometrics, and dynamic stability.	20 wk	Not provided	NMT: 5 min warm-up including aerobic, static and dynamic stretching components to be completed in 5 min, in addition to 10 min of INT (i.e., strength, agility, balance) and a 15-min home-based balance-training program. CG: 15 min standardized warm-up (aerobic, static, and dynamic stretching components) and a home-based stretching program.
Faigenbaum [23]	Strength and dynamic stability training.	8 wk	2x/wk	NMT: 15 min sessions of INT followed by regularly scheduled 43 min PE class. CG: Participants only attended their regular PE class.
Fernández [24]	Plyometrics, agility, and speed.	8 wk	2x/wk	NMT: 8 min dynamic warm-up and combined RS training and ExpS. CG: Usual tennis training sessions.
Filipa [25]	Strength and dynamic stability.	8 wk	2x/wk	NMT: 5 min warm-up on agility ladder, 2 x 45-min increments of lower extremity strength and core stability training, and a 5-min cool-down that included static and dynamic stretches. CG: Instructed to continue only their usual activities.
Klugman [26]	Strength, plyometrics, and dynamic stability.	10 wk	3x/wk	NMT: Six exercises focused on hamstring strength, plyometrics, and dynamic stability, progressing every two weeks. CG: Received no intervention and continued their regular in-season routine.

Table 3. *Cont.*

Author	INT Contents	Duration	Frequency	Program
Lindblom [27]	Strength, dynamic stability, and coordination.	11 wk	2x/wk	NMT: 15 min. Six exercises targeting strength, dynamic stability, landing technique, and proper knee alignment. Each exercise had four different levels of difficulty. CG: Instructed to perform their usual football training without any changes.
Mcleod [28]	Strength, plyometrics, agility, and dynamic stability training.	6 wk	2x/wk	NMT: 90 min/session, 5 min warm-up (jogging, side shuffles, cariocas, and stretches) followed by (a) strengthening (30 min), (b) plyometrics (20 min), (c) agility training (10 min), and balance training (10 min). CG: No training protocol, instructed to continue only their usual activities.
Richmond [29]	Strength and dynamic stability.	12 wk	2/3x/wk	NMT: 15 min warm-up to PE class in junior high school. The warm-up included (a) 10 min NMT aerobic session and (b) 5 min lower extremity strength and dynamic stability component. CG: A variation of a "current standard of practice warm-up" including a low-intensity aerobic component (10 min of low-intensity jogging around the gym) and static and dynamic stretching components (5 min).
Steffen [30]	Strength, plyometrics, and dynamic stability.	8 months	15 consecutive sessions and thereafter 1x/wk until the season completion	NMT: 20 min/session, including 5 min of jogging before starting the exercises. 10 exercises focusing on dynamic stability, strength, and plyometrics. CG: Continued with their normal training.
Steffen [31]	Strength, plyometrics, and dynamic stability.	10 wk	3x/wk	NMT: 15 min. 10 exercises focusing on core stability, neuromuscular control, and eccentric hamstrings strength. CG: Warm up as usual, with jogging and ball-based exercises.
Steffen [32]	Strength, plyometrics, dynamic stability, and agility.	4 months	2/3x/wk	CTG: 20 min warm-up. 15 exercises divided into three parts, including initial and final running exercises with a focus on cutting, jumping, and landing techniques (parts 1 and 3) and strength, plyometrics, agility, and dynamic stability components (part 2). They did warm-up at the beginning of all practices (all parts) and matches (just part 1 and 3). Program was taught and supervised. RG: They received the information, but program was not taught or supervised. CG: Were solely provided with online access to the 11+ program.

NMT: Training group; CG; control group; PLYO: Plyometric training group; Bal-PLYO: Balance and plyometric training group; CORE: Core training group; CTG: Comprehensive group; RG: Regular training group; INT: Integrative neuromuscular training; PE: Physical education; wk: weeks.

4. Discussion

To the best of our knowledge, this is the first systematic review that analyzes the effects of INT in young population considering the contents used in the intervention programs. The primary purpose of this research was to determine which contents of INT provide better results in terms of injury prevention and performance in young athletes. However, there was a desire to advance one step further in the description of the different parameters that determine the application of the training type (i.e., which frequency, intensity, and volume can optimize the benefits of this type of training).

4.1. Characteristics of the Included Studies

Most of the studies in this review included only female athletes [20,25–28,30–32], in addition, there is a lack of studies assessing exclusively male athletes [21,24]. This fact makes it difficult to extrapolate the results to other populations. Furthermore, there is a clear tendency to analyze only soccer players [22,25–27,30–32], suggesting the need to extend the results in other sports. Once the age of the participants was analyzed, we found again that most of the studies focused on post-PHV athletes [19–22,24–28,30–32] and only two studies [23,29] analyzed the effects of INT on pre-PHV athletes. This is an important factor, since we cannot generalize the results obtained from athletes at different maturation stages [34]. In their study, Granacher et al. [34] suggested training different contents depending on the athletes' maturation stage. The study considered that focus on coordination, agility, dynamic stability, and strength is more significant in early stages, which aim to improve the movement techniques at a controlled speed. Whilst in a more advanced stage, youths should be trained more frequently on plyometrics, core, and strength exercises with a greater sport specificity and execution speed. This is a tendency generally followed by the reviewed articles. We observed that in the particular studies that trained pre-PHV athletes only strength and dynamic stability training were utilized [23,29], while the studies performed on post-PHV athletes included alternative contents, such as plyometric training [19–22,24,26,28,30–32].

4.2. Outcomes Measured

Concerning the effects of INT in youth injury prevention and performance, it was suggested that dynamic stability can be improved [19,21,25,28,32] with an INT program. Notwithstanding, most studies reported improvements in this outcome. While two studies [23,27] did not find changes, possibly by the cause of the training, it did not have the magnitude or stimulus required to enhance this outcome [23,27], also possibly due to a low player attendance in the training sessions [27].

When the coordination of performing fundamental movement skills was assessed, we found some contradictory results. While significant changes after an INT protocol were reported in two studies [19,20], one study [26] observed non-significant between-group changes. Again, an insufficient application of exercise could give an explanation to these results as the authors themselves indicated. There might be a dose response relationship for improving the fundamental movement skills with INT [26]. By taking a closer look at this study it was found that the training protocol consisted of only six exercises focused on hamstring strength, plyometrics, and dynamic stability. Nevertheless, there were no specific coordination exercises including fundamental movement skills. Furthermore, the same tendency can be observed with the sport-specific skills. One study [31] did not report between-group differences. Subsequently, an INT program of 15 min per session composed of 10 exercises focusing on strength, plyometric, and dynamic stability training was accomplished. It also concluded that the training volume and intensity for each of the exercises were too low to result in performance improvements. It is necessary that at least 20 min of INT replace the ordinary warm-up exercises used by the team [31].

As previously reported, there is also a controversy in strength and plyometrics after a neuromuscular training protocol. Further, three of five studies found positive changes in strength [21,23,24], while the remaining of two [27,31] did not report any differences between groups.

One possible explanation to this lack of effect could be attributable, at least in one of the studies [27], to a low training adherence (59.6 ± 14.3%). It is assumed that a higher adherence to the training program would be necessary to improve this outcome [27,31]. Along the same line, non-significant changes between groups were found in plyometric tests after a period of INT in two studies [27,32]. This was probably due to a combination of the lack of stimulus and low adherence to the training program [27,32].

Moreover, contradictory results were found once more with respect to agility and speed tests. Most of the studies did not report significant changes between-groups in agility [23,27] or speed [27,29] performance. As it was described above, it appears that the adherence to the training programs and training stimulus should have been greater in order to enhance agility or speed performance [23,27,29]. Notwithstanding, another possible explanation of the discrepancies in this outcome is the number of contents performed. It has been suggested that performing a 15 min INT program, two times per week, training on only strength and dynamic stability contents [23,27,29], are not enough to improve agility and speed performance.

Fatigue resistance was the only INT content that improved in the three studies, which measured this capacity [23,24,29]. Therefore, it can be suggested that an INT program can improve this outcome in young athletes. This can happen after a period of training of 8–12 weeks with two/three sessions per week focused on around 15 min training of dynamic stability and strength contents [23,29] or plyometrics, agility, and speed contents [24].

Finally, regarding the injury surveillance, most of the analyzed studies observed a decrease in the rate of injuries [22,29,30,32]. Exclusively, one study [30] did not report significant results, probably considering the low compliance of the training program. The intervention teams included the INT program (composed of 20 min sessions where 10 exercises were performed focusing on strength, dynamic stability, and plyometrics) in only 60% of their training sessions during the first half of the season [30]. Due to these findings, the injury risk in athletes is likely to be reduced after an INT program with adequate adherence.

4.3. Characteristics of the Intervention Programs

In the literature, strength training is considered as one of the best paradigms to enhance physical performance in youth athletes [35,36]. Together with dynamic stability training, the INT content was used more in the training programs analyzed in the current study [19–23,25–32]. Furthermore, strength deficits have been associated with less neuromuscular control, which increases the occurrence of injuries mainly in the lower limbs [12]. In this systematic review, those studies which included strength training in their INT programs reduced the rates of injury [22,29,32]. However, these changes were also accompanied by improvements in dynamic stability [19,21,25,28,32], functional movement skills [19,20], strength [21,23], agility [21], speed [21], or fatigue resistance [23,29]. Nonetheless, four [26,27,30,31] of the 13 articles [19–23,25–32] that used strength training did not find positive changes in any of the outcomes measured. This generates a controversy due to the program characteristics in these articles, which were very similar to those studies that did find significant improvements.

Numerous explanations might be suggested for these discrepancies: (a) Application of exercise, (b) age of participants, and (c) contents included in the program. It could be speculated that it is the combination of training contents, rather than the sum of them individually, that confers the benefits on these young people. In fact, the four studies that did not find positive changes employed only three contents: Strength, plyometrics, and dynamic stability in three studies [26,30,31] and strength, dynamic stability and coordination in another one [27]. The age should be another key factor when programing this content. While pre-PHV athletes will have a neuromuscular type of adaptation, exercises must focus on techniques, postural control, and circum-PHV. Post-PHV athletes will not have just neuromuscular but also structural adaptations. The proposed task should additionally focus on improving technique in more advanced and sport-specific exercises and increasing execution speed [11]. Among the aforementioned studies with limited results, all were post-PHV (>14 years old) and one [31] presented the highest mean age of all analyzed studies. This suggests the need for an adjustment of adequate training loads and progression in each content.

Dynamic stability was the other main content used in the intervention programs [19–23,25–32]. Evidence supports the need to train this content due to improvements in the sensorimotor system, which can improve neuromuscular control. This leads to a better joint dynamic stability and, therefore, a decrement in the injury risk [37]. In fact, the youth athletes' inability to maintain postural balance in static and dynamic actions has been associated with an increased likelihood of injury [38]. In the reviewed articles, those which included dynamic stability training in their INT programs also included strength training [19–23,25–32]. Thus, similar improvements were found for both contents, which allowed us to suggest that dynamic stability along with strength training might be a good strategy to improve performance and prevent injuries in young athletes. The literature shows that dynamic stability should be trained through exercises focused on the lower limbs and core [39], which challenges the feed-forward mechanism. This is described as anticipated actions that occur before the sensorimotor system detects changes in the environment [40]. This mechanism is the most important factor of training in order to maintain balance in landing, deceleration and cut off maneuvers, which helps to decrease injury risk [41]. Nonetheless, the lack of these types of exercises can be clearly detected in the training programs described in the reviewed articles [19–32].

Plyometric training was the other content of INT widely included in the intervention programs [19–22,24,26,28,30–32], which has been shown to improve performance and to decrease injury risk in young athletes [13]. In the current review, studies which included plyometric training in their INT programs improved dynamic stability [19,21,28,32], functional movement skills [19,20], strength [21,24], agility [21], speed [21,24], fatigue resistance [24] tests, and reduced injury risk [22,32]. Only three [26,30,31] of the 10 studies [19–22,24,26,28,30–32] that used plyometric training did not report significant changes between-groups. One of the particularities of this INT content is the need for an adequate progression. This progression should start with low-intensity exercises performed at a slow velocity with a proper technique and develop at later stages to higher velocity and intensity drills [13].

In various studies [19,20,27], coordination of fundamental movement skills' training was included in their intervention programs. From these studies, we found two reported significant improvements in fundamental movement skills [19,20] and dynamic stability tests [19]. One study [27] could not find significant changes probably due to the low compliance with the training program (mean player attendance at the training sessions was 59.6 ± 14.3%). These skills are particularly important in pre PHV athletes during sport-specific skills development in circum-PHV and post-PHV athletes [42]. Coordination should be trained at early stages due to the greater neural plasticity that children have at this age [43]. Moreover, children are encouraged to be involved in a variety of sports at this age in order to develop neural adaptations for many skills before sport specialization [11]. Likewise, a proper movement competency not only enhances physical ability but, in addition, is suggested to decrease injury risk [44]. However, none of the studies that included coordination in the program subsequently analyzed the injury rates. Furthermore, it should also be highlighted that the protocol of sport-specific skills was not clearly defined in the analyzed studies. Throughout the same line, four articles [22,24,28,32] that included agility training in their protocols were found. Literature has defined agility as the combination of decision-making process and change of direction [45]. Decision-making will improve overall during adolescence based on the experience acquired by the athletes. Nevertheless, the change of direction technique can be enhanced in pre-PHV [15], probably as the result of the great neural plasticity of these youth athletes [43]. From the reviewed articles that used agility training in their protocols, we found improvements in dynamic stability [28,32], strength [24], speed [24], fatigue resistance [24] tests, and injury prevention [22,32]. Hence, it would be interesting to include these outcomes in the youth training programs to enhance performance and to reduce injury risk.

Regarding speed training, evidence suggests developing this INT content across the different stages of growth [15]. From the reviewed literature, we found two studies [20,24] that included speed training in their programs. These studies reported significant improvements in strength [24], speed [24], fatigue resistance [24], and fundamental movement skills [20]. Athletes should start training with the high-velocity capacity and the proper running biomechanics, with plyometric, coordination, and sprint technique exercises. Prior to puberty and before maturation, these parameters are easier to change [46]. However, post-PHV athletes may improve more with a combination of neural basis training mainly focused on strength training [15]. Contrarily, it is remarkable that none of the reviewed studies included detailed fatigue resistance training in their protocols. Since we know that neuromuscular fatigue is an important risk factor in numerous sports injuries [16], central fatigue can decrease the ability to perform complex tasks in sports, which will reduce performance and increase injury risk [47]. Again, at early stages, this content should be focused on the correct technical execution of basic movements (pre PHV) to gradually progress towards small-sided games with higher intensity. Once an athlete has enough competencies, a combination of small-sided games and high-intensity interval training should be included in the training program [11,48].

If the parameters of the exercise are analyzed, the length of the exercise programs varied between six weeks to eight months, with most programs lasting eight weeks [21,23–25]. As it was reported in previous literature, at least four weeks are necessary to generate adaptations [15]. Nonetheless, the goal of INT will be to achieve long-term athlete development, with a multi-year training approach following all stages of growth in youth athletes [15]. Moreover, training frequency varies between one and three sessions per week, with two days per week at the most used frequency [23–25,27,28,31]. According to this finding, manifestly practicing high-intensity INT with two/three days per week on non-consecutive days may be sufficient to produce adaptations in youth athletes [49]. Moreover, the duration of the sessions (which was between 15 min and 60 min or more) was 15 min on average in most of the selected studies [22,23,27,29,31]. With respect to this outcome, the literature suggests sessions of 60 min [50] or between 30 and 90 min for young athletes [51]. However, for physical education classes, it is also scientifically proven that 15 min sessions are effective in reducing injury risk [51]. Further analysis of this parameter in future studies is warranted.

It is interesting to note that while most studies disclosed significant improvements with these characteristics [19–25,28,29,32], some others [26,27,30,31] did not find improvements with similar protocols in terms of length of the program, frequency, duration, or INT contents trained. Therefore, it was hypothesized that the main reason for this controversy is the grade of adherence to the training program. Due to the fact that those studies which could not find significant improvements in any outcome measure, reported low adherence to the protocols as the main cause of their results [26,27,30,31]. Consequently, this makes the training stimulus lower than previously planned.

4.4. Limitations and Strengths of This Review

The main limitation of our systematic review was the lack of a detailed description of the training protocols employed by the included studies. However, in order to overcome this limitation, the authors were contacted to collect the missing data (although the lack of some data persisted in some cases). Moreover, the methodological quality of the included studies is moderate as a result of notable bias in at least six of 14 studies reported. Further, the heterogeneity of the studies (e.g., sample size) limits the discussion of the results. Moreover, only studies published in English language were considered in the current review.

The main strength of this study is the description of the most used INT contents in the literature and their effects on injury prevention and sports performance in the youth population. Additionally, this review analyzed the program characteristics, reporting the most common length, frequency, and duration used by the studies, which accomplished our inclusion criteria.

5. Conclusions

In conclusion, this review presents that INT programs can enhance performance and injury prevention in young athletes, as long as the adherence to the training program is adequate. The most INT contents used in the reviewed studies were strength, dynamic stability, and plyometrics. These contents were followed by coordination of fundamental movement skills, speed, and agility training. However, none of the studies included all the discussed contents in the same protocol. Finally, lack of protocols that included fatigue resistance and specific movement skills training was reported. This is contrary to the recommendations of the literature highlights, with respect to the benefits of discussed INT contents in young athletes.

Future studies are recommended to analyze the effects of a protocol that include all INT contents in young athletes at different growth stages. Researchers need to pay specific attention to pre-PHV athletes and in boys, where a lack of evidence in the literature was found. Needless to say, new lines of research should include INT programs with more drills that challenge the feed-forward mechanisms. This has been described in the literature as one of the most important factors in order to decrease injury risk.

Author Contributions: Conceptualization, B.S. and J.S.-H.; methodology, M.B.-F.; formal analysis, R.T.; investigation, B.S.; writing—original draft preparation, B.S. and J.N.; writing—review & editing, B.S. and J.N.; supervision, M.B.; review & editing, E.A.

Funding: This research received no external funding.

Conflicts of Interest: The authors declare no conflict of interest.

References

1. Rogasch, N.C.; Dartnall, T.J.; Cirillo, J.; Nordstrom, M.A.; Semmler, J.G. Corticomotor plasticity and learning of a ballistic thumb training task are diminished in older adults. *J. Appl. Physiol.* **2009**, *107*, 1874–1883. [CrossRef] [PubMed]
2. Bergeron, M.F.; Mountjoy, M.; Armstrong, N.; Chia, M.; Côté, J.; Emery, C.A.; Faigenbaum, A.; Hall, G.; Kriemler, S.; Léglise, M.; et al. International Olympic Committee consensus statement on youth athletic development. *Br. J. Sports Med.* **2015**, *49*, 843–851. [CrossRef] [PubMed]
3. Hewett, T.E.; Myer, G.D.; Ford, K.R. The influence of growth and pubertal maturation on neuromuscular performance in high-risk female athletes. *Med. Sci. Sports Exerc.* **2002**, *34*, S247. [CrossRef]
4. Quatman-Yates, C.C.; Quatman, C.E.; Meszaros, A.J.; Paterno, M.V.; Hewett, T.E. A systematic review of sensorimotor function during adolescence: A developmental stage of increased motor awkward- ness? *Br. J. Sports Med.* **2012**, *46*, 649–655. [CrossRef] [PubMed]
5. Myer, G.; Faigenbaum, A. Pediatric physical activity exercise is sports medicine in youth: Integrative neuromuscular training to optimize motor development. *Rev. Kronos* **2011**, *10*, 39–48.
6. Myer, G.D.; Ford, K.R.; Foss, K.D.; Goodman, A.; Ceasar, A.; Rauh, M.J.; Divine, J.G.; Hewett, T.E. The incidence and potential pathomechanics of patellofemoral pain in female athletes. *Clin. Biomech.* **2010**, *25*, 700–707. [CrossRef]
7. Gianotti, S.M.; Marshall, S.W.; Hume, P.A.; Bunt, L. Incidence of anterior cruciate ligament injury and other knee ligament injuries: A national population-based study. *J. Sci. Med. Sport* **2009**, *12*, 622–627. [CrossRef]
8. Bloemers, F.; Collard, D.; Paw, M.C.A.; Van Mechelen, W.; Twisk, J.; Verhagen, E. Physical inactivity is a risk factor for physical activity-related injuries in children. *Br. J. Sports Med.* **2012**, *46*, 669–674. [CrossRef]
9. Caine, D.; Purcell, L.; Maffulli, N. The child and adolescent athlete: A review of three potentially serious injuries. *Sports Sci. Med. Rehabil.* **2014**, *6*, 22. [CrossRef]
10. Sport & Fitness Industry Association. *Sports, Fitness, and Leisure Activities Topline Participation Report*; Sport & Fitness Industry Association: Silver Spring, MD, USA, 2013.
11. Fort-Vanmeerhaeghe, A.; Romero-Rodriguez, D.; Lloyd, R.S.; Kushner, A.; Myer, G.D. Integrative Neuromuscular Training in Youth Athletes. Part II: Strategies to Prevent Injuries and Improve Performance. *Strength Cond. J.* **2016**, *38*, 9–27. [CrossRef]
12. Hewett, T.E.; Myer, G.D.; Kiefer, A.W.; Ford, K.R. Longitudinal increases in knee abduction moments in females during adolescent growth. *Med. Sci. Sports Exerc.* **2015**, *47*, 2579–2585. [CrossRef] [PubMed]
13. Lloyd, R.S.; Meyers, R.W.; Oliver, J.L. The natural development and Trainability of plyometric ability during childhood. *Strength Cond. J.* **2011**, *33*, 23–32. [CrossRef]

14. Fort-Vanmeerhaeghe, A.; Romero-Rodriguez, D.; Montalvo, A.M.; Kiefer, A.W.; Lloyd, R.S.; Myer, G.D. Integrative Neuromuscular Training and Injury Prevention in Youth Athletes. Part I: Identifying Risk Factors. *Strength Cond. J.* **2016**, *38*, 36–48. [CrossRef]
15. Lloyd, R.S.; Oliver, J.L. *Strength and Conditioning for Young Athletes: Science and Application*; Routledge: Abingdon, UK, 2013.
16. Brazen, D.M.; Todd, M.K.; Ambegaonkar, J.P.; Wunderlich, R.; Peterson, C. The effect of fatigue on landing biomechanics in single-leg drop landings. *Clin. J. Sport Med.* **2010**, *20*, 286–292. [CrossRef] [PubMed]
17. Lloyd, R.S.; Oliver, J.L.; Faigenbaum, A.D.; Howard, R.; Croix, M.B.; Williams, C.A.; Best, T.M.; Alvar, B.A.; Micheli, L.J.; Thomas, D.P.; et al. Long-term athletic development-part 1: A pathway for all youth. *J. Strength Cond. Res.* **2015**, *29*, 1439–1450. [CrossRef] [PubMed]
18. Higgins, J.P.; Green, S. (Eds.) *Cochrane Handbook for Systematic Reviews of Interventions*; John Wiley & Sons: Hoboken, NJ, USA, 2011; Volume 4.
19. Barendrecht, M.; Lezeman, H.C.; Duysens, J.; Smits-Engelsman, B.C. Neuromuscular training improves knee kinematics, in particular in valgus aligned adolescent team handball players of both sexes. *J. Strength Cond. Res.* **2011**, *25*, 575–584. [CrossRef] [PubMed]
20. Brown, T.N.; Palmieri-Smith, R.M.; McLean, S.G. Comparative adaptations of lower limb biomechanics during unilateral and bilateral landings after different neuromuscular-based ACL injury prevention protocols. *J. Strength Cond. Res.* **2014**, *28*, 2859–2871. [CrossRef]
21. Chaouachi, A.; Othman, A.B.; Hammami, R.; Drinkwater, E.J.; Behm, D.G. The combination of plyometric and balance training improves sprint and shuttle run performances more often than plyometric-only training with children. *J. Strength Cond. Res.* **2014**, *28*, 401–412. [CrossRef]
22. Emery, C.A.; Meeuwisse, W.H. The effectiveness of a neuromuscular prevention strategy to reduce injuries in youth soccer: A cluster-randomised controlled trial. *Br. J. Sports Med.* **2010**, *44*, 555–562. [CrossRef]
23. Faigenbaum, A.D.; Myer, G.D.; Farrell, A.; Radler, T.; Fabiano, M.; Kang, J.; Ratamess, N.; Khoury, J.; Hewett, T.E. Integrative neuromuscular training and sex-specific fitness performance in 7-year-old children: An exploratory investigation. *J. Athl. Train.* **2014**, *49*, 145–153. [CrossRef]
24. Fernandez-Fernandez, J.; Sanz-Rivas, D.; Kovacs, M.S.; Moya, M. In-season effect of a combined repeated sprint and explosive strength training program on elite junior tennis players. *J. Strength Cond. Res.* **2015**, *29*, 351–357. [CrossRef] [PubMed]
25. Filipa, A.; Byrnes, R.; Paterno, M.V.; Myer, G.D.; Hewett, T.E. Neuromuscular training improves performance on the star excursion balance test in young female athletes. *J. Orthop. Sports Phys. Ther.* **2010**, *40*, 551–558. [CrossRef] [PubMed]
26. Klugman, M.F.; Brent, J.L.; Myer, G.D.; Ford, K.R.; Hewett, T.E. Does an in-season only neuromuscular training protocol reduce deficits quantified by the tuck jump assessment? *Clin. Sports Med.* **2011**, *30*, 825. [CrossRef] [PubMed]
27. Lindblom, H.; Waldén, M.; Hägglund, M. No effect on performance tests from a neuromuscular warm-up programme in youth female football: A randomised controlled trial. *Knee Surg. Sports Traumatol. Arthrosc.* **2012**, *20*, 2116–2123. [CrossRef] [PubMed]
28. McLeod, T.C.V.; Armstrong, T.; Miller, M.; Sauers, J.L. Balance improvements in female high school basketball players after a 6-week neuromuscular-training program. *J. Sport Rehab.* **2009**, *18*, 465–481. [CrossRef]
29. Richmond, S.A.; Kang, J.; Doyle-Baker, P.K.; Nettel-Aguirre, A.; Emery, C.A. A school-based injury prevention program to reduce sport injury risk and improve healthy outcomes in youth: A pilot cluster-randomized controlled trial. *Clin. J. Sport Med.* **2016**, *26*, 291–298. [CrossRef] [PubMed]
30. Steffen, K.; Myklebust, G.; Olsen, O.E.; Holme, I.; Bahr, R. Preventing injuries in female youth football—A cluster randomized controlled trial. *Scand. J. Med. Sci. Sports* **2008**, *18*, 605–614. [CrossRef] [PubMed]
31. Steffen, K.; Bakka, H.M.; Myklebust, G.; Bahr, R. Performance aspects of an injury prevention program: A ten-week intervention in adolescent female football players. *Scand. J. Med. Sci. Sports* **2008**, *18*, 596–604. [CrossRef]
32. Steffen, K.; Emery, C.A.; Romiti, M.; Kang, J.; Bizzini, M.; Dvorak, J.; Finch, C.F.; Meeuwisse, W.H. High adherence to a neuromuscular injury prevention programme (FIFA 11+) improves functional balance and reduces injury risk in Canadian youth female football players: A cluster randomised trial. *Br. J. Sports Med.* **2013**, *47*, 794–802. [CrossRef]

33. Malina, R.M.; Bouchard, C.; Bar-Or, O. *Growth, Maturation, and Physical Activity*; Human Kinetics: Champaign, IL, USA, 2004.
34. Granacher, U.; Lesinski, M.; Büsch, D.; Muehlbauer, T.; Prieske, O.; Puta, C.; Gollhofer, A.; Behm, D.G. Effects of Resistance Training in Youth Athletes on Muscular Fitness and Athletic Performance: A Conceptual Model for Long-Term Athlete Development. *Front. Phys.* **2016**, *7*, 164. [CrossRef]
35. Lloyd, R.S.; Faigenbaum, A.D.; Stone, M.H.; Oliver, J.L.; Jeffreys, I.; Moody, J.A. Position statement on youth resistance training: The 2014 international consensus. *Br. J. Sports Med.* **2014**, *48*, 498–505. [CrossRef] [PubMed]
36. Faigenbaum, A.D.; Myer, G.D. Pediatric resistance training: Benefits, concerns, and program design considerations. *Curr. Sports Med. Rep.* **2010**, *9*, 161–168. [CrossRef] [PubMed]
37. Lloyd, R.S.; Oliver, J.L.; Faigenbaum, A.D.; Myer, G.D.; De Ste Croix, M.B.A. Chronological age versus biological maturation: Implications for exercise programming in youth. *J. Strength Cond. Res.* **2014**, *28*, 1454–1464. [CrossRef] [PubMed]
38. Plisky, P. Star excursion balance test as a predictor of lower extremity injury in high school basketball players. *J. Orthop. Sports Phys. Ther.* **2006**, *36*, 911–919. [CrossRef] [PubMed]
39. Gamble, P. *Strength and Conditioning for Team Sports: Sport-Specific Physical Preparation for High Performance*; Routledge: Abingdon, UK, 2013.
40. Riemann, B.L.; Lephart, S.M. The sensorimotor system, part I: The physiologic basis of functional joint stability. *J. Athl. Train.* **2002**, *37*, 71–79. [PubMed]
41. Holmes, A.; Delahunt, E. Treatment of common deficits associated with chronic ankle instability. *Sport Med.* **2009**, *39*, 207–224. [CrossRef]
42. Morgan, P.J.; Barnett, L.M.; Cliff, D.P.; Okely, A.D.; Scott, H.A.; Cohen, K.E.; Lubans, D.R. Fundamental movement skill interventions in youth: A systematic review and meta-analysis. *Pediatrics* **2013**, *132*, e1361–e1383. [CrossRef]
43. Zetou, E.; Vernadakis, N.; Tsetseli, M.; Kampas, A.; Michalopoulou, M. The effect of coordination training program on learning tennis skills. *Sport J.* **2012**, *15*, 1–7.
44. Stodden, D.; Langendorfer, S.; Roberton, M.A. The association between motor skill competence and physical fitness in young adults. *Res. Q. Exerc. Sport* **2009**, *80*, 223–229. [CrossRef]
45. Young, W.B.; James, R.; Montgomery, I. Is muscle power related to running speed with changed of direction? *J. Sports Med. Phys. Fit.* **2002**, *42*, 282.
46. Barlett, R. *Introduction to Sports Biomechanics: Analysing Human Movement Patterns*; Routledge: Abingdon, UK, 2007.
47. Borotikar, B.S.; Newcomer, R.; Koppes, R.; McLean, S.G. Combined effects of fatigue and decision making on female lower limb landing postures: Central and peripheral contributions to ACL injury risk. *Clin. Biomech.* **2008**, *23*, 81–92. [CrossRef] [PubMed]
48. Harrison, C.B.; Gill, N.D.; Kinugasa, T.; Kilding, A.E. Development of aerobic fitness in young team sport athletes. *Sports Med.* **2015**, *45*, 969–983. [CrossRef] [PubMed]
49. Myer, G.D.; Faigenbaum, A.D. Exercise is sports medicine in youth: Integrative neuromuscular training to optimize motor development and reduce risk of sports related injury. *Rev. Kronos* **2011**, *10*, 39–48.
50. World Health Organization. *Reproductive Health. Medical Eligibility Criteria for Contraceptive Use*; World Health Organization: Geneva, Switzerland, 2010.
51. Faigenbaum, A.D.; Bush, J.A.; McLoone, R.P.; Kreckel, M.C.; Farrell, A.; Ratamess, N.A.; Kang, J. Benefits of strength and skill-based training during primary school physical education. *J. Strength Cond. Res.* **2015**, *29*, 1255–1262. [CrossRef] [PubMed]

applied sciences

MDPI

Article

3D Device for Forces in Swimming Starts and Turns

Karla de Jesus [1,2,3,4,*], Luis Mourão [1,2,5], Hélio Roesler [6], Nuno Viriato [2,7], Kelly de Jesus [1,2,3,4], Mário Vaz [2,7], Ricardo Fernandes [1,2] and João Paulo Vilas-Boas [1,2]

1 Centre of Research, Education, Innovation and Intervention in Sport, Faculty of Sport, University of Porto, 91 Dr. Plácido Costa st., 4200-450 Porto, Portugal
2 Porto Biomechanics Laboratory, University of Porto, 91 Dr. Plácido Costa st., 4200-450 Porto, Portugal
3 Human Performance Studies Laboratory, Faculty of Physical Education and Physiotherapy, Federal University of Amazonas, 3000 Gal. Rodrigo Octávio Jordão Ramos ave., South MiniCampus, Coroado I, Manaus 69077-000, Amazonas, Brazil
4 Human Motor Behaviour Studies Laboratory, Faculty of Physical Education and Physiotherapy, Federal University of Amazonas. 3000 Gal. Rodrigo Octávio Jordão Ramos ave., South MiniCampus, Coroado I, Manaus 69077-000, Amazonas, Brazil
5 Superior Institute of Engineering of Porto, Polytechnic Institute of Porto, 431, Dr. António Bernardino de Almeida st., 4249-015 Porto, Portugal
6 Centre of Physical Education, Physiotherapy and Sports, Aquatic Biomechanics Research Laboratory, University of the State of Santa Catarina, 358 Pascoal Simone st., Coqueiros, Florianópolis 88080-350, Santa Catarina, Brazil
7 Institute of Mechanical Engineering and Industrial Management, Faculty of Engineering, University of Porto. Dr. Roberto Frias st., 4200-465 Porto, Portugal
* Correspondence: karladejesus@ufam.edu.br; Tel.: +55-92-3305-1181

Received: 29 June 2019; Accepted: 16 August 2019; Published: 30 August 2019

check for updates

Abstract: Biomechanical tools capable of detecting external forces in swimming starts and turns have been developed since 1970. This study described the development and validation of a three-dimensional (six-degrees of freedom) instrumented block for swimming starts and turns. Seven force plates, a starting block, an underwater structure, one pair of handgrips and feet supports for starts were firstly designed, numerically simulated, manufactured and validated according to the Fédération Internationale de Natation rules. Static and dynamic force plate simulations revealed deformations between 290 to 376 με and 279 to 545 με in the anterior-posterior and vertical axis and 182 to 328.6 Hz resonance frequencies. Force plates were instrumented with 24 strain gauges each connected to full Wheatstone bridge circuits. Static and dynamic calibration revealed linearity (R^2 between 0.97 and 0.99) and non-meaningful cross-talk between orthogonal (1%) axes. Laboratory and ecological validation revealed the similarity between force curve profiles. The need for discriminating each upper and lower limb force responses has implied a final nine-force plates solution with seven above and two underwater platforms. The instrumented block has given an unprecedented contribution to accurate external force measurements in swimming starts and turns

Keywords: sports engineering; biomechanics; ground reaction forces; swimming; performance

1. Introduction

The analysis of swimmers' performance has traditionally examined spatiotemporal variables representing the start, turn and clean swimming distance [1]. The start and turn are fundamentally different skills than free swimming, which does not necessarily indicate a similar level of start or turning performance [1]. Generally, at the elite level, it is not only swimming speed that wins the races but rather the start and turn where most expert swimmers are travelling at their fastest velocity [2,3]. For instance, the start phase of the 50 m men's freestyle at the 2019 FINA Champions Swim Series at

Budapest has showed that 15 m after the start take-off phase, the second-placed swimmer was 0.08 s slower and the final race time difference was 0.15 s. Moreover, over a 200 m event, the turn contributes 21% to total race performance and progressively more as race distance increases [4].

External forces generated during the ventral starts using instrumented starting blocks have been measured since 1971 with a uniaxial force platform placed at the edge of the pool, which had assessed horizontal forces applied by the swimmers' feet during the conventional circular arm swing technique [5]. Cavanagh et al. [6] had optimised Elliot and Sinclair's solution including eight strain gauges in a horizontal bar to measure the horizontal and vertical forces applied by the hands and feet during the grab ventral start technique. Three-dimensional forces were assessed in starts since 2003 by Naemi et al. [7] revealing that the grab start performed with hands in between feet was less stable in preventing platform twist. Horizontal forces applied by feet and hands in the backstroke start were firstly measured in 2011 (de Jesus) using a uniaxial platform and a load cell, being fixed on the starting wall and on the handgrip, respectively.

The changes in the starting block design in 2008 implied adjustments in the previously developed instrumented platforms for start (e.g., [2]) force analyses. Mason and co-authors [2] have presented an instrumented block comprised of four triaxial force sensors placed in a main force plate fixed over the starting block, inclined rear plate, handles for ventral and backstroke start and underwater platform with holes over the front surface. Recently, researchers have shown a new device to measure independently three-axial forces exerted on hands and feet in the ventral kick start technique [8], but limitations have been identified on hand placement dependence. Despite the relevant shortcomings for start analysis that the previous systems have provided, none of them are able to narrow in a unique solution for all ventral, backstroke and relay start force analysis possibilities, assuming laterality effects and independency in hands and feet force assessment. The horizontal and vertical force components determine steering start strategies, being the most assessed. However, lateral responses are essentially a controlling movement in starts [7,9,10] and it can be better understood when assessing each swimmer's limb force contribution.

The description of forces generated during swimming turns also started in the 1970s using uniaxial force plates to assess the horizontal component in tumble and open turns (e.g., [11]). Swimming turns external forces analysis using three-dimensional force plates is still scarce, with the flip technique and its variants being the most commonly investigated (e.g., [12,13]). A double underwater tri-axial force plate solution developed for independent assessment of forces applied by each foot in a backstroke start can also provide a direct measurement of the combined forces exerted during turns and the determination of foot position. Furthermore, in some turn techniques, coaches can assess laterality data if the foot location is compatible with each force platform (e.g., [13]). Mason and co-authors' underwater force platform design has a multitude of holes system to reduce wave effects [2]. However, this configuration limits the possibility to use the same force plate for water wave effect analyses [14].

Coaching and commercial ongoing-instrumented starting blocks are still lacking some final integrated solution for start and turn force analyses that could inspire new biomechanical research directions. In a competitive, rapid uptake market such as sports equipment, it is important to keep searching for new and improved designs and materials at affordable prices [15]. Commercial force plate prices are ~$20,000, which can usually be very expensive for coaches and biomechanists [16]. Beside the high prices, technical assistance issues and spare parts' availability are sources of many difficulties for labs and researchers [17]. Thus, due to the high cost of existing force plate systems, the development of simpler low cost, adjustable and accurate models for biomechanical analysis is desirable for force measurements in swimming start and turn techniques (cf. [18]).

A dynamometric unit composed by independent 3D force plates, a starting block, an underwater structure, ventral and dorsal start handgrips and feet supports is original. It presents crucial potentialities, particularly regarding forces and momenta measurement in individual ventral and backstroke starts, relays and turning techniques. As a versatile and low cost system, the force plates can be used uncoupled from the dynamometric unit to measure active (e.g., [19]) and passive swimmers' drag (e.g., [14]), and underwater gait ground reaction forces (e.g., [20]). The current study aimed to design, construct and validate an instrumented swimming start block, emphasizing geometry description, numerical simulation, sensors bonding, calibration, experimental and ecological validation procedures.

2. Materials and Methods

2.1. 3D Geometric Computer Aided Design

Force plates, the starting block, underwater structure, handgrips and feet supports were 3D designed using a solid modelling computer aided design software (SolidWorks 2012, Dassault Systèmes, SOLIDWORKS Corporation, Waltham, MA, USA). Each force plate and handgrip was framed to achieve proper sensibility with a high rigidity and reduced mass. The start block project prioritized low deformation to support seven to nine force plates, handgrips and an underwater structure. In addition, force plates, the start block, underwater structure, handgrips and feet support dimensions (Figure 1) were conditioned to comply with the Fédération Internationale de Natation rules (FINA; FR 2.7 and FR 2.10).

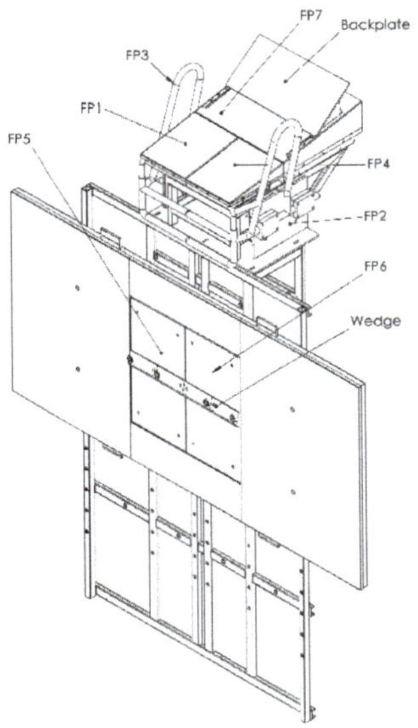

Figure 1. The seven force plates solution distributed in the start block for individual limb force-moment assessments: Each force plate (FP 1 to 7) and feet supports (back plate and wedge) are also shown.

2.2. Force Plates Spatial Layout

Contrarily to arbitrarily multiplying force platforms on the block, it is worth it to consider a minimum number of them. Such a number considers the limbs involved in the starting technique force development and the set of limb positioning on the starting block/wall. The rationale relies on the following: If a sole limb force is measured then the centre of pressure $(\vec{COP})$ lies on the limb contact area. On the other hand, if a sole platform measures simultaneously two-limb forces then the centre of pressure lies between the limbs and perhaps outside each limb contact area. Such measurement can mask postural constraints necessary for maximum performance data interpretation. The five above water force plates configuration was firstly developed for independent swimmers' upper and lower limbs kinetic analysis in individual ventral start techniques (e.g., grab and track start).

Subsets of analogous platforms have been preferable, aiming for the comparison of force plates behaviour, which implied that two pairs of platforms were similar (300 mm × 250 mm above water and 600 mm × 300 mm underwater). This resemblance provides an interchangeable ability, which might be of great importance if mass production or maintenance were considered. The fifth above water force plate is 500 mm × 440 mm, enabling support for the rear foot as used in the track start technique.

2.3. Force Plates Geometry

The main requirements of the load sensor followed Lywood et al.'s geometry [21], which consisted of instrumented rectangular section bars oriented to be most sensitive to x, y and z force and torque components (cf. [22]), obviating most cross-talk. A waterproof sensor choice enables force plates immersion into the bottom of the swimming pool (~2 m). Sensor locus, distant from the centre anchorage points, allows a better interpolation approach. As the Roesler [20] platform design complied with these previous requirements, it was selected to serve as the testing tool.

It has already been mentioned that Roesler's [20] force plate topology allows a more accurate $\overrightarrow{COP}$ determination, as well as direct and independent 3D forces and moment measurements without interference among them [18]. Each force plate core was designed to be manufactured in galvanized steel and is essentially composed by two vertical and two horizontal beams plus two lateral boxes. The beams, contrarily to the ring or pylon, can acquire the applied load with better accuracy and precision, also allowing better minute change capture in the strain throughout the top plate and not just at the corners [23]. Above and underwater top (Figure 2a,b) and bottom (Figure 2c,d) force plates were built in duralumin to minimise force plate mass. The mounting apparatus plays a crucial role in providing accurate and reliable measurements, with the force plate top and core unattached with commercial bushings.

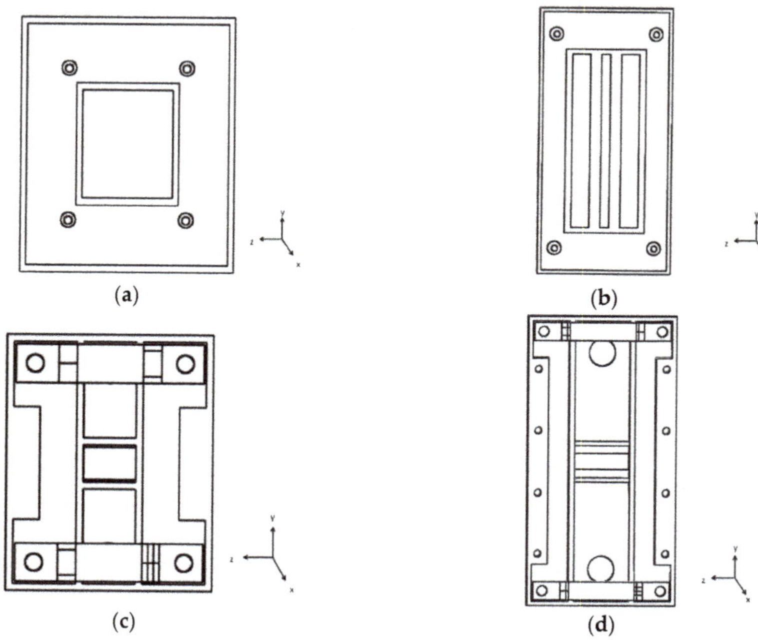

Figure 2. Above and underwater top (**a**,**b** panels) and core (**c**,**d** panels) 300 mm × 250 mm and 600 mm × 300 mm force plates top.

2.4. Start Block, Underwater Structure, Handgrips and Feet Support

Two start block projects were previously designed: (i) a bulky and solid structure and (ii) a lattice block with a declination support, which was replaced by a lattice galvanized steel structure with zero inclination due to excessive mass and reduced stability showed by the previous structures. The starting block was projected to be fixed over an underwater structure attached vertically to the swimming pool wall by front and rear edges, being similar to a previous structure used by Pereira et al. [13,14] and de Jesus et al. [24] for flip turn and backstroke start kinetics assessment. The first underwater structure included a two independent force plates support. The underwater structure evolved from a heavier to a lattice form, with its version being slighter, including holes with 100 mm distance between them placed at different heights on both force plates. It presented a flat rectangular surface for swimming turn analysis (replying the touch pad; FINA, FR 2.13), with a hollow area for underwater force plates embedding, as previously used in turn analysis [13,14].

Handgrips were projected to be independent and framed in galvanized steel, being the first design very versatile to be easily used in ventral and dorsal start technique analysis. However, the first design had shown a handgrip positioning dependency on measured strain signals that did not allow real training and competitive swimmers' movement. In fact, strain gauges would be bonded to the handgrip pipes, obliging swimmers to position their hands on a fixed place to allow comparisons (this was eliminated by the handgrips fixation on each lateral force plate top). With two force plates fixed each one on the start block lateral, forces and moments could be measured and handgrips positioning could be located. The second handgrip project was based on a simple and outdated handgrip version (e.g., [24]), being updated for two horizontal (i.e., the highest and the lowest, 0.43 and 0.56 m above water surface, respectively) and vertical bars, following the OSB11 starting block configurations. The final handgrip prototype received fine arrangements due to the existent pipe profile. The adjustable feet support for ventral and backstroke start was framed in galvanised steel and nylon (FINA, FR 2.7 and 2.10, respectively), allowing the five rear foot authorised positions.

2.5. Finite Element Analysis

We conducted static structural simulations using modelling software for finite element analysis (Ansys v.12.1, ANSYS Inc., Canonsburg, PA, USA), thus enabling predictions on how the dynamometric unit would strain under isolated and integrated conditions. Dynamic simulations were applied to verify resonance frequency, equivalent stress, equivalent strain and deformations. Based on Roesler et al. [20] geometry and sensor location definition, a 8000 N load was vertically and antero-posteriorly applied to confirm the previously determined sensor location (Figure 3a). The 8000 N load was simulated to allow force plate use in other data collection purposes that had depicted increased ground reaction forces (e.g., long distance jump, 15.2 times body weight; [25]).

Static and dynamic force plate simulations were performed with a core, top and mounting apparatus (i.e., polyethylene bushing and screw). Simulations with the starting block and handgrips were conducted with a 2500 N (centrally located; Figure 3b) and 2000 N load (vertical and antero-posterior; Figure 3c) in the commonly used handgrip positioning for ventral and backstroke start [10]. The underwater structure with force plates vertically mounted was simulated with the gravity at sea level (i.e., 9.81 m/s^2) and values of total deformation, equivalent stress and equivalent elastic strain have also been obtained. The most refined mesh for force plates, the starting block, underwater structures and handgrip simulations was composed of pyramids and cobbled with a 1 mm length and 39,197 nodes and 12,431 elements, 76,344 nodes and 15,724 elements, 273,837 nodes and 107,653 elements, and 46,372 nodes and 17,459 elements, respectively.

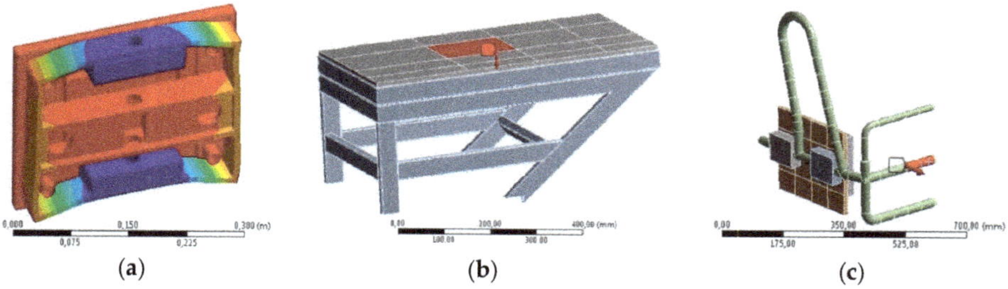

Figure 3. Strain numerical simulation in the assembled force plate (**a** panel), starting block (**b** panel), and handgrips (**c** panel).

2.6. Electrical Circuit

Following structures manufacturing, strain gauges were bonded to the platforms as sensing elements due to the previous research group background and short budget available. Each force plate was instrumented with 24 waterproof strain gauges (Kyowa, Electronic Instruments, KFW-5-120-C1-5M2B, Tokyo, Japan), arranged in six independent full Wheatstone bridges, minimising temperature effects. Those strain gauges were internally bonded to each force plate core and positioned as depicted in Figure 4a–d.

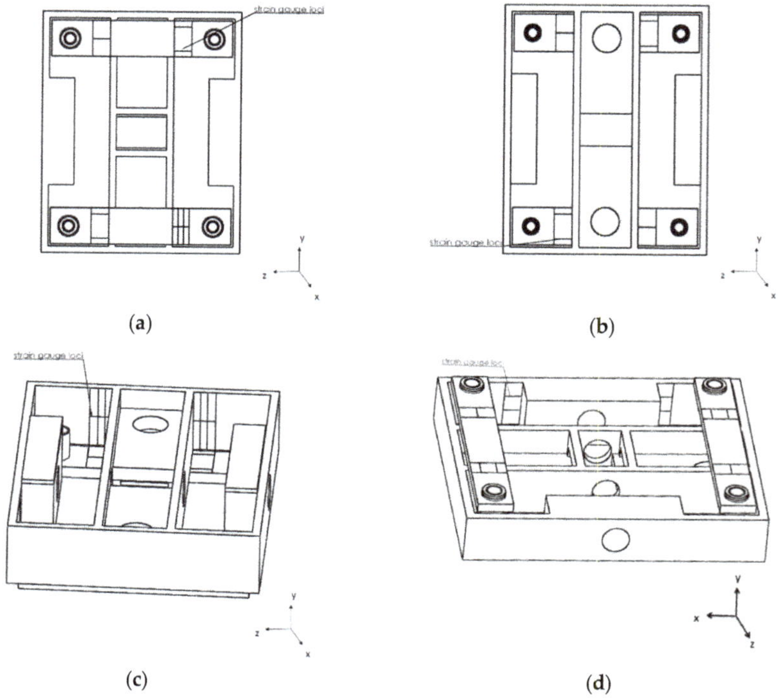

Figure 4. Strain gauge zones: Vertical force (y), antero-posterior (x) and medium-lateral (z) moment (top and bottom view, **a**,**b** panel), antero-posterior (x) force (**c** panel) and medium-lateral (z) force and vertical moment (**d** panel).

After bonding, each strain gauge received an additional protection of a two-part polybutadiene resin designed for re-enterable splice protection (Scotchcast TM re-enterable electrical insulation resin

2123, 3M TM, St. Paul, MN, USA), minimising chlorine wear. Strain gauge wires were brazed in a full Wheatstone bridge configuration (six for each force plate), silicone protected. Each full Wheatstone bridge was connected to a shielded and unfilled cable and provided data from each variable of interest (i.e., horizontal, vertical and lateral forces and momenta).

The six-shielded unfilled cables were connected to an analogue-to-digital converter module to transmit full Wheatstone bridge signals (NI9237 50 kS/s/ch 24 bit-4Channels, National Instruments Corporation, NI™, Austin, TX, USA) and to its scanner chassis (NI CompactDAQ 9172 and 9188 with 8 slots, National Instruments Corporation, NI™, Austin, TX, USA) through RJ50 connectivity, which interfaced with the computer. Strain gauges were identified by digit codes and cables wiring had also been standardised with colour code jacket terminals nearby the RJ50 plug. The dynamometric system specs included 42 strain plus two trigger channels acquiring at a 2000 Hz sampling rate.

Custom-designed data processing software was created in LabView 2013 (SP1, NI™, Austin, TX, USA) to acquire, plot and save the strain readings of each three axial force and moment of force component from each force plate. The executable file was programmed to record data in a total of 8 s (4 s before and after the trigger signal), which was a strategy implemented for the full relay start force acquisitions, anticipatory individual start (e.g., pre-activation) and turn actions (e.g., wave drag). Each force and moment of force curve profile was observed in real data time-acquisition and data acquisition files were later operated to transform strain signals into force-moment (using MatLab R2014a, The MathWorks Incorporated, Natick, MA, USA) routines for the required matricial conversion operations.

2.7. Static Calibration

Static calibration was performed on dry land before force plate proper use and applied both to a load and to an unload sequence with 10 kg individual masses (up to 50 kg in each positioning; cf. [8,18]) allowing the correspondence between strain and applied load. The vertical force component was calibrated at five positions with the use of a tension, compression machine (Instron 8804 Servohydraulic Fatigue Testing System, Instron®, Ilinois Tool Works Inc., Norwood, MA, USA) that laid the right load on each force plate top centre. On antero-posterior and lateral axis force calibration, platforms were vertically fixed on the wall and the load was applied on each centre of interest through a stainless-steel cable connection. A stainless-steel cable was fixed through holes made on the lateral of each force plate top (three per edge). Forces and moment were calibrated using the central and lateral holes (respectively).

2.8. Laboratory Experimental Validation

An experimental validation was completed comparing the results of a rigid body theoretical free rotation around a pivot point $\overrightarrow{COP}$ fall force pattern [26] and the associated strain signal pattern generated. Rigid body inertia moment of the inverted pendulum had to be assessed previously to know mass distribution to allocate its centre of mass and to calculate the centre of mass to the centre of pressure distance. This assessment implies the mass and geometrical dependency of moment of inertia and dynamic behaviour observed in force patterns generated, obviating the use of any force platform except gravity acceleration knowledge.

2.9. Ecological Experimental Validation

The instrumented start block was tested in a 25 m long and 1.90 m deep indoor swimming pool for real data acquisitions, which were then qualitatively compared with starting and turning data previously presented in the literature. All experimental procedures conformed to the requirements stipulated in accordance with The Code of Ethics of the World Medical Association (Declaration of Helsinki) and were approved accordingly by the local research ethics committee. Swimmers and parents and/or guardians (when participants were under 18 years old) provided informed written consent before data collection.

Nine well-trained, healthy and able-bodied male swimmers (mean and standard deviations: 21.09 ± 5.64 years, stature 1.74 ± 0.05 m, and body mass 71.32 ± 11.03 kg) and eight age-group male swimmers engaged on a regular basis in regional- and national-level competitions (mean ± SD: 12.4 ± 0.6 years old, 155.1 ± 13.6 cm of height, 44.6 ± 10.9 kg of body mass, 14.1 ± 5.3% of body fat, with 3.5 ± 1.4 years of experience in competitive swimming and 3.3 ± 0.7 (2–4) of Tanner maturation scale by self-evaluation) volunteered to participate in the start and turn validation study, respectively.

Swimmers performed a familiarisation period with each start and turn technique studied (cf. [27,28]). For the start protocol, each swimmer randomly performed three maximal 15 m trials of backstroke with vertical handgrips, track and one step relay starts. Swimmers performing the turn protocol were also tested in a set of three trials for the open backstroke to the breaststroke medley turning technique. Trials started and finished from the mid-pool (at 12.5 m from the turning wall) and swimmers were instructed to swim in and out at maximum speed until the 12.5 reference [13,28]. Rest periods of 2 min were observed between each repetition in both protocols.

A start signal complying with the FINA SW 4.2 rule was produced through an official device (StartTime IV acoustic start, Swiss Timing Ltd., Corgémento, Switzerland) and instrumented to simultaneously generate an auditory signal and export a pulse to the force plates with convenient signal conditioning. In the one step relay start protocol and open backstroke to breaststroke turn the starter device (without sound) was triggered 7.5 m before the incoming swimmer had touched the wall.

3. Results and Discussion

3.1. Simulations

Table 1 presents strain in antero-posterior and vertical axes (8000 N centred) and resonance frequency of the first, third and final prototypes of above water 300 mm × 250 mm, 500 mm × 440 mm and underwater 600 mm × 300 mm platforms, corroborating Roesler's [20] suggestions for strain values between 100 and 500 με. The force plates design optimisation was a compromise among maximum rigidity, minimum mass and high frequency, which allows small deformations, uncoupling, good linearity and low hysteresis (c.f. [17]). The specific force plate application determined proper resonance frequencies and the maximal vertical load was supported. The waterproof force plate used by Roesler [20], with 500 mm × 500 mm framed in galvanized steel, obtained a 35 Hz resonance frequency, which was considered sufficient for underwater applications due to over damping effects. However, in the current project, force plate versatility has been prioritized and values higher than ~140 Hz were required (cf. [20]) as they can be both used independently of the starting block and out of the water. Commercial force platforms with a natural frequency ≥150 Hz have been well accepted for triple jump ground reaction forces assessment (e.g., [25]).

Table 1. Antero-posterior and vertical strain and their resonance frequency above 300 mm × 250 mm, 600 mm × 300 mm and 500 mm × 440 mm underwater force plates of first, third and final prototypes.

Project Evolution	Static and Dynamic Simulations	300 × 250 mm	600 × 300 mm	500 × 440 mm
First	Antero-posterior (με)	323.0	470.0	297.0
	Vertical (με)	464.0	489.0	473.0
	Resonance frequency (Hz)	235.7	153.0	140.0
Third	Antero-posterior (με)	287.0	357.0	349.0
	Vertical (με)	545.0	521.0	511.0
	Resonance frequency (Hz)	323.7	197.0	177.0
Final	Antero-posterior (με)	290.0	375.0	376.0
	Vertical (με)	545.0	541.0	279.0
	Resonance frequency (Hz)	328.6	199.2	182.0

Table 2 shows results from one example of static simulation from each force plate considering all 24 strain gauge responses, which allow noticing maximal cross-talk less than 5% (cf. [20,21]). In fact, Roesler's [19] findings have shown ~3% of maximal interference among loads when 800 N had been applied. In addition, Lywood et al. [21] reported a cross-talk between orthogonal axes less than 5% for 40 N and 10 N of vertical and horizontal forces applied. The 300 mm × 250 mm force plate laterally positioned on the starting block with fixed handgrip was simulated with a 2000 N load applied vertically on the handgrips, indicating an expectable relevant lateral force (strain gauges 17, 18, 23, 24; cf. Figure 4d). When 8000 N was centrally and vertically applied on a 600 mm × 300 mm force plate, only strain gauges responsible for this measurement responded (i.e., 1 to 12, vertical force, antero-posterior and lateral moment; cf. Figure 4a,b).

Table 2. The 24 strain gauge responses when a 2000 N and 8000 N vertical load was applied on 300 mm × 250 mm and 600 mm × 300 mm force plates.

Strain Gauges	Variable	300 × 250 mm (με)	600 × 300 mm (με)
1	Lateral moment	−4.54	78.13
2	Vertical force	84.66	301.17
3	Lateral moment	−23.73	157.09
4	Lateral moment	−149.79	108.67
5	Vertical force	200.78	316.68
6	Lateral moment	83.29	93.08
7	Antero-posterior moment	−40.01	−215.51
8	Vertical force	133.75	−282.81
9	Antero-posterior moment	−74.94	−205.90
10	Antero-posterior moment	9.51	−160.94
11	Antero-posterior moment	55.70	−188.20
12	Vertical force	−72.43	−238.46
13	Antero-posterior force	−104.08	1.60
14	Antero-posterior force	93.30	0.52
15	Antero-posterior force	−177.25	0.62
16	Antero-posterior force	176.88	0.14
17	Lateral force	249.58	14.55
18	Lateral force	256.12	−10.41
19	Vertical moment	−26.40	−7.71
20	Vertical moment	−45.34	13.11
21	Vertical moment	7.09	13.10
22	Vertical moment	33.94	−8.22
23	Lateral force	−222.66	−10.20
24	Lateral force	−264.68	15.83

The starting block total deformation under a 2500 N centre vertical load was 0.00030553 m. The gravity at sea level (9.81 m/s^2) tested over the underwater structure and two force plates vertically fixed on it revealed a maximal deformation of 0.00012322 m. Moreover, using the same standard gravity at sea level, the underwater structure with force plates has showed a maximum of 0.00000813 Pa and 0.0000409 m/m, considering equivalent von-Mises stress and equivalent von-Mises elastic strain, respectively, indicating short stress gradients in the underwater structure regions. An antero-posterior 2000 N load applied both on the lowest and on the highest horizontal and vertical handgrips revealed 200, 165 and 115 maximal με. A vertical 2000 N load applied on the lowest and on the highest horizontal and vertical handgrips indicated 591, 585 and 205 maximal με.

3.2. Calibrations

The calibration regression equation for the antero-posterior and medium-lateral axis of 300 mm × 250 mm and 600 × 300 mm force plates is depicted in Figure 5a–d. Results evidenced the previously noticed linearity (R^2 ranging between 0.97 and 0.99) and non-meaningful cross-talk between orthogonal axes (small and negligible; <5%) when quantifying any couple of force plate output signals (cf. [20,21]). Calibration results of 300 mm × 250 mm force plates are depicted considering the forces applied on handgrips positioning previously simulated.

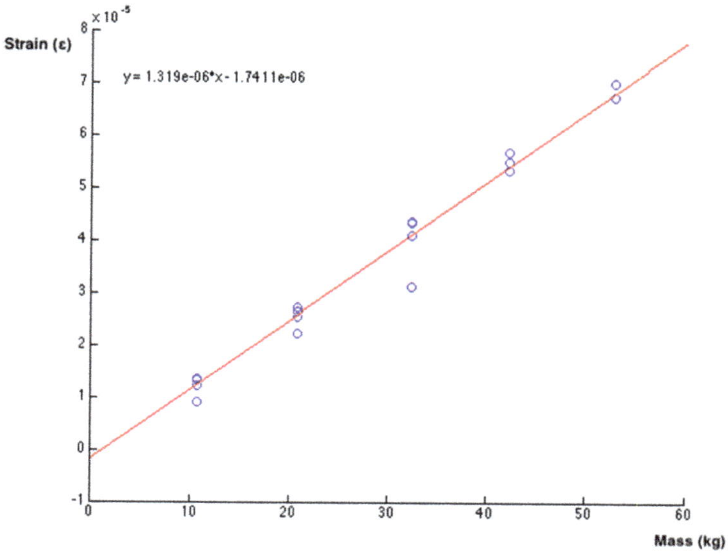

(a)

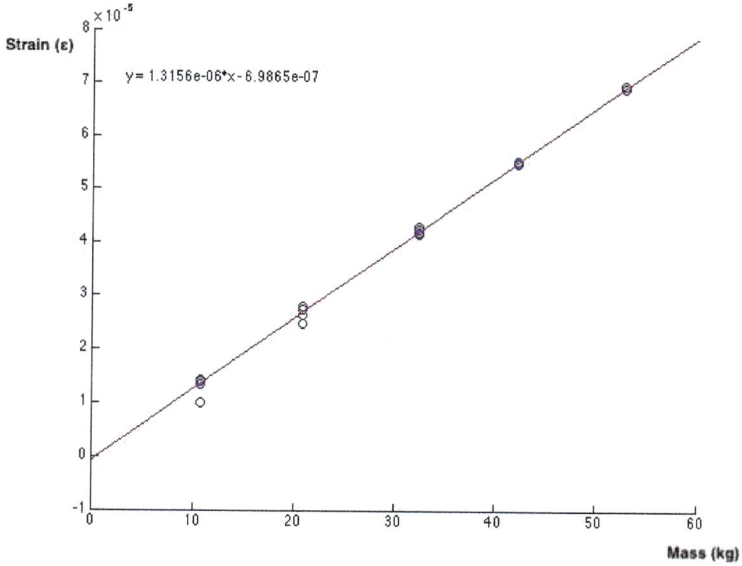

(b)

Figure 5. *Cont.*

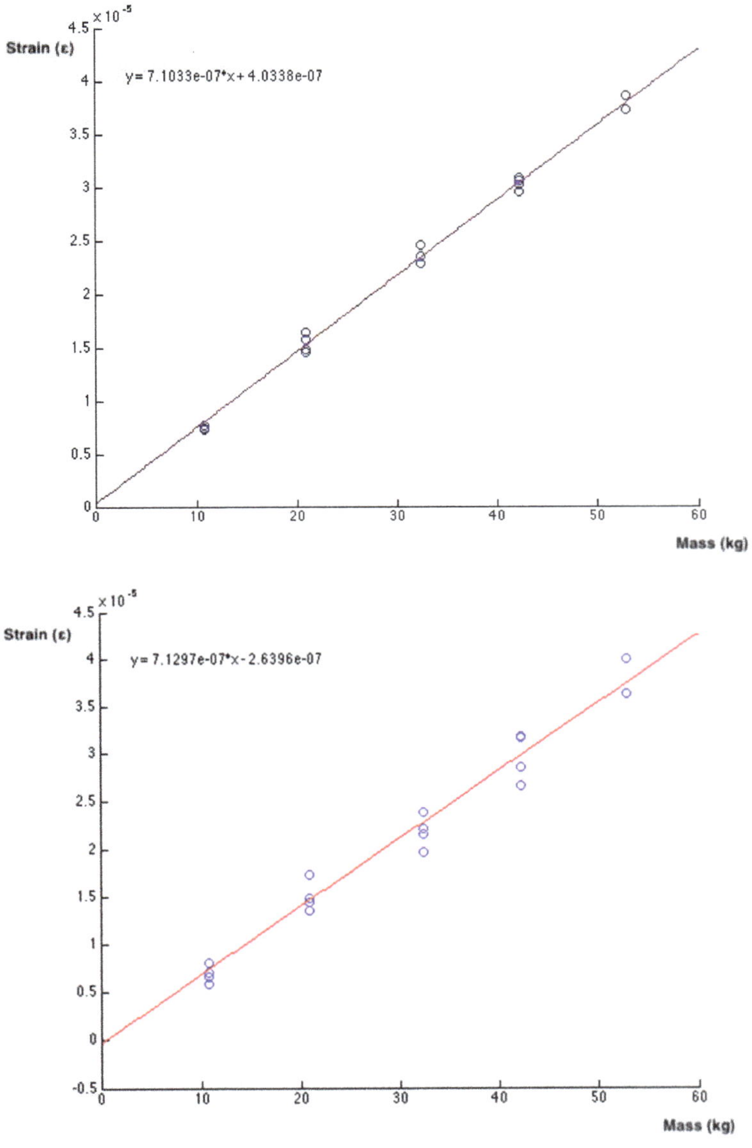

(c)

(d)

Figure 5. Force plates calibration results: 300 mm × 250 mm force plate antero-posterior force (**a** panel), medium-lateral force (**b** panel), 600 mm × 300 mm force plate antero-posterior force (**c** panel), and medium-lateral force (**d** panel).

Figure 6 presents vertical force, antero-posterior and lateral moment calibration (in normalised strain measure) for the underwater 600 mm × 300 mm force plate, evidencing linearization and cross-talk ~5% between orthogonal sensors (cf. [18,20]). The force plate design used has revealed greater sensibility in responses to the vertical loads, corroborating Lywood [21].

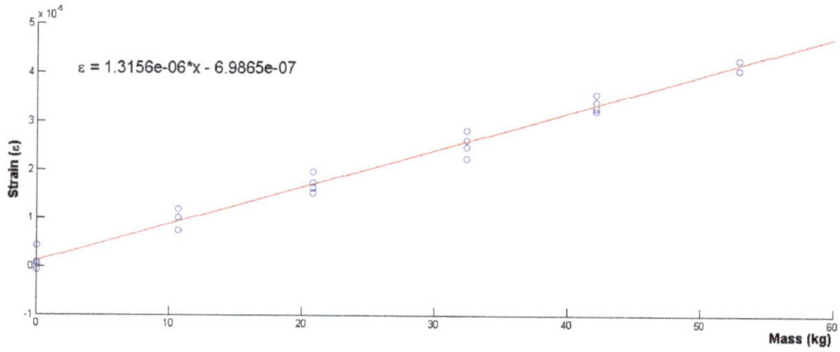

Figure 6. Force plates calibration results: 600 mm × 300 mm force plate vertical force.

Repeatability for loads above 300 N and below 40 N has revealed within 5% and 10–15% with a 95% confidence level. The $\overrightarrow{COP}$ locus was more uncertain farther from the platform centre with a reasonable radial around centre dependency.

3.3. Laboratory Experimental Validation

Due to in situ installation procedures, usage and aging, force plates accuracy may decrease, which can be propagated to the already calculated kinetic quantities. Based on these limitations, some research groups have developed systems to assess force plate accuracy using ad hoc designed in situ devices [8,26]. In the current study, static calibrations were followed by dynamical calibrations performed with a rigid body falling procedure [26], revealing homogeneity of static calibration results, particularly in the time = 0 force time curve of the falling body. The simultaneous correlation coefficient between the posterior strain filtered signal (moving average of 32 samples) and the theory generated by a previous application in MatLab R2014 (MathWorks Inc., Natick, MA,USA) has shown a value of 0.95.

3.4. Ecological Experimental Validation

The horizontal force–time curve exerted by hands and produced during a backstroke start data acquisition is similar to the one peak force profile previously presented (~0.5 BW; [24]). Most of the force–time curves displayed in backstroke start studies have analysed the horizontal component exerted on feet (e.g., [24,29]), which is similar to the dual peak force profile found in the current study (Figure 7a), with swimmers performing ~1.6 BW in the both peak forces. Vertical backstroke start forces exerted by feet have shown similar values to those found in de Jesus et al. (i.e., ~1.0 BW; [10,29]). Lateral force exerted by feet has depicted a peak force instant before the hands-off. Horizontal and vertical force–time curves exerted on feet and hands obtained in the new track start technique performed with the back plate (Figure 8) have registered similar rear (~800 N and ~1000 N) and front foot (~800 N and 600 N; cf. [1,8]) as well as hand force profiles (~150 N and ~800 N; cf. [8]). Previous studies have evidenced that the test–retest reliability of the kinetic gait parameters in the aquatic environment presented poor (medial lateral force component) to excellent reliability (vertical and antero-posterior force component; [30]).

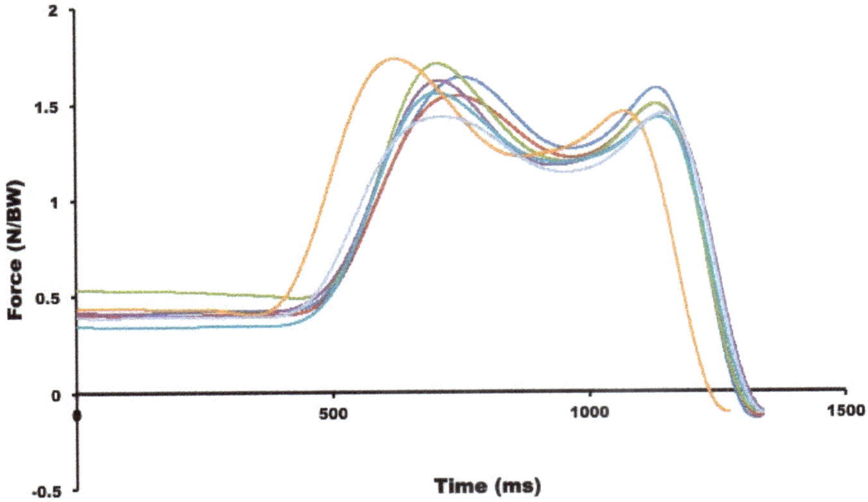

(a)

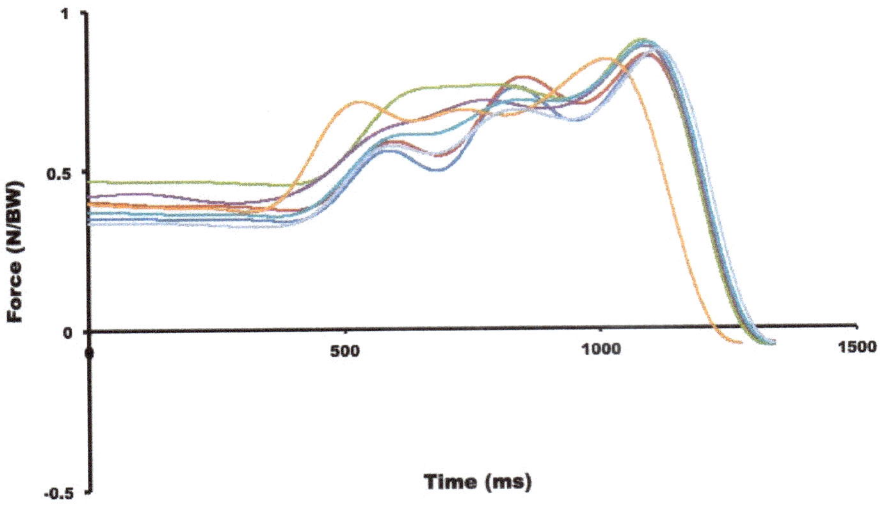

(b)

Figure 7. *Cont.*

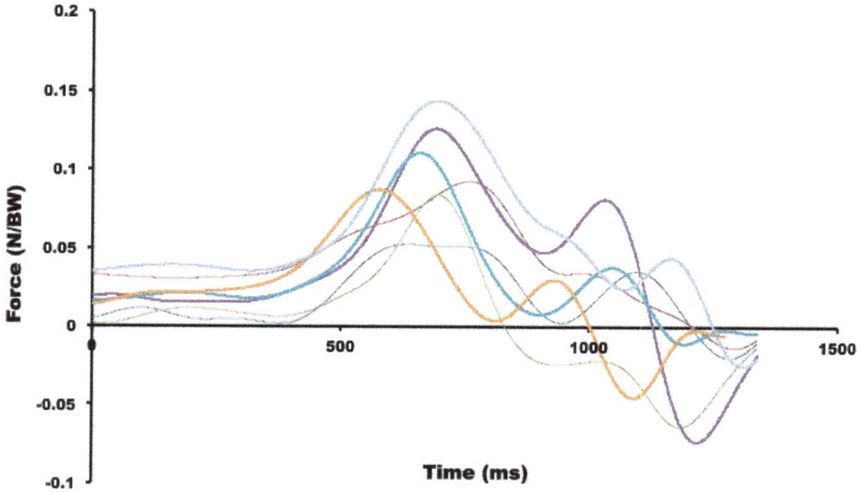

(c)

Figure 7. Backstroke start horizontal (**a** panel), vertical (**b** panel) and lateral (**c** panel) forces from seven trials of one swimmer (**c** panel).

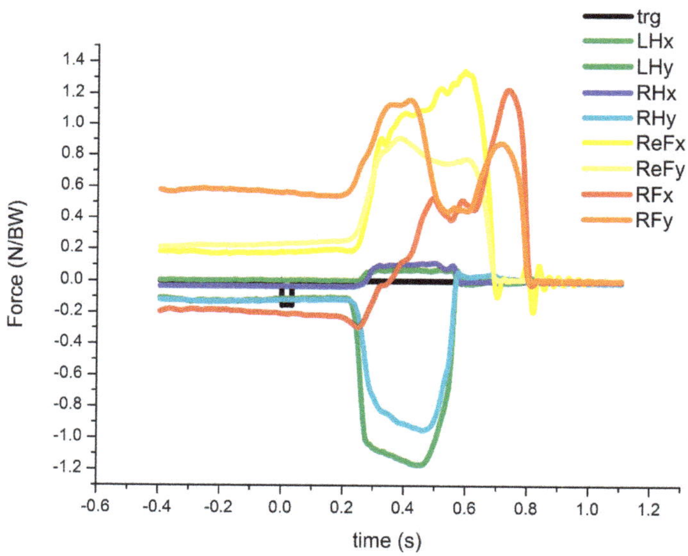

Figure 8. Individual antero-posterior (x) and vertical (y) force—time curves representing forces exerted on left and right hands (LH and RH), as well as rear and front foot (ReF and RF) in the new track start technique synchronised with the trigger (trg).

In relay start techniques, the horizontal force–time curve exerted on the feet and obtained during one step start technique (Figure 9) revealed a similar finding as obtained by Takeda et al. [31], although these authors had not measured right and left forces exerted on each foot.

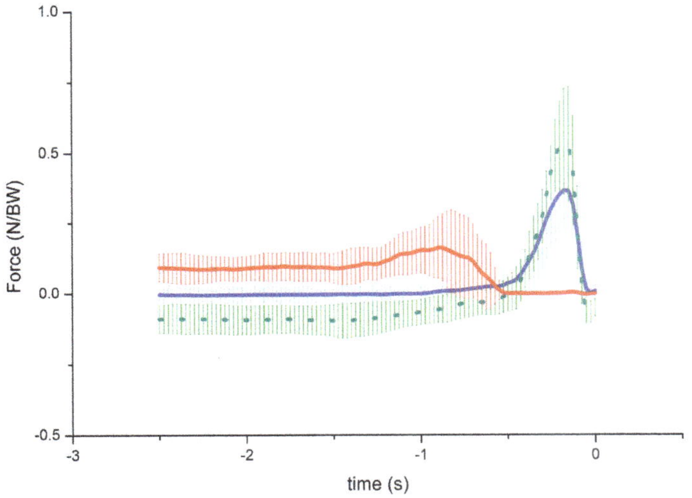

Figure 9. Horizontal force–time curve profiles from eight swimmers at the one step relay start technique: Rear foot (red line), front left and right foot (green and blue line, respectively).

Tri-axial forces measured in one underwater force plate during the open backstroke to breaststroke changeover have evidenced two distinct turn phases, the hand contact (~4.60 to 5.50 s) followed by swimmers' rotation (~4.80 s to 7.30 s) and the push-off (~7.30 to 7.80 s; Figure 10). Previous studies have already mentioned that the push-off phase depicted explosive lower limb movements since the first feet contact (~7.20 s) to the end of push-off (~7.80 s), which has been mainly observed in the horizontal force component [13,28]. Purdy and co-authors [31] have found ~193 N for the peak horizontal ground reaction force in female swimmers performing the open turn technique.

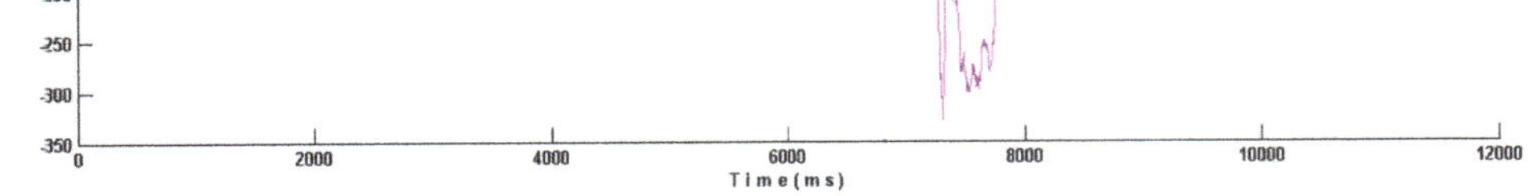

Figure 10. Individual backstroke to breaststroke open medley turning technique force–time curve with horizontal, vertical and medium-lateral components (purple, yellow and red line, respectively).

The current and the previously developed instrumented starting block configurations have shown restrictions regarding independent hand force and momentum measures with natural swimmers' hands and feet placement during all ventral start techniques. (e.g., [2,6,8]). The solution found in the current project was the replication of four more tri-axial force plates of 300 mm × 125 mm designed and numerically simulated (Figure 11a–d) to be placed in the frontal start block edge. An 800 N vertical, and 200 N medium-lateral and horizontal load were applied and registered 280.11, 109.89 and 117.34 με maximal deformation and a maximal 286.45 Hz frequency. The four-force plates solution enables swimmers to grasp the top and bottom of the starting block surface, being versatile and valid to independently measure each swimmer's hand and foot forces and momenta generated regardless of the ventral starting technique used.

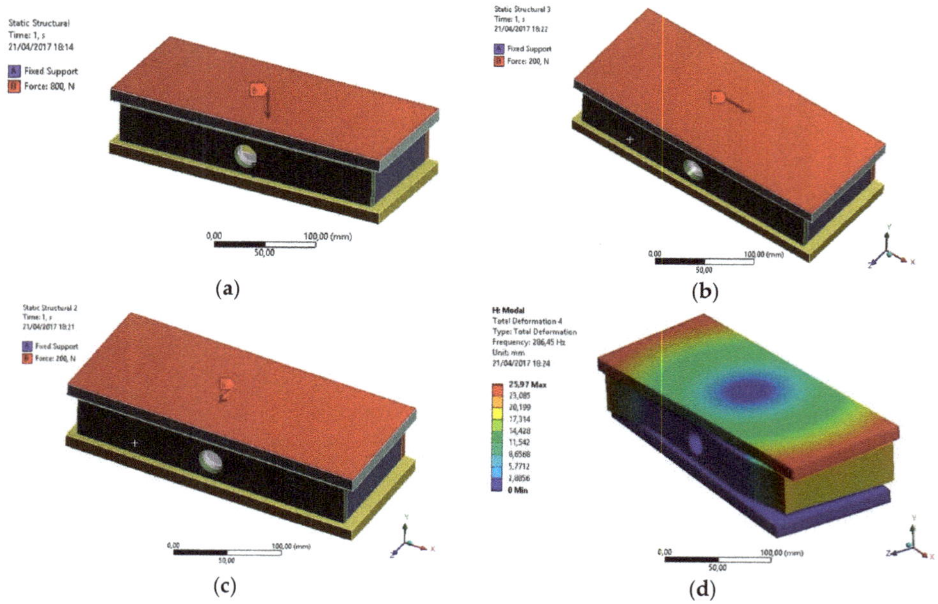

Figure 11. The four 300 mm × 125 mm force plates in static and dynamic simulation conditions for vertical, antero-posterior and medium lateral deformation (**a–c** panels), as well as the frequency analysis (**d** panel).

Notwithstanding the relevance of the dynamometric unit solution developed for start and turn analysis, it is recognised that future work is needed to improve main limitations: (i) finite element analysis revealed important data for project manufacturing. However, irregularities in the starting pool wall cannot be controlled and could affect measurements; (ii) despite static and dynamic calibration procedures, force plate linear responses were revealed, suggesting a less time-consuming in situ calibration procedure should be developed to avoid the detachment of each force plate from the dynamometric central; and (iii) increasing underwater force plate placement possibilities beyond side-by-side but also in an up and down location would cover independent feet force measurements from other turn techniques.

4. Conclusions

The dynamometric central for individual and relay swimming start and turn techniques framed according to FINA facility rules has evidenced reliable and accurate external force data. This device is ecologically valid and versatile, being able to be used in integrated or in future independent force plate analysis, as bow wave measurements in turns, passive drag and jumping techniques.

Author Contributions: Conceptualization, K.d.J. (Karla de Jesus), L.M., H.R., M.V. and J.P.V.-B.; data curation, K.d.J. (Karla de Jesus), L.M., M.V., R.F. and J.P.V.-B.; formal analysis, K.d.J. (Karla de Jesus), L.M., H.R., N.V., M.V. and J.P.V.-B.; funding acquisition, K.d.J. (Karla de Jesus), M.V., R.F. and J.P.V.-B.; investigation, K.d.J. (Karla de Jesus), K.d.J. (Kelly de Jesus), M.V., R.F., J.P.V.-B.; methodology, K.d.J. (Karla de Jesus), LM., H.R., N.V., K.d.J. (Kelly de Jesus), M.V., R.F., J.P.V.-B.; software, L.M., H.R., N.V., M.V., project administration; K.d.J. (Karla de Jesus), R.F., J.P.V.-B.; resources, K.d.J. (Karla de Jesus), R.F., J.P.V.-B.; supervision; K.d.J. (Karla de Jesus), L.M., M.V., R.F., J.P.V.-B.; validation; K.d.J. (Karla de Jesus), L.M., K.d.J. (Kelly de Jesus), J.P.V.-B.; visualization, K.d.J. (Karla de Jesus), R.F., J.P.V.-B.; writing—original draft, K.d.J., (Karla de Jesus), L.M., H.R., N.V., K.d.J. (Kelly de Jesus), R.F., J.P.V.-B.; writing—review and editing, K.d.J. (Karla de Jesus), L.M., M.V., R.F., J.P.V.-B.

Funding: This research was funded by the [Coordination for the Improvement of Higher Education Personnel—CAPES] grant number [BEX 0761/12-5/2012-2015], [Foundation for Science and Technology—FCT] grant number [EXPL/DTP-DES/2481/2013-FCOMP-01-0124-FEDER-041981], [CAPES-FCT] grant number [99999.008578/2014-01], SANTADER grant number [PP/IJUP2011/123] and [Amazonas State Research Support Foundation - FAPEAM] grant number [POSGRAD 2017 FAPEAM, 002/2016]. Funders had no role in study design, data collection and analysis, decision to publish, or manuscript preparation.

Acknowledgments: Leandro dos Santos Coelho, Alexandre Igor Araripe Medeiros and Phornpot Chainok provided valuable feedback on manuscript drafts.

Conflicts of Interest: The authors declare no conflict of interest. The funders had no role in the design of the study; in the collection, analyses, or interpretation of data; in the writing of the manuscript, or in the decision to publish the results.

References

1. Slawson, S.E.; Conway, P.P.; Cossor, J.; Chakravorti, N.; West, A.A. The categorisation of swimming start performance with reference to force generation on the main block and footrest components of the Omega OSB11 start blocks. *J. Sports Sci.* **2013**, *31*, 468–478. [CrossRef] [PubMed]
2. Mason, B.; Mackintosh, C.; Pease, D. The development of an analysis system to assist in the correction of inefficiencies in stars and turns for elite competitive swimming. In Proceedings of the 30th International Symposium on Biomechanics in Sports, Melbourne, Australia, 2–6 July 2012; Bradshaw, E.J., Burnett, A., Hume, P.A., Eds.; International Society of Biomechanics in Sports: Melbourne, Australia, 2012; pp. 249–252.
3. Veiga, S.; Carla, A.; Frutos, P.; Navarro, E. Comparison of starts and turns of national and regional level swimmers by individualized-distance measurements. *Sports Biomech.* **2014**, *13*, 285–295. [CrossRef] [PubMed]
4. Slawson, S.; Conway, P.; Justham, L.; Le Sage, T.; West, A. Dynamic signature for tumble turn performance in swimming. *Procedia Eng.* **2010**, *2*, 3391–3396. [CrossRef]
5. Elliott, G.M.; Sinclair, H. The influence of block angle on swimming sprint starts. In *Swimming I, Proceedings of the I International Symposium of Biomechanics and Swimming, Waterpolo and Divig, Bruxelles, Belgium, 14–16 September 1970*; Lewillie, L., Clarys, J.P., Eds.; Brussel Universite Libre de Bruxelles, Laboratoire de l' effort: Bruxelles, Belgium, 1971; pp. 183–189.
6. Cavanagh, P.R.; Palmgren, J.V.; Kerr, B.A. A device to measure forces at hands during the grab start. In *Swimming II, Proceedings of the II International Symposium of Biomechanics in Swimming, Brussels, Belgium*; Lewillie, L., Clarys, J.P., Eds.; Tokyo Universitaire Park Press: Baltimore, MD, USA; London, UK, 1975; pp. 43–50.
7. Naemi, R.; Arshi, A.R.; Ahadian, A.; Barjasteh, B. 3D kinematic and kinetic analyses of two methods for grab start technique. In Proceedings of the 19th International Symposium on Biomechanics in Sports, Melbourne, Australia, 26–29 June 2001; Blackwell, J.R., Sanders, R.H., Eds.; International Society of Biomechanics in Sports: San Francisco, CA, USA, 2001; pp. 96–99.
8. Takeda, T.; Sakai, S.; Takagi, H.; Okuno, K.; Tsubakimoto, S. Contribution of hand and foot force to take-off velocity for the kick-start in competitive swimming. *J. Sports Sci.* **2017**, *35*, 565–571. [CrossRef] [PubMed]
9. Vantorre, J.; Seifert, L.; Fernandes, R.J.; Vilas-Boas, J.P.; Chollet, D. Comparison of grab start between elite and trained swimmers. *Int. J. Sports Med.* **2010**, *31*, 887–893. [CrossRef] [PubMed]
10. De Jesus, K.; de Jesus, K.; Gonçalves, P.; Vasconcelos, M.O.; Medeiros, A.I.A.; Carvalho, D.A.D.; Fernandes, R.J.; Vilas-Boas, J.P. Lateral kinetic proficiency and asymmetry in backstroke start performed with horizontal and vertical handgrips. *Sports Biomech.* **2018**, 13–15. [CrossRef] [PubMed]
11. Nicol, K.; Krueger, F. Impulses exerted in performing several kinds of swimming turns. In *Swimming III, Proceedings of the III International Symposium of Biomechanics in Swimming, Alberta, AB, Canada*; Terauds, J., Bedingfield, E.W., Eds.; Baltimore University Park Press: Baltimore, MD, USA; London, UK, 1979; pp. 222–232.
12. Puel, F.; Morlier, J.; Avalos, M.; Mesnard, M.; Cid, M.; Hellard, P. 3D kinematic and dynamic analysis of the front crawl tumble turn in elite male swimmers. *J. Biomech.* **2012**, *45*, 510–515. [CrossRef] [PubMed]
13. Pereira, S.M.; Ruschel, C.; Hubert, M.; Machado, L.; Roesler, H.; Fernandes, R.J.; Vilas-Boas, J.P. Kinematic, kinetic and EMG analysis of four front crawl flip turn techniques. *J. Sports Sci.* **2015**, *33*, 2006–2015. [CrossRef] [PubMed]
14. Pereira, S.M.; Gonçalves, P.; Fernandes, R.J.; Machado, L.; Roesler, H.; Vilas-Boas, J.P. Graphic removal of water wave impact in the pool wall during the flip turn. In Proceedings of the XI International Symposium for Biomechanics and Medicine in Swimming, Oslo, Norway, 16–19 June 2010; Kjendlie, P.L., Stallman, R.K., Cabri, J., Eds.; Norwergian School of Sport Science: Oslo, Norway, 2010; pp. 148–150.

15. Duncan, O.; Shepherd, T.; Moroney, C.; Foster, L.; Venkatraman, P.; Winwood, K.; Allen, T.; Alderson, A. Review of Auxetic Materials for Sports Applications: Expanding Options in Comfort and Protection. *Appl. Sci.* **2018**, *8*, 941. [CrossRef]
16. Wardoyo, S.; Hutajulu, P.T.; Togibasa, O. A Development of Force Plate for Biomechanics Analysis of Standing and Walking. *J. Phys. Conf. Ser.* **2016**, *739*. [CrossRef]
17. Silva, M.G.; Moreira, P.V.S.; Rocha, H.M. Development of a low cost force platform for biomechanical parameters analysis. *Res. Biomed. Eng.* **2017**, *33*, 259–268. [CrossRef]
18. Mourão, L.; Jesus, K.; Viriato, N.; Fernandes, R.J.; Vilas-Boas, J.P.; Vaz, M. Design and construction of a 3D force plate prototype for developing an instrumented swimming start block. *J. Biomed. Eng. Inform.* **2016**, *2*, 99. [CrossRef]
19. Formosa, D.; Mason, B.R.; Burkett, B.J. Measuring active drag within the different phases of front crawl swimming. In Proceedings of the XI International Symposium on Biomechanics and Medicine in Swimming, Oslo, Norway, 16–19 June 2010; Kjendlie, L., Stallman, R.K., Cabri, J., Eds.; Norwegian School of Sport Science: Oslo, Norway, 2010; pp. 82–84.
20. Roesler, H. Turning measurements in swimming using underwater force platforms. In Proceedings of the IX International Symposium on Biomechanics and Medicine in Swimming, de Saint-Étienne, France, 16–19 June 2010; Chatard, J.C., Ed.; l' Université de Saint-Étienne: Saint-Étienne, France, 2003; pp. 243–248.
21. Lywood, D.W.; Adams, D.J.; Van Eyken, A.; Macpherson, J.M. Small, triaxial force plate. *Med. Biol. Eng. Comput.* **1987**, *25*, 698–701. [CrossRef] [PubMed]
22. Wu, G.; Siegler, S.; Allard, P. ISB recommendation on definitions of joint coordinate system of various joints for the reporting of human joint motion-part I: Ankle, hip, and spine. International Society of Biomechanics. *J. Biomech.* **2002**, *35*, 543–548. [CrossRef]
23. Wright, D.A. *Development of a Waterproof Force Plate for Pool Applications*; Ohio State University: Columbus, OH, USA, 2011.
24. De Jesus, K.; de Jesus, K.; Figueiredo, P.; Gonçalves, P.; Pereira, S.; Vilas-Boas, J.P.; Fernandes, R. Backstroke start kinematic and kinetic changes due to different feet positioning. *J. Sports Sci.* **2013**, *31*, 1665–1675. [CrossRef] [PubMed]
25. Perttunen, J.O.; Kyrolainen, H.; Komi, P.V.; Heinonen, A. Biomechanical loading in the triple jump. *J. Sports Sci.* **2000**, *18*, 363–370. [CrossRef] [PubMed]
26. Mourão, L.; de Jesus, K.; Roesler, H.; Machado, L.; Fernandes, R.; Vilas-Boas, J.P.; Vaz, M. Effective swimmer's action during the grab start technique. *PLoS ONE* **2015**, *15*, e0123001. [CrossRef] [PubMed]
27. De Jesus, K.; de Jesus, K.; Medeiros, A.; Gonçalves, P.; Figueiredo, P.; Fernandes, R.J.; Vilas-Boas, J.P. Neuromuscular activity of upper and lower limbs during two backstroke swimming start variants. *J. Sport Sci. Med.* **2015**, *14*, 591–601.
28. Chainok, P.; Lauer, J.; Zacca, R.; Barbosa, R.; Gonçalves, P.; Fernandes, R.; Vilas-Boas, J.P. Neuromuscualar activation during rotation and push-off phases of backstroke turning techniques in age-group swimmers. In Proceedings of the XXXIV International Symposium on Biomechanics in Sports, Tsukuba, Japan, 18–22 July 2016; Ae, M., Enomoto, Y., Fujii, N., Takagi, H., Eds.; International Society of Biomechanics in Sports: Tsukuba, Japan, 2016; pp. 541–544.
29. Nguyen, C.; Bradshaw, E.; Pease, D.; Wilson, C. Is starting with the feet out of the water faster in backstroke swimming? *Sports Biomech.* **2014**, *13*, 154–165. [CrossRef] [PubMed]
30. Barreto, M.S.; Dela Bela, L.F.; Dias, J.M.; Pelegrinelli, A.R.; de Campos, R.R.; Carvalho, R.G.; Taglietti, M.; Batista, J.P., Jr.; Silva, M.F.; Olkoski, M.M.; et al. Reliability of ground reaction forces in the aquatic environment. *J. Electromyogr. Kinesiol.* **2016**, *30*, 23–30. [CrossRef] [PubMed]
31. Purdy, E.; Hurley, N.; Bengry, A.; Jensen, R. Force and tie analysis of backstroke to breaststroke turns. In Proceedings of the 30th International Symposium on Biomechanics in Sports, Melbourne, Australia, 2–6 July 2012; Bradsha, E.J., Burnett, A., Hume, P.A., Eds.; International Society of Biomechanics in Sports: Melbourne, Australia, 2012; pp. 196–199.

applied sciences

MDPI

Article

Isokinetic Strength in Peritoneal Dialysis Patients: A Reliability Study

Daniel Collado-Mateo [1,*], Francisco Javier Dominguez-Muñoz [1], Zelinda Charrua [2], José Carmelo Adsuar [1], Eugenio Merellano-Navarro [4] and Armando Manuel Raimundo [2,3]

1 Faculty of Sport Sciences, University of Extremadura, 10003 Cáceres, Spain
2 Departamento de Desporto e Saúde, Escola de Ciência e Tecnologia, Universidade de Évora, 7005-399 Évora, Portugal
3 Comprehensive Health Research Center (CHRC), Universidade de Évora, 7005-399 Évora, Portugal
4 Facultad de Educación, Universidad Autonoma de Chile, Talca 3460000, Chile
* Correspondence: dcolladom@unex.es; Tel.: +34-927257460

Received: 5 July 2019; Accepted: 26 August 2019; Published: 29 August 2019

Abstract: Although there are studies assessing the effects of interventions on the knee strength of patients undergoing dialysis, there are no previous studies investigating the test–retest reliability of isokinetic measures in people undergoing peritoneal dialysis. The objective of this study was to determine the relative and absolute reliability of peak torque and work measurements for isokinetic concentric knee and elbow extension and flexion in peritoneal dialysis patients. Thirty-one patients undergoing peritoneal dialysis (19 males) participated in the current study. All isokinetic tests were performed using a Biodex System 3. Participants performed three concentric repetitions of each test (flexion or extension) with the dominant limb (knee and elbow) at 60°/s. Peak torque (Nm) and work (J) were extracted. The intraclass correlation coefficient (ICC), standard error of measurement (SEM) and smallest real difference (SRD) were calculated. The results showed that all knee peak torque and work measures had an ICC of >0.90. On the other hand, the ICC for peak torque and work in the elbow concentric extension was <0.90, while the remaining elbow-related variables achieved an excellent reliability. Therefore, isokinetic dynamometry is a reliable technique to evaluate peak torque and work for concentric flexion and extension in both the knee and elbow joints in patients undergoing peritoneal dialysis.

Keywords: kidney; torque; exercise; physical fitness; peritoneal dialysis

1. Introduction

Peritoneal dialysis (PD) patients present a condition resulting from multiple physiological and behavioral changes which contribute to muscle wasting [1]. Recent studies showed that PD patients often present decreased levels of physical activity (PA) matching a sedentary lifestyle [2]. These circumstances significantly affect the health condition of those patients and are associated with significant morbimortality [3]. Moreover, the nutritional status and sedentary behavior directly affect muscle function, exercise performance, physical function, strength, and health-related quality of life [4–6]. Although the proportion of patients active during leisure time is low [4], they need adequate levels of strength to accomplish daily life activities such as the need to displace, walking up and down stairs, maintaining a standing position, etc.

Exercise programs are recommended to increase lean body mass, strength, and physical functioning in frail elderly persons and those with chronic diseases, including PD patients [5]. Precise and sensitive strength tests are required to appropriately extract conclusions in studies focused on the evaluation of strength. Among these tests, the hand grip dynamometer and the isokinetic dynamometer are two of the most reliable and widely used devices. In this regard, the handgrip dynamometer has been previously used to evaluate muscle strength of the upper limb in patients undergoing dialysis [4,7]. Isokinetic dynamometry has been previously used to evaluate the effects of different interventions and therapies on knee flexion and extension strength in dialysis patients [8,9], whereas, to our knowledge, there is no study evaluating upper limb strength using an isokinetic dynamometer in this population.

Although there are studies assessing the effects of interventions on the knee strength of patients undergoing dialysis, to our knowledge, no previous study has investigated the reliability of knee isokinetic procedures adapted for use in PD patients with poor muscle strength. In healthy individuals, the use of isokinetic dynamometry to evaluate the knee strength when subjects receive adequate instructions and are familiar with the equipment is sufficiently reliable [10,11], but the reliability in PD patients still remains unknown, which impairs and limits the appropriate interpretation of results. In this regard, reliability is considered an important prerequisite for the correct interpretation of isokinetic dynamometry data, which allows the clinician to identify whether or not a genuine change has occurred [12]. Therefore, the aim of the current study was to determine the relative and absolute intra-session reliability of peak torque and work measurements for isokinetic concentric knee and elbow extension and flexion in PD patients.

2. Materials and Methods

2.1. Participants

Patients were recruited from the Nephrology Unit from the Espírito Santo Hospital of Évora, Portugal. The inclusion criteria to participate in the study were as follows: (1) be a peritoneal dialysis patient for at least 6 months; (2) do not have any impediment to perform the strength tests according to the physician's criteria; and (3) give their informed consent to participate in the study. A total of 49 patients fulfilled the first two inclusion criteria and were invited to participate. Of them, 31 (12 women and 19 men) agreed to participate in the study and signed the informed consent. The University of Évora ethics committee approved the protocol of this study, which was conducted in accordance with the updated World Medical Association's Declaration of Helsinki for human studies [13]. Of the 31 participants, 18 were on continuous ambulatory peritoneal dialysis while 14 were on automated peritoneal dialysis. Patients were having PD for a mean of 943 ± 552 days.

2.2. Instrumentation

All isokinetic tests were performed using a Biodex System 3 quick-set isokinetic dynamometer (Biodex Corp., Shirley, NY, USA) and System 3 software (version 3.40). Body fat and lean percentage were assessed by dual-energy X-ray absorptiometry (DXA—Hologic QDR, Hologic, Inc., Bedford, MA, USA). Finally, physical activity level was assessed using the ActiGraph accelerometer device with dimensions 3.8 × 3.7 × 1.8 cm (27 g). All participants were asked to use an accelerometer on the right hip, near the iliac crest, during seven consecutive days.

2.3. Procedures

The procedure is depicted in Figure 1. The intra-session reliability of the measurements was evaluated. All tests were conducted by the same researcher. Only the dominant arm and knee were tested. The dominant arm was defined as the arm used to write and the dominant knee was defined as that of the preferred kicking leg. Knee and elbow protocols followed the Biodex Isokinetic System 3 quick-set application/operation manual instructions [11]. Prior to the implementation of the protocols, all the subjects performed 15 min of warm-up with joint mobilization and stretching.

1. Knee protocol: Participants were seated in a seatback tilt at 85°. The dynamometer orientation and dynamometer tilt were 45° and 0°, respectively. The participant's axis of rotation of the knee was aligned with the dynamometer shaft. All patients were informed about the tasks they were going to perform and performed two familiarization and warm-up repetitions. The

weight of the leg was recorded using the dynamometer software, and gravity adjustments were made. All participants were asked to perform three concentric movements of the knee involving alternative extension and flexion at 60°/s. Reliability was calculated between the second and the third repetition. The rest interval was 2 min long. This protocol has been used previously in the scientific literature [14,15]. The participants were verbally encouraged during the tests.

2. Elbow protocol: Participants were seated in a seatback tilt at 85°. The seat orientation was 15°. The dynamometer orientation and dynamometer tilt were 15° and 0°, respectively. Participants were stabilized with shoulder, waist, and thigh straps. They were informed about the tasks and performed two repetitions, aimed to warm-up and also to get used to the position, the angular speed, and the proposed task. The weight of the arm was recorded using the dynamometer software, and gravity adjustments were made. All participants were asked to perform three concentric movements of the elbow involving alternative extension and flexion at 60°/s. Reliability was calculated between the second and the third repetition. The rest interval was 2 min long. The participants were verbally encouraged during the tests.

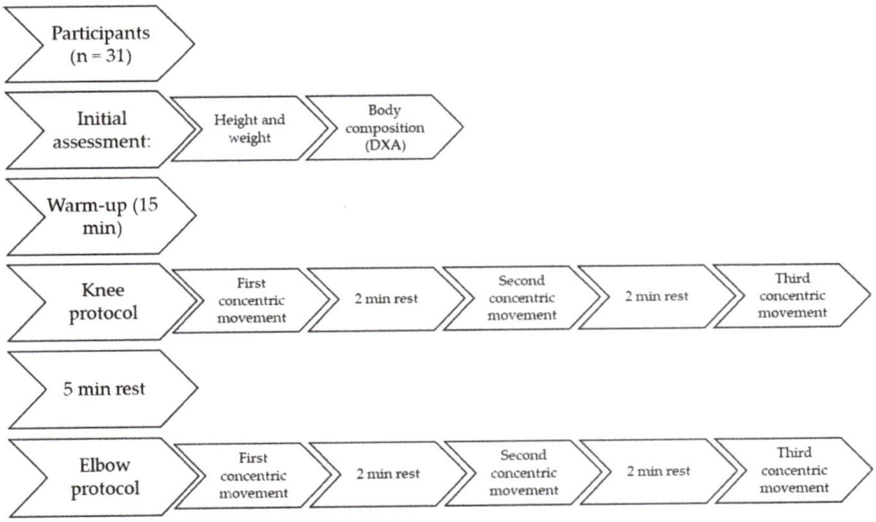

Figure 1. Study procedure. DXA: dual-energy X-ray absorptiometry.

2.4. Measures

Peak torque (Nm) and work (J) were extracted from the System 3 software for knee and elbow flexion/extension. Peak torque is defined as "the single highest torque output recorded throughout the range of motion of each repetition". Work is defined as "the output of mechanical energy" and is represented by the area under the curve of torque versus angular displacement [16].

2.5. Statistical Analysis

The statistical analysis was performed following the criteria used in previous studies [12,14,15]. Absolute values for peak torque and work were obtained as means and standard deviations. Differences in the descriptive characteristics between men and women were evaluated using independent samples *t*-test.

Relative reliability was estimated using the ICC3,1 (intraclass correlation coefficient, two-way mixed single measures) with 95% confidence intervals across the two test repetitions [17]. An ICC higher than 90 was interpreted according to Munro et al. as excellent [18]. A paired sample *t*-test was performed to analyze differences in the mean values of the isokinetic variables between Repetitions 2 and 3.

Absolute reliability was determined by calculating the standard error of measurement (SEM; SEM = SD, where SD is the mean SD of Repetition 2 and Repetition 3) and the smallest real difference (SRD; SRD = 1.96) [19]. Additionally, the SEM and SRD were converted to percentages in order to facilitate the comparability of errors of measurement with those in other studies. These percentages were calculated as follows: SEM% = (SEM/mean peak torque or work of the two repetitions)·100 and SRD% = SRD/(mean peak torque or work of the two repetitions)·100.

Isokinetic variables were also correlated with anthropometric and body composition variables using bivariate Pearson's correlation.

3. Results

The characteristics of patients undergoing PD are reported in Table 1. The mean age was 48.45 (13.39) and the mean body mass index (BMI) was 24.35 (3.67), which is close to the overweight threshold. Differences between men and women were observed in age, body composition, height, and weight. In general terms, there was sedentary behavior, with more than 18 h/day of sedentary time and about 5000 daily steps.

Table 1. Characteristics of patients undergoing peritoneal dialysis (N = 31).

	All	Men (N = 19)	Women (N = 12)
Anthropometric and Body Composition Measurements			
Age (years)	48.45 ± 13.39	52.42 ± 11.54 *	42.16 ± 14.16
Height (cm)	162.19 ± 9.67	167.63 ± 6.80 *	153.58 ± 6.89
Weight (kg)	64.29 ± 9.45	67.40 ± 7.88 *	59.37 ± 9.95
BMI (kg/m^2)	24.35 ± 3.67	23.82 ± 3.37	25.20 ± 4.10
Body fat percentage	27.79 ± 8.49	23.23 ± 5.25	35.01 ± 7.65 *
Body lean percentage	68.58 ± 8.23	72.92 ± 5.14 *	61.70 ± 7.58
Physical Activity Levels			
Sedentary time (h/day)	19.34 ± 1.61	19.44 ± 1.86	19.15 ± 1.11
Light physical activities (h/day)	3.50 ± 1.54	3.43 ± 1.77	3.63 ± 1.09
Moderate and vigorous activities (h/day)	0.28 ± 0.26	0.27 ± 0.26	0.29 ± 0.27
Number of daily steps	5157.18 ± 2483.80	5005.63 ± 2698.69	5429.98 ± 2149.99
Lipid Profile and Other Parameters			
Serum cholesterol (mg/dL)	182.12 ± 31.42	183.00 ± 31.22	180.75 ± 33.09
Serum high-density lipoprotein (mg/dL)	45.35 ± 13.63	45.42 ± 14.08	45.25 ± 13.49
Serum low-density lipoprotein (mg/dL)	107.64 ± 39.05	112.68 ± 34.71	99.66 ± 45.53
Serum triglycerides (mg/dL)	167.29 ± 70.94	171.68 ± 74.31	160.33 ± 67.84
Haemoglobin, (g/dL)	11.66 ± 1.47	11.98 ± 1.51	11.15 ± 1.32
Serum albumin (g/dL)	3.90 ± 0.41	3.94 ± 0.41	3.83 ± 0.43
Serum sodium (mEq/L)	137.58 ± 3.91	137.73 ± 3.44	137.33 ± 4.71
Serum potassium (mEq/L)	4.34 ± 0.75	4.36 ± 0.64	4.30 ± 0.92

Data reported as mean ± SD. * Significantly higher compared to the other group based on results from independent samples t-test. BMI: body mass index.

Table 2 shows the peak torque and work at 60°/s in each of the two repetitions for all variables. There were no statistically significant differences in any test. In general, the mean peak torque and work of male patients were higher than values from female participants.

Table 3 summarizes the ICC values and the 95% confidence intervals, as well as the SEM and SRD in absolute values and percentages. All knee peak torque and work measures had an ICC greater than 0.90, which is excellent according to the classification by Munro, Visintainer, and Page [18]. On the other hand, peak torque in the elbow concentric extension in males and work in the elbow concentric extension in males and the general population showed an ICC lower than the threshold for excellent reliability. The remaining elbow-related variables achieved an excellent reliability. Regarding peak torque, the SRD% in the general population ranged between 18.41% in the elbow concentric flexion and 29.44% in the knee concentric extension. Regarding work, the SRD% in the general population ranged between 26.15% in the elbow concentric extension and 33.63% in the knee concentric flexion.

Table 2. Summary of isokinetic peak torque and work at 60°/s in two repetitions (N = 31).

		Test Measurement	Peak Torque (Nm)			Work (J)		
			Repetition 2	Repetition 3	p *	Repetition 2	Repetition 3	p *
General (n = 31)	Knee	Concentric Extension	77.57 ± 31.40	78.27 ± 32.56	0.747	76.89 ± 34.98	79.06 ± 36.34	0.258
		Concentric Flexion	43.32 ± 21.57	43.43 ± 22.93	0.922	43.80 ± 29.92	44.85 ± 31.17	0.448
	Elbow	Concentric Extension	20.77 ± 8.00	19.92 ± 7.26	0.061	26.86 ± 13.48	25.66 ± 12.64	0.055
		Concentric Flexion	35.69 ± 9.65	35.57 ± 10.18	0.842	51.62 ± 13.39	49.63 ± 15.90	0.170
Men (n = 19)	Knee	Concentric Extension	80.77 ± 34.03	83.86 ± 32.95	0.317	82.25 ± 38.69	87.45 ± 37.62	0.062
		Concentric Flexion	48.67 ± 23.85	49.48 ± 25.56	0.626	50.18 ± 34.26	51.83 ± 36.16	0.437
	Elbow	Concentric Extension	24.84 ± 6.95	23.88 ± 5.78	0.182	33.25 ± 12.38	31.66 ± 11.55	0.116
		Concentric Flexion	39.93 ± 9.41	40.31 ± 9.49	0.697	58.27 ± 11.93	55.77 ± 16.52	0.280
Women (n = 12)	Knee	Concentric Extension	72.50 ± 27.35	69.40 ± 31.24	0.244	68.40 ± 27.57	65.77 ± 31.15	0.203
		Concentric Flexion	34.85 ± 14.50	33.85 ± 14.19	0.364	33.69 ± 18.42	33.80 ± 17.14	0.939
	Elbow	Concentric Extension	14.33 ± 4.65	13.65 ± 4.39	0.063	16.75 ± 7.91	16.15 ± 7.52	0.122
		Concentric Flexion	28.98 ± 5.37	28.07 ± 5.95	0.072	41.09 ± 7.65	39.91 ± 8.63	0.242

Note: Values are mean ± standard deviation; * Paired sample t-test was performed to analyze differences in the mean values of the isokinetic variables between Repetitions 2 and 3.

Table 3. Reliability of isokinetic concentric knee measurements (N = 31).

		Peak Torque (Nm)					
		Test	ICC (95% CI)	SEM (Nm)	SEM (%)	SRD (Nm)	SRD (%)
General (n = 31)	Knee	Concentric Extension	0.933 (0.868–0.967)	8.27	10.62	22.94	29.44
		Concentric Flexion	0.965 (0.930–0.983)	4.16	9.59	11.53	26.60
	Elbow	Concentric Extension	0.945 (0.890–0.973)	1.78	8.79	4.95	24.37
		Concentric Flexion	0.943 (0.887–0.972)	2.36	6.64	6.56	18.41
Men (n = 19)	Knee	Concentric Extension	0.923 (0.816–0.970)	9.29	11.28	25.75	31.29
		Concentric Flexion	0.961 (0.903–0.985)	4.87	9.94	13.52	27.55
	Elbow	Concentric Extension	0.885 (0.730–0.954)	2.15	8.86	5.98	24.56
		Concentric Flexion	0.909 (0.784–0.964)	2.85	7.10	7.90	19.69
Women (n = 12)	Knee	Concentric Extension	0.954 (0.855–0.986)	6.28	8.85	17.41	24.54
		Concentric Flexion	0.968 (0.897–0.990)	2.56	7.47	7.11	20.70
	Elbow	Concentric Extension	0.960 (0.872–0.988)	0.90	6.46	2.50	17.91
		Concentric Flexion	0.952 (0.849–0.986)	1.24	4.27	3.43	11.84
		Work (J)					
		Test	ICC (95% CI)	SEM (J)	SEM (%)	SRD (J)	SRD (%)
General (n = 31)	Knee	Concentric Extension	0.957 (0.913–0.979)	7.39	9.48	20.49	26.28
		Concentric Flexion	0.969 (0.938–0.985)	5.37	12.13	14.90	33.63
	Elbow	Concentric Extension	0.964 (0.927–0.982)	2.47	9.43	6.86	26.15
		Concentric Flexion	0.853 (0.719–0.926)	5.61	11.09	15.56	30.74
Men (n = 19)	Knee	Concentric Extension	0.949 (0.875–0.980)	8.61	10.15	23.88	28.14
		Concentric Flexion	0.968 (0.920–0.987)	6.29	12.34	17.45	34.22
	Elbow	Concentric Extension	0.934 (0.839–0.974)	3.07	9.47	8.52	26.25
		Concentric Flexion	0.767 (0.498–0.903)	6.86	12.04	19.03	33.37
Women (n = 12)	Knee	Concentric Extension	0.972 (0.910–0.992)	4.91	7.32	13.61	20.29
		Concentric Flexion	0.967 (0.894–0.990)	3.22	9.57	8.95	26.53
	Elbow	Concentric Extension	0.985 (0.952–0.996)	0.94	5.74	2.61	15.92
		Concentric Flexion	0.915 (0.743–0.974)	2.37	5.85	6.57	16.24

Abbreviations: CI, confidence interval; ICC, intraclass correlation coefficient; SEM, standard error of measurement; SEM%, standard error of measurement as a percentage; SRD, smallest real difference; SRD%, smallest real difference as a percentage.

Table 4 shows the correlations between isokinetic strength and different variables such as age, height, weight, BMI, fat mass, and lean mass. The variable "height" was significantly correlated with all isokinetic strength outcomes, while BMI was not correlated with any variable.

Table 4. Correlations between isokinetic strength and anthropometric and body composition variables.

	Age	Height	Weight	BMI	Fat Mass %	Lean Mass %
Knee Concentric Extension Peak Torque	−0.314	0.450 *	0.015	−0.327	−0.290	0.279
Knee Concentric Extension Work	−0.199	0.554 **	0.143	−0.279	−0.337	0.319
Knee Concentric Flexion Peak Torque	−0.040	0.436 *	0.268	−0.082	−0.258	0.249
Knee Concentric Flexion Work	−0.071	0.455 *	0.221	−0.150	−0.297	0.285
Elbow Concentric Extension Peak Torque	0.087	0.748 **	0.455 *	−0.185	−0.522 **	0.508 **
Elbow Concentric Extension Work	0.075	0.647 **	0.396 *	−0.129	−0.414 *	0.393 *
Elbow Concentric Flexion Peak Torque	0.266	0.506 **	0.529**	0.041	−0.269	0.274
Elbow Concentric Flexion Work	0.132	0.634 **	0.509 **	−0.021	−0.363 *	0.342

* $p < 0.05$; ** $p < 0.01$; BMI, Body mass index.

4. Discussion

The main finding of this study was that the test–retest reliability of elbow and knee concentric flexion is good or excellent in PD patients. These results were similar to those observed in sit-to-stand-to-sit, six-minute walk, one-leg heel-rise, and handgrip strength tests in people undergoing hemodialysis [20] or in the incremental shuttle walk test, the estimated maximum repetition for quadricep strength, and VO_2peak by cardiopulmonary exercise testing in non-dialysis chronic kidney disease [21]. However, little was known about the reliability of physical function tests in PD patients and, to our knowledge, this is the first study aimed to evaluate the test–retest reliability of isokinetic measures in this population.

The evaluation of physical function is relevant since it is strongly related to quality of life, independence, and the ability to perform activities of daily living. Patients suffering from chronic kidney disease may have a higher risk of having low strength levels and low muscular mass, which is commonly associated with a higher risk of mortality [22]. However, most of the studies aimed to evaluate physical function in patients undergoing dialysis have been conducted with hemodialysis patients [23], while further research is needed in PD patients.

According to Zuo et al. [24], exercise capacity may be reduced in about 96% of PD patients. This reduction in exercise tolerance could be partially determined by age, sex, and body composition and is strongly associated with health-related quality of life in this population [25]. In this regard, the reduced quality of life and physical function may be similar in patients undergoing hemodialysis or PD [26]. However, Kang et al. [27] observed more favorable mental and physical components in hemodialysis patients compared with PD patients. Although the differences between the two dialysis modalities in terms of physical function and quality of life still remain unclear, patient satisfaction is commonly higher in patients undergoing PD [26,27]. Therefore, physical function is a relevant measure that should be included in comprehensive health assessments in dialysis patients. The current study provides test–retest reliability parameters that should be used to interpret health-related physical function evaluations in this population.

Isokinetic dynamometry is considered the gold standard for dynamic muscle performance testing [28,29]; thus, clinicians and researchers should be encouraged to use this device to conduct their physical function evaluations. Previous studies have reported the test–retest reliability of isokinetic measures in several populations different from patients undergoing dialysis. In this regard, reliability results obtained in the current study are similar to those obtained in postmenopausal women with osteopenia [30], women with fibromyalgia [15], patients with knee osteoarthritis [31], or persons with chronic stroke [32]. Furthermore, the current study not only reports the reliability parameters of measures of the lower limb but also assesses the reliability of the elbow flexion and extension strength tests. Therefore, future research may use both the absolute and the relative reliability parameters reported here to evaluate whether an observed change in lower or upper limb strength represents a true change or not.

Almost every measure achieved an excellent reliability according to the classification by Munro, Visintainer, and Page [18]. The ICC in elbow flexion in men was lower than 0.90; thus, it was classified as "good" but not "excellent". On the other hand, the reliability of this measure was excellent among women. Regarding the SRD, which is the parameter that defines the threshold to consider an observed change as "true" or "real", the values for the knee peak torque ranged between 20.70% in the knee concentric flexion of women and 31.29% in the knee concentric extension of men. Regarding the elbow's peak torque, it ranged between 11.84% in the concentric flexion of women and 24.56% in the concentric extension of men. Overall, higher reliability was observed in women compared to men, and the SRD was lower in the concentric flexion compared to the concentric extension. This was also observed when assessing the work. Although hypothetical, these sex differences might be related to the significant differences between men and women observed at baseline, i.e., age, height, weight, fat mass, and lean mass. However, given the relatively low sample size, this finding must be taken with caution.

The current study has some limitations. First, the participants did not undergo a familiarization session because of time and financial constraints. Second, although the sample size was similar to or even higher than those from previous studies [15,30,31], results from the stratification by men and women must be taken with caution. Third, the mean age of males was 52.42, which means that they were more than 10 years older than the women; thus, comparison between males and females could be influenced by that age difference. Despite these three limitations, the current study provides useful information about the reliability of isokinetic measures in people undergoing PD.

5. Conclusions

Isokinetic dynamometry is a reliable technique to evaluate peak torque and work for concentric flexion and extension in both the knee and the elbow joints in patients undergoing PD. Although results from division by sex must be taken with caution, higher reliability was observed in women compared to men and in the concentric flexion compared to the concentric extension. In addition to the reported good or excellent reliability based on ICC values, the present study provided novel SRD data which are associated to measurement error and individual variability and will assist healthcare professionals in interpreting treatment effects on isokinetic strength in this population.

Author Contributions: Conceptualization, N.B. and A.M.R.; Data curation, Z.C., N.B. and A.M.R.; Formal analysis, D.C.-M., F.J.D.-M. and J.C.A.; Investigation, D.C.-M., Z.C., J.C.A. and E.M.-N.; Methodology, N.B.; Supervision, J.C.A., N.B., E.M.-N. and A.M.R.; Writing—original draft, D.C.-M.; Writing—review and editing, D.C.-M., F.J.D.-M., Z.C., J.C.A., N.B., E.M.-N. and A.M.R

Funding: This research received no external funding.

Acknowledgments: The authors acknowledge the support provided by Manuel Amoedo in the evaluations of the patients.

Conflicts of Interest: The authors declare no conflict of interest.

References

1. Carrero, J.J.; Stenvinkel, P.; Cuppari, L.; Ikizler, T.A.; Kalantar-Zadeh, K.; Kaysen, G.; Mitch, W.E.; Price, S.R.; Wanner, C.; Wang, A.Y. Etiology of the protein-energy wasting syndrome in chronic kidney disease: A consensus statement from the International Society of Renal Nutrition and Metabolism (ISRNM). *J. Renal Nutr.* **2013**, *23*, 77–90. [CrossRef] [PubMed]
2. Broers, N.J.; Martens, R.J.; Cornelis, T.; van der Sande, F.M.; Diederen, N.M.; Hermans, M.M.; Wirtz, J.J.; Stifft, F.; Konings, C.J.; Dejagere, T. Physical Activity in End-Stage Renal Disease Patients: The Effects of Starting Dialysis in the First 6 Months after the Transition Period. *Nephron* **2017**, *137*, 47–56. [CrossRef] [PubMed]
3. Oliveira, E.A.; Cheung, W.W.; Toma, K.G.; Mak, R.H. Muscle wasting in chronic kidney disease. *Pediatr. Nephrol.* **2018**, *33*, 789–798. [CrossRef] [PubMed]
4. Rosa, C.S.; Bueno, D.R.; Souza, G.D.; Gobbo, L.A.; Freitas, I.F.; Sakkas, G.K.; Monteiro, H.L. Factors associated with leisure-time physical activity among patients undergoing hemodialysis. *BMC Nephrol.* **2015**, *16*, 192. [CrossRef] [PubMed]

5. Rhee, C.M.; Kalantar-Zadeh, K. Resistance exercise: An effective strategy to reverse muscle wasting in hemodialysis patients? *J. Cachexia Sarcopenia Muscle* **2014**, *5*, 177–180. [CrossRef] [PubMed]
6. Van Den Ham, E.C.; Kooman, J.P.; Schols, A.M.; Nieman, F.H.; Does, J.D.; Franssen, F.M.; Akkermans, M.A.; Janssen, P.P.; Van Hooff, J.P. Similarities in skeletal muscle strength and exercise capacity between renal transplant and hemodialysis patients. *Am. J. Transplant.* **2005**, *5*, 1957–1965. [CrossRef] [PubMed]
7. Wang, A.Y.-M.; Sea, M.M.-M.; Ho, Z.S.-Y.; Lui, S.-F.; Li, P.K.-T.; Woo, J. Evaluation of handgrip strength as a nutritional marker and prognostic indicator in peritoneal dialysis patients. *Am. J. Clin. Nutr.* **2005**, *81*, 79–86. [CrossRef] [PubMed]
8. Dziubek, W.; Bulińska, K.; Rogowski, Ł.; Gołębiowski, T.; Kusztal, M.; Grochola, M.; Markowska, D.; Zembroń-Łacny, A.; Weyde, W.; Klinger, M. The effects of aquatic exercises on physical fitness and muscle function in dialysis patients. *Bio. Med. Res. Int.* **2015**, *2015*, 912980. [CrossRef]
9. Blake, C.; O'meara, Y.M. Subjective and objective physical limitations in high-functioning renal dialysis patients. *Nephrol. Dial. Transplant.* **2004**, *19*, 3124–3129. [CrossRef]
10. Sole, G.; Hamrén, J.; Milosavljevic, S.; Nicholson, H.; Sullivan, S.J. Test-retest reliability of isokinetic knee extension and flexion. *Arch. Phys. Med. Rehabil.* **2007**, *88*, 626–631. [CrossRef]
11. Perrin, D.H. *Isokinetic Exercise and Assessment*; Human Kinetics: Charlottesville, VA, USA, 1993.
12. Adsuar, J.C.; Olivares, P.R.; Parraca, J.A.; Hernández-Mocholí, M.A.; Gusi, N. Applicability and test-retest reliability of isokinetic shoulder abduction and adduction in women fibromyalgia patients. *Arch. Phys. Med. Rehabil.* **2013**, *94*, 444–450. [CrossRef] [PubMed]
13. Association, W.M. World Medical Association Declaration of Helsinki: Ethical principles for medical research involving human subjects. *JAMA* **2013**, *310*, 2191.
14. Collado-Mateo, D.; Dominguez-Munoz, F.J.; Batalha, N.; Parraca, J.; Tomas-Carus, P.; Adsuar, J.C. Test-Retest Reliability of Isokinetic Arm Strength Measurements in Competitive Swimmers. *J. Hum. Kinet.* **2018**, *65*, 5–11. [CrossRef] [PubMed]
15. Adsuar, J.C.; Olivares, P.R.; del Pozo-Cruz, B.; Parraca, J.A.; Gusi, N. Test-retest reliability of isometric and isokinetic knee extension and flexion in patients with fibromyalgia: Evaluation of the smallest real difference. *Arch. Phys. Med. Rehabil.* **2011**, *92*, 1646–1651. [CrossRef] [PubMed]
16. Kannus, P.; Beynnon, B. Peak torque occurrence in the range of motion during isokinetic extension and flexion of the knee. *Int. J. Sports Med.* **1993**, *14*, 422–426. [CrossRef]
17. Shrout, P.E.; Fleiss, J.L. Intraclass correlations: Uses in assessing rater reliability. *Psychol. Bull.* **1979**, *86*, 420–428. [CrossRef]
18. Munro, B.H.; Visintainer, M.A.; Page, E.B. *Statistical Methods for Health Care Research*; JB Lippincott: Philadelphia, PA, USA, 1986.
19. Weir, J.P. Quantifying test-retest reliability using the intraclass correlation coefficient and the SEM. *J. Strength Cond. Res.* **2005**, *19*, 231–240. [CrossRef]
20. Segura-Orti, E.; Martinez-Olmos, F.J. Test-retest reliability and minimal detectable change scores for sit-to-stand-to-sit tests, the six-minute walk test, the one-leg heel-rise test, and handgrip strength in people undergoing hemodialysis. *Phys. Ther.* **2011**, *91*, 1244–1252. [CrossRef]
21. Wilkinson, T.J.P.; Xenophontos, S.M.; Gould, D.W.P.; Vogt, B.P.P.; Viana, J.L.P.; Smith, A.C.P.; Watson, E.L.P. Test-retest reliability, validation, and "minimal detectable change" scores for frequently reported tests of objective physical function in patients with non-dialysis chronic kidney disease. *Physiother. Theor. Pract.* **2019**, *35*, 565–576. [CrossRef]
22. Pereira, R.A.; Cordeiro, A.C.; Avesani, C.M.; Carrero, J.J.; Lindholm, B.; Amparo, F.C.; Amodeo, C.; Cuppari, L.; Kamimura, M.A. Sarcopenia in chronic kidney disease on conservative therapy: Prevalence and association with mortality. *Nephrol. Dial. Transplant.* **2015**, *30*, 1718–1725. [CrossRef]
23. Cheema, B.S.; Singh, M.A. Exercise training in patients receiving maintenance hemodialysis: A systematic review of clinical trials. *Am. J. Nephrol.* **2005**, *25*, 352–364. [CrossRef] [PubMed]
24. Zuo, M.L.; Yue, W.S.; Yip, T.; Ng, F.; Lam, K.F.; Yiu, K.H.; Lui, S.L.; Tse, H.F.; Siu, C.W.; Lo, W.K. Prevalence of and associations with reduced exercise capacity in peritoneal dialysis patients. *Am. J. Kidney Dis.* **2013**, *62*, 939–946. [CrossRef] [PubMed]
25. Uchiyama, K.; Washida, N.; Muraoka, K.; Morimoto, K.; Kasai, T.; Yamaki, K.; Miyashita, K.; Wakino, S.; Itoh, H. Exercise Capacity and Association with Quality of Life in Peritoneal Dialysis Patients. *Peritoneal. Dial. Int.* **2019**, *39*, 66–73. [CrossRef] [PubMed]

26. Iyasere, O.U.; Brown, E.A.; Johansson, L.; Huson, L.; Smee, J.; Maxwell, A.P.; Farrington, K.; Davenport, A. Quality of Life and Physical Function in Older Patients on Dialysis: A Comparison of Assisted Peritoneal Dialysis with Hemodialysis. *Clin. J. Am. Soc. Nephrol.* **2016**, *11*, 423–430. [CrossRef] [PubMed]
27. Kang, S.H.; Do, J.Y.; Lee, S.Y.; Kim, J.C. Effect of dialysis modality on frailty phenotype, disability, and health-related quality of life in maintenance dialysis patients. *PLOS ONE* **2017**, *12*, e0176814. [CrossRef] [PubMed]
28. Dvir, Z.; Muller, S. Multiple-Joint Isokinetic Dynamometry: A Critical Review. *J. Strength Cond. Res.* **2019**. [CrossRef] [PubMed]
29. Stark, T.; Walker, B.; Phillips, J.K.; Fejer, R.; Beck, R. Hand-held dynamometry correlation with the gold standard isokinetic dynamometry: A systematic review. *PM R J.* **2011**, *3*, 472–479. [CrossRef]
30. Eitzen, I.; Hakestad, K.A.; Risberg, M.A. Inter- and intrarater reliability of isokinetic thigh muscle strength tests in postmenopausal women with osteopenia. *Arch. Phys. Med. Rehabil.* **2012**, *93*, 420–427. [CrossRef]
31. Kean, C.O.; Birmingham, T.B.; Garland, S.J.; Bryant, D.M.; Giffin, J.R. Minimal detectable change in quadriceps strength and voluntary muscle activation in patients with knee osteoarthritis. *Arch. Phys. Med. Rehabil.* **2010**, *91*, 1447–1451. [CrossRef]
32. Ekstrand, E.; Lexell, J.; Brogardh, C. Isometric and isokinetic muscle strength in the upper extremity can be reliably measured in persons with chronic stroke. *J. Rehabil. Med.* **2015**, *47*, 706–713. [CrossRef]

applied sciences

MDPI

Article

Relationship between Kinematic Variables of Jump Throwing and Ball Velocity in Elite Handball Players

Abdel-Rahman Akl [1], Ibrahim Hassan [2], Amr Hassan [3,*] and Phillip Bishop [4]

1 Faculty of Physical Education-Abo Qir, Alexandria University, Alexandria 21913, Egypt
2 Faculty of Physical Education, Zagazig University, Zagazig 44519, Egypt
3 Department of Sports Training, Faculty of Sports Education, Mansoura University, Mansoura 35516, Egypt
4 Department of Kinesiology, University of Alabama, Tuscaloosa, AL 35487, USA
* Correspondence: amrahh@mans.edu.eg; Tel.: +20-436506870621

Received: 14 July 2019; Accepted: 16 August 2019; Published: 20 August 2019

check for updates

Abstract: The purpose of this pilot study was to evaluate the relationship between the kinematic variables of the right hand and left leg with ball velocity during jump-throwing phases in handball for better-informed training. We investigated ball velocity and the key kinematic variables of jump throwing during different throwing phases in three strides. Ten right-handed male handball professional players who had competed in the Egyptian Handball Super League participated in this study. Jump throwing performance was divided into three phases (cocking, acceleration and follow-through), which included eight events during the throwing. Five trials were captured for each player, and a 3D analysis was performed on the best trial. Results indicated that the velocity of the throwing hand was the most important variable during jump throwing, which was correlated with ball velocity during the three phases of performance in four events: Initial contact (IC) (r = 0.65*), initial flight (IF) (r = 63*), maximum height of the throwing hand (Max-HH) (r = 0.78*) and ground contact (GC) (r = 0.83*). In addition, the initial flight was the most important event in which players need to be using the best angles during performance, particularly the shoulder angle.

Keywords: jump throwing ability; 3D motion analysis; acceleration

---

1. Introduction

Team handball is an Olympic sport played worldwide at a professional level in several countries. Recently, handball has received increased attention in research studies, especially in biomechanics [1–8].

Jump throwing is an essential task in handball and is used frequently from different positions when a player shoots at the goal. As has been mentioned in earlier studies, handball players perform about 48,000 throws during the season, with the mean throwing speed of 130 km·h^{-1} [9], they and commonly use (73–75% of the time) jump throwing throws during the competition [5,7]. Hence, jump throwing and players' ability to accelerate the ball with the arm throw are vital demands during a handball game.

During jump throwing in handball, ball velocity and jump height range are key factors for better throwing performance [10]. Thus, most studies have investigated ball velocity [5,11,12].

Kinematic variables contribute to the velocity of the ball in order to find the fundamental procedures for improving handball players. As such, previous studies have documented the kinematic results of jump throwing performance in handball. Wagner, Buchecker, von Duvillard and Muller [7] found a significant difference of ball velocity, body height and weight between elite and lower performance levels. As well as, Wagner, Pfusterschmied, von Duvillard and Müller [5] compared the ball velocity and throwing accuracy between jump and standing throws. In addition, van den Tillaar and Ettema [8] investigated the contribution of upper-extremity, trunk, and lower-extremity movements, and van den

Tillaar [2] compared the range of motion with throwing kinematic variables. Plummer and Oliver [1] investigated the effects of fatigue on kinematic and kinetic changes of upper-extremity jump throwing.

Many kinematic experiments have studied ball throwing for team handball players and investigated the movement phases during the ball throwing [7,8,13,14]. Few studies have examined the kinematic variables effects and established relationships with ball velocity using 3D biomechanical tools. The handball coaches tend to focus on improving the throwing velocity for players [15].

Thus, through the investigation in the biomechanics of throwing velocity in elite handball players for the purpose of providing guidance to coaches and players desiring to improve this key aspect of performance, previous studies have shown a distinctive contribution of a velocity–accuracy trade-off and have concluded that kinematic variable as displacement along the velocity accuracy is achieved as an important contributed factor of decision-making. Therefore, ball velocity is considered the main factor for high quality jump throwing towards the goal [2,16,17].

In this regard, Stirn, et al. [18] opined that there are many factors contributing to the final velocity of the ball at release. To evaluate these factors, different demanding and time-consuming acquisition and analysis methods are required, including kinematic and electromyography assessments. Wagner, Pfusterschmied, von Duvillard and Müller [5] argued that the lower extremity in this type of skill plays an important role to drive the upper-extremity during performance. Therefore, throwing performance is considered to be the final outcome of an efficient kinetic chain.

The drive leg step before the take-off during the throwing jump skill could be an essential part of throwing to provide a support for the transfer of momentum through the pelvis and trunk to the dominant throwing arm. The kinematic investigation of the current study should be helpful for identifying the optimum jumping throwing mechanics. Consequently, we evaluated the relationship between ball velocity and some kinematic variables to identify key variables which should contribute to improving jump throwing performance.

Few studies have examined kinematic variable effects and established relationships with ball velocity using 3D biomechanical tools. Our goal for this study was to evaluate the correlation between key variables and throwing velocity, with the goal of providing information useful to coaches and players for improving jump-throwing performance. Therefore, we hypothesized that there would be positive and strong relationships between the kinematic variables of the right hand and left leg with ball velocity, and we determined key kinematic variables of jump throwing.

2. Materials and Methods

2.1. Subjects

Ten right-handed male handball players participated in current study (age: 20.8 ± 1.21 years; body mass: 82.8 ± 8.57 kg; height: 189.6 ± 8.65 cm, training experience: 9 years). They were part of a professional team which competed in the Egyptian Handball Super League. The study was approved by the institutional ethics committee of studies and research, and each player's consent was obtained.

2.2. Procedures

After a 15-min warm-up including general and shoulder-specific mobility exercises, as well as stretching and familiarization with the protocol, participants performed jump-throws after three running steps while positioned in front of the goal. Players were instructed to throw the ball as fast as possible. A total of five successful attempts were recorded for each player, with a one-minute rest between attempts. The best attempt, according to velocity of the ball, was selected for 3D analysis. The jump throwing skill was broken into three phases: The cocking phase, the acceleration phase, and the follow-through phase. Eight events were identified, beginning with the touch down (this event was determined when the jump leg touched the force platform starting the take-off).

The second event was the maximum ground reaction force (this event was designated as when the peak ground reaction force was achieved). Next was take-off (this event was designated as when the jump leg left the force platform). Maximum arm cocking was designated as when the arm reached the maximum back swing. Maximum height of center of mass was defined as when the center of mass achieved maximum height.

The maximum height of throwing arm was defined as when the throwing hand was at the maximum height. Ball release was defined as when the ball was released from the throwing arm, and the last event was landing, which was designated as when the player touched down from flight during performance (Figure 1).

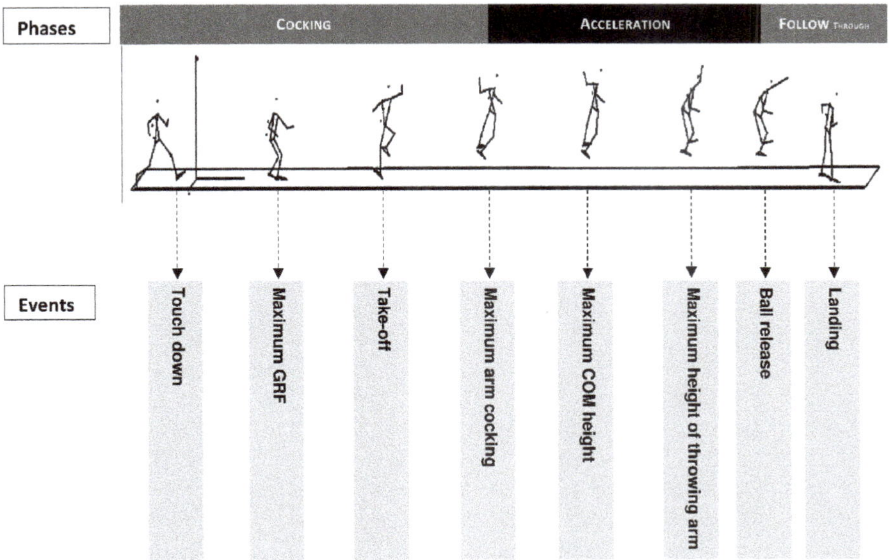

Figure 1. Jump throwing phases (cocking phase, acceleration phase, follow-through phase), and events (touch down, maximum ground reaction force (GRF), take-off, maximum arm cocking, maximum height of center of mass (COM), maximum height of throwing arm, ball release, and landing) during performance.

The kinematic variables of the right-hand (throwing arm) were measured using a 3-D motion capture system (Simi Reality motion analysis V. 9.0.6, eight synchronized Basler scA640-120gc-High-Speed Cameras were used at a 100 Hz frequency) that tracked the position of the reflective markers on anatomical landmarks according to the Hanavan model [19,20], (Figure 2). The 3D coordinates were the X (medio–lateral), Y (anterior–posterior) and Z (vertical) directions. Throwing performance was evaluated by the ball velocity [21], which was calculated by creating a rigid body of the ball using software and tracked. The angular kinematics of the joints were derived from relative angles between the two relevant segments (sagittal plane). A strain gage force platform (MP4060®, Bertec Corporation, Columbus, OH, USA) was used to determine the touch down and maximum ground reaction force events. The standard setting of software for filtering were used, and differentiation was used to calculate the kinematics variables [22].

2.3. Statistical Analysis

The Pearson coefficient of correlation was used to determine the strength of the relationships between each kinematic variable and ball velocity during each of the jump throwing phases (Table 1). Asterisk signs above the number represent significant differences between kinematic variables and ball velocity, and (*) indicates $p \leq 0.05$ for statistical procedures (SPSS 21, V. 21 Statistics for Windows. IBM Corp, Armonk, NY, USA).

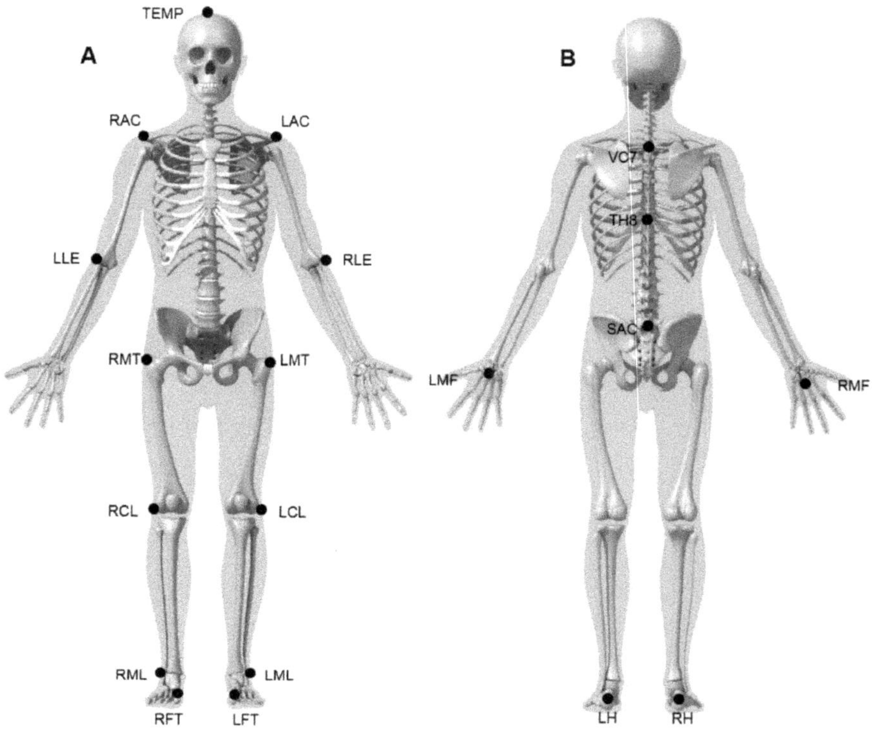

Figure 2. Full body marker set in (**A**) anterior (**B**) posterior view. The markers were placed at tempus head (TEMP), acromion left/right (LAC, RAC), epicondylus lateralis (elbow lateral) left/right (LLE, RLE), trochanter major left/right (LMT, RMT), Condylus lateralis left/right (LCL, RCL), malleolus lateralis left/right (LML, RML), foot tip left/right (LFT, RFT), vertebra C7 (VC7), vertebra Th8 (TH8), mid spina Iliaca posterior superior (SAC), middle finger base joint left/right (LMF, RMF), and heel left/right (LH, RH).

3. Results

The relationships between kinematic variables and ball velocity are shown in Tables 1 and 2 during each jump throwing phase (see Figure 1).

Table 1 provides the Pearson correlation coefficients between ball velocity and kinematic variables (displacement, velocity, and acceleration) at three investigated phases for eight events. No significant correlation was seen between ball velocity and right-hand kinematic variables during the cocking phase (displacement) and the acceleration phase (acceleration).

During the cocking phase, the ball velocity showed moderate correlations with the right arm velocity at initial contact (IC) (Y and R coordinates) ($r = 0.66$, $r = 0.63$) and initial flight (IF) (Y) ($r = 0.63$). Furthermore, there was a relatively strong correlation ($r = 0.72$) between the right arm acceleration at maximum ground reaction force (Max-GRF) (Y) and ball velocity. Through the acceleration phase, right arm displacement at maximum height of body (GC—ground contact) during flight (Max-GCH) (Y) had the highest correlation coefficients during this phase ($r = 0.83$) with ball velocity, and at release ball (RB) (Y) a strong correlation ($r = 0.77$) was observed. Likewise, ball velocity had moderate and strong correlations with the right arm displacement at maximum height of the throwing hand (Max-HH) (R) and RB (R), ($r = 0.67$ and $r = 0.79$, respectively). In addition, strong and moderate correlations were seen during the acceleration phase between velocity of the right hand and the ball velocity at Max-HH (Y) and Max-HH (R), ($r = 0.78$ and $r = 0.70$, respectively). As shown in Table 1 during the follow-through phase, the kinematic variables of the right hand such as the displacement at GC (Y, R) and velocity at GC (Y) had the highest correlation coefficients (r = 0.84, r = 0.88, and r = 0.83, respectively). A moderate correlation was observed between ball velocity and the right-hand acceleration (r = 0.68) at GC (Y) during this phase.

Table 1. Correlation between right-hand center of mass kinematic variables and ball velocity during jump throwing phases in handball. N = 10.

Variables (Unit)	Coordinates	Cocking Phase			Acceleration Phase				Follow-Through Phase
		IC	Max-GRF	IF	Max-BC	Max-GCH	Max-HH	RB	GC
Displacement (m)	X	−0.23	0.10	0.15	0.21	0.05	0.14	−0.04	−0.11
	Y	−0.27	0.32	0.34	0.57	0.83 *	0.63	0.77 *	0.84 *
	Z	0.56	−0.43	−0.27	−0.00	0.13	0.04	−0.25	−0.04
	R	0.49	−0.54	−0.21	0.04	0.22	0.67 *	0.79 *	0.88 *
Velocity (m/s)	X	−0.41	−0.05	0.15	−0.18	0.27	−0.28	−0.37	−0.39
	Y	0.66 *	0.20	0.63 *	0.22	0.51	0.78 *	−0.40	0.83 *
	Z	0.22	−0.00	−0.18	0.21	0.27	0.43	−0.51	0.50
	R	0.63 *	−0.04	0.09	0.26	0.41	0.70 *	0.04	−0.04
Acceleration (m/s^2)	X	0.56	0.62	−0.02	0.01	−0.05	−0.50	0.22	−0.34
	Y	0.40	0.79 *	0.24	−0.35	0.24	0.35	−0.57	0.68 *
	Z	−0.59	0.31	0.28	−0.40	0.26	0.08	0.43	0.30
	R	0.17	−0.54	−0.22	−0.42	0.11	−0.01	0.44	0.17

Note: Coordinates: X (medio–lateral), Y (anterior–posterior), Z (vertical), and R (resultant); IC: Initial contact (end of last step); Max-GRF: Maximum Ground Reaction force; IF: Initial flight; Max-BC: Maximum back cooking; Max-GCH: Maximum height of body GC during flight; Max-HH: Maximum height of the throwing hand GC during flight; RB: Release ball; GC: Ground contact; * Correlation were considered significant at $p < 0.05$.

Table 2. Correlation between right hand and left leg angular variables and ball velocity during jump throwing phases in handball.

Joint (Unit)	Variables (Unit)	Cocking Phase			Acceleration Phase				Follow-Through Phase
		IC	Max-GRF	IF	Max-BC	Max-GCH	Max-HH	RB	GC
Right elbow	Angle (°)	−0.21	0.22	0.08	0.03	−0.07	−0.64 *	−0.29	−0.33
	Velocity (°/s)	−0.20	−0.30	0.05	−0.41	−0.07	−0.11	−0.56	−0.03
	Acceleration (°/s^2)	0.44	−0.05	−0.05	0.25	0.20	0.55	−0.28	0.19
Right shoulder	Angle (°)	0.49	−0.51	−0.32	−0.38	−0.45	−0.31	−0.32	−0.63
	Velocity (°/s)	0.30	−0.32	0.69 *	−0.18	0.13	−0.37	−0.67 *	0.27
	Acceleration (°/s^2)	−0.59	0.31	0.67 *	0.55	0.28	−0.07	0.25	−0.21
Left hip	Angle (°)	0.07	0.33	0.41	0.32	0.28	0.36	0.22	0.73 *
	Velocity (°/s)	0.44	−0.27	−0.51	−0.05	−0.37	0.35	−0.21	−0.26
	Acceleration (°/s^2)	−0.30	−0.65 *	−0.67 *	0.24	−0.44	−0.23	−0.26	−0.14
Left knee	Angle (°)	0.16	−0.57	−0.01	−0.29	0.23	0.28	−0.07	−0.34
	Velocity (°/s)	0.46	−0.26	0.13	−0.02	0.28	−0.26	−0.09	−0.67 *
	Acceleration (°/s^2)	−0.15	−0.51	−0.18	0.22	−0.50	−0.33	0.08	−0.52
Left ankle	Angle (°)	−0.71 *	−0.80 *	0.30	0.01	−0.17	−0.10	−0.50	−0.57
	Velocity (°/s)	−0.03	0.06	0.38	−0.09	0.36	−0.61	0.08	−0.45
	Acceleration (°/s^2)	0.56	0.16	−0.53	−0.01	0.13	−0.19	0.37	−0.30

Note: IC: Initial contact (end of last step); Max-GRF: Maximum ground reaction force; IF: Initial flight; Max-BC: Maximum back cooking; Max-CGH: Maximum height of body GC during flight; Max-HH: Maximum height of the throwing hand GC during flight; RB: Release ball; GC: Ground contact; * Correlation were considered significant at $p < 0.05$.

The Pearson correlation coefficients between ball velocity with the right hand and left leg angular variables (angle, velocity, acceleration) during the three phases can be seen in Table 2. There were no significant correlation coefficients observed between the ball velocity and the kinematic variables of right elbow during all phases except a moderate negative performance score (r = −0.64) of angle at Max-HH during the acceleration phase. Likewise, during the follow-through phase, the velocity of the left knee had a moderate negative correlation (r = −0.67) at GC. Right shoulder velocity and acceleration had a moderate correlation with ball velocity at IF during the cocking phase (r = 0.69, r = 0.67, respectively). At RB during the acceleration phase, the right shoulder velocity had a moderate negative correlation (r = −0.67). Left hip acceleration during the cocking phase had a moderate negative correlation with ball velocity at Max-GRF and IF (r = −0.65, r = −0.67, respectively), and the angle of the left hip had a strong correlation (r = 0.73) at GC during the follow-through phase. The angle of left ankle during the cocking phase had a strong negative correlation with performance at IC and Max-GRF (r = −0.71, r = −0.80, respectively).

4. Discussion

The current study examined the relationships between ball velocity and kinematic variables during the jump throwing of elite handball players. The main aim was to determine the relationships between ball velocity and right hand (throwing hand) linear kinematics, as well as right arm and left leg (take-off leg) angular kinematics during three investigated phases over eight events.

The results in Table 1 show that the kinematic variables at maximum ground reaction force, maximum height of body center of gravity during flight, release ball and ground contact events were correlated strongly with ball velocity. These findings support a previous study which indicated important kinematic variables of the standing throw in handball [23].

The analysis of the cocking phase in our study indicated a moderate correlation between ball velocity and the velocity of the right hand at IC (Y, R) and IF (Y). This finding indicates that the primary velocity of elbow and the initial flight with the ball increased the ball velocity. During the same phase, the right-hand acceleration had a strong correlation with ball velocity at Max-GRF (Y). This strong correlation indicates that the Max-GRF during the jump throwing improved the horizontal velocity of the right hand (elbow and wrist joint) before ball release, which led to improving the arm swing during throwing.

The displacement of the right hand during the acceleration phase was strongly correlated with ball velocity at Max-CGH (Y), RB (Y) and RB (R). In addition, the Max-HH (R) during this phase correlated moderately with the ball velocity. This finding implies that maximum height of body GC and the dominant throwing hand increase the kinematic requirements of the hip and shoulder to maintain a high velocity of ball and performance; van den Tillaar and Ettema [24] showed some kinematic changes during overarm throwing related to the elbow extension and internal rotation.

Furthermore, the velocity of the right hand correlated strongly with Max-HH (Y) and moderately with Max-HH (R). The positive correlation with ball velocity and maximum heights of the throwing hand (Y, R) coordinates during the acceleration phase implies that arm motion continues until ball release time, which generates a ball rotation around the central axis, increases and generates the required release ball velocity. This supports the study of Werner, et al. [25], who suggested that the arm acceleration phase is the dynamic phase between the maximum external shoulder rotation and the instant of ball release.

During the follow-through phase, the results indicated that a higher correlation between ball velocity and the displacement of the right hand at GC (Y and R), and the velocity of the right hand had a high correlation at GC (Y). However, a moderate correlation was seen during this phase with the acceleration of the right hand at GC (Y). A strong correlation of displacement and velocity were found during this phase, so we suggest that when players perform the jump throw, they must increase their movement horizontally to enable forward movement at the landing. Hirashima, et al. [26] indicated that the ability of skilled throwers to optimize the throwing arm event of inertia during the arm cocking and arm acceleration phases highlights the importance of the arm movement to reach a high ball velocity.

Our findings in Table 2 indicate that the moderate negative correlation between ball velocity and the right elbow angle during the Max-HH event. This finding indicates that the smaller angle at the elbow joint assists in the improvement of a higher ball velocity at release for longer throwing. This finding supports previous studies [8,15,23], which indicated that the elbow extension and internal rotation velocity were factors for fast shoulder throwing, wherein about 73% of the contribution of ball velocity was based on the maximal internal rotation velocity of the shoulder and maximal elbow extension during the throw.

A moderate correlation was seen between the ball velocity and the right shoulder velocity and acceleration variables at IF during the cocking phase. In addition, a moderate correlation was observed between the right shoulder velocity at RB and ball velocity during the acceleration phase. This finding indicates that player who moves forward at take-off also maintains a higher projection angle with the wrist hyperextension at the ball release, which leads to a higher flight time style. This result is seen in the correlation between ball velocity and the releasing of ball during the acceleration phase. During the acceleration phase, the maximal angular velocities of the elbow extension, wrist flexion, and the internal rotation of the shoulder joint make a substantial contribution to overarm throwing in team handball [23].

The left hip and right hand had moderate correlations with ball velocity at Max-GRF and with IF events during the cocking phase. During the follow-through phase at GC, a strong correlation was observed between the angle of the left hip and ball velocity. These negative correlations with ball velocity during the cocking phase indicated that handball players were using a greater angle of projection to reach for optimum position before releasing the ball and generate a required power with the throwing arm. Thus, higher velocities of the arm joints could be produced [23], which could result in higher external rotation while throwing. This finding confirmed that the movement of the large body segments such as legs and trunk assist with the reduction of energy in the throwing arm in order to reduce the loads, especially on the shoulder and elbow [25,27,28].

The ball velocity had a moderate correlation with the velocity of the left knee at GC during the follow-through phase and the left ankle angle at the IC and Max-GRF events during the cocking phase (Table 2). The negative correlation between left knee velocity with ball velocity implies that players had a greater knee flexion in the follow-through phase for preparation for landing. Olsen, et al. [29] and Koga, et al. [30] reported that landing after a jump throwing with the knee near full extension was important to prevent ACL injuries in team handball players.

It should be noted that these findings were derived from a small sample of homogenous elite handball players. We selected elite players because we wanted data from high-level performance. A small sample was used in this pilot study in order to evaluate the utility of this approach. Finding strong and moderate correlations in a small, homogenous sample suggests that these findings are very conservative. We hope that future investigations will expand on these preliminary findings.

A limitation was that the sample size in the present study was small (ten right-handed male handball players), and we analyzed the best attempt of the five successful performed attempts. All our participants were elite, meaning that our sample was homogenous, making correlations more challenging to identify. We were willing to accept these limitations in this pilot study because our goal was to identify the correlations for top performers, since it was assumed that these participants represent the best performances. However, future investigations should recruit a larger number of participants and/or increase the number of attempts.

5. Conclusions

Based on the main findings from this study, the velocity of the right hand at IC and IF events (anterior–posterior direction) correlated with ball velocity during the cocking phase as a preparation phase that accelerates the arm throwing. The increases in right-hand acceleration at Max-GRF, Max-HH events (anterior-posterior direction) led to the improvement of the horizontal velocity of the right hand (elbow and wrist joint) before the RB event achieved the highest ball velocity. Furthermore, the results indicated that the change in angular variables of shoulder joint are greater than those of the elbow. Thus, we recommend that, in training programs, coaches focus on increasing their player's movement horizontally to enable forward movement at the landing when evaluating the performance level of handball players.

Also, we recommend that coaches to use special power and reaction time exercises to reduce the dominant foot contact time at the jump moment and before throwing the ball. This would allow the player to jump higher before the ball release event and keep the maximum velocity of the ball while throwing performance.

Author Contributions: Conceptualization, A.-R.A. and I.H.; methodology, A.-R.A., I.H. and A.H.; software, A.-R.A.; validation, A.-R.A., I.H., P.B. and A.H.; formal analysis, A.-R.A.; investigation, A.-R.A. and I.H.; resources, A.-R.A.; data curation, A.-R.A. writing—original draft preparation, A.-R.A., I.H. and A.H.; writing—review and editing, P.B.; visualization, A.H.; supervision, P.B.; project administration, A.-R.A.

Funding: This research received no external funding.

Conflicts of Interest: The authors declare no conflict of interest.

References

1. Plummer, H.A.; Oliver, G.D. The effects of localised fatigue on upper extremity jump shot kinematics and kinetics in team handball. *J. Sports Sci.* **2017**, *35*, 182–188. [CrossRef] [PubMed]
2. Van den Tillaar, R. Comparison of range of motion tests with throwing kinematics in elite team handball players. *J. Sports Sci.* **2016**, 1–7. [CrossRef] [PubMed]
3. Serrien, B.; Clijsen, R.; Blondeel, J.; Goossens, M.; Baeyens, J.P. Differences in ball speed and three-dimensional kinematics between male and female handball players during a standing throw with run-up. *BMC Sports Sci. Med. Rehabil.* **2015**, *7*, 27. [CrossRef] [PubMed]
4. Setuain, I.; Millor, N.; Gonzalez-Izal, M.; Gorostiaga, E.M.; Gomez, M.; Alfaro-Adrian, J.; Maffiuletti, N.A.; Izquierdo, M. Biomechanical jumping differences among elite female handball players with and without previous anterior cruciate ligament reconstruction: A novel inertial sensor unit study. *Sports Biomech. Int. Soc. Biomech. Sports* **2015**, *14*, 323–339. [CrossRef] [PubMed]
5. Wagner, H.; Pfusterschmied, J.; von Duvillard, S.P.; Müller, E. Performance and kinematics of various throwing techniques in team-handball. *J. Sports Sci. Med.* **2011**, *10*, 73–80.
6. Wagner, H.; Buchecker, M.; von Duvillard, S.P.; Müller, E. Kinematic Comparison of Team Handball Throwing with Two Different Arm Positions. *Int. J. Sports Physiol. Perform.* **2010**, *5*, 469–483. [CrossRef] [PubMed]
7. Wagner, H.; Buchecker, M.; von Duvillard, S.P.; Muller, E. Kinematic description of elite vs. Low level players in team-handball jump throw. *J. Sports Sci. Med.* **2010**, *9*, 15–23.
8. Van den Tillaar, R.; Ettema, G. A three-dimensional analysis of overarm throwing in experienced handball players. *J. Appl. Biomech.* **2007**, *23*, 12–19. [CrossRef]
9. Almeida, G.P.; Silveira, P.F.; Rosseto, N.P.; Barbosa, G.; Ejnisman, B.; Cohen, M. Glenohumeral range of motion in handball players with and without throwing-related shoulder pain. *J. Shoulder Elb. Surg. Am. Shoulder Elb. Surg.* **2013**, *22*, 602–607. [CrossRef]
10. Bayios, I.A.; Anastasopoulou, E.; Sioudris, D.; Boudolos, K.D. Relationship between isokinetic strength of the internal and external shoulder rotators and ball velocity in team handball. *J. Sports Med. Phys. Fit.* **2001**, *41*, 229.
11. García, J.A.; Sabido, R.; Barbado, D.; Moreno, F.J. Analysis of the relation between throwing speed and throwing accuracy in team-handball according to instruction. *European Journal of Sport Science* **2013**, *13*, 149–154. [CrossRef]
12. Wagner, H.; Pfusterschmied, J.; Von Duvillard, S.P.; Muller, E. Skill-dependent proximal-to-distal sequence in team-handball throwing. *J. Sports Sci.* **2012**, *30*, 21–29. [CrossRef] [PubMed]
13. Meister, K. Injuries to the shoulder in the throwing athlete. Part one: Biomechanics/pathophysiology/classification of injury. *Am. J. Sports. Med.* **2000**, *28*, 265–275. [CrossRef] [PubMed]
14. Wagner, H.; Finkenzeller, T.; Wurth, S.; von Duvillard, S.P. Individual and team performance in team-handball: A review. *J. Sports Sci. Med.* **2014**, *13*, 808–816. [PubMed]

15. Debanne, T.; Laffaye, G. Coaches' Beliefs and Knowledge: Training Programs Used by French Professional Coaches to Increase Ball-throwing Velocity in Elite Handball Players. *Int. J. Sports Sci. Coach.* **2013**, *8*, 557–570. [CrossRef]
16. Van den Tillaar, R.; Ettema, G. A comparison between novices and experts of the velocity-accuracy trade-off in overarm throwing. *Percept. Mot. Ski.* **2006**, *103*, 503–514. [CrossRef] [PubMed]
17. Van den Tillaar, R.; Ettema, G. Influence of instruction on velocity and accuracy of overarm throwing. *Percept Mot. Ski.* **2003**, *96*, 423–434. [CrossRef] [PubMed]
18. Stirn, I.; Carruthers, J.; Sibila, M.; Pori, P. Frequent Immediate Knowledge of Results Enhances the Increase of Throwing Velocity in Overarm Handball Performance. *J. Hum. Kinet.* **2017**, *56*, 197–205. [CrossRef]
19. Akl, A.-R.; Salem, M. Effects of center of mass kinematics on ball velocity during jump throwing in handball. *MOJ Appl. Bionics Biomech.* **2018**, *2*, 219–221. [CrossRef]
20. Hanavan, E.P., Jr. *A mathematical model of the human body*; USAF Institute of Technology: Greene County, OH, USA, 1964.
21. Rivilla-Garcia, J.; Grande, I.; Sampedro, J.; van den Tillaar, R. Influence of opposition on ball velocity in the handball jump throw. *J. Sports Sci. Med.* **2011**, *10*, 534–539.
22. Hanley, B.; Tucker, C.B.; Bissas, A. Differences between motion capture and video analysis systems in calculating knee angles in elite-standard race walking. *J. Sports Sci.* **2017**, 1–6. [CrossRef]
23. Van den Tillaar, R.; Ettema, G. A force-velocity relationship and coordination patterns in overarm throwing. *J. Sports Sci. Med.* **2004**, *3*, 211–219.
24. Van den Tillaar, R.; Ettema, G. A comparison of kinematics between overarm throwing with 20% underweight, regular, and 20% overweight balls. *J. Appl. Biomech.* **2011**, *27*, 252–257. [CrossRef] [PubMed]
25. Werner, S.L.; Fleisig, G.S.; Dillman, C.J.; Andrews, J.R. Biomechanics of the elbow during baseball pitching. *J. Orthop. Sports Phys. Ther.* **1993**, *17*, 274–278. [CrossRef]
26. Hirashima, M.; Yamane, K.; Nakamura, Y.; Ohtsuki, T. Kinetic chain of overarm throwing in terms of joint rotations revealed by induced acceleration analysis. *J. Biomech.* **2008**, *41*, 2874–2883. [CrossRef]
27. Fleisig, G.S.; Andrews, J.R.; Dillman, C.J.; Escamilla, R.F. Kinetics of baseball pitching with implications about injury mechanisms. *Am. J. Sports Med.* **1995**, *23*, 233–239. [CrossRef]
28. Grezios, A.K.; Gissis, I.T.; Sotiropoulos, A.A.; Nikolaidis, D.V.; Souglis, A.G. Muscle-contraction properties in overarm throwing movements. *J. Strength Cond. Res. Natl. Strength Cond. Assoc.* **2006**, *20*, 117–123.
29. Olsen, O.E.; Myklebust, G.; Engebretsen, L.; Bahr, R. Injury mechanisms for anterior cruciate ligament injuries in team handball: A systematic video analysis. *Am. J. Sports Med.* **2004**, *32*, 1002–1012. [CrossRef]
30. Koga, H.; Nakamae, A.; Shima, Y.; Iwasa, J.; Myklebust, G.; Engebretsen, L.; Bahr, R.; Krosshaug, T. Mechanisms for noncontact anterior cruciate ligament injuries: Knee joint kinematics in 10 injury situations from female team handball and basketball. *Am. J. Sports Med.* **2010**, *38*, 2218–2225. [CrossRef]

applied sciences

MDPI

Article

Morphology-Related Foot Function Analysis: Implications for Jumping and Running

Peimin Yu [1], Liangliang Xiang [1], Minjun Liang [1], Qichang Mei [2], Julien S. Baker [3] and Yaodong Gu [1,2,*]

1 Faculty of Sports Science, Ningbo University, Ningbo 315211, China
2 Auckland Bioengineering Institute, University of Auckland, Auckland 1142, New Zealand
3 School of Science, University of the West of Scotland, Hamilton ML3 OJB, Scotland, UK
* Correspondence: guyaodong@nbu.edu.cn; Tel.: +86-574-87600456

Received: 14 July 2019; Accepted: 5 August 2019; Published: 8 August 2019

Abstract: Barefoot and shod running has received increased attention in recent years, however, the influence of morphology-related foot function has not been explored. This study aimed to investigate morphology-related jumping and running biomechanical functions in habitually barefoot and shod males. A total of 90 barefoot males (Indians) and 130 shod males (Chinese), with significant forefoot and toe morphology differences, participated in a vertical jump and running test to enable the collection of kinematic and kinetic data. The difference of pressure distribution in the hallux and forefoot was shown while jumping and running. The unrestricted forefoot and toes of the barefoot group presented flexible movement and leverage functions to expand the forefoot loading area during performance of the two tasks. Findings related to morphology functions, especially in the forefoot and toe may provide useful information for footwear design.

Keywords: foot morphology; toes function; biomechanics; barefoot; jumping; running

1. Introduction

Human feet are the basic terminal structures that support human walking, running, jumping, and other locomotion. The foot is a complex structure that controls balance and movement [1,2]. Foot morphology has been studied since the early 20th century [3]. Previous studies have demonstrated that the foot differs significantly between habitually barefoot and shod people [4–6], and differences in the kinetics of walking, running, and jumping have been observed [7–9].

Different foot morphology may also be a contributory factor for injury during motion [10], and may also influence physical activity performance [11,12]. There are many reasons for morphological differences in humans, which include disease, foot malfunctions, genetics, and deformity [13]. Research findings have indicated that external factors, such as footwear, may deform foot structure, and result in conditions such as hallux valgus (HV) [3,14]. HV could induce foot dysfunction [15], influence foot morphology [16], and may impair quality of life [17], which may result in depression and pain [18].

In addition, when compared to habitually shod populations, habitually barefoot populations demonstrate more toe separation [3,4,14]. Studies on foot morphology have focused on the width and length of the foot [6], and several studies have investigated the morphological differences between the hallux and other toes [4,14]. However, whether these differences influence the motions needed for physical activity is unclear.

Jumping is a typical movement in many sports, and has attracted much attention from the research community [19,20]. Jumping performance has been evaluated using a one-foot and a two-feet jump [19], and toe flexor function has also been examined [20]. Furthermore, the countermovement jump has been important to support clinicians in the medical diagnosis of muscle power during prolonged recovery periods following ankle injuries [21]. The contribution of the forefoot and toes has been evaluated while performing the vertical jump, and kinematics, kinetics, and spatiotemporal parameters have been recorded and analyzed [22].

Lieberman et al. [23] indicated that habitually barefoot populations and shod populations present different foot strike patterns. Habitually barefoot populations land on the forefoot, then bring down the heel, and have been observed landing with a flat foot, but seldom on the heel. Habitually shod populations mostly land with a rearfoot strike. The elevated and cushioned heel of the modern running shoe may be a contributory factor that has facilitated the differences in the strike patterns observed. However, strike patterns have been observed to be variant, even between shod or barefoot populations, in recent studies [2,7,8,24]. In spite of the conflicting opinions about barefoot locomotion, it has gained in popularity in recent years, and is now included in athletic training [25], recreational running [26], and rehabilitation [27]. A previous study has revealed the foot shape and function differences in native barefoot walkers [5] and runners [24]. The morphological differences between habitually barefoot and shod runners were found to exist in the forefoot and toe regions [4]. However, morphology based on the functions of the forefoot and toes while performing vertical jumping and running has not been investigated.

Therefore, the purpose of this study was to examine morphology-related performance differences while conducting vertical jumping and running tasks between habitually barefoot males and shod males. A further aim was to explore any functional differences in the forefoot and toes, based on foot morphological characteristics. It was hypothesized that the lower extremity kinematics and plantar forefoot loading distribution would be different due to the morphological differences in the forefoot and toes region.

2. Materials and Methods

2.1. Participants

The sample size was calculated prior to this study using the power package in R-3.6.1 (effect size = 0.5, α level = 0.05, power value = 0.9, type: two-sample, alternative: two sided). A total of 90 barefoot males (Indians) and 130 shod males (Chinese), who presented significant forefoot and toe morphology differences in a previously published study [4], volunteered to participate in the vertical jumping and running test to enable collection of kinematic and kinetic data. All participants were students in the University and had a history of running or other physical activities. Participants of Indian ethnicity originated from South India (Kerala state), were running or taking part in physical activities barefoot since birth, and wore slippers during daily life. Participants of Chinese ethnicity were shod runners since birth and wore different kinds of shoes in daily life. Participants with hallux valgus, high-arched foot, flat foot, diabetic foot, or any other foot deformities were excluded via foot scan prior the test. None of the participants had sustained injuries or surgeries to their lower limbs in the previous half year.

Data for 62 barefoot males (age: 22 ± 1.9 years; weight: 65 ± 8.6 kg; height: 1.69 ± 0.16 m), presenting with a forefoot strike during running, and 112 shod males (age: 23 ± 2.8 years; weight: 66 ± 7.8 kg; height: 1.71 ± 0.11 m), presenting with a rearfoot strike during running, were included for analysis via post data procession. This study, with detailed guidelines for participants' safety and experimental protocols, was approved by the Human Ethics Committee at the Research Institute of Ningbo University ARGH20160819. The study was conducted in accordance with the declaration of Helsinki. Prior to the test, all subjects gave informed consent, with full knowledge of test procedures and requirements.

2.2. Experiment Protocol

The test protocol was consistent with a previously reported experiment [23], which was published from our laboratory recently [1,24]. After completion of foot scanning, participants revisited the motion capture lab for experimental vertical jumping and running tests. Participants were instructed to warm up and to familiarize themselves with the lab environment for 5 min prior to data collection. Before data collection, three familiarization trials were performed for each task.

While performing the vertical jump, participants stood on the ground in an akimbo position (right foot on the force platform) to reduce the interference from the upper body during performance of a maximal vertical jump. Each participant completed six trials with the right foot on the force platform (Model 9281B, Winterthur, Switzerland).

Running tests were conducted on a runway in the lab. Subjects performed barefoot running with the right foot striking the force platform, which was located in the middle of the runway and was used for kinetic data collection. The force platform and pressure data were used to assist in the definition of striking patterns following a previously established protocol [28,29]. Each participant performed six trials of running using a self-selected running speed, to present natural strike patterns during running and a collection of biomechanical characteristics. For both jumping and running sessions, there were 30 s rest intervals between each trial to minimize the effect of fatigue.

The pressure platform (Novel EMED System, Munich, Germany) was reported to have high reliability correlations (>0.7) [30], and the insole (Novel Pedar System, Munich, Germany) plantar pressure distribution system also displayed excellent reliability correlations (>0.9) [31]. The pressure plate was used to record barefoot jumping and running plantar pressure distribution data with a frequency of 100 Hz. The in-shoe plantar pressure measurement system was placed in the shoes for collection of the shod jumping and running plantar pressure distribution data among habitually shod males, with a frequency of 100 Hz. The habitually shod males (shod) performed shod running wearing shoes that were the same brand and model, for consistency.

2.3. Data Acquisition

Previous studies have outlined data collected from insole pressure sensors and pressure plates and show high reliability [32]. All the anatomical region division analysis was performed in the Novel Database in the data post-procession based on an auto-masking algorithm [33]. For trials of barefoot and shod vertical jumping, only the data in the forefoot and toes were included. The collected plantar pressure data while performing vertical jumping were separated into the push-off and landing phases for analysis. Thus, the plantar surface was divided into five anatomical regions: medial forefoot (MF), central forefoot (CF), lateral forefoot (LF), hallux (H), and other toes (OT), as this study mainly focused on the instant push-off and landing phase of the vertical jump. For trials using barefoot and shod running, the plantar surface was divided into eight anatomical regions, including medial rearfoot (MR), lateral rearfoot (LR), medial midfoot (MM), lateral midfoot (LM), medial forefoot (MF), lateral forefoot (LF), hallux (H), and other toes (OT). The variables for jumping and running included peak pressure, contact area, and the pressure–time integral of each anatomical region.

The kinematic test used the eight-camera Vicon motion analysis system (Oxford Metric Ltd., Oxford, UK) to collect the lower extremity kinematic data, with a frequency of 200 Hz. Sixteen reflective points (diameter: 14 mm) were attached with adhesive tape on the lower limbs of subjects, following a previously published protocol [34]. The anatomical landmarks included the anterior–superior iliac spine, posterior–superior iliac spine, lateral mid-thigh, lateral knee, lateral mid-shank, lateral malleolus, second metatarsal head, and calcaneus. A Kistler Force Platform (Model 9281B, Winterthur, Switzerland) was used to record ground reaction forces (GRFs), with a frequency set at 1000 Hz, to define the running foot striking patterns and contact time. The force platform was zero-levelled prior to testing each participant. The on and off force platform was defined from the value of vertical GRF as 20 N. Participants were required to strike the force platform with the right foot while performing the running and jumping tests on the force platform. The variables of running included spatiotemporal parameters, such as stride length, stride time and contact time, peak angles during stance, and joints range of motion (ROM) in a gait cycle. The spatiotemporal parameters were generated from the Workstation in the Vicon Nexus software (v1.8.5), including hip, knee, and ankle angles in the sagittal plane, coronal plane, and horizonal plane, computed from the Vicon Plug-in-Gait Model using established protocols [20,30]. Vertical jump height was calculated by Equation (1) [35]:

$$\text{Jumpheight(m)} = \frac{9.80\text{m}\cdot\text{s}^{-2} \times \text{flighttime(s)}^2}{8}. \tag{1}$$

2.4. Statistical Analysis

Normal distribution was checked for all variables, including jump height, peak pressure, pressure time integral, and contact area of vertical jumping, and running spatiotemporal parameters, such as stride length, stride time and contact time, running peak angles during stance, and joints range of motion in a gait cycle. Independent-sample *T* tests were used to analyze the significance of kinematic, plantar loading, and spatiotemporal variables between the barefoot and shod group. SPSS 18.0 (SPSS Inc., Chicago, IL, USA) software was used for the analysis, with statistical significance set at $p < 0.05$.

3. Results

After calculation and comparison of jump height, there were no significant differences between the height of the barefoot jump (386.4 ± 13.6 mm) and shod jump (408.2 ± 12.9 mm), with $p > 0.05$.

As shown in Figure 1, during the take-off phase (left), significant differences ($p < 0.05$) were found between barefoot and shod jumping in H ($p = 0.02$ and 0.01), MF ($p = 0.018$ and 0.029), and CF ($p = 0.026$ and 0.03) for peak pressure and the pressure–time integral. Significance for the contact area was also found in H ($p = 0.032$).

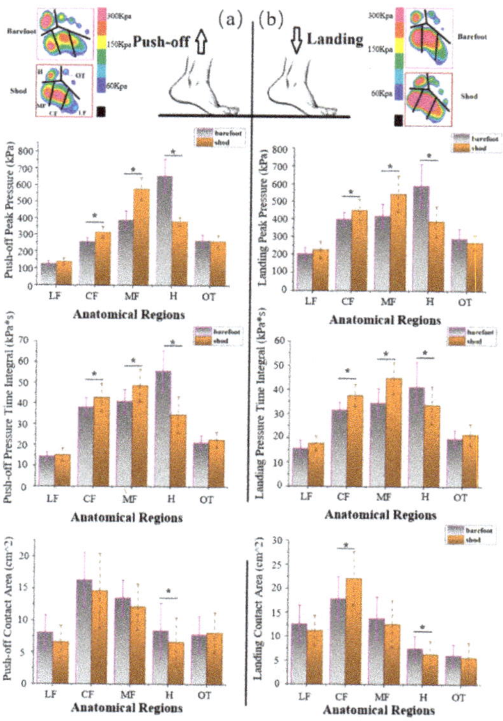

Figure 1. The peak pressure, pressure time integral, and contact area in the anatomical regions during the push-off (**a**) and landing (**b**) phases of the vertical jump. lateral forefoot (LF), central forefoot (CF), medial forefoot (MF), hallux (H), and other toes (OT). * indicates significance between variables, $p < 0.05$.

During the landing phase (right), significant differences were found between barefoot and shod jumping in H ($p = 0.016$ and 0.021), MF ($p = 0.026$ and 0.031), and CF ($p = 0.04$ and 0.033) for peak pressure and the pressure time integral. For contact area, significant differences were found in H ($p = 0.034$) and CF ($p = 0.02$).

As measured from the running test, participants' running speeds were self-selected as comfortable from the generated spatiotemporal parameter. The comparison of collected spatiotemporal parameters, including stride length, stride time, and contact time, in one gait cycle between barefoot and shod running, are presented in Table 1.

Table 1. The spatiotemporal parameters between barefoot and shod running.

	Stride Length (m)	Stride Time (s)	Contact Time (s)
Barefoot	2.35 ± 0.19 *	0.76 ± 0.027 *	0.252 ± 0.018 *
Shod	2.46 ± 0.21	0.794 ± 0.032	0.298 ± 0.013

Note: * Significance between barefoot and shod runners, $p < 0.05$.

As shown in Figure 2, the stances of barefoot and shod running were highlighted with solid (33.2 ± 0.7%) and dashed (37.5 ± 0.8%) vertical lines, which were calculated from the percentage of contact time in stride time. The peak angles during the stance were thus obtained between barefoot and shod running, and statistical significance was highlighted with a red dotted line with an asterisk (*), with Table 2 presenting detailed values.

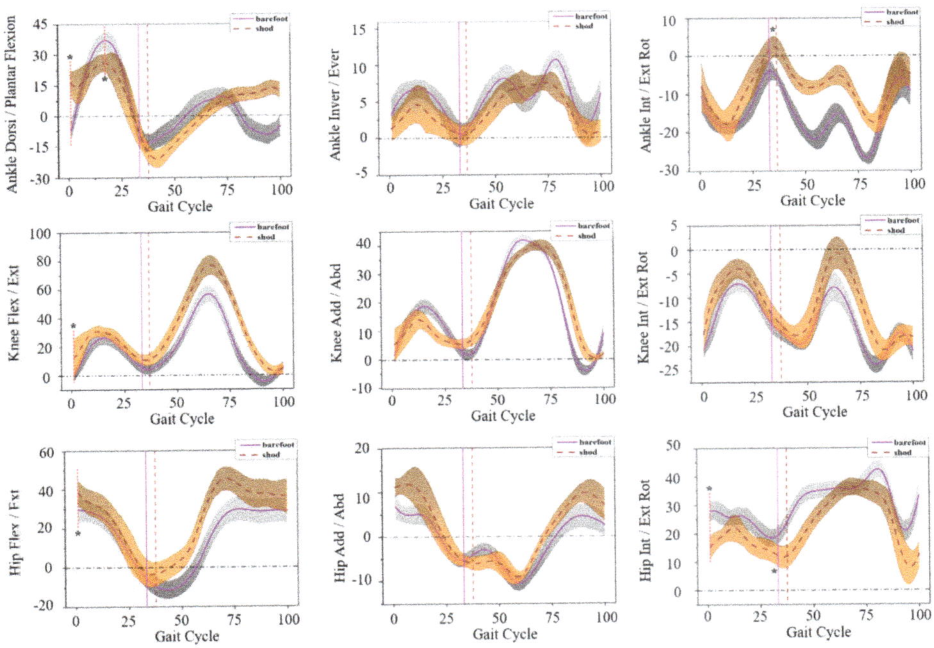

Figure 2. Joint angles curves of the ankle, knee, and hip in sagittal, frontal, and horizontal planes during one gait cycle. Red dotted lines with * indicate a significant difference, $p < 0.05$.

There was a significant difference between the foot strike angle of the ankle between shod and barefoot running, with the foot strike angle of shod running at 17.1 ± 4.3°, and barefoot running at −7.2 ± 3.9° (minus indicates plantarflexion), $p = 0.00$. Internal and external ankle rotation also showed a significant difference, $p < 0.05$. The maximal rotation angle during the push-off phase of the stance was 3.24 ± 2.26° (shod running) and −3.76 ± 1.5° (barefoot running). Barefoot running showed significantly larger ankle ROM than shod running, $p = 0.00$ (Table 3).

The knee joint contact angles while foot landing were 12.33 ± 8.45° (shod) and 0.1 ± 2.3° (barefoot), showing significance ($p = 0.012$) (highlighted in Figure 2). Smaller knee joint ROM in the sagittal plane was also observed, with $p = 0.021$ (Table 3).

For hip movement, shod running (38.79 ± 7.81°) presented a larger flexion angle than barefoot running (30.12 ± 5.66°) while landing ($p = 0.03$) (Table 2). A greater internal rotation angle was observed for barefoot running (27.21 ± 3.66°) than shod running (14.21 ± 2.66°), as the foot was landing ($p = 0.32$) (Figure 2). Shod running presented significantly larger ROM than barefoot running in the gait cycle (Table 3).

Table 2. Peak joints' angles between barefoot and shod running during stance.

		Barefoot (Mean ± SD)		Shod (Mean ± SD)	
		Max.	Min.	Max.	Min.
Ankle	Sagittal	37.12 ± 2.8° *	−11.22 ± 5.4°	28.47 ± 2.6°	−18.7 ± 6.3°
	Coronal	6.98 ± 2.1°	0.46 ± 1.5°	4.13 ± 2.5°	0.73 ± 1.23°
	Horizontal	−3.76 ± 1.5° *	−19.14 ± 1.66°	3.24 ± 2.26°	−19.73 ± 3.5°
Knee	Sagittal	26.33 ± 4.45° *	0.1 ± 2.3°	32.2 ± 3.1°	2.57 ± 3.6°
	Coronal	18.75 ± 3.1°	1.98 ± 2.1°	14.39 ± 3.1°	4.28 ± 3.99°
	Horizontal	−6.98 ± 1.99°	−20.4 ± 3.1°	-3.1 ± 1.9°	−18.2 ± 2.2°
Hip	Sagittal	30.12 ± 5.66° *	−7.04 ± 2.99°	38.79 ± 7.81°	−5.57 ± 4.88°
	Coronal	6.9 ± 2.89°	−5.37 ± 2.33°	13.3 ± 4.1°	−6.1 ± 1.68°
	Horizontal	28.57 ± 4.1°	18.3 ± 3.6°	22.86 ± 4.5°	11.17 ± 4.17°

* Significance between barefoot and shod runners.

Table 3. Lower extremity joints' range of motion (ROM) between barefoot and shod running in gait cycle.

		Barefoot (Mean ± SD)	Shod (Mean ± SD)
Ankle	Sagittal	50.93 ± 3.81°	49.67 ± 5.12°
	Coronal	10.58 ± 3.56° *	3.89 ± 1.66°
	Horizontal	23.86 ± 5.22°	24.09 ± 7.6°
Knee	Sagittal	61.67 ± 8.26° *	74.67 ± 9.15°
	Coronal	46.11 ± 7.55°	39.9 ± 5.45°
	Horizontal	17.56 ± 2.3°	21.05 ± 4.3°
Hip	Sagittal	42.62 ± 9.59°	42.44 ± 11.2°
	Coronal	17.39 ± 5.66°	22.09 ± 7.58°
	Horizontal	24.43 ± 6.89° *	31.61 ± 9.16°

* Significance between barefoot and shod runners.

Peak pressure, contact area, and the pressure–time integral are shown in Figure 3. For peak pressure, MR, LR, LM, LF, and H showed significant differences between shod and barefoot running. Specifically, barefoot running demonstrated less peak pressure in MR ($p = 0.00$) and LR ($p = 0.00$) than shod running. In contrast, barefoot running showed larger peak pressure in LM ($p = 0.028$), LF ($p = 0.019$), and H ($p = 0.005$) than shod running. For the pressure–time integral, shod running showed a larger pressure–time integral in MR ($p = 0.00$), LR ($p = 0.00$), and MF ($p = 0.02$) than barefoot running. In contrast, barefoot running indicated a larger pressure–time integral in LM ($p = 0.03$), LF ($p = 0.009$), and H ($p = 0.028$) than shod running. For the contact area, shod running presented a larger area in MR ($p = 0.00$), LR ($p = 0.00$), and MM ($p = 0.00$) than barefoot running.

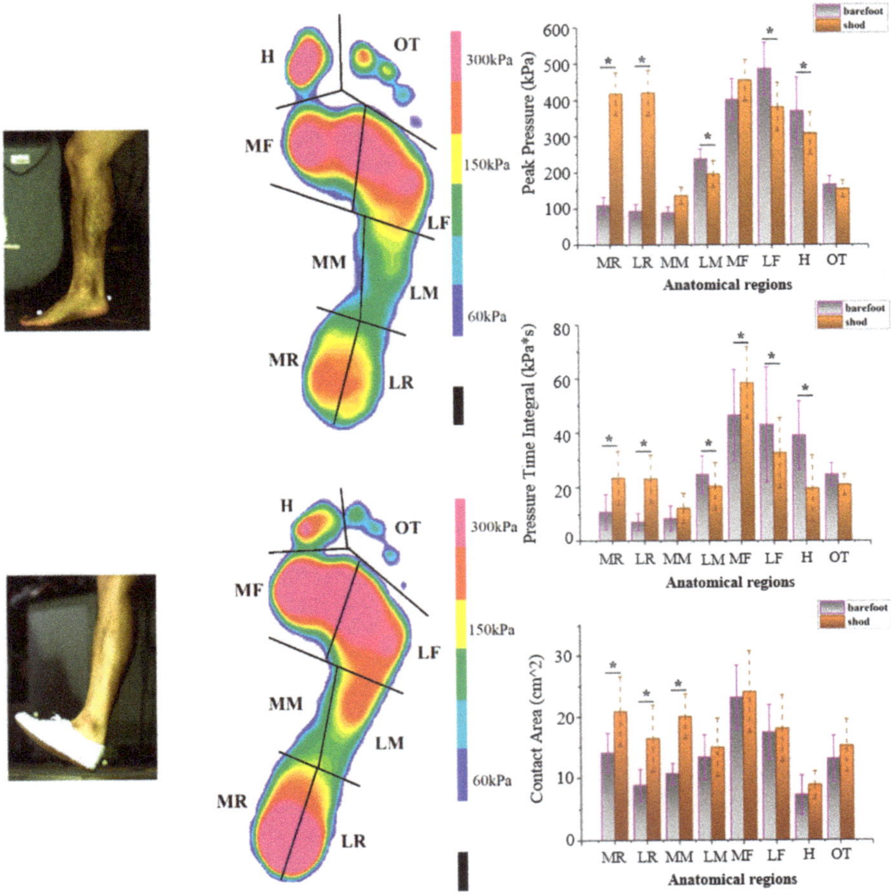

Figure 3. Foot pressure of barefoot and shod running. Medial rearfoot (MR), lateral rearfoot (LR), medial midfoot (MM), lateral midfoot (LM), medial forefoot (MF), lateral forefoot (LF), hallux (H), and other toes (OT). * indicates significance, $p < 0.05$.

4. Discussion

This study aimed to analyze foot morphology-related jumping and running biomechanics and evaluate any potential functional differences. Participants in this study were from different parts of Asia, with a barefoot group from Indian ethnicity and a shod group from Chinese ethnicity. The main findings were that, (i) during the push-off and landing phases of the vertical jump, the separate hallux of barefoot individuals shared loading from the metatarsals, and thus expanded the loading concentrated region; (ii) during the push-off phase of running, there were plantar pressure differences in the hallux and forefoot of barefoot individuals compared with shod individuals; (iii) barefoot individuals with separate toes presented a flexible range of motion, particularly in the coronal plane of the ankle, sagittal plane of the knee, and horizontal plane of the hip.

Hallux angle has been reported to be different among populations of different ethnicities [11]. However, few studies have focused on the minimal distance between the hallux and the interphalangeal joint of the second toe. Compared with results of our previous study [4], barefoot groups in this study had a larger distance and smaller hallux angle, while the shod group had a larger hallux angle and smaller distance. It may be concluded that the barefoot group had more flexible hallux than the shod group [1,3,36]. Lambrinudi et al. [36] reported that if the separate hallux has ambulatory and prehensile functions, it could work fundamentally the same way as the fingers. However, wearing shoes may block these prehensile and separate functions of the toes, due to the sharp-headed or ill-fitted space restrictions [3,6,14]. The hallux angle and minimal distance are the basis of the morphological differences for the vertical jumping and running test in this study.

Results from the vertical jump test indicated that the hallux presented larger plantar loading in the barefoot group compared with the shod group (Figure 1). During the push-off phase, the plantar loading of the barefoot group was larger under the hallux, while the pressure of the shod group was larger under the medial forefoot and central forefoot. The same pressure time integrals were presented in these anatomical regions, which may imply that the hallux of the barefoot group was used predominantly, while the forefoot of shod group was used primarily. Moreover, the peak pressure of the barefoot reduced in the forefoot regions while landing. This suggests that the gripping function of the hallux could firm and expand the supporting base during the push-off and landing phases by the separate toes [3,36]. In addition, large loading under the hallux could reduce the impact force to the forefoot [24]. Previous research has reported that excessive loading under the metatarsal head area (forefoot) would lead to forefoot injuries [7]. These findings imply that the foot morphology related to toe gripping functions may have a possible link with forefoot metatarsal stress injuries. However, this study did not investigate the injury risks between the two population groups. The jumping height showed no significance, which implies that the morphological differences may not be linked with jumping performance, or there may be a limitation in the akimbo position. Further research is needed and should focus on jumping performance via comprehensive kinematic analysis.

Research pertaining to habitually barefoot and shod people has received increased attention in recent years. Different ethnicities [4,5,24], pathological factors [37], and different forms of sport participation [10] could influence foot morphological differences. Among all the barefoot and shod participants, biomechanical data for the forefoot strike barefoot running, and rearfoot strike shod running were included in this study. Barefoot running was reported to be different to minimalist, racing, or regular shoe conditions in a previous biomechanical study of experienced runners [38]. The extrinsic muscles of the foot presented reduced muscular activity during barefoot running, for instance, peroneus longus [39,40]. Other controversial opinions were proposed between the minimalist and barefoot running when compared to traditional shoe running [41,42]. This study focused on the biomechanics from the forefoot and toe morphology, thus, shod running from the barefoot group and barefoot running from the shod group were not performed during the test, which was aimed to reduce the acute response of altering to shod (for the barefoot group) or barefoot (for the shod group) running.

During the running test, each participant performed running at a comfortable speed so as to present natural running biomechanics [7,43]. The results indicated significant differences in strike length, strike time, and contact time between the shod and barefoot groups, which are consistent with previous studies [2,7,8]. In previous barefoot running studies, the barefoot group was observed to reduce these spatiotemporal parameters [8,44,45]. The stride time of the barefoot group was significantly less than the shod [46]. The running performance of the barefoot group was characterized by landing on the forefoot, and the ankle changed from plantarflexion to dorsiflexion in the sagittal plane, which contributed to the greater dorsiflexion angle during the stance. The shod running resulted in landing with the rearfoot, and the ankle in the dorsiflexion position, which could explain the ankle angle difference as the foot strikes. Different foot strike patterns could be a reason for the strike time differences observed [8].

The observed knee contact flexion angle and peak flexion angle difference of shod running in this study may be a compensatory movement (with larger sagittal knee ROM), resulting from the previously established greater knee impact while rearfoot shod running [9,45,47]. As shown by the hip flexion angle, the contact flexion angle was larger than that of barefoot running, and this could explain the increased stride length of shod running and, although not significant, an about hip flexion-extension ROM was observed.

In terms of the plantar pressure distribution, the barefoot group showed smaller peak pressure and pressure–time integral than the shod group in the rearfoot. This may have been caused by the rear foot landing during shod running, which results in a larger contact area in MR and LR. The difference in contact area in MM may be related to the uppers and soles of the footwear while shod running. Owing to the forefoot strike of barefoot running, the larger peak pressure and pressure time integral to the LM and LF may result from the landing impact. This finding could explain the previously reported forefoot metatarsus fatigue injury due to the repetitive impact and lack of cushioning protection from footwear [1,24]. The hallux showed an increased peak pressure and pressure–time integral during barefoot running, and, while not significant, this was also observed for the contact area. This may result from the active gripping motion of the separate hallux "leverage function" expanding the push-off supporting area (fulcrum) [1,36]. Thus, less loading was found in the MF compared with shod running, which presented greater MF loading and smaller H loading. The greater ankle ROM in the coronal plane and peak angle while pushing off may be kinematic evidence for the active toes function related to the morphological differences in this study. As reported, the function of the remaining toes may also be used for balance and stability control under static and dynamic conditions [48], and it may also be useful during running and jumping performances. Further benefits from the toes in relation to balance and coordination, especially contributions to long distance endurance racing and related events [49], needs further investigation.

Several limitations should be considered in this study. Firstly, participants were physically active males in their early twenties, which may be a limiting factor for generalizing findings from this study to different age groups and genders. Secondly, this study lacked information on the vertical jump biomechanics between the two-population groups, which should be a future research project to investigate potential differences in jumping performance. Thirdly, the entire test was conducted in a lab-based environment, and it is possible the jumping and running biomechanical performances may be different in an outdoor environment.

Previous research revealed that running injuries are a multifactorial issue, including systemic factors, training (experience), health factors, and lifestyle factors [50]. The foot type, structure, or morphology were considered for musculoskeletal injuries in several studies [10,51–53]; however, morphology-related foot functions have been rarely investigated. This needs investigation as a potential contributory factor for injury research.

5. Conclusions

This study analyzed the morphology-related jumping and running biomechanical functions of habitually barefoot and shod males. The unrestricted forefoot and toes of the barefoot group presented flexible movements and a leverage function to expand the forefoot loading area during jumping and running. Findings from the study in relation to morphology-related functions, especially the contribution of the forefoot and toes, may provide useful information for footwear design and injury prevention.

Author Contributions: Conceptualization, P.Y., M.L., Q.M. and Y.G.; methodology, P.Y., L.X., Q.M., J.S.B. and Y.G..; validation, L.X., M.L. and Q.M.; investigation, P.Y. and L.X.; writing—original draft preparation, P.Y., L.X and Q.M.; writing—review and editing, J.B. and Y.G.

Funding: This research was funded National Natural Science Foundation of China (No.81772423), National Key R&D Program of China (2018YFF0300903) and K.C. Wong Magna Fund in Ningbo University.

Conflicts of Interest: The authors declare no conflict of interest.

References

1. Mei, Q.; Fernandez, J.; Hume, P.; Gu, Y. Investigating biomechanical function of toes through external manipulation integrating analysis. *Acta Bioeng. Biomech.* **2016**, *18*, 87–92. [CrossRef]

2. Franklin, S.; Grey, M.J.; Heneghan, N.; Bowen, L.; Li, F.X. Barefoot vs common footwear: A systematic review of the kinematic, kinetic and muscle activity differences during walking. *Gait Posture* **2015**, *42*, 230–239. [CrossRef] [PubMed]
3. Hoffmann, P. Conclusions drawn from a comparative study of the feet of barefooted and shoe-wearing peoples. *Am. J. Orthop. Surg.* **1905**, *3*, 105–163.
4. Shu, Y.; Mei, Q.; Fernandez, J.; Li, Z.; Feng, N.; Gu, Y. Foot morphological difference between habitually shod and unshod runners. *PLoS ONE* **2015**, *10*, e0131385. [CrossRef] [PubMed]
5. D'Août, K.; Pataky, T.C.; De Clercq, D.; Aerts, P. The effects of habitual footwear use: foot shape and function in native barefoot walkers. *Footwear Sci.* **2009**, *1*, 81–94. [CrossRef]
6. Kadambande, S.; Khurana, A.; Debnath, U.; Bansal, M.; Hariharan, K. Comparative anthropometric analysis of shod and unshod feet. *Foot* **2006**, *16*, 188–191. [CrossRef]
7. Tam, N.; Astephen Wilson, J.L.; Noakes, T.D.; Tucker, R. Barefoot running: an evaluation of current hypothesis, future research and clinical applications. *Br. J. Sports Med.* **2014**, *48*, 349–355. [CrossRef] [PubMed]
8. Lohman, E.B.; Balan Sackiriyas, K.S.; Swen, R.W. A comparison of the spatiotemporal parameters, kinematics, and biomechanics between shod, unshod, and minimally supported running as compared to walking. *Phys. Ther. Sport* **2011**, *12*, 151–163. [CrossRef]
9. Yeow, C.H.; Lee, P.V.S.; Goh, J.C.H. Shod landing provides enhanced energy dissipation at the knee joint relative to barefoot landing from different heights. *Knee* **2011**, *18*, 407–411. [CrossRef]
10. Cain, L.E.; Nicholson, L.L.; Adams, R.D.; Burns, J. Foot morphology and foot/ankle injury in indoor football. *J. Sci. Med. Sport* **2007**, *10*, 311–319. [CrossRef]
11. Gurney, J.K.; Kersting, U.G.; Rosenbaum, D. Dynamic foot function and morphology in elite rugby league athletes of different ethnicity. *Appl. Ergon.* **2009**, *40*, 554–559. [CrossRef] [PubMed]
12. Mei, Q.; Graham, M.; Gu, Y. Biomechanical analysis of the plantar and upper pressure with different sports shoes. *Int. J. Biomed. Eng. Technol.* **2014**, *14*, 181–191. [CrossRef]
13. Ledoux, W.R.; Shofer, J.B.; Ahroni, J.H.; Smith, D.G.; Sangeorzan, B.J.; Boyko, E.J. Biomechanical differences among pes cavus, neutrally aligned, and pes planus feet in subjects with diabetes. *Foot Ankle Int.* **2003**, *24*, 845–850. [CrossRef] [PubMed]
14. Ashizawa, K.; Kumakura, C.; Kusumoto, A.; Narasaki, S. Relative foot size and shape to general body size in Javanese, Filipinas and Japanese with special reference to habitual footwear types. *Ann. Hum. Biol.* **1997**, *24*, 117–129. [CrossRef] [PubMed]
15. Xiang, L.; Mei, Q.; Fernandez, J.; Gu, Y. Minimalist shoes running intervention can alter the plantar loading distribution and deformation of hallux valgus: A pilot study. *Gait Posture* **2018**, *65*, 65–71. [CrossRef]
16. Calvo-Lobo, C.; Garrido-Marín, A.; Rodríguez-Sanz, D.; López-López, D.; Palomo-López, P.; Romero-Morales, C.; Sanz-Corbalán, I. Ultrasound evaluation of intrinsic plantar muscles and fascia in hallux valgus: A case-control study. *Medicine* **2016**, *95*, e5243. [CrossRef]
17. López-López, D.; Callejo-González, L.; Losa-Iglesias, M.E.; Saleta-Canosa, J.L.; Rodríguez-Sanz, D.; Calvo-Lobo, C.; Becerro-de-Bengoa-Vallejo, R. Quality of life impact related to foot health in a sample of older people with hallux valgus. *Aging Dis.* **2016**, *7*, 1–8. [CrossRef]
18. López-López, D.; Vilar-Fernández, J.M.; Losa-Iglesias, M.E.; Álvarez-Castro, C.; Calvo-Lobo, C.; Ramos-Galván, J.; Becerro-de-Bengoa-Vallejo, R. Influence of depression in a sample of people with hallux valgus. *Int. J. Ment. Health Nurs.* **2016**, *25*, 574–578. [CrossRef]
19. Vint, P.F.; Hinrichs, R.N. Differences between one-foot and two-foot vertical jump performances. *J. Appl. Biomech.* **1996**, *12*, 338–358. [CrossRef]
20. Goldmann, J.P.; Sanno, M.; Willwacher, S.; Heinrich, K.; Brüggemann, G.P. The potential of toe flexor muscles to enhance performance. *J. Sports Sci.* **2013**, *31*, 424–433. [CrossRef]
21. Sman, A.D.; Hiller, C.E.; Rae, K.; Linklater, J.; Black, D.A.; Refshauge, K.M. Prognosis of ankle syndesmosis injury. *Med. Sci. Sports Exerc.* **2014**, *46*, 671–677. [CrossRef]
22. Shu, Y.; Zhang, Y.; Fu, L.; Fekete, G.; Baker, J.S.; Li, J.; Gu, Y. Dynamic loading and kinematics analysis of vertical jump based on different forefoot morphology. *SpringerPlus* **2016**, *5*, 1–9. [CrossRef] [PubMed]
23. Lieberman, D.E.; Venkadesan, M.; Werbel, W.A.; Daoud, A.I.; D'Andrea, S.; Davis, I.S.; Mang'Eni, R.O.; Pitsiladis, Y. Foot strike patterns and collision forces in habitually barefoot versus shod runners. *Nature* **2010**, *463*, 531–535. [CrossRef] [PubMed]

24. Mei, Q.; Fernandez, J.; Fu, W.; Feng, N.; Gu, Y. A comparative biomechanical analysis of habitually unshod and shod runners based on a foot morphological difference. *Hum. Mov. Sci.* **2015**, *42*, 38–53. [CrossRef] [PubMed]
25. Krabak, B.J.; Hoffman, M.D.; Millet, G.Y.; Chimes, G.P. Barefoot running. *PM&R J.* **2011**, *3*, 1142–1149. [CrossRef]
26. De Cock, A.; Vanrenterghem, J.; Willems, T.; Witvrouw, E.; De Clercq, D. The trajectory of the centre of pressure during barefoot running as a potential measure for foot function. *Gait Posture* **2008**, *27*, 669–675. [CrossRef] [PubMed]
27. Hall, J.P.L.; Barton, C.; Jones, P.R.; Morrissey, D. The biomechanical differences between barefoot and shod distance running: A systematic review and preliminary meta-analysis. *Sport. Med.* **2013**, *43*, 1335–1353. [CrossRef] [PubMed]
28. Santuz, A.; Ekizos, A.; Arampatzis, A. A pressure plate-based method for the automatic assessment of foot strike patterns during running. *Ann. Biomed. Eng.* **2016**, *44*, 1646–1655. [CrossRef] [PubMed]
29. Santuz, A.; Ekizos, A.; Janshen, L.; Baltzopoulos, V.; Arampatzis, A. The influence of footwear on the modular organization of running. *Front. Physiol.* **2017**, *8*, 1–10. [CrossRef] [PubMed]
30. Hafer, J.F.; Lenhoff, M.W.; Song, J.; Jordan, J.M.; Hannan, M.T.; Hillstrom, H.J. Reliability of plantar pressure platforms. *Gait Posture* **2013**, *38*, 544–548. [CrossRef]
31. Godi, M.; Turcato, A.M.; Schieppati, M.; Nardone, A. Test-retest reliability of an insole plantar pressure system to assess gait along linear and curved trajectories. *J. Neuroeng. Rehabil.* **2014**, *11*, 1–8. [CrossRef] [PubMed]
32. Razak, A.; Hadi, A.; Zayegh, A.; Begg, R.K.; Wahab, Y. Foot plantar pressure measurement system: A review. *Sensors* **2012**, *12*, 9884–9912. [CrossRef] [PubMed]
33. Ellis, S.J.; Stoecklein, H.; Yu, J.C.; Syrkin, G.; Hillstrom, H.; Deland, J.T. The accuracy of an automasking algorithm in plantar pressure measurements. *HSS J.* **2011**, *7*, 57–63. [CrossRef] [PubMed]
34. Gu, Y.; Lu, Y.; Mei, Q.; Li, J.; Ren, J. Effects of different unstable sole construction on kinematics and muscle activity of lower limb. *Hum. Mov. Sci.* **2014**, *36*, 46–57. [CrossRef] [PubMed]
35. Bosco, C.; Luhtanen, P.; Komi, P.V. A simple method for measurement of mechanical power in jumping. *Eur. J. Appl. Physiol.* **1983**, *50*, 273–282. [CrossRef]
36. Lambrinudi, C. Use and abuse of toes. *Postgrad. Med. J.* **1932**, *8*, 459–464. [CrossRef] [PubMed]
37. Hillstrom, H.J.; Song, J.; Kraszewski, A.P.; Hafer, J.F.; Mootanah, R.; Dufour, A.B.; Chow, B.S.; Deland, J.T. Foot type biomechanics part 1: Structure and function of the asymptomatic foot. *Gait Posture* **2013**, *37*, 445–451. [CrossRef] [PubMed]
38. Bonacci, J.; Saunders, P.U.; Hicks, A.; Rantalainen, T.; Vicenzino, B.T.; Spratford, W. Running in a minimalist and lightweight shoe is not the same as running barefoot: a biomechanical study. *Br. J. Sports Med.* **2013**, *47*, 387–392. [CrossRef] [PubMed]
39. Roca-Dols, A.; Losa-Iglesias, M.E.; Sánchez-Gómez, R.; Becerro-de-Bengoa-Vallejo, R.; López-López, D.; Palomo-López, P.; Rodríguez-Sanz, D.; Calvo-Lobo, C. Electromyo-graphy activity of triceps surae and tibialis anterior muscles related to various sports shoes. *J. Mech. Behav. Biomed.* **2018**, *86*, 158–171. [CrossRef]
40. Roca-Dols, A.; Losa-Iglesias, M.E.; Sánchez-Gómez, R.; López-López, D.; Becerro-de-Bengoa-Vallejo, R.; Calvo-Lobo, C. Electromyography comparison of the effects of various footwear in the activity patterns of the peroneus longus and brevis muscles. *J. Mech. Behav. Biomed.* **2018**, *82*, 126–132. [CrossRef]
41. Squadrone, R.; Gallozzi, C. Biomechanical and physiological comparison of barefoot and two shod conditions in experienced barefoot runners. *J. Sports Med. Phys. Fitness* **2009**, *49*, 6–13. [PubMed]
42. Paquette, M.; Zhang, S.; Baumgartner, L. Acute effects of barefoot, minimal shoes and running shoes on lower limb mechanics in rear and forefoot strike runners. *Footwear Sci.* **2013**, *5*, 9–18. [CrossRef]
43. Hatala, K.G.; Dingwall, H.L.; Wunderlich, R.E.; Richmond, B.G. Variation in foot strike patterns during running among habitually barefoot populations. *PLoS ONE* **2013**, *8*, 4–9. [CrossRef] [PubMed]
44. Altman, A.R.; Davis, I.S. A kinematic method for footstrike pattern detection in barefoot and shod runners. *Gait Posture* **2013**, *35*, 298–300. [CrossRef] [PubMed]
45. Daoud, A.I.; Geissler, G.J.; Wang, F.; Saretsky, J.; Daoud, Y.A.; Lieberman, D.E. Foot strike and injury rates in endurance runners: A retrospective study. *Med. Sci. Sports Exerc.* **2012**, *44*, 1325–1334. [CrossRef] [PubMed]

46. Roca-Dols, A.; Losa-Iglesias, M.E.; Sánchez-Gómez, R.; Becerro-de-Bengoa-Vallejo, R.; López-López, D.; Rodríguez-Sanz, D.; Martínez-Jiménez, E.M.; Calvo-Lobo, C. Effect of the cushioning running shoes in ground contact time of phases of gait. *J. Mech. Behav. Biomed.* **2018**, *88*, 196–200. [CrossRef] [PubMed]
47. Cheung, R.T.H.; Rainbow, M.J. Landing pattern and vertical loading rates during first attempt of barefoot running in habitual shod runners. *Hum. Mov. Sci.* **2014**, *34*, 120–127. [CrossRef]
48. Ku, P.X.; Abu Osman, N.A.; Yusof, A.; Abas, W.A.B. The effect on human balance of standing with toe-extension. *PLoS ONE* **2012**, *7*, e41539. [CrossRef]
49. Rolian, C.; Lieberman, D.E.; Hamill, J.; Scott, J.W.; Werbel, W. Walking, running and the evolution of short toes in humans. *J. Exp. Biol.* **2009**, *212*, 713–721. [CrossRef]
50. Van Gent, R.N.; Siem, D.; van Middelkoop, M.; van Os, A.G.; Bierma-Zeinstra, S.M.A.; Koes, B.W. Incidence and determinants of lower extremity running injuries in long distance runners: A systematic review. *Br. J. Sports Med.* **2007**, *40*, 16–29. [CrossRef]
51. Kaufman, K.R.; Brodine, S.K.; Shaffer, R.A.; Johnson, C.W.; Cullison, T.R. The effect of foot structure and range of motion on musculoskeletal overuse injuries. *Am. J. Sports Med.* **1999**, *27*, 585–593. [CrossRef] [PubMed]
52. Barnes, A.; Wheat, J.; Milner, C. Association between foot type and tibial stress injuries: a systematic review. *Br. J. Sports Med.* **2008**, *42*, 93–98. [CrossRef] [PubMed]
53. Sánchez-Sáez, J.M.; Palomo-López, P.; Becerro-de-Bengoa-Vallejo, R.; Calvo-Lobo, C.; Losa-Iglesias, M.E.; López-Del-Amo-Lorente, A.; López-López, D. Stability of Three Different Sanitary Shoes on Healthcare Workers: A Cross-Sectional Study. *Int. J. Environ. Res. Public Health* **2019**, *16*, 2126. [CrossRef] [PubMed]

MDPI
St. Alban-Anlage 66
4052 Basel
Switzerland
Tel. +41 61 683 77 34
Fax +41 61 302 89 18
www.mdpi.com

Applied Sciences Editorial Office
E-mail: applsci@mdpi.com
www.mdpi.com/journal/applsci

www.ingramcontent.com/pod-product-compliance
Lightning Source LLC
Chambersburg PA
CBHW061602220926
51850CB00005B/403
* 9 7 8 3 0 3 9 3 6 3 9 6 4 *